10th Workshop on Computational Linguistics and Clinical Psychology (CLPsych 2025)

Albuquerque, New Mexico, USA
3 May 2025

ISBN: 979-8-3313-1915-1

Printed from e-media with permission by:

Curran Associates, Inc.
57 Morehouse Lane
Red Hook, NY 12571

Some format issues inherent in the e-media version may also appear in this print version.

Association for Computational Linguistics
209 N. Eighth Street
Stroudsburg, Pennsylvania 18360

Phone: 1-570-476-8006
Fax: 1-570-476-0860

acl@aclweb.org

Additional copies of this publication are available from:

Curran Associates, Inc.
57 Morehouse Lane
Red Hook, NY 12571 USA
Phone: 845-758-0400
Fax: 845-758-2633
Email: curran@proceedings.com
Web: www.proceedings.com

CLPsych 2025

The 10th Workshop on Computational Linguistics and Clinical Psychology

Proceedings of the Workshop

May 3, 2025

The CLPsych organizers gratefully acknowledge the support from the following sponsors.

ARIEL UNIVERSITY

Introduction

Mental health remains a critical issue. Globally, mental health conditions rank among the top causes of disability [4, 6], and the economic burden of mental health issues, including neurological and substance use disorders, is expected to exceed $16 trillion from 2011 to 2030 [1]. In the United States alone, suicide ranked among the top nine leading causes of death for individuals aged 10-64 in 2020, and was the second leading cause of death for those aged 10-14 and 25-34 [3]. The COVID-19 pandemic has further exacerbated these mental health challenges. Research by Sheridan et al. [5] indicates that suicide attempts among children aged 10-12 have increased more than five-fold from 2010 to 2020. Recent advancements in large language models (LLMs) have demonstrated significant potential in mental health. They are not only used for diagnostic purposes but have also been shown to provide valuable explanations for prediction outcomes [2]. In light of this, for The Tenth Workshop on Computational Linguistics and Clinical Psychology (CLPsych), we adopt the theme "understanding the mental health state – going beyond classification".

CLPsych was a hybrid workshop that accommodated both in-person and remote participation. It was collocated with NAACL'25, which took place in Albuquerque, New Mexico, USA on May 3rd, 2025. Since 2014, CLPsych has been successful in bringing together people from different backgrounds (e.g. mental health experts, clinicians, and computational linguists), to share and discuss their work and results. Its central goal is to build bridges so that these different disciplines can integrate to improve our understanding of mental health issues, and to deliver better mental health treatments and diagnoses to everybody.

The CLPsych 2025 Shared Task focused on capturing mental health dynamics from social media timelines, through a novel multi-task framework grounded in the transtheoretical MIND approach. Building on CLPsych 2022's longitudinal modeling approach, it combines monitoring mental states with evidence and summary generation through four subtasks: (A.1) Evidence Extraction, highlighting text spans reflecting adaptive or maladaptive self-states; (A.2) Well-Being Score Prediction, assigning posts a 1 to 10 score based on social, occupational, and psychological functioning; (B) Post-level Summarization of the interplay between adaptive and maladaptive states within individual posts; and (C) Timeline-level Summarization capturing temporal dynamics of self-states over posts in a timeline. Overall, 14 teams completed the shared task, proposing solutions from traditional machine learning methods with domain-specific features to LLM pipelines with demonstration and retrieval. The results shed light on the complexity of capturing mental health states beyond static classification and offer directions for future work.

Our program committee included mental health and technological experts, in order to provide all the papers with more informative feedback that addresses both aspects. CLPsych'25 received a total of 31 papers for the main workshop, of which 15 were accepted; all 11 submitted shared task papers were also accepted. The organizing committee, with the help of the program committee scores, and feedback chose 6 main workshop papers and 4 shared task papers as oral presentations, and the rest were presented in the poster session.

CLPsych'25 also hosted excellent invited speakers and panelists. Our keynote speakers were Zac Imel (University of Utah), and Zohar Elyoseph (University of Haifa). Additionally, we hosted a panel that included short talks and discussion by Philip Resnik (University of Maryland, College Park), Sunny Tang (Hofstra/Northwell University), and moderated by Steven Bedrick (Oregon Health & Science University).

The CLPsych organizing committee would like to extend special thanks to all the people that helped make the workshop a success. This includes and is not limited to our authors, shared task participants and organizers, program committee members. We also would like to thank the North American chapter of the Association for Computational Linguistics for making this workshop possible and to Philip Resnik who assisted with general advice. Special thanks to our generous sponsors: Ariel University. Their funds helped to support the workshop's program, and provided support for attendees that don't have the financial means to cover their registration costs.

Ayah Zirikly, Andrew Yates, Kfir Bar, Bart Desmet, Molly Ireland, Sean MacAvaney, Yaakov Ophir, and Steven Bedrick.

References

[1] David E Bloom, Elizabeth Cafiero, Eva Jané-Llopis, Shafika Abrahams-Gessel, Lakshmi Reddy Bloom, Sana Fathima, Andrea B Feigl, Tom Gaziano, Ali Hamandi, Mona Mowafi, et al. 2012. The global economic burden of noncommunicable diseases. Technical report, Program on the Global Demography of Aging.

[2] Jenny Chim, Adam Tsakalidis, Dimitris Gkoumas, Dana Atzil-Slonim, Yaakov Ophir, Ayah Zirikly, Philip Resnik, and Maria Liakata. 2024. Overview of the clpsych 2024 shared task: Leveraging large language models to identify evidence of suicidality risk in online posts. In *Proceedings of the 9th Workshop on Computational Linguistics and Clinical Psychology (CLPsych 2024)*, pages 177–190.

[3] Daniel C Ehlman. 2022. Changes in suicide rates—united states, 2019 and 2020. *MMWR. Morbidity and Mortality Weekly Report*, 71.

[4] Christopher JL Murray, Alan D Lopez, World Health Organization, et al. 1996. *The global burden of disease: a comprehensive assessment of mortality and disability from diseases, injuries, and risk factors in 1990 and projected to 2020: summary*. World Health Organization.

[5] David C Sheridan, Sara Grusing, Rebecca Marshall, Amber Lin, Adrienne R Hughes, Robert G Hendrickson, and B Zane Horowitz. 2022. Changes in suicidal ingestion among preadolescent children from 2000 to 2020. *JAMA pediatrics*.

[6] Ma-Li Wong and Julio Licinio. 2001. Research and treatment approaches to depression. *Nature Reviews Neuroscience*, 2(5):343–351.

Organizing Committee

Workshop Co-chairs

Ayah Zirikly, Johns Hopkins University
Andrew Yates, Johns Hopkins University

Organizing Committee

Bart Desmet, National Institutes of Health
Molly Ireland, Receptiviti
Steven Bedrick, Oregon Health & Science University
Sean MacAvaney, University of Glasgow
Kfir Bar, Reichman University
Yaakov Ophir, Ariel University & University of Cambridge

Shared Task Organizers

Jenny Chim, Queen Mary University of London
Talia Tseriotou, Queen Mary University of London
Dana Atzil Slonim, Bar-Ilan University
Ayal Klein, Bar-Ilan University
Aya Shamir, Bar-Ilan University
Ayah Zirikly, Johns Hopkins University
Maria Liakata, Queen Mary University of London & The Alan Turing Institute

Program Committee

Program Committee

Mosab Alfaqeeh, Queen's University
Nick Allen, University of Oregon
Dana Atzil-Slonim, Psychology Department Bar-Ilan University
Alon Aviram, The Hebrew University of Jerusalem
Kfir Bar, Reichman University
Steven Bedrick, Oregon Health & Science University
Nan Bernstein Ratner, University of Maryland
Hoda Bidkhori, George Mason University
Laura Biester, Middlebury College
Daniel Chechelnitsky, Carnegie Mellon University
Jenny Chim, Queen Mary University of London
Trevor Cohen, University of Washington
Hadar Fisher, Harvard Medical Scool
Karina Halevy, Carnegie Mellon University
Patrick Healey, Queen Mary, University of London
Heyuan Huang, Johns Hopkins University
Zac Imel, University of Utah
Loring Ingraham, George Washington University
Molly Ireland, Receptiviti
Zehan Li, UTHealth
Maria Liakata, Queen Mary University of London
Siyang Liu, university of michigan
Antolin Llorente, Depts Pediatrics/Psychiatry, Penn State Hershey, College of Medicine
Sarah Morgan, King's College London
Thong Nguyen, University of Amsterdam
Yaakov Ophir, Technion - Israel Institute of Technology
Debjyoti Paul, Amazon Research
Baruch Perlman, Hebrew University of Jerusalem
David Piterman, NATAL
Emily Prud'hommeaux, Boston College
Mehrdad Rafiepour, University of Kashan
Philip Resnik, University of Maryland
Rebecca Resnik, Rebecca Resnik and Associates LLC
Kaushik Roy, AI Institute, UofSC
Frank Rudzicz, Dalhousie University
Jonathan Schler, HIT
H. Andrew Schwartz, Stony Brook University
Siddharth Singh, University of Amsterdam
Marija Stanojevic, WinterLightLabs
Shabnam Tafreshi, UMD:ARLIS
Ahmad Tafti, University of Pittsburgh
Talia Tseriotou, Queen Mary University of London
Adithya V Ganesan, Stony Brook University
Bram Van Dijk, Leiden University
Vasudha Varadarajan, Stony Brook University
Bo Wang, Massachusetts General Hospital

Maria Wolters, University of Edinburgh
Yumeng Yang, The University of Texas Health Science Center at Houston
Refael Yonatan-Leus, The College of Management Academic Studies
Moreah Zisquit, Reichman University

Table of Contents

Program

Saturday, May 3, 2025

08:45 - 08:55	*Workshop Intro*
08:55 - 09:30	*Keynote 1*
09:30 - 10:30	*Paper Session 1*
10:30 - 11:00	*Break*
11:00 - 12:00	*Panel*
12:00 - 13:00	*Shared Task Session*
13:00 - 14:15	*Lunch*
14:15 - 14:50	*Keynote 2*
14:50 - 15:30	*Poster Session*
15:30 - 16:00	*Break*
16:00 - 17:00	*Paper Session 2*
17:00 - 17:15	*Closing Remarks*

Assessing the Reliability and Validity of GPT-4 in Annotating Emotion Appraisal Ratings

Deniss Ruder
Institute of Computer Science
University of Tartu
deniss.ruder@ut.ee

Andero Uusberg
Institute of Psychology
University of Tartu
andero.uusberg@ut.ee

Kairit Sirts
Institute of Computer Science
University of Tartu
kairit.sirts@ut.ee

Abstract

Appraisal theories suggest that emotions arise from subjective evaluations of events, referred to as appraisals. The taxonomy of appraisals is quite diverse, and they are usually given ratings on a Likert scale to be annotated in an experiencer-annotator or reader-annotator paradigm. This paper studies GPT-4 as a reader-annotator of 21 specific appraisal ratings in different prompt settings, aiming to evaluate and improve its performance compared to human annotators. We found that GPT-4 is an effective reader-annotator that performs close to or even slightly better than human annotators, and its results can be significantly improved by using a majority voting of five completions. GPT-4 also effectively predicts appraisal ratings and emotion labels using a single prompt, but adding instruction complexity results in poorer performance. We also found that longer event descriptions lead to more accurate annotations for both model and human annotator ratings. This work contributes to the growing usage of LLMs in psychology and the strategies for improving GPT-4 performance in annotating appraisals.

1 Introduction

According to appraisal theories, emotions emerge from the individual's subjective appraisals of significant events (Scherer, 2009). Appraisals are the person's evaluations of what situations mean for their needs, goals, and other concerns (Moors et al., 2013). Appraisals consist of values on abstract dimensions representing key aspects of a situation, such as how important, desirable, self-caused, certain, and controllable it is. Appraisals orchestrate changes in other components of an emotional episode, including tendencies to act in some way (motivational component), visceral preparations for these actions (somatic component), facial and bodily expressions (motor component), and a conscious feeling (experiential component) (Moors et al., 2013). Even as all of these components can influence each other, appraisals are often considered as pivotal for initiating and shaping the dynamic interactions that underlie a person's experience of emotion.

Despite their centrality in emotion theory, there have been only few attempts to apply NLP methods to automatically extracting appraisals from text (Hofmann et al., 2020, 2021; Troiano et al., 2023). In those previous works, appraisals have been of interest as a component of emotions, with the goal of eventually predicting the emotion themselves (Hofmann et al., 2020; Troiano et al., 2023). These early results are promising, but more research is needed to further improve the precision and robustness of appraisal prediction models.

Advancing the NLP research on appraisal requires suitable annotated datasets. The most accurate way to annotate appraisals is the so-called experiencer-annotator paradigm, where a person provides both a textual description of some emotional event and ratings on appraisal dimensions. However, the experiencer-annotator method can only be used on texts collected specifically for this purpose. In an alternative reader-annotator paradigm, the appraisal ratings are provided by another person reading the textual descriptions of emotional events. The reader-annotator method is more flexible, as it enables using existing datasets of emotional event descriptions, even those that have not been specifically collected for that purpose but rather have been generated spontaneously by people, for instance in social media.

Generating appraisal ratings with the reader-annotator procedure requires human labor that can become prohibitively expensive for large real-world datasets such as social media and blog posts. Therefore, we are interested whether this process can be automated. Previously, Hofmann et al. (2021) adopted a rule-based approach to assign appraisal labels to texts, given that the emotion

Proceedings of the 10th Workshop on Computational Linguistics and Clinical Psychology (CLPsych 2025), pages 1–11

label of the text was known. However, this approach requires the emotion labels which are often unavailable in real-world datasets. Also, Hofmann et al. (2021) considered only six appraisals and represented them as binary variables. Thus, the deterministic rule-based approach is clearly not feasible with a larger number of appraisal dimensions assessed on Likert scales.

The goal of this work is to assess the suitability of Large Language Models (LLMs) to act as an annotator in the reader-annotator paradigm for emotional appraisals. Several previous works have shown the utility of using LLMs as a viable alternative for human annotators for labeling data for NLP tasks using both the proprietary GPT models (Ding et al., 2023) as well as open-source LLMs (Alizadeh et al., 2023). Other studies have found that using GPT-4 is generally a more reliable annotator than crowdworkers for social science related text annotations (Gilardi et al., 2023), and it is better than vocabulary-based methods for annotating several psychological constructs (Rathje et al., 2024). Although several studies have assessed the ability of GPT-4 to detect discrete emotion labels (Niu et al., 2024; Lian et al., 2024), Kocoń et al. (2023) found that GPT-4 was performing worse than fine-tuned classification models, especially on emotion prediction tasks.

Our study focuses on adopting GPT-4 for annotating emotion appraisals in the reader-annotator paradigm to assess the viability of the generative LLM to act as an alternative to human reader-annotators. Few studies have explored the capabilities of GPT-4 and similar LLMs in this context. Tak and Gratch (2023) analyzed the emotional reasoning abilities of GPT models, finding that while they align well with human appraisals, they struggled with predicting emotion intensity and coping response. Yongsatianchot et al. (2023) studied how LLMs perceive emotions using appraisal theory, demonstrating that these models can effectively generate context-specific emotional appraisals. However, these studies focused more on the rationales of LLMs than appraisal rating predictions and used a limited list of appraisal dimensions. Zhan et al. (2023) evaluated annotating emotion appraisals by multiple LLMs, showing that they can produce human-like emotional appraisals. However, the dataset used in this research was relatively small and, more importantly, did not contain original experience-annotator ratings, which prevents comparison between the event experiencer and reader annotations.

For our study, we use the crowd-enVent dataset collected by Troiano et al. (2023), which consists of crowd-sourced emotional event descriptions, supplied by the appraisal rating of both the experiencers and readers. We formulate research questions about the *reliability* of GPT-4 in generating appraisal ratings (Q1) and the *accuracy* of GPT-4 ratings compared to both experiencer-annotators and human reader-annotators (Q2). In addition, we test whether the application of *a majority voting algorithm* and a model confidence tiebreaker improves accuracy (Q3). Also, we examine the impact of adding *the emotion prediction* on the accuracy of the appraisal ratings (Q4). Finally, we study the impact of the event *description length* on the accuracy of the ratings by both the GPT-4 and human reader-annotators (Q5).

2 Research Questions

We pose two main research questions to evaluate the reliability and accuracy of GPT-4 to act as an annotator for emotion appraisals in the reader-annotator paradigm. Also, we pose three additional research questions that study the impact of a majority voting algorithm, the prediction of emotions, and the length of the event description on the accuracy of the appraisals. This section outlines our research questions.

Q1: Is GPT-4 reliable in generating emotion appraisal ratings? Reliability is a prerequisite for validity. For the GPT-4 to act as a valid annotator of the emotion appraisals, the reliability of its ratings needs to be high in terms of good inter-annotator agreement between several independent GPT-4 runs. We assessed the reliability of the appraisal ratings in a subsample of 108 randomly sampled texts and measured the reliability using Spearman correlation coefficients and root mean squared errors (RMSE). For GPT-4 to be considered reliable, the ratings of different runs should show at least the same level of agreement as different human reader-annotators.

Q2: How accurate are the GPT-4 appraisal ratings compared to the ratings given by the human reader-annotators? To validate GPT-4 as an accurate annotator of emotion appraisals based on event descriptions, we examined the difference between the ratings assigned by GPT-4 and those assigned by experiencer-annotators. We compared these differences with the rating differences between experiencer-annotators and reader-annotators. For GPT-4 to be considered a valid emotion appraisal annotator, its ratings should remain at least close to those of the human reader-annotators ratings.

Q3: Can majority voting with or without a confidence tiebreaker improve GPT-4 accuracy? Applying a majority voting algorithm on a human reader-annotators dataset greatly affected the accuracy of reader guesses (Troiano et al., 2023). The authors also noticed that a substantial number of votes required tiebreakers and proposed breaking the ties by assigning a higher weight to the annotators with stronger confidence. We assumed that majority voting might similarly impact GPT-4 performance, and we can improve it with a model confidence tiebreaker.

Q4: Does adding an emotion prediction task to the prompt impact appraisal ratings accuracy? During the crowd-enVent dataset collection process, human reader-annotators were first asked to select an emotion that the event experiencer-annotator likely felt (anger, boredom, disgust, fear, guilt, joy, pride, relief, sadness, shame, surprise trust, no emotion) as well as to rate how confident they were about their chosen emotion (1-5) and the intensity of emotion (1-5). This procedure makes sense both intuitively and theoretically, as different emotions are expected to have different signature appraisal profiles (Moors et al., 2013). Thus, knowing/guessing the emotion might help to predict/guess also the appraisal ratings more accurately. We tested if reproducing this exact sequence of questions improves GPT-4 appraisal prediction accuracy. We also attempted to use the newly generated confidence and intensity ratings as majority voting tiebreakers.

Q5: How does the length of the event description affect the accuracy of the appraisal ratings for both the GPT-4 and the human reader-annotators? The length of the description of the emotional event is one practical consideration in the reader-annotator paradigm. While the experiencer-annotator will have direct access to the emotion and the related appraisals, the reader, whether an LLM or a human, has to make inferences based on the written text only. It seems intuitive that longer descriptions would give rise to more accurate appraisal ratings. In this research question, we tested this intuition and sought to identify a minimum required length of text to make accurate predictions or guesses.

3 Method

3.1 Data

We use the crowd-enVent data collected by Troiano et al. (2023)[1] who explored the application of appraisal theories to the analysis of emotions in the text. They investigated whether human reader-annotators and an automatic RoBERTa-based text classifier can reproduce appraisal ratings of the original event experiencers and whether these appraisal ratings can assist in labeling emotions.

The data collection process included two phases. During the first phase, the authors collected event descriptions, appraisal ratings, and the categorical emotion experienced in relation to the event from the experiencer-annotators. In total, 6600 event descriptions were collected via crowd-sourcing. In the second phase, a subset of 1200 event descriptions was subsequently annotated by five human crowd-sourced reader-annotators. In addition to guessing the appraisal ratings of the experiencers of the event, the reader-annotators also were asked to guess the emotion label related to the event, and rate their confidence in the 5-point Likert scale in that guess. Thus, the final appraisal-annotated crowd-enVent corpus includes two datasets that we call experiencer-annotator and reader-annotator datasets, respectively. Both datasets are annotated with the following 21 appraisal dimensions on a 1-5 Likert scale.

1. **Suddenness**: The event was sudden or abrupt.
2. **Familiarity**: The event was familiar.
3. **Event Predictability**: I could have predicted the occurrence of the event.
4. **Pleasantness**: The event was pleasant.
5. **Unpleasantness**: The event was unpleasant.
6. **Goal Relevance**: I expected the event to have important consequences for me.
7. **Situational Responsibility**: The event was caused by chance, special circumstances, or natural forces.
8. **Own Responsibility**: The event was caused by my own behavior.
9. **Others' Responsibility**: The event was caused by someone else's behavior.

[1]Available in `https://github.com/sarnthil/crowd-enVent-modeling`

10. **Anticipated Consequences**: I anticipated the consequences of the event.
11. **Goal Support**: I expected positive consequences for me.
12. **Urgency**: The event required an immediate response.
13. **Own Control**: I was able to influence what was going on during the event.
14. **Others' Control**: Someone other than me was influencing what was going on.
15. **Situational Control**: The situation was the result of outside influences that no one had control.
16. **Accepted Consequences**: I accepted that I would easily live with the unavoidable consequences of the event.
17. **Internal Norms**: The event clashed with my standards and ideals.
18. **External Norms**: The actions that produced the event violated laws or socially accepted norms.
19. **Attention**: I had to pay attention to the situation.
20. **Not consider**: I tried to shut the situation out of my mind.
21. **Effort**: The situation required me a great deal of energy to deal with it.

In addition, the experiencer-annotators and the reader-annotators had to choose one emotion they or the experiencer felt from a list of 13 emotions: anger, boredom, disgust, fear, guilt, joy, pride, relief, sadness, shame, surprise, trust and no emotion.

3.2 Model Setup

To generate appraisal ratings, we used the GPT-4[2] via the Azure OpenAI REST API service.[3] We adopted the default parameters of temperature 0.7 and top_p 0.95, with both presence and frequency penalty 0. For Q1, we conducted 5 independent runs of a single prompt. For Q2, we performed only 1 run. For research questions Q2, Q3, and Q4, we set the number of completions GPT-4 parameter to 5.

3.3 Prompt

Our prompt to the GPT-4 consists of three parts: the general context **C**, more detailed instructions **I** and the description of the event **D**.

[2]Model version: 0613

[3]API version: 2023-07-01-preview

Context: The context part of the prompt gives an overall instruction to the model to act as an expert:

[C] *"You are an expert in human psychology. You will read descriptions people have written about an event or situation eliciting an emotional reaction. You will see a series of questions about how people think about these events."*

Instruction: The instruction part of the prompt asks the model to give integer ratings to the listed questions and specifies the output format. We used two versions of the instruction. The first instruction, I1, used for Q1 and Q2, asks GPT-4 to assign ratings to appraisal dimensions. Q3 used the same instruction with one additional question, "How confident is the model about chosen ratings?". The second instruction, I2, used for Q4, asks for the prediction of the emotion label and the confidence and intensity ratings for the emotion.

[I1] *Instruction: Based on event descriptions, assign integer ratings varying from 1 to 5 to the list of questions.*
Desired format: plain values in this order: <list of appraisal titles>
List of questions: <list of appraisals>

[I2] *Instruction: Based on descriptions, choose one emotion from the list of emotions and assign integer ratings varying from 1 to 5 to the list of questions.*
Desired format: plain values in this order: emotion, confidence, intensity, <list of appraisal titles>
List of emotions: <list of emotion titles>
List of questions: How confident are you about your emotion?, "How intense do you think the emotion was?", <list of appraisals>

Description: For event descriptions, we used the version of the dataset in which explicit words of the target emotion were masked, to avoid the model using these words as superficial heuristics for predicting evaluation ratings. The same masked version of the text was used by Troiano et al. (2023) to collect the appraisal ratings by the annotators.

[D] *"People get under my skin. Like for example if an entitled customer shows up at my work and demands to speak to my manager for a simple issue that I can resolve. This happens on* *almost a daily occurrence and it really makes me <masked>."*

4 Results

4.1 Q1 Reliability of GPT-4

We studied the Q1 on a subsample of the reader-annotator dataset. We aimed for a subsample of at least 100 event descriptions stratified to match the emotion category distribution in the full dataset. In order to account for potential losses due to Azure policy and other model and format generation errors, we opted for taking a subsample of 9% of the reader-annotator dataset. During appraisal rating generation, we encountered no policy errors, nor partial or empty outcomes, resulting in a subset on 108 event descriptions. To assess GPT-4 reliability in generating appraisal ratings, we calculated Root Mean Squared Errors (RMSE) and Spearman correlation coefficients between all pairs of appraisal rating vectors from five GPT-4 runs[4] for each appraisal dimension. We then averaged these metrics across all pairs.

The average pairwise RMSE (APRMSE) values, macro-averaged over all appraisal dimensions, were very close to each other, ranging from 0.6 to 0.63 over the five runs, with a mean of 0.61 (see Table 1. Across individual appraisal dimensions, the lowest (e.g., the most accurate) APRMSE value was observed for External Norms, while the model struggled the most with Anticipated Consequences, for which it obtained the highest (e.g., the least accurate) APRMSE score of 0.79.

Analyzing Spearman correlation coefficients, we observed strong and very strong correlations between responses from different GPT-4 runs. Mean Spearman coefficients ranged from 0.67 for Attention to 0.97 for Pleasantness, with a mean of 0.87 for the macro-average over all appraisal dimensions (1). This compares favorably with common guidelines in psychometrics that consider test-retest correlations above 0.8 to indicate good reliability and above 0.90 to indicate excellent reliability.

Based on these results, we conclude that GPT-4 is reliable in generating appraisal ratings, showing statistically significant Spearman correlations and average pairwise RMSE scores falling into a small tight range.

[4]In total $\frac{n\cdot(n-1)}{2} = \frac{5\cdot 4}{2} = 10$ pairs.

Appraisal	APRMSE	ρ
Suddenness	0.69	0.89
Familiarity	0.77	0.84
Event predictability	0.69	0.87
Pleasantness	0.38	0.97
Unpleasantness	0.42	0.96
Goal Relevance	0.58	0.89
Situational Responsibility	0.72	0.77
Own Responsibility	0.62	0.91
Others' Responsibility	0.62	0.91
Anticipated Consequences	0.79	0.68
Goal Support	0.59	0.93
Urgency	0.69	0.86
Own Control	0.66	0.87
Others' Control	0.66	0.92
Situational Control	0.76	0.81
Accepted Consequences	0.58	0.84
Internal Norms	0.57	0.93
External Norms	0.37	0.92
Attention	0.65	0.67
Not Consider	0.51	0.94
Effort	0.57	0.85
Macro average	0.61	0.87

Table 1: Average Pairwise RMSE (APRMSE) and Spearman correlation coefficients (ρ) of 108 random samples of five GPT-4 runs. All Spearman correlations were statistically significant at the level of $p < 0.001$.

4.2 Q2 Accuracy of GPT-4

After we ensured that GPT-4's responses were reliable on a small subset, we examined how well GPT-4 performed compared to the experiencer-annotators and human reader-annotators on more extensive data. We prompted GPT-4, using the prompt *I1*, to generate appraisal ratings to all 1200 event descriptions in the reader-annotator dataset. Recall, that this data is a subset of the experiencer-annotator dataset, so it has ratings from both the original event experiencer-annotators as well as human reader-annotators.

We calculated the RMSE between the GPT-4 predictions and the experiencer-annotators' responses, which shows how accurately the model can predict the appraisal ratings as a model reader-annotator. We also calculated the RMSE between the experiencer-annotators' responses and human reader-annotators' guesses, which enables to compare how well the model reader-annotator compares to human reader-annotators. Each text in the reader-annotator dataset has five human-reader annotations; therefore, we aggregated them by computing the mean RMSE of five appraisal ratings.

The results of this experiment are shown in the left-most section of Table 2. We found that the GPT-4 predictions were slightly closer to the hu-

	Q2 - Accuracy		Q3 - Majority Voting and Ties					Q4 - Emotion
Appraisal	GPT-4	$\text{Human}_{\text{avg}}$	$\text{GPT-4}_{\text{avg}}$	$\text{GPT-4}_{\text{maj}}$	$\text{Human}_{\text{maj}}$	$\text{GPT-4}_{\text{conf}}$	$\text{Human}_{\text{conf}}$	$\text{GPT-4}_{\text{emo}}$
Suddenness	1.50	1.58	1.58	1.21	1.03	1.25	1.08	1.37
Familiarity	1.51	1.58	1.96	1.53	1.04	1.61	1.16	1.31
Event predictability	1.44	1.56	1.70	1.31	1.08	1.38	1.16	1.14
Pleasantness	1.11	1.23	1.15	0.59	0.57	0.61	0.63	0.59
Unpleasantness	1.20	1.33	1.32	0.73	0.68	0.72	0.71	0.69
Goal Relevance	1.54	1.63	1.61	1.19	1.08	1.22	1.14	1.20
Situational Responsibility	1.43	1.65	1.67	1.10	1.03	1.12	1.14	1.50
Own Responsibility	1.34	1.45	1.60	1.03	0.83	1.06	0.91	2.26
Others' Responsibility	1.51	1.60	1.86	1.26	1.01	1.24	1.06	1.68
Anticipated Consequences	1.60	1.69	1.68	1.32	1.21	1.35	1.28	1.57
Goal Support	1.32	1.50	1.48	0.96	0.90	0.96	0.95	1.07
Urgency	1.56	1.76	1.75	1.33	1.28	1.32	1.25	1.32
Own Control	1.44	1.56	1.43	1.04	1.05	1.12	1.16	1.06
Others' Control	1.61	1.63	2.01	1.54	1.04	1.52	1.13	1.46
Situational Control	1.58	1.65	1.66	1.10	0.99	1.12	1.14	1.10
Accepted Consequences	1.65	1.84	1.64	1.25	1.39	1.27	1.43	1.39
Internal Norms	1.45	1.57	1.60	1.14	0.92	1.29	0.96	0.89
External Norms	1.08	1.31	1.34	0.79	0.59	0.87	0.66	0.81
Attention	1.45	1.53	1.50	1.08	1.01	1.07	1.02	1.06
Not Consider	1.59	1.64	1.52	0.98	1.03	1.00	1.07	1.18
Effort	1.40	1.61	1.48	1.10	1.09	1.13	1.16	1.05
Macro average	1.45	1.57	1.61	1.12	0.99	1.15	1.06	1.22

Table 2: RMSE results for research questions Q2, Q3, and Q4. GPT-4 stands for the GPT-4 annotator, Human for the human reader-annotator, *avg*: average of five GPT-4 completions/human guesses, *maj*: majority vote of five GPT-4 completions/human guesses, *conf*: majority vote of five GPT-4 completions/human guesses with confidence rating as a tiebreaker, *emo*: majority vote of five GPT-4 completions with the emotion prediction task in a prompt. Bold/underline marks the appraisal dimensions that are consistently predicted better/worse than the macro average.

man experiencer-annotators' guesses compared to human reader-annotators with an average RMSE over all appraisal dimension of 1.45 compared to 1.57. The results for all distinct appraisals (also in Table 2) reveal that for all appraisal dimensions, the model preditions were closer to human experiencer-annotators compared to human reader-annotators.

We note that the accuracy of the GPT-4-generated appraisal ratings is in the same range to the results reported by Troiano et al. (2023) obtained with a fine-tuned RoBERTa model (RMSE = 1.40), with a difference that our results in this section are obtained from just one prediction, whereas the results reported by Troiano et al. (2023) are the average over five runs. Thus, we conclude that GPT-4 is an effective tool for predicting appraisal ratings, performing very close or even slightly better than human reader-annotators.

4.3 Q3 Effect of Majority Voting

In previous work (Troiano et al., 2023), the accuracy of human reader-annotators guesses was considerably improved when their ratings were aggregated using the majority voting over the guesses of five annotators. Encouraged by this, we used GPT-4 to generate five completions for each text in the reader-annotator dataset, and tested whether the majority voting has an impact on prediction accuracy. If there was a clear majority vote, that rating was selected as the final prediction. In case of ties (e.g., either two ratings were predicted twice or all five ratings were different), we adopted the following essentially random procedure. While each text had five independent predictions from the GPT-4 model, we pretended to have done just a single completion and chose the rating from the first of the five runs. In total, we generated predictions for 1200×5 data samples using a prompt *I1* with added question "How confident is the model about chosen ratings?".

The application of the majority voting algorithm improved the RMSE on average by about 30% from 1.61 to 1.12 (columns $\text{GPT-4}_{\text{avg}}$ and $\text{GPT-4}_{\text{maj}}$ in Table 2). However, compared to the results shown in previous Section 4.2 (also shown in Table 2), generating five runs and taking the average increased the RMSE considerably, which is now in the same range to the average over the human reader-annotator ratings (1.61 vs 1.57; columns $\text{GPT-4}_{\text{avg}}$ and $\text{Human}_{\text{avg}}$ in Table 2). At the same time, the RMSE obtained using the majority voting is also considerably lower than the results shown in the previous subsection from just one run (1.12 vs 1.45; columns GPT-4_{maj} and GPT-4 in Table 2). Similar or even better results were obtained by applying majority voting to the guesses of the human reader-annotators, showing an improvement from 1.57 to 0.99 (columns $Human_{avg}$ and $Human_{maj}$ in Table 2).

Next, we analyzed how often a tie needs to be resolved in the ratings of human reader-annotators and GPT-4. We found that on average, 22% of human reader-annotators guesses required tie-breaking, while for GPT-4, only 9% required that. Since the number of ties was not negligible, we considered using a method similar to the one adopted by Troiano et al. (2023) based on rating the confidence. Specifically, we added to the prompt's questions list the question "How confident is the model about chosen ratings?" as an alternative to the question posed to the human reader-annotator's by Troiano et al. (2023) "How confident are you about your answer?".[5] Similar to other questions, this question was rated on the 5-point Likert scale with values ranging from 1-5. To break a tie, the value with a highest average confidence rating was chosen.

The results of this experiment are shown in Table 2 columns GPT-4_{conf} and $Human_{conf}$. We can see that using the confidence rating to break the ties did not improve the RSME neither for the model nor for the human reader-annotators compared to the random choice. The mean RMSE for GPT-4 from 1.12 to 1.15 (columns GPT-4_{maj} and GPT-4_{conf} in Table 2), and the mean RMSE for human reader-annotators increased from 0.99 to 1.06 (columns $Human_{maj}$ and $Human_{conf}$ in Table 2). When looking at individual appraisal dimensions, we can see that the RMSE's are in most cases lower in the random tie-breaking setting compared to using the confidence rating, and that applies to both the model and humans. Thus, we conclude that using the model-generated confidence rating is a not a useful cue for breaking the ties when aggregating several ratings via majority voting.

4.4 Q4 Effect of Adding Emotion Prediction

To study Q4, we again generated five appraisal ratings for each of the 1200 event description in the reader-annotator dataset, using the prompt *I2* that added the task of identifying the target emotion. That means, the prompt asked to pick one emotion from a set of given list of 13 emotion categories (including the no emotion category) most likely felt by the author of the text. This addition made the generation process more error-prone, requiring significantly more runs for entries with shorter descriptions.

In addition, while generating data for Q4, we again generated ratings (confidence and intensity) to be used as tiebreakers, but this time, we prompted exactly the same questions that were used by Troiano et al. (2023) in the human reader-annotator questionnaire. Instead of "How confident is the model about chosen ratings?" we asked, "How confident are you about your answer?"[6] and for the intensity rating, we asked, "How intense do you think the emotion was?". However, we got almost identical results to the random choice voting, which leaves open the question of how to effectively break ties.

The results of adding emotion to a prompt are shown in the last column of Table 2. The average RMSE scores for five completions applying the majority voting algorithm was 1.22 (see column GPT-4_{emo}), which is worse than the same task without the emotion label prediction studied in relation to research question Q3. Thus, we conclude that simply adding the target emotion identification task does not improve the appraisal rating predictions. We note also that the emotion prediction accuracy was ca 55%, and in 24 cases, GPT-4 generated emotions not present in the predefined list.[7]

4.5 Q5 Impact of Event Description Length

From the beginning of the research, we noted that GPT-4 struggled with shorter event descriptions by often producing inconsistent or empty responses, which suggests that the predictions can be less accurate for shorter texts. Therefore, we analyzed the correlations between event description length (in characters) and the RMSE of the appraisal predictions and examined if the correlation pattern is similar between GPT-4 and human reader-annotators.

We split the reader-annotator dataset into ten bins, each containing even descriptions grouped by length of 100 character intervals and compared the average RMSE scores of each bin for both GPT-4 predictions and human reader-annotator guesses. For that experiment, we used that GPT-4 predictions obtained for Q2 in Section 4.2.

[5] "Answer" refers to "emotion" used in the questionnaire

[6] Referring to the emotion label prediction.

[7] anxiety, disappointment, embarrassment, frustration, jealousy, betrayal, pain, distraction, and indifference

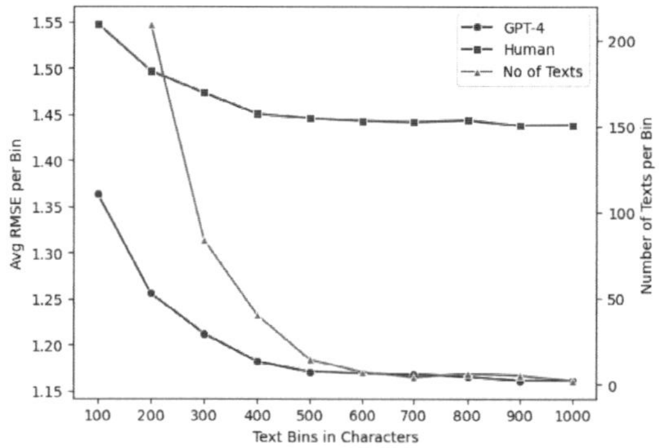

Figure 1: Average RMSE of texts with different lengths. The x-axis labels show the end of the bin in characters: the first bin contains texts with length up to 100 characters, the second between 100 and 200 characters, etc. The secondary y-axis plots the number of texts in each bin, except for the first bin, where the number of texts was ca 800 and thus did not fit to the plot.

We calculated Spearman's correlation coefficients between the text length and the macro-averaged RMSE for both the model predictions and human reader-annotator guesses. The correlations were negative and very close for the model and humans on average: -0.79 for GPT-4 and -0.75 for human reader-annotators, indicating that for both the model and humans, it is harder to predict appraisal ratings for shorter texts. The relation between the text length and the RMSE plotted in Figure 1 show that while the human annotator scores show consistently higher RMSE values, the shape of both curves is similar.

We also looked at the correlations between the text length and the RMSE of each appraisal dimension separately. Two appraisal dimensions indicating the valence of the event (Pleasantness and Unpleasantness) showed near perfect correlations for both GPT-4 and human annotators. Few other dimensions (Suddenness, Own responsibility, Others' responsibility, Situational control) also showed very strong correlations (in the range of -0.99 – -0.80). Most other dimensions fell into the strong correlation range (-0.79 – -0.60), but some were moderate (-0.59 – -0.40).

We also observed that GPT-4 and human annotators had moderate coefficients for different appraisals: Familiarity, Anticipated Consequences, and Not Consider for GPT-4, and Goal Relevance, Urgency, Internal and External Norms for human reader-annotators. Finally we note that, with few exceptions, the GPT-4 correlations for individual appraisals tend to be stronger than for human annotators, which means that the accuracy of the GPT-4 predictions is somewhat more dependent on the text length.

Overall, we conclude that there is, on average, a strong negative correlation between the event description length and the accuracy of predicting appraisal rating for both GPT-4 and human reader-annotators. The sweet spot seems to be around 400-500 characters where the RMSE starts to plateau for both GPT-4 and humans.

5 Discussion

In analyzing appraisal dimensions across all experiments, we looked for patterns by comparing the RMSE of individual appraisal dimensions to the macro-averaged RMSE. Appraisals consistently showing better accuracy across both GPT-4 models and human annotators include Pleasantness, Unpleasantness, Goal Support, and External Norms (marked as bold in Table 2), suggesting that these are the easiest to infer based on text. Own Responsibility, Own Control, Not Consider, and Effort also generally performed well, although Own Responsibility is predicted remarkably worse in the GPT-4_{emo} setting. In contrast, dimensions such as Familiarity, Others' Responsibility, Anticipated Consequences, Urgency, Others' Control, and Accepted Consequences consistently show RMSE values equal to or higher than the macro average for both GPT-4 models and human reader-annotators (marked with underscore in Table 2), meaning that the rating of these dimensions were more difficult to infer accurately.

In contrast to our expectations, asking the GPT-4 model to predict the emotion first and then the appraisal ratings (Q4) did not improve the overall prediction, although the emotion label was expected give useful information about the appraisal values. This might be due to the accuracy of the emotion prediction task being moderate, only at 55%, and incorrect emotion category predictions may have confused the model. To check that, we split the data into two groups by correctly and incorrectly predicted emotion labels. Although texts with correctly predicted labels demonstrated better macro average RMSE compared to the texts with incorrect labels (1.21 vs 1.24 RMSE), the difference is not large. However, a closer look at individual appraisal dimensions revealed that most appraisals showed significantly better or comparable accuracy for correctly predicted emotions, with few exceptions of Situational Responsibility (1.57 vs 1.42), Others' Responsibility (1.70 vs 1.65), Anticipated Consequences (1.65 vs 1.48), and Not Consider (1.25 vs 1.10). A more fine-grained analysis of why the predictions of these dimensions were more accurate with incorrect emotion labels remains for future research.

We also found that shorter event descriptions generally exhibit lower RMSE values. This result is expected as very short texts convey too little information to predict various appraisal aspects accurately. We found that the optimal event description length that appears to start from the range of roughly 400 to 500 characters and although the improvement in prediction accuracy is more steep for the GPT-4 model, a similar pattern can be observed also for human-readers. This finding has implications for collecting data for emotion appraisal research, as it suggests that researchers should aim for eliciting event descriptions at least 500 characters long. Finally, throughout the research, we observed that adding more complexity to a prompt resulted in less consistent responses as well as higher RMSE scores. This observation is in line with the reports of other researchers (Herderich et al., 2024).

We conclude that GPT-4 is an effective tool for annotating appraisals, though the reliability and accuracy vary across different appraisal dimensions. Our expenses of generating more than 14000 data points to test different strategies were around €200, significantly lower than the £2188 reported by Troiano et al. (2023) for 6000 entries annotated by human readers. Thus, GPT-4 annotations can be a viable alternative to the more costly human reader-annotator ratings in studies requiring large datasets or for generating enough synthetic data for training smaller, local models.

6 Conclusions

In this paper, we studied the reliability and accuracy of GPT-4 in annotating 21 emotion appraisals as an alternative to human reader-annotators. The results showed that GPT-4 annotations are highly reliable across several independent runs and it can annotate appraisals with near-human accuracy. Moreover, these results can be considerably improved with using majority voting algorithm over five model completions, which increased the accuracy of both GPT-4 and human reader-annotators by more than 30%. Although we tried using predicted confidence rating to resolve the ties, it did not lead to lower RMSE. Thus, there is room for further improvements by finding a better way to resolve the ties.

Impact

In our research, GPT-4 predictions performed similarly to human reader-annotators in annotating appraisal ratings and thus could be applied in psychological research and practice. Predicting user appraisal profile of emotional events could help to identify behavioral and emotional patterns, support therapeutic interventions, or have other practical applications.

Our study also contributes to the set of effective strategies in predicting appraisal ratings by GPT-4 or potentially similar LLMs. We show that adopting majority voting algorithm based on five completions can considerably improve the performance of this subjective task. Moreover, we empirically establish an minimum optimal event description length below which both human readers and GPT-4 model prediction accuracy starts to degrade, thus providing a practical guideline for appraisal researchers interested in using predictive models.

The results of this work can inform further research in developing automated reappraisal self-help systems or similar applications, offering practical tools for emotional reframing and appraisal annotation. However, in those settings, the requirement to submit sensitive and private user content to the GPT-4 API might not be desirable. An alternative would be to use GPT-4 predictions to augment the training data to improve the accuracy of smaller, local, models that could be subsequently applied in more sensitive data settings.

Limitations

An important limitation of our research is the limited generalizability of the results. We used only one model, GPT-4, and thus our results cannot be generalized to other available LLMs, although we anticipate that open-source models would probably show poorer results. We also used the specific default configuration of the GPT-4 model. LLMs are sensitive to hyperparameters tuning, and our findings can be applicable only to the used settings.

Another significant limitation arises from the dataset characteristics. The dataset used in this study reflects limited demographic and socioeconomic diversity, and the emotional events reported by crowd-source participants may be synthetic, potentially reducing their relevance to real-world contexts.

Finally, GPT-4 and other LLM models can only act as reader-annotators and are limited in the experiencer-annotators roles. Thus, our results might have limited impact in psychological research aiming to study the role of appraisals in emotional experiences, as this subjective information can only be provided by human experiencers.

Ethical Considerations

Using GPT-4 in a reader-annotator context raises ethical questions regarding the accuracy and bias of the annotations. Therefore, GPT-4 annotations should always be validated by both actual event experiencers and human reader-annotators.

Another ethical aspect that has to be taken into account is the privacy and sensitivity of the data. Our research used an open and freely available dataset that does not contain sensitive or private content, collected by Troiano et al. (2023) for research purposes.

However, the use this dataset raises some considerations regarding its intended scope and limitations. The dataset was crowd-sourced and as such, we can assume that the participants were a sample from a generally healthy population. This constraint needs to be kept in mind when using the results of our research in designing systems for clinical domains, such as self-help systems intended to aid in emotional reappraisal for people with clinical issues.

References

Meysam Alizadeh, Maël Kubli, Zeynab Samei, Shirin Dehghani, Juan Diego Bermeo, Maria Korobeynikova, and Fabrizio Gilardi. 2023. Open-source large language models outperform crowd workers and approach chatgpt in text-annotation tasks. *arXiv preprint arXiv:2307.02179*.

Bosheng Ding, Chengwei Qin, Linlin Liu, Yew Ken Chia, Boyang Li, Shafiq Joty, and Lidong Bing. 2023. Is GPT-3 a good data annotator? In *Proceedings of the 61st Annual Meeting of the Association for Computational Linguistics (Volume 1: Long Papers)*, pages 11173–11195, Toronto, Canada. Association for Computational Linguistics.

Fabrizio Gilardi, Meysam Alizadeh, and Maël Kubli. 2023. Chatgpt outperforms crowd workers for text-annotation tasks. *Proceedings of the National Academy of Sciences*, 120(30):e2305016120.

Alina Herderich, Jana Lasser, Mirta Galesic, Segun Taofeek Aroyehun, David Garcia, and Joshua Garland. 2024. Measuring complex psychological and sociological constructs in large-scale text.

Jan Hofmann, Enrica Troiano, and Roman Klinger. 2021. Emotion-aware, emotion-agnostic, or automatic: Corpus creation strategies to obtain cognitive event appraisal annotations. In *Proceedings of the Eleventh Workshop on Computational Approaches to Subjectivity, Sentiment and Social Media Analysis*, pages 160–170.

Jan Hofmann, Enrica Troiano, Kai Sassenberg, and Roman Klinger. 2020. Appraisal theories for emotion classification in text. In *Proceedings of the 28th International Conference on Computational Linguistics*, pages 125–138.

Jan Kocoń, Igor Cichecki, Oliwier Kaszyca, Mateusz Kochanek, Dominika Szydło, Joanna Baran, Julita Bielaniewicz, Marcin Gruza, Arkadiusz Janz, Kamil Kanclerz, et al. 2023. Chatgpt: Jack of all trades, master of none. *Information Fusion*, 99:101861.

Zheng Lian, Licai Sun, Haiyang Sun, Kang Chen, Zhuofan Wen, Hao Gu, Bin Liu, and Jianhua Tao. 2024. Gpt-4v with emotion: A zero-shot benchmark for generalized emotion recognition. *Information Fusion*, 108:102367.

Agnes Moors, Phoebe C Ellsworth, Klaus R Scherer, and Nico H Frijda. 2013. Appraisal theories of emotion: State of the art and future development. *Emotion Review*, 5(2):119–124.

Minxue Niu, Mimansa Jaiswal, and Emily Mower Provost. 2024. From text to emotion: Unveiling the emotion annotation capabilities of llms. In *Proc. Interspeech 2024*, pages 2650–2654.

Steve Rathje, Dan-Mircea Mirea, Ilia Sucholutsky, Raja Marjieh, Claire E Robertson, and Jay J Van Bavel. 2024. Gpt is an effective tool for multilingual psychological text analysis. *Proceedings of the National Academy of Sciences*, 121(34):e2308950121.

Klaus R Scherer. 2009. The dynamic architecture of emotion: Evidence for the component process model. *Cognition and emotion*, 23(7):1307–1351.

Ala N Tak and Jonathan Gratch. 2023. Is gpt a computational model of emotion? detailed analysis. *arXiv preprint arXiv:2307.13779*.

Enrica Troiano, Laura Oberländer, and Roman Klinger. 2023. Dimensional modeling of emotions in text with appraisal theories: Corpus creation, annotation reliability, and prediction. *Computational Linguistics*, 49(1):1–72.

Nutchanon Yongsatianchot, Parisa Ghanad Torshizi, and Stacy Marsella. 2023. Investigating large language models' perception of emotion using appraisal theory. In *2023 11th International Conference on Affective Computing and Intelligent Interaction Workshops and Demos (ACIIW)*, pages 1–8.

Hongli Zhan, Desmond Ong, and Junyi Jessy Li. 2023. Evaluating subjective cognitive appraisals of emotions from large language models. In *Findings of the Association for Computational Linguistics: EMNLP 2023*, pages 14418–14446, Singapore. Association for Computational Linguistics.

AutoPsyC: Automatic Recognition of Psychodynamic Conflicts from Semi-structured Interviews with Large Language Models

Sayed Muddashir Hossain[1], Simon Ostermann[1], Patrick Gebhard[1], Cord Benecke[2], Josef van Genabith[1], Philipp Müller[1]

[1]DFKI, Saarbrücken, Germany
[2]University of Kassel, Kassel, Germany

{sayed_muddashir.hossain,simon.ostermann,patrick.gebhard}@dfki.de
{josef.van_genabith,philipp.mueller}@dfki.de

Abstract

Psychodynamic conflicts are persistent, often unconscious themes that shape a person's behaviour and experiences. Accurate diagnosis of psychodynamic conflicts is crucial for effective patient treatment and is commonly done via long, manually scored semi-structured interviews. Existing automated solutions for psychiatric diagnosis tend to focus on the recognition of broad disorder categories such as depression, and it is unclear to what extent psychodynamic conflicts which even the patient themselves may not have conscious access to could be automatically recognised from conversation. In this paper, we propose AutoPsyC, the first method for recognising the presence and significance of psychodynamic conflicts from full-length Operationalized Psychodynamic Diagnostics (OPD) interviews using Large Language Models (LLMs). Our approach combines recent advances in parameter-efficient fine-tuning and Retrieval-Augmented Generation (RAG) with a summarisation strategy to effectively process entire 90 minute long conversations. In evaluations on a dataset of 141 diagnostic interviews we show that AutoPsyC consistently outperforms all baselines and ablation conditions on the recognition of four highly relevant psychodynamic conflicts.

1 Introduction

Accurate and detailed analysis of clinical interviews is essential for effective psychodynamic diagnostics. In particular, Operationalized Psychodynamic Diagnostics (OPD) interviews (Force, 2008) serve as a cornerstone in psychodynamic assessment. A key aspect of OPD is the assessment of the patient's life-determining, often unconscious inner conflicts, such as conflicts relating to Dominance or Submissiveness, or to Self-value/esteem. Automated analysis of psychodynamic conflicts from clinical interviews has the potential to support clinicians, reduce manual work, enhance objectivity, and may even lay the groundwork for the delivery of diagnostic interviews by artificial agents. However, due to their long duration, low level of standardisation, and richness of information, semi-structured interviews pose unique challenges (Adams, 2010; Magaldi and Berler, 2020). Prior natural language processing (NLP) work has often focused on short interview excerpts and broad diagnostic categories (Low et al., 2020; Milintsevich et al., 2023). To the best of our knowledge, no previous work has addressed the recognition of fine-grained psychodynamic concepts from long semi-structured diagnostic interviews.

In this work, we introduce a novel approach for recognising the presence and significance of psychodynamic conflicts as classified in the OPD from full-length interviews. Our method combines recent advancements in parameter-efficient fine-tuning (Hu et al., 2022) and Retrieval-Augmented Generation (RAG) (Lewis et al., 2020) with a summarisation approach in order to process and classify long (90 min) semi-structured psychodynamic diagnostic interviews. In particular, we make use of a RAG framework to let the LLM access full-length interviews. To allow the model to effectively reason about the interview to be scored, we additionally prompt it with a summary of the interview. The classification is performed by an ensemble of four models, each of which was fine-tuned to analyse a specific temporal portion of an interview. To evaluate our approach, we make use of a dataset of 141 OPD interview recordings (Bock et al., 2016). Our approach consistently improves over baselines and ablation conditions. It is able to reach weighted F1 scores of 0.78 and 0.81 for the conflicts *Self-dependency and Dependency on Others*, and *Dominance or Submissiveness*. For the more challenging conflicts *Self-sufficiency* and *Self-value/esteem*, it is able to reach 0.59 and 0.58 F1, respectively.

Our specific contributions are threefold:

Proceedings of the 10th Workshop on Computational Linguistics and Clinical Psychology (CLPsych 2025), pages 12–25
May 3, 2025 ©2025 Association for Computational Linguistics

1. We present AutoPsyC[1], the first LLM-based method for the recognition of presence and severity of psychodynamic conflicts from full-length OPD interviews, thereby bridging the fields of psychodynamic diagnostics and advanced NLP.

2. We evaluate AutoPsyC on a dataset of 141 90 minute long OPD interviews, showing that AutoPsyC consistently outperforms all baselines and in-depth ablation comparisons.

3. We demonstrate that information contained in the middle sections of interviews is particularly informative for classifier training.

2 Related Work

2.1 Diagnostic Interviews in Psychotherapy and Psychiatry

Structured interviews, using standardized questions and scoring, improve psychiatric diagnosis reliability by reducing clinician bias. Tools like the Structured Clinical Interview for DSM-5 (SCID-5, First et al. (2016)) ensure DSM-aligned accuracy but require extensive training, while the Mini-International Neuropsychiatric Interview (MINI, Sheehan et al. (1998)) offers efficient screening at the cost of some diagnostic precision. The Structured Interview of Personality Organization (STIPO) is a structured interview designed to assess personality functioning based on Kernberg's object relations theory (Clarkin et al., 2007).

Unstructured interviews emphasize patient narratives and clinical intuition, enabling the exploration of unique experiences (Nordgaard et al., 2013). While fostering rapport and uncovering insights, their lack of standardization introduces variability and reduces reliability (Shea, 2016; Corbin and Morse, 2003; O'Brien and Tabaczynski, 2007; Widiger, 2008; Fava et al., 2024; Lenouvel et al., 2022). In this context, the PDM-2 provides a multi-dimensional diagnostic framework that emphasizes psychological functioning and personality organization over categorical symptom-based diagnosis (Lingiardi et al., 2015). Likewise, Malan's triangles offer a conceptual model for understanding intrapsychic conflict and resistance, rather than a formalized interview procedure (Malan, 1979).

Semi-structured interviews blend the structure of standardized questions with the flexibility to address emergent themes (Fava et al., 2024; Lenouvel et al., 2022; Adams, 2010; Brinkmann, 2014; Magaldi and Berler, 2020; Adeoye-Olatunde and Olenik, 2021). They have been shown to be particularly useful in complex cases like major depressive disorder (Dupuy et al., 2020). One example for a sem-structured format is the Core Conflictual Relationship Theme (CCRT) method to identify recurring interpersonal themes (Luborsky and Crits-Christoph, 1998). Operationalized Psychodynamic Diagnosis (OPD) uses semi-structured methods rooted in psychodynamic theory to assess self-experience, interpersonal relationships, and unconscious conflicts (Force, 2008; Cierpka et al., 2007). Unlike symptom-focused tools, OPD provides in-depth insights into personality organization and internal dynamics, aiding personalized therapeutic interventions. Research has demonstrated the clinical relevance of OPD within therapeutic settings (Cierpka et al., 2007; Benecke, 2024; Cierpka et al., 2001; Rudolf et al., 2004). Despite their importance, the automatic analysis of semi-structured interviews remains under-explored. In particular no previous work has attempted to automatically score OPD interviews.

2.2 Large Language Models for Psychiatric Diagnosis

The integration of NLP and machine learning in different aspects of mental health is a rapidly growing field of research (Le Glaz et al., 2021; Lindsay et al., 2021; Hossain et al., 2024). One particular focus of attention is the automated diagnosis of conditions such as depression or schizophrenia by analysing text, speech, and facial expressions (Barzilay et al., 2019; Low et al., 2020; Kishimoto et al., 2022; Oh et al., 2024; Milintsevich et al., 2023; Ettore et al., 2023). Tools like *Diagnostisches Expertensystem für psychische Störungen* (DIA-X-5) are being tested for consistency (Hoyer et al., 2020), while patient involvement is emphasized to ensure ethical use (Brederoo et al., 2021).

Recent work has shown that LLMs can be utilised to analyse complex human affect expression in conversation (Broekens et al., 2023; Müller et al., 2024), making them a promising candidate for applications in psychiatric disorders. Indeed, LLMs are increasingly applied in psychiatry, identifying linguistic markers of disorders from social media posts and clinical transcripts (Farruque et al., 2024; Lan et al., 2024; Zhang et al., 2024b). These models also assist in parsing unstructured EHR

[1] Code available at `https://git.opendfki.de/philipp.mueller/autopsyc`

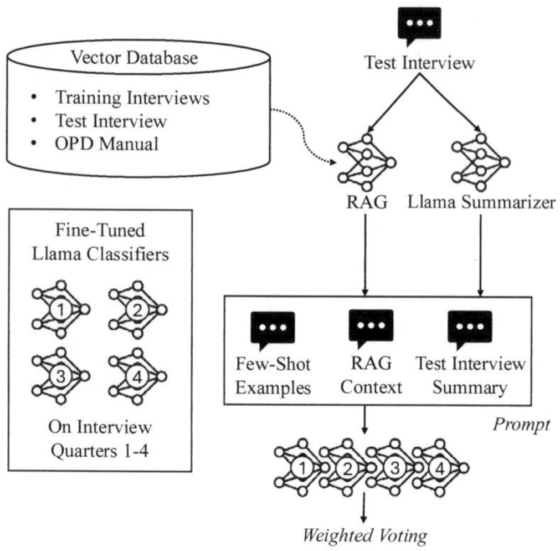

Figure 1: Overview over AutoPsyC.

notes for early diagnosis (Zhang et al., 2024b).

However, current automatic methods mainly detect broad disorder categories without capturing the more detailed and often unconscious factors explored in psychodynamic assessments. In particular, we are not aware of any approach to automatically recognise psychodynamic conflicts from clinical interviews.

2.3 Integrating Domain Knowledge in Large Language Models

The integration of domain-specific knowledge into LLMs, particularly through techniques like Retrieval-Augmented Generation (RAG), is transforming psychiatric applications by enabling models to access and apply current, relevant external information in real time (Lewis et al., 2020). Building on RAG, RAFT (Retrieval-Augmented Facilitation for Text) further optimizes domain-specific knowledge integration by prioritizing the most relevant medical literature and clinical guidelines (Zhang et al., 2024a). Other methods, such as parameter-efficient domain knowledge integration and the use of human-annotated features, also improve LLM performance in biomedical contexts (Ke et al., 2021; Kim et al., 2024). In our work, we utilise RAG techniques to present the first LLM-based system able to recognise psychodynamic conflicts expressed in full-length OPD interviews.

3 Method

A schematic overview of our method is shown in Figure 1. The classification of a test interview consists of three steps. In the first step, we employ LLaMA 3.1 (8B) (Grattafiori et al., 2024) to generate a summary of the interview. In the second step, we use this summary to build a prompt for the final classification, which is performed by an ensemble of LLaMa 3.1 (8B) models that are fine-tuned to different interview segments. The prompts for these specialised models contain 5 interview summaries from the training set, including ground truth (few-shot examples). Via a RAG framework (Lewis et al., 2020), each specialised model also has access to relevant sections of the OPD manual, as well as to the full test interview and all full interviews from the training set. In the third step, we combine the individual classifications obtained from the specialised models using a weighted voting scheme driven by a multinomial logistic regression.

3.1 Summarization Method

To obtain a focused representation of each interview, we first generate a summary using a LLaMA 3.1 8b model (Grattafiori et al., 2024). The summarization prompt includes an example summary excerpted from the OPD Manual, instructing the model to adhere to a consistent style that reflects the diagnostic criteria. In this way, the generated summaries capture the diagnostically relevant information while filtering out extraneous details.

3.2 Training Data Integration and RAG Setup

In addition to the test interview summary, the classification prompt also includes few-shot examples in the form of summaries of interviews from the training set with associated ground truth labels. We include one interview summary per ground truth class. To further ground the classification in a domain-specific context, we employ a Retrieval-Augmented Generation (RAG) framework. In particular, we upload the following information into the RAG vector database.

1. Training Interviews: We upload the full transcripts of all interviews from the training set without ground truth into the RAG knowledge base, pointing the model to them for retrieval. Adding ground truth information did not lead to improvements in preliminary experiments.

2. Test Interview: We also upload the full transcript of the test interview.

3. OPD Manual: The OPD-2 manual (Force, 2008) is organized into chapters correspond-

ing to its axes, providing detailed descriptions and examples of OPD tasks and classifications. For our purposes, we included excerpts from the chapter on conflicts (Axis III) and the introductory section where the axes are defined and explained. Preliminary tests showed that incorporating the entire OPD-2 manual into the retrieval-augmented generation (RAG) system did not improve model performance compared to using only the relevant excerpts.

This integration ensures that the classification model benefits from exemplars of each diagnostic class and explicit domain knowledge. The model is also able to access detailed information present in the full interview transcripts in case the summaries are inconclusive.

3.3 Classification Stage: Interview Segmentation and Finetuning

Given that OPD interviews are semi-structured—with diagnostic cues distributed throughout—we split each interview (or its summary) into k segments, where in our implementation $k = 4$ (each segment being roughly 5,000 words). In this way, each segments represents one quarter of the interview. For each segment, we fine-tune a separate Llama 3.1 (8B) model using LoRA (Hu et al., 2022), which allows for parameter-efficient adaptation. During training, we provide the model with a prompt including the segment summary, the RAG-augmented context (i.e., training interview summaries and manual excerpts), and an example for each of the five classes. The model is trained to output a probability distribution over the diagnostic classes. This process allows each fine-tuned model to capture the specific semantic and contextual nuances present in its corresponding interview segment.

3.4 Result Aggregation

After obtaining classification probabilities from each of the four fine-tuned models, we combine their outputs using a weighted voting scheme. Specifically, we train a multinomial logistic regression model that assigns a weight w_i to the prediction $p_i(c)$ of the i-th segment for class c. The final predicted class $\hat{y}$ is computed as:

$$\hat{y} = \arg\max_{c} \sum_{i=1}^{k} w_i \, p_i(c) \tag{1}$$

where $k = 4$ in our implementation.

4 Dataset

The Kassel dataset (Bock et al., 2016), utilized in this study, comprises 141 participants recorded during Operationalized Psychodynamic Diagnostics (OPD) interviews.

4.1 Participants

The dataset includes both male (n = 21) and female (n = 120) participants, aged between 18 and 57 years. Among them, 64 were inpatients diagnosed with at least one DSM-IV (Association, 2000) disorder, while 20 were healthy controls. The remaining participants had diverse diagnostic categories, including somatoform disorders (n = 22), borderline personality disorder (n = 19), depression (n = 18), and eating disorders such as anorexia (n = 14) and bulimia (n = 14). Anxiety disorders were observed in 13 participants. The inclusion criteria required informed consent, age above 18 years, and the absence of acute psychosis or schizophrenia.

4.2 Data Collection

Each participant underwent a clinical interview based on the framework of Operationalized Psychodynamic Diagnostics (Force, 2008). The interviews were carried out by a team of two male and two female interviewers (Bock et al., 2016), all of whom were certified and trained in OPD application. The sessions, with an average duration of approximately 90 minutes, were recorded using split-screen technology to capture both the participants and interviewers. Both the interviewer and the interviewee were equipped with microphones to ensure clear audio capture. The audio from each session was transcribed verbatim into text by research staff or trained transcribers (Bock et al., 2016; Vierl et al., 2023). These recordings provided the foundational data for subsequent analyses of behavioural and contextual elements.

4.3 Clinical Ground Truth

The dataset includes scores for Axes I-V of the OPD system. In the scope of this paper, we focus on Axis III, which captures the patient's life-determining (un)conscious inner conflicts. In

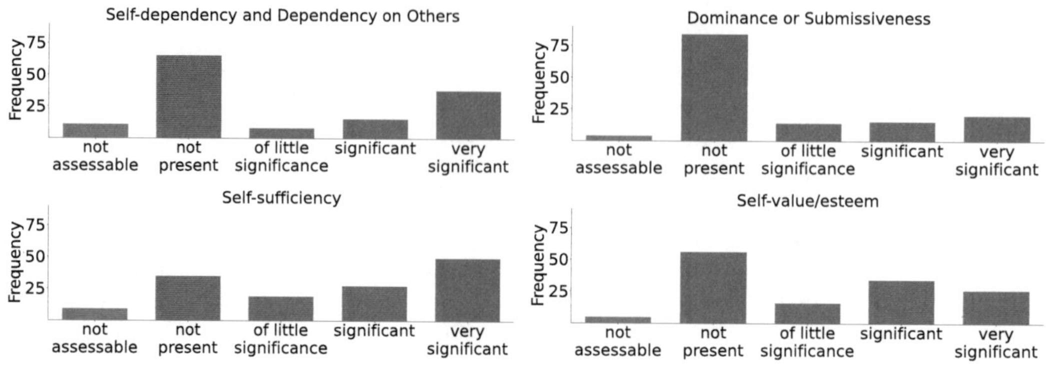

Figure 2: Class distributions for the psychodynamic conflicts investigated in this study.

particular, these are: *Conflicts related to self-dependency and dependency on others*, *Conflicts associated with dominance or submissiveness*, *Conflicts revolving around self-sufficiency*, *Self-value and self-esteem conflicts*, *Oedipal conflicts*, *Identity-related conflicts*. Each of these conflicts is rated with a five-class classification scheme. The classes are *Not assessable, Not present, Of little significance, Significant, Very significant*. Their detailed description can be found in A.3.

We decided to omit the conflicts *Oedipal Conflict* and *Conflicts Related to Identity* from further analysis, as these conflicts were diagnosed in only a few instances, making a robust evaluation of predictions infeasible. For instance, in the case of *Oedipal Conflict*, 120 out of 141 instances were labeled as *not present* (see Appendix for further details). Figure 2 illustrates the class distribution of the remaining conflicts. While *not present* is the most prevalent class for every conflict except Self-sufficiency, in all cases a significant portion of participants exist for whom the respective conflict is at least present with little significance. In the following we provide a concise explanation of the four conflicts included in our analysis.

Conflicts related to self-dependency and dependency on others refer to the tension between striving for autonomy and seeking support from others, often leading to struggles between independence and fear of isolation. *Conflicts associated with dominance or submissiveness* involve power dynamics in interpersonal relationships, where individuals may oscillate between asserting control and yielding to authority, potentially resulting in power struggles or passivity. *Conflicts revolving around self-sufficiency* pertain to the balance between the need for care and the desire for independence, with individuals experiencing inner turmoil when their reliance on others contradicts their self-reliance. *Self-value and self-esteem conflicts* center on an individual's sense of worth, encompassing struggles with feelings of inadequacy or inferiority, often manifesting in compensatory behaviours aimed at reinforcing self-image. The conflicts were classified based on the overall interview. When binarizing the conflicts by treating *not assessable* and *not present* as "No," and all other categories as "Yes," only three cases did not present with any of the conflicts.

5 Experiments

In this section, we describe the evaluation protocol and baselines.

5.1 Evaluation Protocol

The dataset was partitioned into five fixed folds using stratified 5-fold cross-validation to maintain the proportional representation of key demographic and diagnostic variables. This was achieved by utilizing the `StratifiedKFold` module from the `scikit-learn` library (Pedregosa et al., 2011), with stratification based on *Gender* and *Diagnosis*. The stratification process guarantees a fair distribution of these attributes across all folds. This consistency was preserved across all experiments involving the 5-fold cross-validation framework.

To evaluate all models and baselines, we make use of the weighted F1 score. The weighted F1 score accounts for class imbalance by weighting classes proportionally to their prevalence, ensuring robust evaluation of both frequent and rare diagnostic categories. It balances precision (avoiding over-

	Self-dep. & others-dep.	Dom. or sub.	Self-suff.	Self-val. & self-est.
Naive Baselines				
Demographic	0.31 (±0.01)	0.46 (±0.02)	0.31 (±0.03)	0.26 (±0.01)
Random	0.30 (±0.00)	0.33 (±0.00)	0.20 (±0.00)	0.23 (±0.00)
No Training Data in VDB, No Fine-tuning				
w/o Manual	0.51 (±0.01)	0.63 (±0.01)	0.39 (±0.02)	0.43 (±0.03)
w/o Test Interv. in VDB	0.46 (±0.01)	0.60 (±0.01)	0.46 (±0.02)	0.48 (±0.01)
w/o Test Interv. Summary	0.53 (±0.01)	0.64 (±0.01)	0.42 (±0.02)	0.46 (±0.01)
w/o Few-shot Examples	0.65 (±0.01)	0.68 (±0.01)	0.53 (±0.03)	0.48 (±0.02)
AutoPsyC	0.68 (±0.01)	0.70 (±0.02)	0.55 (±0.01)	0.48 (±0.02)
Training Data in VDB (Unlabelled), No Fine-tuning				
w/o Manual	0.48 (±0.03)	0.61 (±0.02)	0.43 (±0.02)	0.49 (±0.02)
w/o Test Interv. in VDB	0.50 (±0.01)	0.60 (±0.01)	0.45 (±0.01)	0.50 (±0.01)
w/o Test Interv. Summary	0.62 (±0.01)	0.69 (±0.02)	0.47 (±0.01)	0.52 (±0.02)
w/o Few-shot Examples	0.68 (±0.02)	0.73 (±0.01)	0.57 (±0.01)	0.50 (±0.04)
AutoPsyC	0.70 (±0.01)	0.74 (±0.02)	0.58 (±0.01)	0.50 (±0.02)
Training Data in VDB (Unlabelled), Fine-tuning				
w/o Test Interv. Summary & Manual & Train Interv. in VDB	0.65 (±0.04)	0.68 (±0.01)	0.49 (±0.02)	0.47 (±0.02)
w/o Test Interv. Summary & Manual	0.69 (±0.02)	0.72 (±0.03)	0.50 (±0.01)	0.49 (±0.01)
w/o Test Interv. Summary & Weighted Voting	0.73 (±0.01)	0.75 (±0.02)	0.56 (±0.01)	0.53 (±0.02)
w/o Test Interv. Summary & Ensemble	0.72 (±0.02)	0.74 (±0.01)	0.55 (±0.02)	0.52 (±0.01)
w/o Manual & Train Interv. in VDB	0.68 (±0.02)	0.72 (±0.01)	0.51 (±0.02)	0.49 (±0.01)
w/o Weighted Voting	0.75 (±0.02)	0.78 (±0.01)	0.55 (±0.02)	0.57 (±0.01)
w/o Ensemble	0.71 (±0.02)	0.74 (±0.01)	0.56 (±0.02)	0.55 (±0.01)
w/o Train Interv. in VDB	0.74 (±0.02)	0.77 (±0.01)	0.56 (±0.02)	0.55 (±0.01)
w/o Manual	0.72 (±0.01)	0.74 (±0.02)	0.54 (±0.01)	0.52 (±0.02)
w/o Test Interv. in VDB	0.70 (±0.02)	0.73 (±0.01)	0.53 (±0.02)	0.50 (±0.01)
w/o Test Interv. Summary	0.75 (±0.01)	0.80 (±0.02)	0.57 (±0.01)	0.57 (±0.02)
w/o Few-shot Examples	0.73 (±0.01)	0.74 (±0.02)	0.55 (±0.01)	0.51 (±0.02)
AutoPsyC	**0.78 (±0.02)**	**0.81 (±0.01)**	**0.59 (±0.02)**	**0.58 (±0.01)**

Table 1: Average Weighted F1-Scores with 95% Confidence Intervals.

pathologizing) and recall (preventing missed conflicts), aligning with clinical priorities. To robustly estimate performance, we repeated all experiments several times and report the average weighted F1 score across all runs. In the case of experiments involving LLMs, we average across 100 runs, and in the case of the computationally less expensive baseline experiments, we average across 1000 runs. In addition to the averages, we also report their 95% confidence interval.

5.2 Baselines

We implement two simple baselines: a *Demographic Baseline* and a *Random Baseline*.

Demographic Baseline. This baseline employs a neural network classifier using demographic attributes such as gender, clincial diagnosis group, and binned age as input features. Numerical features were normalized by subtracting the mean and dividing by the standard deviation, while categorical features were converted into numerical representations. The neural network, implemented in PyTorch (Paszke et al., 2019), consists of three fully connected layers with ReLU activations. It was trained for 30 epochs using cross-entropy loss and the Adam optimizer.

Random Baseline. The random baseline leverages the `DummyClassifier` module from `scikit-learn`, configured with the `stratified` strategy. This classifier generates predictions by randomly assigning labels based on the class distribution of the training set. This random classifier serves as a naive baseline, highlighting the minimum expected performance for the classification task.

6 Results and Discussion

6.1 Overview

Table 1 summarises our weighted F_1 scores across the four psychodynamic conflicts. To more easily navigate the table, we partition the different ablation conditions into three cases, based on whether unlabelled training data is encorporated in the RAG framework and based on whether fine-tuning is performed with labelled training data. Ablations are always named relative to the partition their are in. For example, the ablation *w/o Manual* in the partition *No Training Data in VDB, No Fine-tuning* describes an ablation condition without training data integration into the RAG framework, without fine-tuning, and without integration of the OPD Manual in the RAG framework.

We observe that our full method (AutoPsyC), which combines our summarisation strategy with weighted voting across fine-tuned, temporally specialised models, as well as domain knowledge integration into the RAG framework, consistently outperforms all baselines and ablation conditions. As illustrated in Figure 3, models fine-tuned on the middle segments of the interviews consistently outperform those focusing on earlier or later sections. Moreover, Figure 4 indicates that deviating from four total models or partitioning the interviews into fewer or more than four segments leads to a noticeable drop in overall performance. Finally, we present an analysis of gender fairness of our model. Overall low values of Conditional Demographic Disparity (CDD) indicate no major gender-related biases (Table 2).

6.2 Which Model Configuration Works Best?

Our experimental results demonstrate the effectiveness of combining our summarisation strategy with Retrieval-Augmented Generation (RAG), instruction tuning and section-wise model specialization for psychodynamic conflict classification in clinical interviews. As can be seen in (Table 1), AutoPsyC achieves superior performance across all four conflict categories, with weighted F1-scores ranging from 0.58 to 0.81. This represents a substantial improvement over both naive baselines (Demographic: 0.26–0.46; Dummy: 0.20–0.33) and non-instruction-tuned variants (0.50–0.74).

Our detailed ablation experiments indicates that AutoPsyC effectively integrates all available information. We can observe a large decrease in performance when the test interview transcript is removed from the vector database (0.50-0.73 F1). This indicates that our model indeed makes use of the full test interview transcript that is provided via the RAG framework to fill in information missing in the interview summary. At the same time, we see that it does profit from the test interview summary, with losses of up to 0.04 F1 when the summary is removed. We furthermore observe a clear loss in performance when the OPD manual is removed from the vector data base (0.52-0.74 F1), and a slightly lower loss in performance when the training set interviews are removed from the database (0.55-0.77 F1). This indicates that even when using fine-tuned classification model, domain knowledge integration via the RAG setup is still helpful. The weighted voting mechanism using multinomial logistic regression provides moderate but consistent performance gains (0.58–0.81 F1 vs. 0.55–0.78 for unweighted aggregation), suggesting that different interview sections contribute asymmetrically to conflict identification.

One general observation we can make is that fine-tuning leads to greater robustness w.r.t. other ablation conditions. E.g. removing the OPD Manual from the vector database leads only to a moderate loss in performance when fine-tuned classification models are used (0.58-0.81 F1 vs. 0.52-0.74 F1). In contrast, for the case of no fine-tuning, the losses are more dramatic (0.50-0.74 F1 vs. 0.43-0.61 F1).

6.3 Which Interview Section is most useful?

To further investigate which sections of the interviews are most informative fine-tuning classification models, we investigate the performance of our four pretrained models singled out across all conflicts (see Figure 3). The results indicate that the models fine-tuned using the middle sections of the interviews outperform those tuned with other sections. After a careful examination of the interviews, we observed that, in the quarter 2 & 3, the interviewees often share information more closely related to their condition and situation. An excerpt of the interview can be found in Appendix A.1.

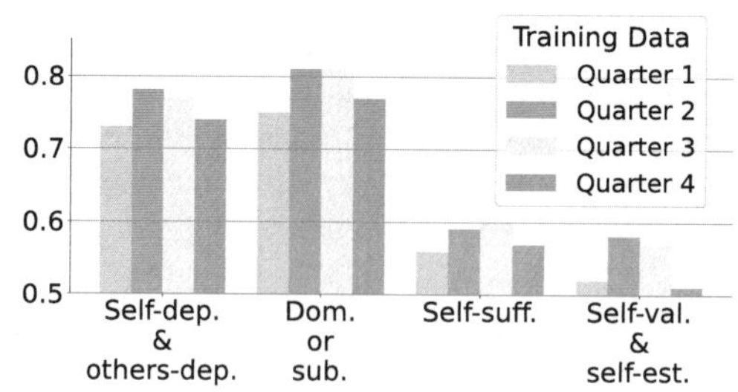

Figure 3: Performance of the four models across all classes.

6.4 Additional Experiments: How Fair is the Model?

Fairness concerning gender is a critical issue in psychiatric diagnoses when using machine learning algorithms, as biases in training data or model predictions can lead to systematic disparities in diagnostic outcomes. For instance, a study by Mosteiro et al.

	Self-dep. & others-dep.	*Dom. or sub.*	*Self-suff.*	*Self-val. & self-est.*
not assessable	0.0031	0.0008	0.0053	0.0044
not present	0.0042	0.0023	0.0034	0.0014
of little significance	0.0018	0.0011	0.0019	0.0020
significant	0.0025	0.0021	0.0054	0.0009
very significant	0.0029	0.0015	0.0010	0.0027

Table 2: One-vs.-rest CDD values for each class across four conflicts.

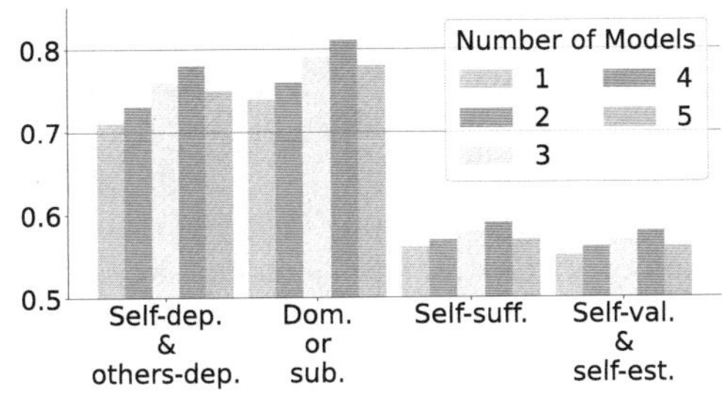

Figure 4: Performance of Different Number of Models

(2022) found that gender played an unexpected role in predictions related to benzodiazepine administration, potentially biasing the model's decisions.

We evaluate fairness with respect to the gender attribute using Conditional Demographic Disparity (CDD) (Wachter et al., 2021). CDD quantifies the difference in expected outcomes across demographic groups, with values closer to zero indicating fairer conditions.

$$\text{CDD} = \mathbb{E}[\hat{y} \mid \text{male}, y] - \mathbb{E}[\hat{y} \mid \text{female}, y] \quad (2)$$

Since we have more than two classes we compute CDD in one-vs.-rest fashion. We compute

$$\text{CDD}_k = \mathbb{E}\big[\hat{p}^k \mid \text{male},\ y\big] \ - \ \mathbb{E}\big[\hat{p}^k \mid \text{female},\ y\big] \quad (3)$$

for class k. This formulation reduces the multi-class problem to a "one vs. rest" scenario by focusing on a single predicted probability $\hat{p}^k$. If class k is deemed the "positive" class, the binary-based fairness thresholds from (Wachter et al., 2020) can be applied to CDD_k.

Overall, the CDD values reported in Table 2 are relatively low (all below 0.006), as indicated by prior studies (Wachter et al., 2021; Koumeri et al., 2023; Wachter et al., 2020), suggesting minimal gender-based disparity across the different classes. Wachter et al. (2021) suggest that CDD values below 0.01 indicate minimal demographic disparity, while Koumeri et al. (2023) and Wachter et al. (2020) provide empirical evidence that values above 0.02-0.05 often indicate fairness concerns. It is important to note that this fairness evaluation is not able to account for potential biases that are already present in the ground truth annotations on the dataset.

6.5 Ethical Considerations and Impact

The automation of psychodynamic diagnostics using NLP and machine learning presents both opportunities and ethical challenges. While enhancing objectivity, efficiency, and accessibility, its implementation requires careful ethical scrutiny to ensure responsible use in mental health care. Psychodynamic interviews contain sensitive data, necessitating strong anonymization and compliance with privacy regulations such as GDPR and HIPAA. Additionally, automated diagnostics may reflect biases present in training data, leading to disparities across demographic groups. Continuous bias auditing and fairness assessments are essential to mitigate these risks and ensure equitable model performance.

Automated tools should complement, not replace, human expertise. AutoPsyC could serve as a supplementary tool for therapists during psychodynamic interviews, acting as a "second-eye" to enhance clinical decision-making (American Psychological Association, 2025). Additionally, AutoPsyC could be integrated into social interactive agents, chat applications, and telepsychiatry platforms, providing complementary therapeutic support (Smith et al., 2019). Furthermore, it could be utilized in psychological training tools to enhance the proficiency of conducting psychodynamic interviews (American Psychological Association, 2023).

Psychodynamic diagnostics involve complex interpretations that extend beyond text-based pattern recognition. Thus, model outputs must be interpretable, allowing clinicians to integrate them into their assessments. Future research should prioritize explainability and transparency in AI-driven diagnostics. As AI applications in mental health expand, concerns arise regarding consent, misuse, and potential stigmatization in non-clinical settings. Interdisciplinary collaboration among clinicians, ethicists, and policymakers is needed to safeguard patient autonomy and well-being.

7 Conclusion

We present a novel framework for automated conflict classification in psychodynamic interviews, achieving clinically relevant performance through three key innovations: (1) domain-adapted instruction tuning using segmented interview data, (2) RAG-enhanced contextual understanding through OPD Manual and other interview integration, and (3) confidence-weighted aggregation of specialized section models.

These results suggest that LLMs can be effectively adapted for complex psychiatric coding tasks when combined with domain-specific knowledge retrieval and structured interview analysis. The demonstrated technique for identifying diagnostically salient interview segments (quarters 2 & 3) offers methodological insights for computational psychiatry research. Future work should explore applications to other OPD Axes and integration with multimodal clinical data.

8 Limitations

While promising, our approach has several limitations. First, the dataset size (n=141 interviews) may limit generalizability, particularly for rare conflict subtypes. Second, the complex pipeline (RAG, summarization, 4 specialized models) incurs significant computational costs compared to monolithic models. Third, performance variation across conflict categories (0.58–0.81 F1) suggests task-specific optimization needs, particularly for *Self-sufficiency* classification.

The reliance on manual OPD Manual examples for summarization introduces potential annotation bias, and the gender fairness analysis does not account for non-binary identities. Additionally, our stratified sampling based on diagnosis and gender may not fully capture all confounding demographic factors. We focus on a single summarization approach, as our primary goal is to establish a proof of concept for automated OPD scoring. While alternative summarization methods could be explored in future work, this choice allows us to maintain methodological consistency and provide a clear baseline for comparison.

We split the interview into four parts based on word counts, which does not fully account for the semi-structured nature of our interviews. Future work could focus on automatically detecting interview segments for the fine-tuning process.

Our fairness analysis showed minimal gender-based disparity in predictions. There are however many other ways in which our model may be biased. On the interview dataset we utilised, we did not have access to e.g. information on socioeconomic status or education. Further variations are such as cultural background are not sufficiently covered by the dataset as it was recorded with German-speaking people in Europe. This geographic and cultural constraint represents another key limitation of our study. It remains unclear, whether our approach would also work in very different cultural contexts.

Future research should address these limitations through larger multicentre datasets (König et al., 2022), lightweight model architectures, and explicit modeling of clinician raters' variance. The current implementation also requires further validation for real-time clinical deployment, including robustness testing against speech recognition errors and patient dialect variations. A more detailed investigation of how AutoPsyC handles defensive processes (Freud, 1936) remains an area for future research.

References

Eike Adams. 2010. The joys and challenges of semi-structured interviewing. *Community Practitioner*, 83(7):18–22.

Omolola A Adeoye-Olatunde and Nicole L Olenik. 2021. Research and scholarly methods: Semi-structured interviews. *Journal of the american college of clinical pharmacy*, 4(10):1358–1367.

American Psychological Association. 2023. Ai is changing every aspect of psychology. here's what to watch for. *Monitor on Psychology*.

American Psychological Association. 2025. Artificial intelligence in mental health care. *American Psychological Association*.

American Psychiatric Association. 2000. *Diagnostic and Statistical Manual of Mental Disorders (4th ed., Text Revision)*. American Psychiatric Association, Washington, DC.

Ran Barzilay, Nadav Israel, Amir Krivoy, Roi Sagy, Shiri Kamhi-Nesher, Oren Loebstein, Lior Wolf, and Gal Shoval. 2019. Predicting affect classification in mental status examination using machine learning face action recognition system: a pilot study in schizophrenia patients. *Frontiers in Psychiatry*, 10:446117.

Cord Benecke. 2024. 30 jahre operationalisierte psychodynamische diagnostik – neuerungen in der opd-3. *Psychotherapeutenjournal*, 23(1):36–46.

Astrid Bock, Eva Huber, and Cord Benecke. 2016. Levels of structural integration and facial expressions of negative emotions. *Zeitschrift für Psychosomatische Medizin und Psychotherapie*, 62:224–238.

SG Brederoo, FG Nadema, FG Goedhart, AE Voppel, JN De Boer, J Wouts, S Koops, and IEC Sommer. 2021. Implementation of automatic speech analysis for early detection of psychiatric symptoms: what do patients want? *Journal of psychiatric research*, 142:299–301.

Svend Brinkmann. 2014. Unstructured and semi-structured interviewing. *The Oxford handbook of qualitative research*, 2:277–299.

Joost Broekens, Bernhard Hilpert, Suzan Verberne, Kim Baraka, Patrick Gebhard, and Aske Plaat. 2023. Fine-grained affective processing capabilities emerging from large language models. In *2023 11th international conference on affective computing and intelligent interaction (ACII)*, pages 1–8. IEEE.

M. Cierpka, C. Grande, J. Rudolf, B. Stasch, and M. von der Tann. 2001. The operationalized psychodynamic diagnostics system: Clinical relevance, reliability, and validity. *Psychopathology*, 34(4):209–220.

Manfred Cierpka, Tilman Grande, Gerd Rudolf, M Von Der Tann, and Michael Stasch. 2007. The operationalized psychodynamic diagnostics system: clinical relevance, reliability and validity. *Psychopathology*, 40(4):209–220.

John F. Clarkin, Eve Caligor, Barry L. Stern, and Otto F. Kernberg. 2007. The structured interview of personality organization (stipo): A preliminary report. *Journal of Personality Assessment*, 88(1):69–83.

Juliet Corbin and Janice M Morse. 2003. The unstructured interactive interview: Issues of reciprocity and risks when dealing with sensitive topics. *Qualitative inquiry*, 9(3):335–354.

Lucile Dupuy, Jean-Arthur Micoulaud-Franchi, Hélène Cassoudesalle, Orlane Ballot, Patrick Dehail, Bruno Aouizerate, Emmanuel Cuny, Etienne de Sevin, and Pierre Philip. 2020. Evaluation of a virtual agent to train medical students conducting psychiatric interviews for diagnosing major depressive disorders. *Journal of Affective Disorders*, 263:1–8.

Eric Ettore, Philipp Müller, Jonas Hinze, Matthias Riemenschneider, Michel Benoit, Bruno Giordana, Danilo Postin, Rene Hurlemann, Amandine Lecomte, Michel Musiol, et al. 2023. Digital phenotyping for differential diagnosis of major depressive episode: narrative review. *JMIR mental health*, 10:e37225.

Nawshad Farruque, Randy Goebel, Sudhakar Sivapalan, and Osmar R Zaïane. 2024. Depression symptoms modelling from social media text: an llm driven semi-supervised learning approach. *Language Resources and Evaluation*, pages 1–29.

Giovanni A Fava, Nicoletta Sonino, David C Aron, Richard Balon, Carmen Berrocal Montiel, Jianxin Cao, John Concato, Ajandek Eory, Ralph I Horwitz, Chiara Rafanelli, et al. 2024. Clinical interviewing: an essential but neglected method of medicine. *Psychotherapy and psychosomatics*, 93(2):94–99.

Michael B. First, Janet B. W. Williams, Rhonda S. Karg, and Robert L. Spitzer. 2016. Structured clinical interview for dsm-5® disorders—clinician version (scid-5-cv).

OPD Task Force. 2008. *Operationalized psychodynamic diagnosis OPD-2: Manual of diagnosis and treatment planning*. Hogrefe Publishing GmbH.

Anna Freud. 1936. Das ich und die abwehrmechanismen.

Aaron Grattafiori, Abhimanyu Dubey, Abhinav Jauhri, Abhinav Pandey, Abhishek Kadian, and Ahmad Al-Dahle et al. 2024. The llama 3 herd of models. *Preprint*, arXiv:2407.21783.

Sayed Muddashir Hossain, Jan Alexandersson, and Philipp Müller. 2024. M3TCM: Multi-modal multi-task context model for utterance classification in motivational interviews. In *Proceedings of the Joint International Conference on Computational Linguistics, Language Resources and Evaluation (LREC-COLING)*.

Jana Hoyer, Catharina Voss, Jens Strehle, John Venz, Lars Pieper, Hans-Ulrich Wittchen, Stefan Ehrlich, and Katja Beesdo-Baum. 2020. Test-retest reliability of the computer-assisted dia-x-5 interview for mental disorders. *BMC psychiatry*, 20:1–16.

Edward J Hu, Yelong Shen, Phillip Wallis, Zeyuan Allen-Zhu, Yuanzhi Li, Shean Wang, Lu Wang, and Weizhu Chen. 2022. LoRA: Low-rank adaptation of large language models. In *International Conference on Learning Representations*.

Pengfei Ke, Xiaoman Ji, Weizhi Wang, and Min Sun. 2021. Parameter-efficient domain knowledge integration from multiple sources for biomedical pre-trained language models. In *Findings of the Association for Computational Linguistics: EMNLP 2021*, pages 3876–3887.

Hyunji Kim, Byeongchang Choi, Hyeongu Cho, Junwon Park, Eunji Kim, Sungwon Kim, and Jiyeon Kang. 2024. Towards understanding counseling conversations: Domain knowledge and large language models. In *Findings of the Association for Computational Linguistics: EACL 2024*, pages 1675–1687.

Taishiro Kishimoto, Hironobu Nakamura, Yoshinobu Kano, Yoko Eguchi, Momoko Kitazawa, Kuo-ching Liang, Koki Kudo, Ayako Sento, Akihiro Takamiya, Toshiro Horigome, et al. 2022. Understanding psychiatric illness through natural language processing (underpin): Rationale, design, and methodology. *Frontiers in Psychiatry*, 13:954703.

Alexandra König, Philipp Müller, Johannes Tröger, Hali Lindsay, Jan Alexandersson, Jonas Hinze, Matthias Riemenschneider, Danilo Postin, Eric Ettore, Amandine Lecomte, et al. 2022. Multimodal phenotyping of psychiatric disorders from social interaction: Protocol of a clinical multicenter prospective study. *Personalized Medicine in Psychiatry*, 33:100094.

L.K. Koumeri, M. Legast, Y. Yousefi, and K. Vanhoof. 2023. Compatibility of fairness metrics with EU non-discrimination laws: Demographic parity & conditional demographic disparity. *arXiv preprint*.

Xiaochong Lan, Yiming Cheng, Li Sheng, Chen Gao, and Yong Li. 2024. Depression detection on social media with large language models. *arXiv preprint arXiv:2403.10750*.

Aziliz Le Glaz, Yannis Haralambous, Deok-Hee Kim-Dufor, Philippe Lenca, Romain Billot, Taylor C Ryan, Jonathan Marsh, Jordan Devylder, Michel Walter, Sofian Berrouiguet, et al. 2021. Machine learning and natural language processing in mental health: systematic review. *Journal of medical Internet research*, 23(5):e15708.

Eric Lenouvel, Camelia Chivu, Janet Mattson, John Q Young, Stefan Klöppel, and Severin Pinilla. 2022. Instructional design strategies for teaching the mental status examination and psychiatric interview: a scoping review. *Academic Psychiatry*, 46(6):750–758.

Patrick Lewis, Ethan Perez, Aleksandra Piktus, Fabio Petroni, Vladimir Karpukhin, Naman Goyal, Heinrich Küttler, Mike Lewis, Wen-tau Yih, Tim Rocktäschel, et al. 2020. Retrieval-augmented generation for knowledge-intensive nlp tasks. *Advances in Neural Information Processing Systems*, 33:9459–9474.

Hali Lindsay, Philipp Mueller, Nicklas Linz, Radia Zeghari, Mario Magued Mina, Alexandra König, and Johannes Tröger. 2021. Dissociating semantic and phonemic search strategies in the phonemic verbal fluency task in early dementia. In *Proceedings of the Seventh Workshop on Computational Linguistics and Clinical Psychology: Improving Access*, pages 32–44.

Vittorio Lingiardi, Nancy McWilliams, Robert F. Bornstein, Francesco Gazzillo, and Robert M. Gordon. 2015. The psychodynamic diagnostic manual version 2 (pdm-2): Assessing patients for improved clinical practice and research. *Psychoanalytic Psychology*, 32(1):94–115.

Daniel M Low, Kate H Bentley, and Satrajit S Ghosh. 2020. Automated assessment of psychiatric disorders using speech: A systematic review. *Laryngoscope investigative otolaryngology*, 5(1):96–116.

Lester Luborsky and Paul Crits-Christoph. 1998. *Who Will Benefit from Psychotherapy?: Predicting Therapeutic Outcomes*. Basic Books, New York, NY.

Danielle Magaldi and Matthew Berler. 2020. Semi-structured interviews. *Encyclopedia of personality and individual differences*, pages 4825–4830.

David H. Malan. 1979. *Individual Psychotherapy and the Science of Psychodynamics*. Butterworth-Heinemann, Oxford, UK.

Kirill Milintsevich, Kairit Sirts, and Gaël Dias. 2023. Towards automatic text-based estimation of depression through symptom prediction. *Brain Informatics*, 10(1):4.

Pablo Mosteiro, Jesse Kuiper, Judith Masthoff, Floortje Scheepers, and Marco Spruit. 2022. Bias discovery in machine learning models for mental health. *Information*, 13(5):237.

Philipp Müller, Alexander Heimerl, Sayed Muddashir Hossain, Lea Siegel, Jan Alexandersson, Patrick Gebhard, Elisabeth André, and Tanja Schneeberger. 2024. Recognizing emotion regulation strategies from human behavior with large language models. In *Proceedings of the International Conference on Affective Computing and Intelligent Interaction (ACII)*.

Julie Nordgaard, Louis A. Sass, and Josef Parnas. 2013. The psychiatric interview: validity, structure, and subjectivity. *Eur Arch Psychiatry Clin Neurosci*, 263:353–64.

Jihoon Oh, Taekgyu Lee, Eun Su Chung, Hyonsoo Kim, Kyongchul Cho, Hyunkyu Kim, Jihye Choi, Hyeon-Hee Sim, Jongseo Lee, In Young Choi, et al. 2024. Development of depression detection algorithm using text scripts of routine psychiatric interview. *Frontiers in psychiatry*, 14:1256571.

William H O'Brien and Tracy Tabaczynski. 2007. Unstructured interviewing. *Handbook of clinical interviewing with children*, pages 16–29.

Adam Paszke, Sam Gross, Francisco Massa, Adam Lerer, James Bradbury, Gregory Chanan, Trevor Killeen, Zeming Lin, Natalia Gimelshein, Luca Antiga, et al. 2019. Pytorch: an imperative style.

F. Pedregosa, G. Varoquaux, A. Gramfort, V. Michel, B. Thirion, O. Grisel, M. Blondel, P. Prettenhofer, R. Weiss, V. Dubourg, J. Vanderplas, A. Passos, D. Cournapeau, M. Brucher, M. Perrot, and E. Duchesnay. 2011. Scikit-learn: Machine learning in Python. *Journal of Machine Learning Research*, 12:2825–2830.

J. Rudolf, C. Grande, and M. Cierpka. 2004. Operationalized psychodynamic diagnosis in planning and conducting psychotherapy. *Psychotherapy Research*, 14(3):295–308.

Shawn Christopher Shea. 2016. *Psychiatric interviewing: The art of understanding: A practical guide for psychiatrists, psychologists, counselors, social workers, nurses, and other mental health professionals*. Elsevier Health Sciences.

David V Sheehan, Yves Lecrubier, K Harnett Sheehan, Patricia Amorim, Juris Janavs, Emmanuelle Weiller, Thierry Hergueta, Roxy Baker, Geoffrey C Dunbar, et al. 1998. The mini-international neuropsychiatric interview (mini): the development and validation of a structured diagnostic psychiatric interview for dsm-iv and icd-10. *Journal of clinical psychiatry*, 59(20):22–33.

Anthony C. Smith, Penny M. N. Probst, Len Gray, Richard G. C. Johnston, Peter C. Wootton, and David G. C. Williams. 2019. Role of artificial intelligence within the telehealth domain. *Yearbook of Medical Informatics*, 28(1):162–167.

Larissa Vierl, Charlotte Von Bremen, York Hagmayer, Cord Benecke, and Christian Sell. 2023. How are psychodynamic conflicts associated with personality functioning? a network analysis. *Frontiers in Psychology*, 14:1152150.

Sandra Wachter, Brent Mittelstadt, and Chris Russell. 2020. Bias preservation in machine learning: the legality of fairness metrics under EU non-discrimination law. *West Virginia Law Review*, 123:765–810.

Sandra Wachter, Brent Mittelstadt, and Chris Russell. 2021. Why fairness cannot be automated: Bridging the gap between eu non-discrimination law and ai. *Computer Law & Security Review*, 41:105567.

Thomas A Widiger. 2008. Clinical interviews. *Evidence-based outcome research: A practical guide to conducting randomized controlled trials for psychosocial interventions*, pages 47–65.

Tianjun Zhang, Shishir G Patil, Naman Jain, Sheng Shen, Matei Zaharia, Ion Stoica, and Joseph E Gonzalez. 2024a. Raft: Adapting language model to domain specific rag.

Xiangyu Zhang, Hexin Liu, Kaishuai Xu, Qiquan Zhang, Daijiao Liu, Beena Ahmed, and Julien Epps. 2024b. When llms meets acoustic landmarks: An efficient approach to integrate speech into large language models for depression detection. *arXiv preprint arXiv:2402.13276*.

A Appendix

A.1 Excerpt from the Interview Transcript

Beginning

Interviewer: So, let us begin with the second part. I will ask you various questions from different areas.

Interviewee: Hmm.

Interviewer: About your present life, your past, relationships, work life, etc. Yes. I would like to start by asking you to describe what is currently the most burdensome for you.

Interviewee: At the moment?

Interviewer: Yes, it can be anything.

Interviewee: (Exhales) Ahm. (-) In general, I am actually doing quite well. However, I must say that things that have burdened me in the past, especially over the past three years, have now become less significant. What currently affects me the most is the situation at home. My parents are about to get divorced, and that has been the most difficult thing in my life so far. I must say, it has also been very stressful for me, but I am slowly managing it quite well.

Middle

Interviewee: The period was simply long. I was 20 when it all started, and I would say that only in the last few months have I truly felt lighter inside. For about a year, things have been steadily improving, but before that, I felt terrible. At home, it was a crisis. My mother was struggling—she barely ate, she was just existing. That made me very sad because I am someone who tries to keep everyone together. Given my age, I was able to grasp everything more clearly. I always spoke with everyone, tried to mediate, and made sure we somehow lived through it. But it was simply too much. (Claps hands on the table.)

I do not regret anything, or at least not much, except for the lingering aftereffects, which sometimes scare me. But otherwise, I would do it all over again. It just went too far. There were long periods where I barely met anyone or made any plans. If someone invited me out, I would always say no because I had to check on my mother to see if she was alone.

It was a responsibility that suddenly fell upon me. I would not say that it was forced upon me—I took it on willingly. That is simply the kind of person I am. If I see someone struggling, I cannot ignore it. I am very attached to my family.

There were times, for example, at Easter, when my mother just drove off. I could see in her eyes that she did not want to live anymore. She says the same thing even now. Back then, it was even stronger—she simply did not want to go on. She just got into the car and drove away. (Shocked and saddened.) It was simply terrible.

At first, I wanted to prevent the separation, of course. As a child, you never want your parents to separate. But later, it was just about minimizing the damage. I lost count of how many times I sat there listening, trying to mediate. I took on the role of always being there. But at some point, it was just too much.

I still managed to get through it, though sometimes I look back and wonder how I did it. I held up well, except for my university studies, where I had some setbacks. That was where the burden really showed. The emotional toll and the time commitment were simply too much.

End

Interviewee: I actually feel much better now. I have accepted everything as it is. I am a realistic person. I do not try to convince myself of things that do not exist. I walk through life with open eyes. I see what is happening around me. I know the divorce statistics.

Interviewer: But until recently, they did not matter to you.

Interviewee: No.

Interviewer: (Laughs.)

Interviewee: (Laughs as well.) Yes, because within those four walls, everything was fine. That was my foundation, my roots, where I came from. It was intact.

Interviewer: But now that has changed.

Interviewee: Yes. And I know that no matter how well things may seem to be going, there is always the risk that it could fall apart. That belief, that certainty I once had, is gone. I used to truly believe in lasting relationships. But now, if you ask me whether I think a relationship will last a lifetime, I no longer believe that. It is a sad realization.

Interviewer: It sounds as if a vision or a dream has been lost.

Interviewee: Yes, definitely. No doubt about it.

A.2 Example Prompt for *Self-dependency and Dependency on Others*

Context: Relationships and attachments are of existential importance in every person's life. They span the opposing poles of striving for close relationships and symbiotic proximity (dependency) and striving for well-developed independence and clear distance (powerful individuation). Individuation and dependency are fundamental elements of human life and experience, present in all areas of life. A life-defining conflict arises when this fundamental bipolar tension turns into a conflictual polarization. An individuation-dependency conflict is present only if this constellation is of existential importance and formative for a person's life history: This conflict involves the activation of experiences that either seek or avoid closeness, not the shaping of relationships in terms of caregiving or avoiding caregiving. The theme of individuation-dependency deals with the question of being alone or the ability to be with others. In its pathological conflict version, it concerns the necessity of being alone or being with others as an existential requirement.

Task: Based on this context, classify the following interview excerpt regarding the theme of "autonomy-dependency" into one of the following categories: *"not present"*, *"not assessable"*, *"of little significance"*, *"significant"*, *"very significant"*. For tasks where the interviews were summarized prior to classification, the model was first instructed to generate a summary of the interview based on a provided example. This example was derived from the OPD Manual.

A.3 Classification Classes

- **Not assessable** – The category cannot be determined due to insufficient or ambiguous information. There may be a lack of relevant content, unclear statements, or methodological limitations preventing a reliable assessment.

- **Not present** – There is no indication of the characteristic or phenomenon being evaluated. The available information does not support its existence or relevance in the given context.

- **Of little significance** – The characteristic or phenomenon is present but plays only a minor role. It appears occasionally but does not have a substantial influence on behavior, emotions, or interactions.

- **Significant** – The characteristic or phenomenon is clearly identifiable and has a notable impact. It influences thoughts, emotions, or interactions and is relevant to the overall assessment.

- **Very significant** – The characteristic or phenomenon is a dominant feature. It strongly shapes experiences, interactions, or coping mechanisms and is central to the evaluation.

A.4 Extra Plots

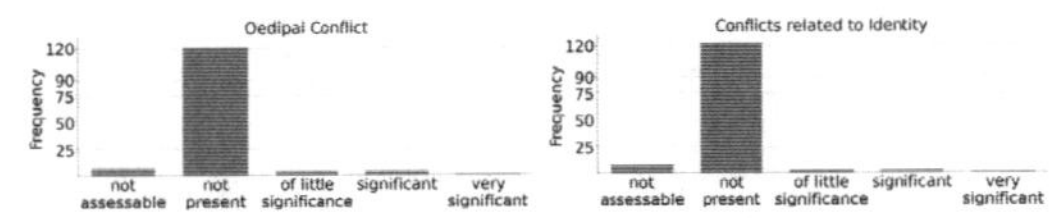

Figure 5: Class distributions for Oedipal Conflict and Conflicts related to Identity

The Emotional Spectrum of LLMs: Leveraging Empathy and Emotion-Based Markers for Mental Health Support

Alessandro De Grandi *
Università della Svizzera italiana
alessandro.de.grandi@usi.ch

Federico Ravenda *
Università della Svizzera italiana
federico.ravenda@usi.ch

Andrea Raballo
Università della Svizzera italiana
Cantonal Sociopsychiatric Organisation
andrea.raballo@usi.ch

Fabio Crestani
Università della Svizzera italiana
fabio.crestani@usi.ch

Abstract

The increasing demand for mental health services has highlighted the need for innovative solutions, particularly in the realm of psychological conversational AI, where the availability of sensitive data is scarce. In this work, we explored the development of a system tailored for mental health support with a novel approach to psychological assessment based on explainable emotional profiles in combination with empathetic conversational models, offering a promising tool for augmenting traditional care, particularly where immediate expertise is unavailable. Our work can be divided into two main parts, intrinsecaly connected to each other. First, we present RACLETTE, a conversational system that demonstrates superior emotional accuracy compared to considered benchmarks in both understanding users' emotional states and generating empathetic responses during conversations, while progressively building an emotional profile of the user through their interactions. Second, we show how the emotional profiles of a user can be used as interpretable markers for mental health assessment. These profiles can be compared with characteristic emotional patterns associated with different mental disorders, providing a novel approach to preliminary screening and support.

1 Introduction

Empathetic chatbots represent a significant evolution in the field of conversational AI, designed not just to understand commands or queries, but to perceive and interpret the emotional states of their users. These advanced agents leverage Natural Language Processing (NLP) approaches to analyze text for emotional content, enabling them to engage in interactions that feel more human-like. By recognizing and responding to a wide range of emotions, empathetic chatbots can tailor their responses to

*The first two authors contributed equally to this work.

Figure 1: An example of how our 3-turns coversation structure has been implemented.

provide comfort, advice, or support, thereby enhancing the user experience. This capability is particularly valuable in applications ranging from customer service and mental health support to personal assistants and social companions, where understanding and addressing the emotional needs of users can significantly impact satisfaction and outcomes (Crestani et al., 2022; Cena et al., 2023).

With the advent of advanced large language models (LLMs), the interaction experience with conversational agents has seen remarkable improvements. These new models exhibit enhanced understanding of natural language, greater contextual awareness, and the ability to generate more coherent and contextually appropriate responses. This technological leap has not only transformed how conversational agents interact with users but has also opened new avenues for analyzing and understanding human emotional expressions (Sekulić et al., 2021).

The motivation behind this research is rooted in the understanding that mental health support can be augmented through the use of empathetic conversational AI. For this purpose, we developed a conversational system called RACLETTE (***R**esponsive **A**nalysis with **C**hatbot **L**LMs for **E**motional and **T**herapeutical **T**racking and **E**valuation*).

The key highlights of this paper include the de-

Proceedings of the 10th Workshop on Computational Linguistics and Clinical Psychology (CLPsych 2025), pages 26–43
May 3, 2025 ©2025 Association for Computational Linguistics

velopment of a conversational model capable of detecting, understanding, and responding to emotional cues similar to human empathy (see Figure 1 for a visual example). This model is based on a novel approach to create emotion embeddings, which allows for the gradual construction of a user's emotional profile through interaction with the empathetic conversational model. We show how the user's emotional profile can be compared with known, pre-computed emotional profiles extracted from specialized datasets where individuals discuss their own experiences on specific mental health issues, with the rationale of potentially obtaining an explainable assessment of the mental state of the user engaging with the system. The contributions and findings of this work are twofold: **(1.)** We define a method to tailor a chatbot, RACLETTE, for reacting empathetically to a specific user. RACLETTE uses an unconventional 3-turn structure where the model is trained to predict the user's emotion as a next-token prediction, leveraging the generative capabilities of the underlying Mistral 7B model, and responds empathetically based on the predicted emotion. During the conversation, the user's emotional profile is updated, making the chatbot aware of the user's emotional condition in real-time. This updating allows the system to refine its understanding of the user's emotional state, enabling more precise and empathetic responses. **(2.)** We demonstrate how different mental disorders can be viewed as mixtures of specific emotions, guided by psychological theory.

This framework suggests that emotional states are interconnected components forming distinct patterns linked to various mental health conditions. Emotional profiles for specific disorders can be pre-calculated and compared with users' emotional profiles to differentiate between conditions, potentially aiding in early detection and diagnosis. These emotional profiles can be viewed as markers for identifying different groups of mental disorders.

The paper is structured as follows. Section 2 discusses the Related Work. Section 3 presents the methodology, data, and provides a psychological rationale supporting our approach. Section 4 presents the results of our model on the task of correct emotion classification and the quality of empathetic response generation. Section 5 discusses the explainable method for generating embeddings associated with various mental disorders and shows qualitative results. In Section 6 we present the results of an experiment using emotional profiles to discriminate users from different subreddit communities. Finally, Section 7 presents the conclusions.

2 Related Works

Significant research efforts have been devoted to developing sophisticated conversational models capable of understanding human emotions and generating empathic responses. The detection of sentiment and emotions has been recognized as crucial for the development of empathetic chatbots, as highlighted in (Felbo et al., 2017; Xu et al., 2018; Shin et al., 2019; Zhou et al., 2020). These works underline the importance of integrating emotional understanding capabilities into automatic dialogue systems to enhance human-computer interaction. In (Morris et al., 2018), authors demonstrated the feasibility of using corpus-based approaches to enable conversational agents to simulate subtle empathy. Recent research has focused on developing personalized conversational systems that can maintain coherence and user engagement throughout interactions (Madotto et al., 2019; Cho et al., 2022). These systems aim to create more natural and personalized dialogue experiences by adapting their responses to specific user characteristics and conversation contexts. Furthermore, the comprehensive scoping review by (Abd-Alrazaq et al., 2021) sheds light on patient perceptions of mental health chatbots, revealing a positive outlook but emphasizing the need for enhanced linguistic capabilities and personalized interactions.

Recently, there has been a significant rise in the application of NLP techniques within the field of psychology (Le Glaz et al., 2021). This growing interest stems from the ability of NLP to extract valuable linguistic markers from both spoken and written communication, offering crucial insights into various mental health disorders (Agurto et al., 2023; Corcoran et al., 2020; Corona Hernández et al., 2023; He et al., 2024).

Research has shown, for example, that measures of language coherence can serve as strong predictors of psychotic symptoms in individuals at high clinical risk (Just et al., 2020). Clearer language production deficits are typically observed during the first episode of psychosis (Gargano et al., 2022). One of the core symptoms, language disorganization, can be evaluated by analyzing the coherence and logical consistency of speech. For example, topic models (Blei et al., 2003) have been used to assess psychotic symptoms during patient interviews. In this context, the use of markers proves valuable for identifying differences within patient groups in an interpretable way. Our method aligns with the trend of leveraging the representational power of large language models to create useful markers for identifying trends within populations. This approach extends the current research in NLP applications for mental health, where language patterns serve as indicators of psychological states. By using emotion embeddings as markers, our method offers a novel way to quantify and analyze the emotional content of language, offering a computational framework for understanding mental health through affective patterns. Similar to how language coherence and organization have been used to predict psychotic symptoms, these emotional markers could potentially serve as early indicators or diagnostic aids for a range of mental disorders.

Building upon these foundational works, this study draws inspiration from CAiRE's empathetic neural chatbot model by (Lin et al., 2020), and the innovative approach of using grayscale labels for emotion recognition as suggested by "*The Emotion is Not One-hot Encoding*" by (Lee, 2022).

3 Methodology

This work proposes a novel methodology, guided by the intuition that one of the fundamental qualities of a therapist is *empathy*. This direction aims to synthesize empathetic responses based on a broader understanding of affective language, circumventing the need for sensitive, real-world conversational data, enabling the model to detect emotions, and create explainable emotional profiles that can be useful for mental health assessment.

3.1 A Psychological Rationale for our Approach

Empathy has two main components:
(1.) *Cognitive Empathy*, the intellectual ability to understand another person's emotions, thoughts, and motives. It involves the ability to comprehend someone else's mental state and why they might be feeling a certain way, which is essential for effective communication and social interaction.
(2.) *Affective Empathy*, the ability to physically feel another person's emotions, often leads to emotional responses such as compassion or concern. For a more detailed discussion, see (Decety, 2005).

This research focuses on Cognitive Empathy, aiming to classify the emotional state of a patient and enable the system to respond appropriately. Teaching machines to physically feel the emotions of others raises deep ethical and philosophical questions about the nature of consciousness and emotion in artificial systems, a topic that will likely remain at the forefront of futuristic research.

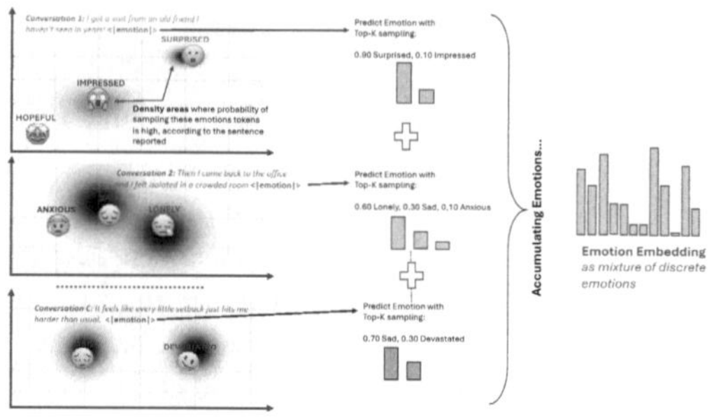

Figure 2: A visual explanation of how the emotional profile of a user is updated across a conversation and how to extract the final emotion embedding.

The approach aims to leverage and enhance the capabilities of empathetic LLMs by integrating emotion embeddings into their framework, guided by the intuition that the emotional spectrum is complex, and many emotions may coexist in a single sentence or piece of text.

We define an *emotion embedding* as a high-dimensional vector representing an emotional state. Unlike word embeddings (Allen and Hospedales, 2019), which capture semantic meaning, emotion embeddings synthesize an individual's emotional state within a conversation by encoding affective information. A distribution is generated by sampling and normalizing multiple predictions from a probabilistic classifier. These distributions can also be interpreted as embeddings, enabling meaningful algebraic operations. Complex emotions are encoded and represented by sequentially accumulating through the summation of many emotion embeddings, e.g., by accumulating over the many interactions that occur over an entire conversation (see Figure 2 for a visual explanation).

An emotion embedding can be defined as:

$$Emotion\ Embedding = \sum_{j}^{K} \alpha_j e_j$$

$\sum_j^K \alpha_j = 1$, K is the total number of all the different type of emotions considered and e_j is a specific emotion.

The use of emotional profiles to assess whether patients suffer from mental disorders is not entirely new in psychometrics. This approach aligns with established psychological assessment methods, such as the Beck Depression Inventory-II (BDI-II) (Beck et al., 1996), a widely used tool for measuring depression severity. The BDI-II includes items evaluating various emotional states and symptoms, like sadness, pessimism, guilt, agitation, irritability, and indecisiveness. Each item contributes to an overall score, aiding in the formulation of a final diagnosis.

For example, in the BDI-II, a patient might score high on sadness and pessimism, while moderate on guilty feelings and irritability. Similarly, our approach creates an emotional profile capturing the interplay of various emotions, offering a comprehensive view of an individual's mental state.

Thus, our work reveals a key insight: emotions act as indicators of deeper, complex mental states.

This study aims to demonstrate that mental states can be represented as collections of different emotions. Therefore, explainable emotion embeddings can be useful not only in identifying individuals in need of assistance but also as a potentially effective tool for automated diagnosis.

This multidimensional approach to emotional assessment acknowledges the complexity of human psychology, where a single emotional label cannot fully capture an individual's experience. By analyzing the emotional distribution, our system provides insights that align with the subtle understanding of mental states in clinical psychology, potentially enabling more accurate and personalized mental health support.

3.2 Datasets

In this work, three main sources of open-source data were used:

Empathetic Dialogues Dataset (Rashkin et al., 2018): it has been used to train the RACLETTE model to identify emotions and respond empathically. This dataset is a large-scale multi-turn empathetic dialogue dataset collected on the Amazon Mechanical Turk, containing 24,850 one-to-one open-domain conversations. This dataset was selected for this task because, other than its high quality and appropriate size, it considers a much wider range of emotions compared to all other available datasets, which usually consider only a very limited subset (5-8) of fundamental emotions (Zahiri and Choi, 2017; Li et al., 2017).

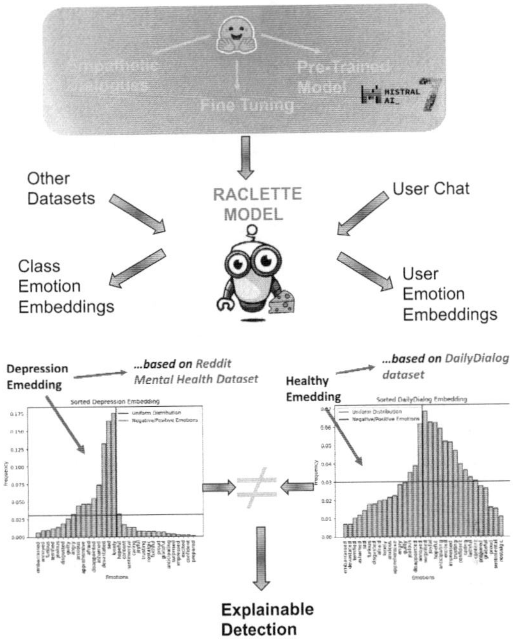

Figure 3: Overview of the main steps of RACLETTE pipeline.

Reddit Mental Health Dataset (Low et al., 2020): a collection of posts from specific Reddit forums (also called subreddits, Table 3 shows all the subreddits considered) have been used to construct the discrete distributions associated with each mental disorder to extract emotion embeddings.

DailyDialog Dataset (Li et al., 2017): a collection of posts used to establish a control group for the emotional profiles assessment. It contains 13,118 dialogues split into a training set with 11,118 dialogues and validation and test sets with 1,000 dialogues each.

Figure 3 illustrates the RACLETTE workflow pipeline, highlighting the specific use of each dataset at various stages of the process. This visual representation provides a clear overview of how the different datasets are integrated into the system's architecture, from training the empathetic model to extracting emotion embeddings and conducting comparative analyses.

3.3 Tailoring an LLM to React Empathetically

For this study, we chose to fine-tune the Mistral 7B model (Jiang et al., 2023), a 7-billion-parameters state-of-the-art LLM, known for its great performance combined with both computational and memory efficiency. This approach aligns with recent findings from (Sekulić et al., 2024), which demonstrate that fine-tuning LLMs on task-oriented dialogue data can reduce hallucinations.

The Empathetic Dialogues Dataset is purposefully formatted using a specific structure that allows to fully leverage the causal attention mask of the transformer decoder model to generate tokens for both the empathic response prediction and dialogue emotion detection tasks. Placing prompts before emotion labels enforces the autoregressive property of the model during training and inference (Sun et al., 2023), allowing the generative model to be used both as a classifier and a conversational agent. The model learns to predict the next tokens by only attending to previous positions in the sequence in order to generate predictions sequentially. Let $P(y_{<emotion>}|y_1, y_2, \ldots, y_N)$ be the probability of the model predicting the emotion token $y_{<emotion>}$ given the sequence of previous tokens $y_1, y_2, \ldots, y_N$, then the model's objective can be defined as:

$$\max \sum_{t=1}^{N} \log P(y_{<emotion>}|y_1, y_2, \ldots, y_N)$$

where N is the length of the sequence.

When generating the prediction of an emotion, the model iteratively produces the tokens that are more likely, given the previous tokens (see Figure 2). Unlike deterministic methods, this process can be guided to generate a diverse set of emotions by iteratively sampling over the predicted probability distribution of all tokens in the vocabulary. In this implementation, Top-K Sampling is used (Holtzman et al., 2019), which limits the sampling pool to the top-K most probable tokens, in this case, *top-10*, balancing diversity with relevance. Then to generate multiple emotions, this process is repeated *10* times independently for each prompt. Let V be the vocabulary and K be the sampling parameter:

$$TopK(P(y_t|y_{<t}), K) = y_i \in V : P(y_i|y_{<t})$$

is among top-k probabilities.

These empirical distributions are aggregated across the entire conversation to obtain the emotional profile of the speaker. Let C be the set of all prompts in a conversation and $e_{i,k}$ the sampled emotion ($K = 10$ samples in total) for prompt i :

$$EmotionalProfile = \frac{1}{|C|} \sum_{i \in C} \frac{1}{K} \sum_{k=1}^{K} e_{i,k}$$

For this study, an unconventional 3-turns structure was implemented (see Figure 1 for an example). It can be summarized as Prompt, Emotion, and Response, separated by the special tokens: $< |prompter| >, < |emotion| >, < |assistant| >$ and $< endoftext >$.

When predicting empathetic responses, the model will attend to the previous tokens in its context, **(1)** the whole history of the conversation, **(2)** the current prompt followed by the emotion, and learn to generate the appropriate reply as seen in the training dataset.

4 Emotion Recognition and Empathetic Response

Table 1 shows the results of RACLETTE in detecting the correct emotion for each conversation both at prompt and conversation levels. The results related to individual prompts refer to the correct classification of emotions for single conversation turns. Regarding the conversation, this approach progressively concatenates each prompt, its predicted emotion, and subsequent response, thus enriching the model's contextual awareness with each conversational turn. The accumulated emotion distributions for each prompt contribute to a more precise classification, resulting in a 3% increase in accuracy (from 56% to 59%). Notably, this methodology enhances accuracy, as the expanding conversational context provides more information for discerning the speaker's emotional profile.

Out of approximately 10.9k utterances present in the test set, the report categorizes 5,242, as the classification is made solely on the speaker's contributions. In addition, to evaluate empathic replies to each of the speaker prompts, this table includes the BERTSCORE (Zhang et al., 2019), an automatic evaluation metric for text generation. Unlike traditional metrics that rely on exact word matches or n-grams, BERTSCORE evaluates the similarity between predicted and target replies by analyzing contextual embeddings of tokens obtained with the BERT model. This approach allows for a semantical understanding of the model's performance, capturing the comparison of empathetic responses beyond mere lexical matching. Notably, a BERTSCORE of 0.87 indicates high semantic similarity between the responses given by the model and the target replies contained in the test set that were given by the human listeners.

Table 2 compares the overall emotional accuracy

Emotion	Precision	Recall	F1-Score	Support
Individual Prompts (5,242 Prompts)				
Macro avg	0.56	0.56	0.55	
Weighted avg	0.56	0.56	0.55	
BERTSCORE	0.873	0.865	0.869	0.87
Accuracy				0.56
Conversations (2,472 Conversations)				
Macro avg	0.59	0.59	0.58	
Weighted avg	0.59	0.59	0.58	
Accuracy				0.59

Table 1: Emotion Classification Report: Evaluated on *Individual prompts* and *Conversations* from the Empathetic Dialogues test set.

Model	Emotional Accuracy
(Gao et al., 2021)	0.42
(Li et al., 2022)	0.46
(Chen et al., 2024)	0.53
CAiRE	0.51
RACLETTE	0.59

Table 2: Emotional accuracy comparison between RACLETTE and other benchmarks (best results highlighted in green).

of RACLETTE with the accuracy of CAiRE, as reported by (Lin et al., 2020). For completeness, in addition to CAiRE, we present other baselines from the literature that have used the same dataset to evaluate their performance. As benchmarks for emotion classification accuracy, we consider the following approaches: **(1.)** In (Chen et al., 2024), authors propose a cause-aware empathetic generation method using Chain-of-Thought fine-tuning on Large Language Models. **(2.)** In (Li et al., 2022), authors introduced a knowledge-enhanced empathetic dialogue generation method incorporating external knowledge and emotional signals. **(3.)** The approach from (Gao et al., 2021), who proposed incorporating emotion cause recognition into empathetic response generation using an emotion reasoner and gated attention mechanism.

We report the accuracy value as presented in works that have used the same dataset, as shown in their respective manuscripts.

Table 2 shows how RACLETTE outperforms the benchmarks considered. Also in this case, it can be observed that the choice of fine-tuning a generative model, leveraging its autoregressive characteristics for classification, leads to the best results.

5 Mental State as Mixtures of Emotions

This experiment shows a possible novel approach to create explainable mental state embeddings based on emotions, expanding the conversational model's role from empathetic response generation to emotion analysis and diagnostic tool. The approach involves leveraging the fine-tuned model from the previous experiment, primarily as an emotion classifier. The idea is to extract emotional embeddings, used as *markers* that are indicative of specific mental disorders from specialized corpora, in this case from social media interactions in mental health forums. The goal is to later compare the distinctive emotional profiles obtained in this experiment to the profiles obtained from users interacting with the model in a conversation.

For this experiment, considering the lack of professionally labeled data, various datasets are gathered from Reddit, a social news website and forum where content is socially curated and promoted by site members through voting. It must be acknowledged that the data obtained from Reddit or other social media platforms may not accurately represent the broader population with mental illnesses, as it only captures those who choose to discuss their experiences online.

Reddit is organized into forums known as "subreddits". Each subreddit focuses on a specific topic, interest, or theme, creating a unique community within the broader Reddit platform. In Table 3 all the considered subreddits obtained from (Low et al., 2020) are reported (for a more in-depth discussion we refer to Section E in Appendix), together with a graphical visualization in Figures 5 and 6 in Appendix of the relative mental state embeddings based on emotions obtained by processing 1,000 posts from each subreddit.

In the approach described in this section, embeddings for each mental disorders were generated by processing posts from the respective subreddits. The methodology involves the empathetic conversational model obtained in the previous experiment. However, rather than responding with both emotion and a reply, the posts are segmented into individual phrases. For each phrase, the model predicts a set of 10 emotions. These predicted emotions are then aggregated across all posts by summation and subsequently normalized. This process results in a characteristic emotional distribution profile for each mental disorder.

This experiment yielded interesting results: the obtained emotion embeddings show significant differences across a spectrum of Reddit communities. Also, similarities across related disorders were to be expected. For instance, depression and suicide, or addiction and alcoholism, show consistent similarities. Overall, these distributions can provide insights into how individuals discussing their experiences with similar conditions might perceive and express their feelings. In Figure 4(**A**), we observe the mental disorder representations in a two-dimensional reduced space after applying t-SNE on top of the emotional embeddings. We can observe how mental disorders such as *alcoholism*, *addiction*, and *eating disorder* are close in the embedding space, as are *depression* and *loneliness* or *schizophrenia* and *post-traumatic stress disorder (PTSD)*. We can also observe interesting properties of our representations, for example, by summing the depression and schizophrenia's embeddings, a new representation can be obtained, that is very close, in the embedding space, to bipolar.

The assumption that mentally distressed individuals show identifiable skewed patterns of emotions must be addressed, by first establishing a normal distribution for comparison. For this, the Daily Dialogue Dataset (Li et al., 2017), a high-quality multi-turn open-domain English dialogue dataset, was chosen as a control group. On average there are around 8 speakers turns per dialogue with around 15 tokens per turn where people discuss their daily lives, the whole training set was used to extract the embedding for this dataset.

The order of the emotional features in the embeddings can be arbitrary. But as an example, for clarity and visual comparison, the control group embeddings and the depression embeddings can be ordered according to what are most commonly considered positive and negative emotions: Figure 4(**B-C**) clearly shows the contrast in emotional profiles, emphasizing the marked disparities in how emotions are manifested and experienced by those within the reddit depression community, exhibiting an extremely skewed distribution towards negative emotions, compared to individuals engaging in daily dialogues.

6 Reddit's Emotion embeddings applied to the Detection of Suicide Risk

This experiment aims to evaluate the use of emotion to create mental state embedding as a mechanism for diagnosing the potential risk of suicide. For lack of a professionally labeled dataset, this experiment, like the previous one, focuses on Reddit's users. To compare our approach with other related tasks and methods, we have built a dataset for binary classification of general conversation text versus suicidal text. We used the two subreddits CasualConversation and SuicideWatch, where CasualConversation is a subreddit for general conversation, and has generally been used by other methods as data for a clinically healthy class in other works (Haque et al., 2021; Shen and Rudzicz, 2017). This dataset is part of a larger collection available on Kaggle[1], which has been carefully cleaned to ensure the reliability of the data. We select 5% of the initial samples at random as a test set ($\approx$ 10,585 samples). After inspecting posts for anomalous length deviations from the average, those lacking informative content are removed.

To compute the sample embeddings, each post is divided into sentences, with 10 emotions predicted per sentence. The final embedding aggregates these emotions across all sentences in the post, following a similar approach as used in Section 5 to extract reference embeddings.

The methodology encompasses the use of three metrics for comparing embeddings: Kullback–Leibler (KL) divergence, Jensen–Shannon (JS) divergence, and Cosine Similarity (CS). Our approach is based on the use of emotional profiles embeddings that are most closely associated with an elevated risk of suicide and match them against the user's emotional profile. Table 3 compares various emotion embeddings, focusing on their differences from the suicide embedding, measured by KL, JS, and CS. The results show emotional proximity between Suicide and Depression, as well as Borderline Personality Disorder, followed by Bipolar Disorder, Addiction, PTSD, and Schizophrenia. This pattern aligns with psychological insights that these mental disorders are often linked to a risk of suicidal tendencies (Song et al., 2020).

For these reasons, this experiment will focus on the use of these specific emotion embeddings, considering the embedding that is most similar to what is obtained by processing the sample post, and mapping it to the predicted label, in an unsupervised fashion, as follows:

Positive labels: suicide, depression, borderline personality disorder (BPD), bipolar disorder, PTSD, addiction, and schizophrenia. By combining these particular embeddings, the study aims to capture a spectrum of characteristic emotional patterns that are potentially indicative of an elevated risk of suicide.

Negative labels: normal and uniform distributions, where normal is obtained from the Daily

[1]https://www.kaggle.com/datasets/suicide-watch

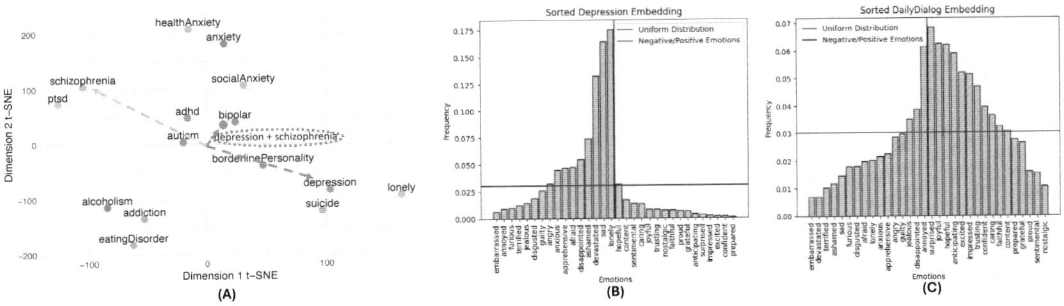

Figure 4: **(A)** 2-Dimensional representation of mental disorders distribution after applying t-SNE dimensionality reduction. **(B)** Sorted emotion embedding of depression. **(C.)** Sorted emotion embedding of DailyDialog.

Emotion	KL	JS	CS
suicide	0.000	0.000	1.000
depression	0.062	0.124	0.969
bpd	0.201	0.226	0.852
bipolar	0.451	0.332	0.637
addiction	0.519	0.349	0.592
ptsd	0.545	0.355	0.594
alcoholism	0.567	0.355	0.586
schizophrenia	0.745	0.407	0.519
eatingDisorder	0.822	0.409	0.496
socialAnxiety	0.830	0.432	0.463
uniform	0.850	0.449	0.540
autism	0.855	0.435	0.467
adhd	1.016	0.454	0.428
anxiety	inf	0.481	0.303
lonely	inf	0.429	0.579
healthAnxiety	inf	0.591	0.260

Table 3: Comparison of Emotion Embeddings: KL Divergence (KL), JS Divergence (JS), and Cosine Similarity (CS) w.r.t. suicide embedding.

Models	Prec	Rec	F1	ACC
RACLETTE				
KL Divergence	0.71	0.90	0.79	0.77
JS Divergence	0.67	0.93	0.78	0.76
Cosine Similarity	0.65	0.93	0.77	0.74
Combined	0.63	0.95	0.76	0.72
Benchmark Models				
RoBERTa + KM	0.72	0.84	0.78	0.77
BERT + KM	0.65	0.80	0.71	0.69

Table 4: Classification results for different methods of comparing embeddings to detect risk of suicide. The higher the score (a.k.a. the greener), the better.

Dialogue dataset, and embedding obtained from Casual Conversation's subreddit are also used as the control group.

6.1 Results for Mental Health Classification

Table 4 shows the performance metrics: precision, recall, F1 score and accuracy. For each similarity metrics — Kullback–Leibler divergence, Jensen–Shannon divergence, and Cosine Similarity — we also introduce a combined method, where if any of these methods detect a risk of suicide, the label is assigned as positive. This experiment is designed to maximize recall, a critical metric in scenarios where missing a positive instance has severe consequences, thus reducing the emphasis on precision and false positives. As the results show, this approach achieves high recall at the cost of other metrics. Additionally, these methods are compared with state-of-the-art unsupervised approaches based on RoBERTa (Liu et al., 2019) and BERT's (Devlin et al., 2019) embedding representations, before grouping them into two classes using a K-Means clustering approach, as done in (Subakti et al., 2022). The results show RACLETTE's Combined method achieving the highest recall of 0.95, indicating superior ability to identify relevant cases, though this comes with a trade-off in precision at 0.63. Conversely, RoBERTa leads in precision at 0.72, but with lower recall at 0.84. The KL Divergence variant of RACLETTE stands out for its balanced performance, maintaining strong scores across all metrics (precision: 0.71, recall: 0.90, F1: 0.79, accuracy: 0.77). Both JS Divergence and Cosine Similarity methods show similar patterns, with high recall (0.93) but lower precision. The color intensity (green shading) in Table 4 indicates better performance, visually highlighting that RACLETTE's approaches generally outperform the benchmark models.

A key advantage of this method is that it generates explainable representations and emotion embeddings, which can be visually inspected, providing insights into an individual's emotional profile.

7 Conclusions and Future Works

This paper introduces the RACLETTE system, which addresses two critical challenges in mental health support: the need for empathetic conversational system and reliable assessment tools. By fine-tuning a LLM, we demonstrate that it's possible to create effective conversational agents that can accurately recognize users' emotional states while generating high-quality empathetic responses, all while avoiding the use of sensitive clinical data. The system not only achieves state-of-the-art performance in emotion recognition, but also introduces a novel methodology for creating emotional profiles. These profiles, generated by aggregating emotion distributions from user interactions, serve as interpretable markers that can be compared with characteristic patterns associated with various mental health conditions. Our experimental results demonstrate both the system's effectiveness in maintaining empathetic conversations and its potential as a preliminary screening tool through the analysis of emotional embeddings.

Future work should extend RACLETTE's capabilities beyond purely supportive interactions to incorporate a broader range of therapeutic approaches. While RACLETTE currently focuses on emotional markers, a more comprehensive model would incorporate cognitive symptoms (memory, attention, perception, etc.) alongside affective components. Our framework can be adapted to such extensions, potentially offering a more holistic assessment that better reflects the complex nature of mental health conditions. By expanding our emotional embeddings to include markers of cognitive functioning, future works could provide more comprehensive profiles that better align with the multifaceted nature of mental health assessment in clinical practice.

8 Limitations

One of the main critical point of this work is represented by the quality and reliability of the emotional data used for training. Emotional data must be diverse and accurately labeled to ensure the model can understand and respond to a wide range of emotional expressions. This data collection process is complex and time-consuming, often requiring manual annotation by experts to maintain high standards.

Furthermore, findings have revealed that individuals affected by mental disorders commonly turn to social media to share their personal experiences, seek out information about mental health and treatment options, and either offer or gain support from others who are dealing with similar challenges (Naslund et al., 2020; Dodemaide et al., 2022). However, noise in the data is another significant limitation. For example, individuals may seek advice on behalf of others, such as family members, which can introduce inaccuracies. Self-reported information, while valuable, may not always be as reliable or accurate as clinically diagnosed conditions due to personal biases, misunderstandings, or intentional misreporting. Additionally, online self-expression can vary greatly between individuals, influenced by factors such as cultural differences, personal communication styles, and the specific context of the interaction.

Confounding factors, such as comorbidities, must also be taken into account. Individuals with multiple overlapping conditions may exhibit complex emotional and psychological profiles that are difficult for the model to parse accurately. Also, the authenticity and accuracy of self-reported conditions cannot be verified, as users may misattribute symptoms or self-diagnose without professional confirmation. Moreover, the way individuals express themselves online can differ significantly from in-person interactions, adding another layer of complexity to the model's ability to interpret and respond appropriately. These limitations necessitate caution when generalizing findings to clinical contexts and highlight the need for validation against professionally assessed populations.

Despite these challenges, the methodology addresses crucial privacy and confidentiality issues that are particularly important in the mental health domain. However, it does not fully address the ethical implications of using AI as a clinical tool, including the potential for misuse and the need for safeguards against harmful or biased behaviors in the conversational model. Continuous improvements and validation against clinical standards are essential to ensure that these tools effectively integrate into traditional care pathways, enhancing rather than disrupting the therapeutic process.

9 Ethical Considerations

The proposed methodology for mental health support and assessment, while innovative, brings several ethical considerations to the forefront that must be addressed to ensure responsible deployment.

There is a potential for AI to be misused as a clinical tool. Without proper safeguards, these models could exhibit harmful or biased behaviors, leading to adverse outcomes for users.

Implementing ethical safeguards is crucial to mitigate the risks associated with AI in mental health. Developing clear guidelines on the appropriate use of AI, managing sensitive data protocols, and ensuring transparency in operations are essential steps. Involving ethicists, and clinicians in the development process will help create a balanced and ethical approach.

It is also crucial to clearly communicate the supplementary nature of these tools and the necessity of professional evaluation and treatment. There is a risk that users may become overly reliant on automated mental health support, potentially neglecting the importance of seeking help from qualified professionals. By ensuring that these tools are integrated into traditional care pathways, they can enhance the therapeutic process, providing additional support while maintaining the central role of professional mental health care providers.

References

Alaa A Abd-Alrazaq, Mohannad Alajlani, Nashva Ali, Kerstin Denecke, Bridgette M Bewick, and Mowafa Househ. 2021. Perceptions and opinions of patients about mental health chatbots: scoping review. *Journal of medical Internet research*, 23(1):e17828.

Carla Agurto, Guillermo Cecchi, Sarah King, Elif K Eyigoz, Muhammad A Parvaz, Nelly Alia-Klein, and Rita Z Goldstein. 2023. Speak and you shall predict: speech at initial cocaine abstinence as a biomarker of long-term drug use behavior. *bioRxiv*.

Carl Allen and Timothy Hospedales. 2019. Analogies explained: Towards understanding word embeddings. In *International Conference on Machine Learning*, pages 223–231. PMLR.

Aaron T Beck, Robert A Steer, and Gregory Brown. 1996. Beck depression inventory–ii. *Psychological assessment*.

David M. Blei, Andrew Y. Ng, and Michael I. Jordan. 2003. Latent dirichlet allocation. *J. Mach. Learn. Res.*, 3(null):993–1022.

Louise Brådvik. 2018. Suicide risk and mental disorders.

Federica Cena, Luca Console, Marta Micheli, and Fabiana Vernero. 2023. Combining genetic algorithms and temporal constraint satisfaction for recommending personalized tourist itineraries. In *International Conference of the Italian Association for Artificial Intelligence*, pages 441–452. Springer.

Xinhao Chen, Chong Yang, Man Lan, Li Cai, Yang Chen, Tu Hu, Xinlin Zhuang, and Aimin Zhou. 2024. Cause-aware empathetic response generation via chain-of-thought fine-tuning. *arXiv preprint arXiv:2408.11599*.

Itsugun Cho, Dongyang Wang, Ryota Takahashi, and Hiroaki Saito. 2022. A personalized dialogue generator with implicit user persona detection. *arXiv preprint arXiv:2204.07372*.

Cheryl M Corcoran, Vijay A Mittal, Carrie E Bearden, Raquel E Gur, Kasia Hitczenko, Zarina Bilgrami, Aleksandar Savic, Guillermo A Cecchi, and Phillip Wolff. 2020. Language as a biomarker for psychosis: a natural language processing approach. *Schizophrenia research*, 226:158–166.

Hugo Corona Hernández, Cheryl Corcoran, Amélie M Achim, Janna N De Boer, Tessel Boerma, Sanne G Brederoo, Guillermo A Cecchi, Silvia Ciampelli, Brita Elvevåg, Riccardo Fusaroli, et al. 2023. Natural language processing markers for psychosis and other psychiatric disorders: emerging themes and research agenda from a cross-linguistic workshop. *Schizophrenia bulletin*, 49(Supplement_2):S86–S92.

Fabio Crestani, David E Losada, and Javier Parapar. 2022. *Early Detection of Mental Health Disorders by Social Media Monitoring: The First Five Years of the ERisk Project*, volume 1018. Springer Nature.

Jean Decety. 2005. Perspective taking as the royal avenue to empathy. *Other minds: How humans bridge the divide between self and others*, 143:157.

Tim Dettmers, Artidoro Pagnoni, Ari Holtzman, and Luke Zettlemoyer. 2023. Qlora: Efficient finetuning of quantized llms. *arXiv preprint arXiv:2305.14314*.

Jacob Devlin, Ming-Wei Chang, Kenton Lee, and Kristina Toutanova. 2019. Bert: Pre-training of deep bidirectional transformers for language understanding. *Preprint*, arXiv:1810.04805.

Paul Dodemaide, Mark Merolli, Nicole Hill, and Lynette Joubert. 2022. Do social media impact young adult mental health and well-being? a qualitative study. *The British Journal of Social Work*, 52(8):4664–4683.

Bjarke Felbo, Alan Mislove, Anders Søgaard, Iyad Rahwan, and Sune Lehmann. 2017. Using millions of emoji occurrences to learn any-domain representations for detecting sentiment, emotion and sarcasm. In *Proceedings of the 2017 Conference on Empirical Methods in Natural Language Processing*, pages 1615–1625.

Jun Gao, Yuhan Liu, Haolin Deng, Wei Wang, Yu Cao, Jiachen Du, and Ruifeng Xu. 2021. Improving empathetic response generation by recognizing emotion cause in conversations. In *Findings of the association for computational linguistics: EMNLP 2021*, pages 807–819.

Giulia Gargano, Elisabetta Caletti, Cinzia Perlini, Nunzio Turtulici, Marcella Bellani, Carolina Bonivento, Marco Garzitto, Francesca Marzia Siri, Chiara

Longo, Chiara Bonetto, et al. 2022. Language production impairments in patients with a first episode of psychosis. *Plos one*, 17(8):e0272873.

Ayaan Haque, Viraaj Reddi, and Tyler Giallanza. 2021. Deep learning for suicide and depression identification with unsupervised label correction. In *Artificial Neural Networks and Machine Learning–ICANN 2021: 30th International Conference on Artificial Neural Networks, Bratislava, Slovakia, September 14–17, 2021, Proceedings, Part V 30*, pages 436–447. Springer.

Rui He, Claudio Palominos, Han Zhang, Maria Francisca Alonso-Sánchez, Lena Palaniyappan, and Wolfram Hinzen. 2024. Navigating the semantic space: Unraveling the structure of meaning in psychosis using different computational language models. *Psychiatry Research*, 333:115752.

Ari Holtzman, Jan Buys, Li Du, Maxwell Forbes, and Yejin Choi. 2019. The curious case of neural text degeneration. *arXiv preprint arXiv:1904.09751*.

Albert Q Jiang, Alexandre Sablayrolles, Arthur Mensch, Chris Bamford, Devendra Singh Chaplot, Diego de las Casas, Florian Bressand, Gianna Lengyel, Guillaume Lample, Lucile Saulnier, et al. 2023. Mistral 7b. *arXiv preprint arXiv:2310.06825*.

Sandra A Just, Erik Haegert, Nora Kořánová, Anna-Lena Bröcker, Ivan Nenchev, Jakob Funcke, Andreas Heinz, Felix Bermpohl, Manfred Stede, and Christiane Montag. 2020. Modeling incoherent discourse in non-affective psychosis. *Frontiers in Psychiatry*, 11:846.

Sidney H Kennedy. 2008. Core symptoms of major depressive disorder: relevance to diagnosis and treatment. *Dialogues in clinical neuroscience*, 10(3):271–277.

Filiz Kulacaoglu and Samet Kose. 2018. Borderline personality disorder (bpd): in the midst of vulnerability, chaos, and awe. *Brain sciences*, 8(11):201.

Aziliz Le Glaz, Yannis Haralambous, Deok-Hee Kim-Dufor, Philippe Lenca, Romain Billot, Taylor C Ryan, Jonathan Marsh, Jordan Devylder, Michel Walter, Sofian Berrouiguet, et al. 2021. Machine learning and natural language processing in mental health: systematic review. *Journal of medical Internet research*, 23(5):e15708.

Joosung Lee. 2022. The emotion is not one-hot encoding: Learning with grayscale label for emotion recognition in conversation. *arXiv preprint arXiv:2206.07359*.

Qintong Li, Piji Li, Zhaochun Ren, Pengjie Ren, and Zhumin Chen. 2022. Knowledge bridging for empathetic dialogue generation. In *Proceedings of the AAAI conference on artificial intelligence*, volume 36, pages 10993–11001.

Yanran Li, Hui Su, Xiaoyu Shen, Wenjie Li, Ziqiang Cao, and Shuzi Niu. 2017. Dailydialog: A manually labelled multi-turn dialogue dataset. *arXiv preprint arXiv:1710.03957*.

Zhaojiang Lin, Peng Xu, Genta Indra Winata, Farhad Bin Siddique, Zihan Liu, Jamin Shin, and Pascale Fung. 2020. Caire: An end-to-end empathetic chatbot. In *Proceedings of the AAAI Conference on Artificial Intelligence*, volume 34, pages 13622–13623.

Yinhan Liu, Myle Ott, Naman Goyal, Jingfei Du, Mandar Joshi, Danqi Chen, Omer Levy, Mike Lewis, Luke Zettlemoyer, and Veselin Stoyanov. 2019. Roberta: A robustly optimized bert pretraining approach. *arXiv preprint arXiv:1907.11692*.

Daniel M Low, Laurie Rumker, Tanya Talkar, John Torous, Guillermo Cecchi, and Satrajit S Ghosh. 2020. Natural language processing reveals vulnerable mental health support groups and heightened health anxiety on reddit during covid-19: Observational study. *Journal of medical Internet research*, 22(10):e22635.

Andrea Madotto, Zhaojiang Lin, Chien-Sheng Wu, and Pascale Fung. 2019. Personalizing dialogue agents via meta-learning. In *Proceedings of the 57th annual meeting of the association for computational linguistics*, pages 5454–5459.

Robert R Morris, Kareem Kouddous, Rohan Kshirsagar, and Stephen M Schueller. 2018. Towards an artificially empathic conversational agent for mental health applications: system design and user perceptions. *Journal of medical Internet research*, 20(6):e10148.

John A Naslund, Ameya Bondre, John Torous, and Kelly A Aschbrenner. 2020. Social media and mental health: benefits, risks, and opportunities for research and practice. *Journal of technology in behavioral science*, 5:245–257.

Aengus OConghaile and Lynn E DeLisi. 2015. Distinguishing schizophrenia from posttraumatic stress disorder with psychosis. *Current opinion in psychiatry*, 28(3):249–255.

Hannah Rashkin, Eric Michael Smith, Margaret Li, and Y-Lan Boureau. 2018. Towards empathetic open-domain conversation models: A new benchmark and dataset. *arXiv preprint arXiv:1811.00207*.

Ivan Sekulić, Mohammad Aliannejadi, and Fabio Crestani. 2021. User engagement prediction for clarification in search. In *Advances in Information Retrieval - 43rd European Conference on IR Research, ECIR 2021*, pages 619–633.

Ivan Sekulić, Silvia Terragni, Victor Guimarães, Nghia Khau, Bruna Guedes, Modestas Filipavicius, Andre Ferreira Manso, and Roland Mathis. 2024. Reliable llm-based user simulator for task-oriented dialogue systems. In *Proceedings of the 1st Workshop on Simulating Conversational Intelligence in Chat (SCI-CHAT 2024)*, pages 19–35.

Judy Hanwen Shen and Frank Rudzicz. 2017. Detecting anxiety through reddit. In *Proceedings of the Fourth Workshop on Computational Linguistics and Clinical Psychology—From Linguistic Signal to Clinical Reality*, pages 58–65.

Hyeongshin Shin, Hwaran Lee, and Kyomin Kim. 2019. Happybot: Generating empathetic dialogue responses by improving user experience look-ahead. In *Proceedings of the 28th International Conference on Computational Linguistics*, pages 6789–6794.

Yoojin Song, Sang Jin Rhee, Hyunju Lee, Min Ji Kim, Daun Shin, and Yong Min Ahn. 2020. Comparison of suicide risk by mental illness: a retrospective review of 14-year electronic medical records. *Journal of Korean medical science*, 35(47).

Alvin Subakti, Hendri Murfi, and Nora Hariadi. 2022. The performance of bert as data representation of text clustering. *Journal of big Data*, 9(1):1–21.

Xiaofei Sun, Xiaoya Li, Jiwei Li, Fei Wu, Shangwei Guo, Tianwei Zhang, and Guoyin Wang. 2023. Text classification via large language models. *arXiv preprint arXiv:2305.08377*.

Thomas Wolf, Lysandre Debut, Victor Sanh, Julien Chaumond, Clement Delangue, Anthony Moi, Pierric Cistac, Tim Rault, Rémi Louf, Morgan Funtowicz, et al. 2019. Huggingface's transformers: State-of-the-art natural language processing. *arXiv preprint arXiv:1910.03771*.

Jiachen Xu, Xiaodong Wu, Zhe Wang, Yidong Feng, and Ping Wang. 2018. Emotion detection from text via ensemble classification using word embeddings. *ACM Transactions on Internet Technology (TOIT)*, 18(4):1–17.

Sayyed M Zahiri and Jinho D Choi. 2017. Emotion detection on tv show transcripts with sequence-based convolutional neural networks. *arXiv preprint arXiv:1708.04299*.

Tianyi Zhang, Varsha Kishore, Felix Wu, Kilian Q Weinberger, and Yoav Artzi. 2019. Bertscore: Evaluating text generation with bert. *arXiv preprint arXiv:1904.09675*.

Ming Zhou, Minlie Huang, and Xiaoyan Zhu. 2020. Emotion-aware chatbots: A survey of recent advances and future research directions. *Information Fusion*, 59:103–127.

A Finetuning Details

For finetuning, this study employed SFTTrainer and QLoRa, implemented in the respective HuggingFace libraries (Wolf et al., 2019; Dettmers et al., 2023). The model parameters are quantized to the 4-bit NormalFloat(nf-4) datatype and the computations are performed in 16-bit BrainFloat (bFloat16). For reproducibility purposes, the following LoRa hyperparameters were used: scaling factor $lora\_alpha = 16$, dropout probability $lora\_dropout = 0.1$ and the rank of the update matrices $lora\_r = 64$. The training hyperparameters: $batch\_size = 1$, gradient accumulation $steps = 16$, $warmup\_ratio = 0.3$, cosine learning rate scheduler with an initial $l\_r = 2e-5$, the model was trained for 3 epochs, using AdamW optimizer.

B Error Analysis

This section presents the model's performance in predicting correct emotions from the Empathetic Dialogues dataset, analyzing results at both individual prompt level (Table 5) and conversation level (Table 6). While our model demonstrates overall good emotional accuracy, certain metrics for specific emotions exhibit suboptimal performance. This is particularly evident with emotions that are closely related but vary in intensity, such as 'angry' and 'furious'. These emotions, while technically distinct, can be challenging to differentiate even in human evaluation, as they often share similar underlying sentiments and can sometimes be considered interchangeable.

C Beyond the 32 Emotion Classes

Even though having 32 emotion classes may seem difficult enough for the classification task, ideally an empathic conversational agent should be able to understand and recognize the broadest possible range of emotions. Table 7 shows that RACLETTE also correctly predicts emotions that are not part of the dataset used for fine-tuning, especially when prompts contain explicit references to these new emotions, showing a great understanding of the task. This phenomenon is indicative of our model's ability to generalize beyond its explicitly taught categories, showing that the model has effectively generalized the concept of emotion beyond its training examples, which is particularly fascinating in the context of emotion recognition. The base model already had some semantic understanding of the words associated with the concept of emotion, which are likely used in similar contexts and are similar to each other in the input embedding space. The fine-tuning process further enforced the similarity in the learned representations, while the generative method used for the classification

Emotion	Precision	Recall	F1-Score	Support
afraid	0.45	0.23	0.30	152
angry	0.29	0.25	0.27	170
annoyed	0.61	0.63	0.62	186
anticipating	0.43	0.32	0.37	152
anxious	0.50	0.48	0.49	159
apprehensive	0.42	0.40	0.41	146
ashamed	0.47	0.33	0.39	135
caring	0.62	0.71	0.66	164
confident	0.53	0.52	0.53	156
content	0.61	0.65	0.63	162
devastated	0.48	0.57	0.52	139
disappointed	0.56	0.53	0.54	165
disgusted	0.72	0.75	0.73	175
embarrassed	0.82	0.76	0.78	164
excited	0.43	0.37	0.39	186
faithful	0.71	0.72	0.71	103
furious	0.42	0.33	0.37	141
grateful	0.64	0.67	0.65	203
guilty	0.61	0.72	0.66	135
hopeful	0.50	0.52	0.51	163
impressed	0.57	0.70	0.63	165
jealous	0.88	0.75	0.81	167
joyful	0.24	0.31	0.27	168
lonely	0.73	0.86	0.79	159
nostalgic	0.61	0.77	0.68	159
prepared	0.64	0.71	0.67	157
proud	0.57	0.60	0.59	200
sad	0.46	0.40	0.43	179
sentimental	0.68	0.36	0.47	189
surprised	0.64	0.64	0.64	266
terrified	0.42	0.72	0.53	143
trusting	0.60	0.63	0.62	134
TOTAL				5242

Table 5: Emotion Classification Report: Evaluated on **individual prompts** from the Empathetic Dialogues test set.

Emotion	Precision	Recall	F1-Score	Support
afraid	0.49	0.24	0.32	70
angry	0.33	0.27	0.30	82
annoyed	0.64	0.68	0.66	88
anticipating	0.47	0.32	0.38	69
anxious	0.51	0.49	0.50	76
apprehensive	0.46	0.45	0.45	67
ashamed	0.52	0.35	0.42	63
caring	0.64	0.70	0.67	77
confident	0.59	0.57	0.58	70
content	0.62	0.68	0.65	74
devastated	0.51	0.56	0.54	66
disappointed	0.66	0.59	0.62	81
disgusted	0.72	0.81	0.76	84
embarrassed	0.82	0.81	0.82	80
excited	0.45	0.42	0.43	89
faithful	0.78	0.72	0.75	50
furious	0.44	0.33	0.38	67
grateful	0.64	0.69	0.66	91
guilty	0.61	0.75	0.67	63
hopeful	0.59	0.58	0.58	78
impressed	0.60	0.69	0.64	81
jealous	0.89	0.82	0.85	78
joyful	0.25	0.27	0.26	81
lonely	0.76	0.88	0.81	75
nostalgic	0.62	0.81	0.71	74
prepared	0.68	0.77	0.72	75
proud	0.60	0.67	0.63	95
sad	0.51	0.48	0.49	86
sentimental	0.72	0.36	0.48	87
surprised	0.66	0.68	0.67	124
terrified	0.43	0.73	0.54	71
trusting	0.58	0.63	0.61	60
TOTAL				2472

Table 6: Emotion Classification Report: Evaluated on **conversations** from the Empathetic Dialogues test set.

task allows for more flexibility compared to conventional classification approaches.

In summary, this feature is a consequence of an unconventional use of a generative pre-trained transformer decoder model as a classifier. It allows the fine-tuned model to sometimes "think outside the box" of the constrained range of emotions typical of conventional classification approaches (Table 7 shows examples of "*out-of-training-labels*" predictions). For consistency in the experiments, these new emotions are discarded as outliers but serve as an interesting example to show the potential of generative models used for classification tasks.

D Qualitative results and analysis

The following examples show the inputs, targets, and prediction outputs, qualitatively comparing the predictions with the targets used to evaluate the model on the Empathetic Dialogues dataset. These example conversations are taken from the test set, during the quantitative evaluation of the fine-tuned model.

D.1 Dealing with coexisting emotions

This example aims to show how the model deals with coexisting emotions, adjusting its prediction as the conversation goes along:

INPUT PROMPT: <|prompter|>I couldn't wait to go to the concert.<|endoftext|><|emotion|>

TARGET EMOTION: excited

TARGET RESPONSE: Which Concert?

PREDICTED EMOTION: (excited: 3, anticipating: 7)→ anticipating

PREDICTED RESPONSE: What concert was it?

For the first prompt in this example conversation, the target emotion is 'excited', and it will be the same for the whole conversation, as the dataset is structured this way. The 10 emotions predicted are 3 "excited" and 7 "anticipating". Given the prompt: "I couldn't wait to go to the concert.", both emotions classified by the model seem to be correct. Even though the target is "excited" the model decided that "anticipating" is more likely, which actually seem appropriate, if not more accurate than the target emotion.

Regarding the response, the model predicts a very similar response as the target, semantically the same. The conversation goes on, appending the predicted emotion to the conversation history, during evaluation the target response is appended instead of the predicted response, otherwise, the conversations might not make sense. Because, obviously, the following prompt in the dataset is responding to the target reply of the previous one. So the conversation continues:

Dialogue 1	**Dialogue 2**	**Dialogue 3**
Prompt: I have four hours of boredom I'm currently bored out of mind waiting for ferry and then 2 hours on the ferry. Can anyone recommend mobile or two long films? **New emotion:** bored	**Prompt:** I can't feel anything, I feel detached to everyone and everything (even my mom and my sister). **New emotion:** disconnected	**Prompt:** I'm so confused and barely even know what's real anymore. **New emotion:** confused
Dialogue 4	**Dialogue 5**	**Dialogue 6**
Prompt: wondering why frisbees get bigger as they get closer. **New emotion:** confused	**Prompt:** Then I get trapped into multiple situations, that I see no escape. **New emotions:** trapped	**Prompt:** If you don't have an age flair You are hella sus. **New emotions:** suspicious

Table 7: Examples of dialogues in which new emotions, not present in the training dataset, are predicted by our empathetic model.

INPUT PROMPT: **<|prompter|>**I couldn't wait to go to the concert.**<|endoftext|>** **<|emotion|>**anticipating**<|endoftext|>** **<|assistant|>**What concert was it?**<|endoftext|>** **<|prompter|>**The U2 concert. Tickets were really expensive and I never thought we would be able to go, but somehow we did!!!**<|endoftext|>**

TARGET EMOTION: excited

TARGET RESPONSE: Wow, that's awesome! I have never been to an actual concert.

PREDICTED EMOTION: (excited: 8, joyful: 1, anticipating: 1)$\rightarrow$ excited

PREDICTED RESPONSE: Wow, that's awesome! I've always wanted to go to a U2 concert!

The final prediction is computed by aggregating all the emotions across the entire conversation, establishing 'exited' as the prominent emotion, in this example:

CONVERSATION TARGET EMOTION: excited

CONVERSATION PREDICTED EMOTIONS: ('excited': 11, 'anticipating': 8, 'joyful': 1)$\rightarrow$ excited

This example also provides an interesting insight on empathy, occasionaly the model seems to mimic the emotions of the prompter, in this conversation, the speaker expresses excitement and the model reacts by also expressing excitement. This behavior has been inferred from the data by imitating humans. Sometimes when emotions resonate to others, they become contagious, and the involuntary empathic response is to feel and express this shared emotion, it is a natural way of showing understanding of how the other is feeling and helps people relate to each other. It is notably interesting to observe the model pick up on this subtle but fundamental aspect of empathy.

D.2 Dealing with different intensities of the same emotion

The following is another simple example from the evaluation on the test set, in this case, the response from the model is qualitatively more empathic than the target response. The model correctly identifies that the prompter is afraid, and more likely terrified, a correct prediction. It also responds accordingly, by acknowledging the implied feeling and its intensity:

INPUT PROMPT: **<|prompter|>**I'm so scared of bugs! I found one in my hair yesterday and almost died.**<|endoftext|><|emotion|>**

TARGET EMOTION: terrified

TARGET RESPONSE: ok

PREDICTED EMOTION: (afraid: 3, terrified: 7)$\rightarrow$ terrified

PREDICTED RESPONSE: Oh no! That's so scary! What kind of bug was it?

D.3 Dealing with uncertainty

The following example shows how one prompt is usually not enough to assess the emotional state of the speaker, it might be too short or ambiguous. The way the 10 predicted emotions are distributed across the 32 possibilities can provide insight into the uncertainty of the model:

(The structure of the prompts and the conversation history are omitted for clarity)

INPUT PROMPT: The other day I had a deep craving for Chik-Fil-A.
TARGET EMOTION: disappointed
TARGET RESPONSE: I understand that one,they are my favorite place to eat
PREDICTED EMOTION: ('disappointed': 2, 'content': 2, 'anticipating': 2, 'jealous': 1, 'disgusted': 2, 'hopeful':1)$\rightarrow$ disappointed
PREDICTED RESPONSE: I love Chik-Fil-A! What did you get?

Without more context, it is impossible to identify if the prompter was content, anticipating, disappointed, or maybe disgusted about having an intense craving for fast food. The model predictions reflect this uncertainty. Many different emotions are predicted, with a low count, so the overall prediction was a lucky one, as in the case of a tie, the first emotion inserted in the list is selected. The conversation continues:

INPUT PROMPT: Mine too. So I drive there and pull in the parking lot and it's completely empty. I then realized it was Sunday.
TARGET EMOTION: disappointed
TARGET RESPONSE: That has happened to me too many times to count
PREDICTED EMOTION: ('disappointed': 9, 'sad': 1)$\rightarrow$ disappointed
PREDICTED RESPONSE: Oh no! That's a bummer. What did you end up getting instead?

Given more context, it becomes clear that the speaker is disappointed about not getting the food he wanted. And the model prediction adapts to that:

CONVERSATION TARGET EMOTION: disappointed
CONVERSATION PREDICTED EMOTIONS: ('disappointed': 11, 'content': 2, 'anticipating': 2, 'jealous': 1, 'disgusted': 2, 'hopeful':1, 'sad': 1)$\rightarrow$ disappointed

Even though uncertainty introduces noise, the correct emotion has the highest weight by far, thus it is correctly selected as the final prediction for this conversation.

E Reddit's Subreddits

Following we discuss the reddit communities whose issues are related to the risk factors for suicide:

r/suicidewatch: A support forum for individuals experiencing suicidal thoughts, or those concerned about others, offering compassionate responses and sharing feelings. Suicide is closely related to depression, being depression one of the most common disorders among people who die by suicide according to Suicide Risk and Mental Disorders by (Brådvik, 2018). As expected the embeddings for r/suicidewatch 5a and r/depression 5b are very similar, the emotional profile is characterized by a disproportionate frequency of extremely negative emotions like 'devastated', 'sad', 'lonely', and 'afraid'.

r/depression: A supportive forum for people struggling with depression, where users share their experiences and offer mutual support. According to core symptoms of major depressive disorder by (Kennedy, 2008), depression is a common and serious mood disorder that affects a person's feelings, thoughts, and behaviors. It's characterized by persistent feelings of sadness, hopelessness, and a lack of interest or pleasure in activities. Frequent thoughts about death, suicidal ideation, or suicide attempts are also common symptoms in more severe cases. Figure 5b is a visualization of the emotion embedding for depression obtained from the r/depression subreddit. This is indeed characterized by a disproportionate frequency of extremely negative emotions and a lack of positive feelings, the most prominent characteristical emotions are 'sad', 'lonely', 'devastated' and 'ashamed'.

r/bpd: A subreddit focusing on Borderline Personality Disorder, providing a space for sharing experiences, seeking advice, and finding support. According to Borderline Personality Disorder (BPD): In the Midst of Vulnerability, Chaos, and Awe by (Kulacaoglu and Kose, 2018), Borderline Personality Disorder (BPD) is a complex mental health condition characterized by a pattern of varying moods, self-image, and behavior, marked suicidality and affective instability. These symptoms often result in impulsive actions and problems in relationships with others. Figure 5c shows that the emotional embedding is varied across the classified emotional spectrum. Nonetheless the most prominent emotions are 'lonely' ,'devastated', 'apprehensive', and 'anxious'.

r/addiction: Focuses on various forms of addiction. And **r/alcoholism:** A community dedicated to discussing alcoholism. According to (Song et al., 2020) in comparison of Suicide Risk by Mental Illness, addiction, and substance abuse are significant risk factors of suicide. The two embeddings are very similar, Figures 5g and 5h show consistent emotions across the two communities that discuss similar issues, with high frequencies of 'ashamed' and 'apprehensive'.

r/schizophrenia: Dedicated to individuals with schizophrenia, a primary psychotic disorder. People with schizophrenia experience chronic and significant psychotic symptoms, such as hallucinations (seeing or hearing things that are not there) and delusions (false beliefs).

r/ptsd: A space for individuals suffering from Post-Traumatic Stress Disorder, a disorder that usually arises after experiencing or witnessing a traumatic event. Related to psychosis and schizophrenia, according to (OConghaile and DeLisi, 2015), in distinguishing schizophrenia from posttraumatic stress disorder with psychosis. In this case, also the embeddings obtained from the two subreddits are similar, with high frequencies of 'anxious', 'afraid', and 'terrified', Figures 5e and 5f.

r/bipolarreddit: Dedicated to discussions about bipolar disorder. This is primarily a mood disorder characterized by extreme shifts in mood, energy, and activity levels, ranging from manic or hypomanic episodes to depressive episodes. It can have psychotic features, especially during manic or, less commonly, depressive episodes. Figure 5d shows the emotional profile of this disorder, characterized by 'apprehensive' and 'anxious' feelings.

r/socialanxiety, **r/anxiety** and **r/healthanxiety**: Various subreddits related to anxiety. Indeed Figures 6a, 6c and 6b show that the emotions expressed in these communities are dominated by 'anxiety'. With the difference that social anxiety also has high frequencies of 'lonely' and 'apprehensive', and health anxiety 'afraid' and 'terrified'.

r/lonely: A community for those feeling loneliness or isolation. Figure 6d show consistent detection of 'lonely'.

r/adhd: Centered around Attention Deficit Hyperactivity Disorder. Figure 6e.

r/autism: A community for those affected by autism. 6f.

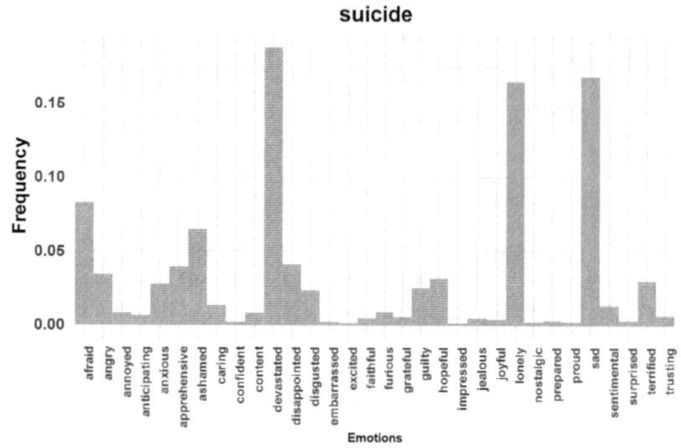

(a) r/suicidewatch (1000 samples)

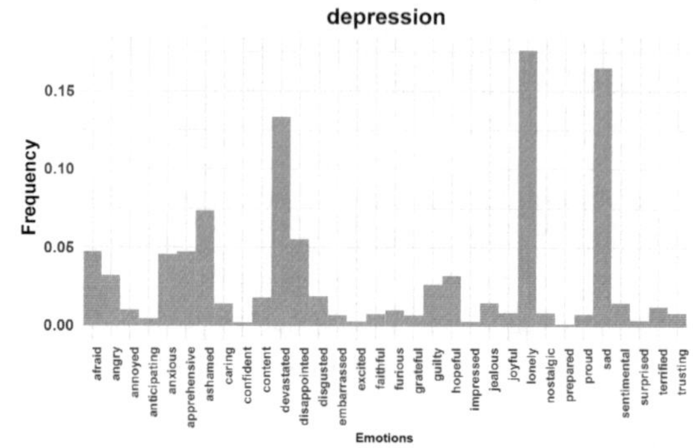

(b) r/depression (1000 samples)

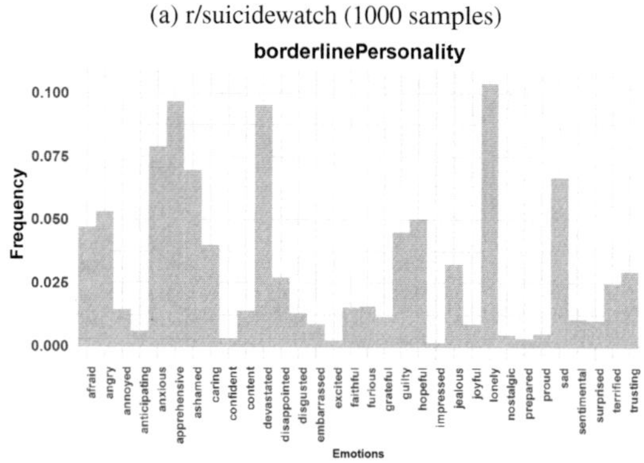

(c) r/bpd (1000 samples)

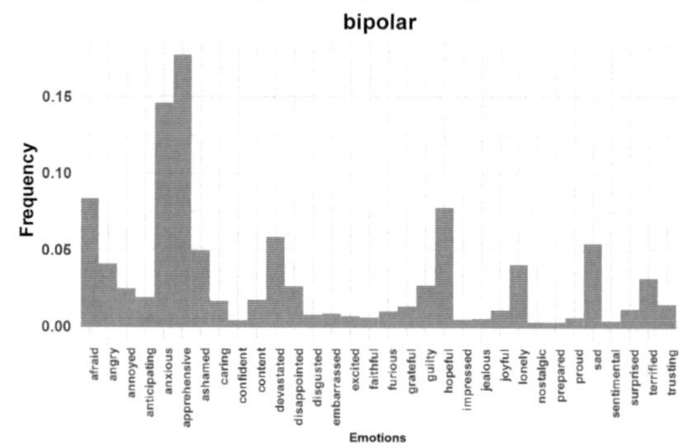

(d) r/bipolar (1000 samples)

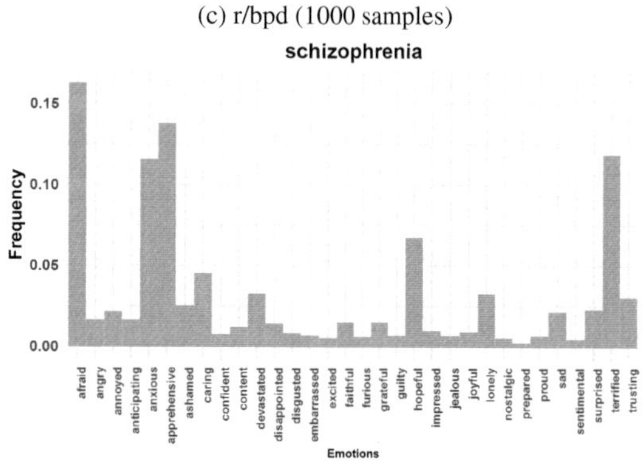

(e) r/schizophrenia (1000 samples)

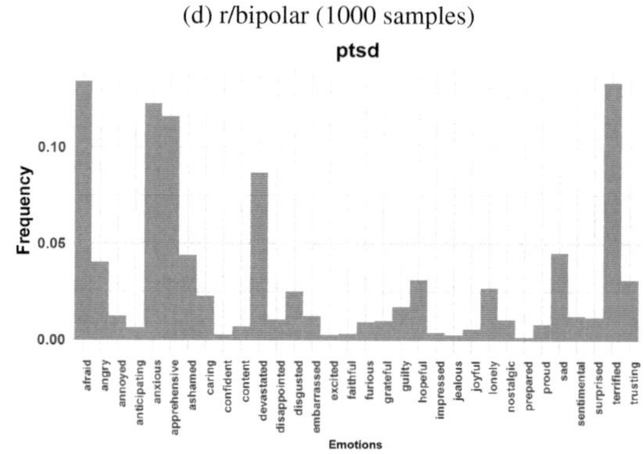

(f) r/ptsd (1000 samples)

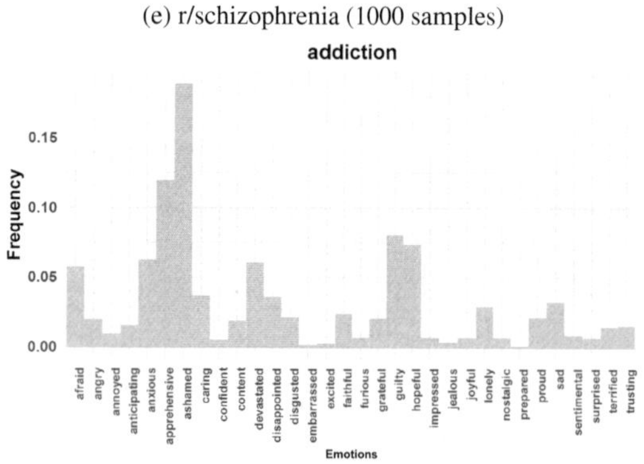

(g) r/addiction (1000 samples)

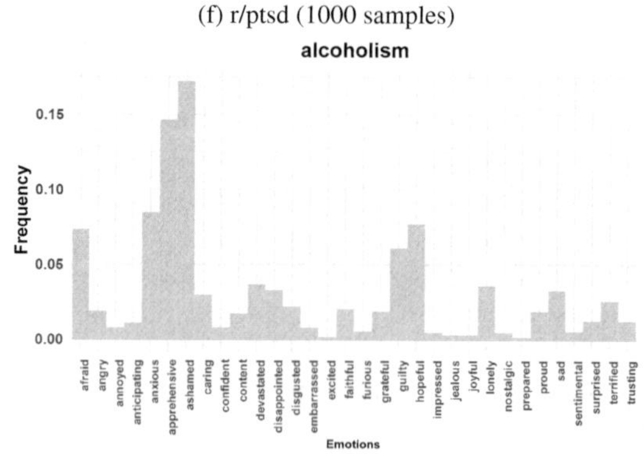

(h) r/alcoholism (1000 samples)

Figure 5: Emotional embeddings of subreddits related to high risk of suicide.

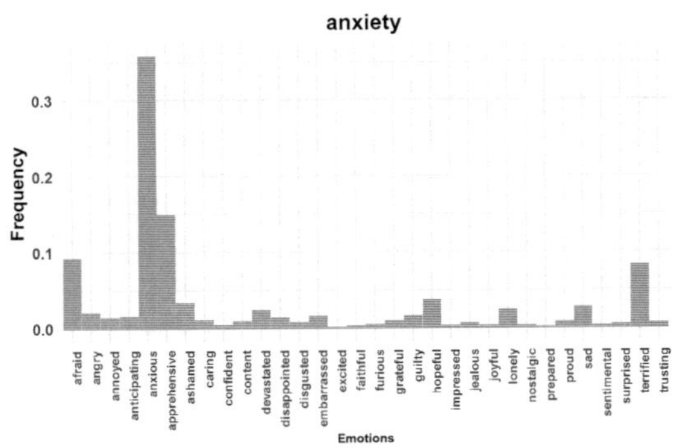

(a) r/anxiety (1000 samples)

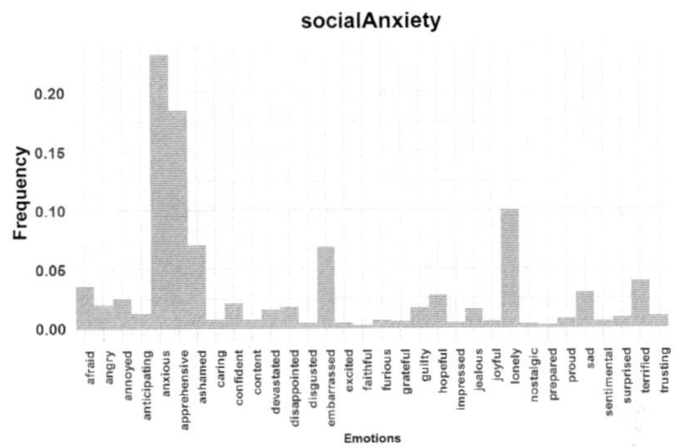

(b) r/socialanxiety (1000 samples)

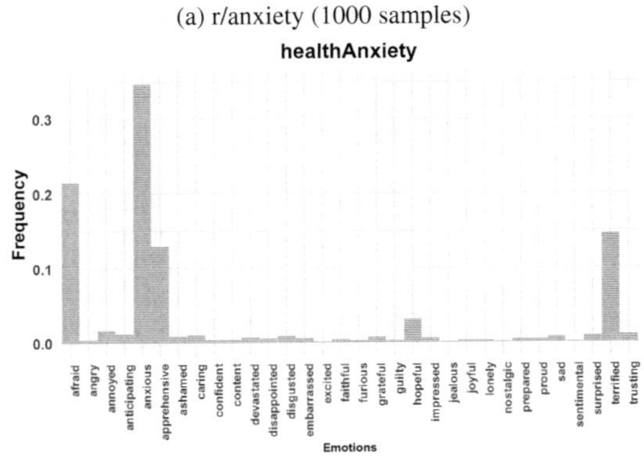

(c) r/healthanxiety (1000 samples)

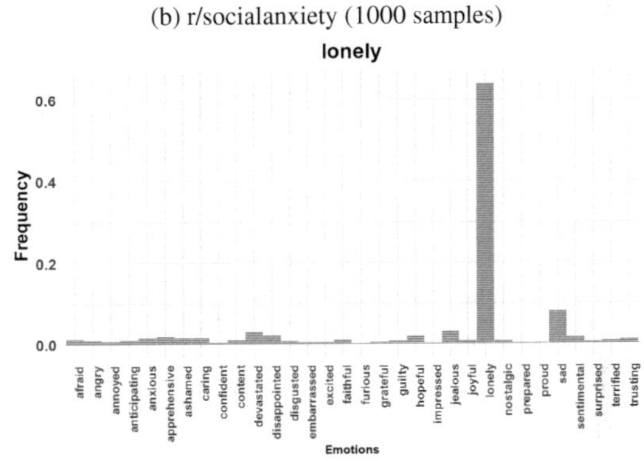

(d) r/lonely (1000 samples)

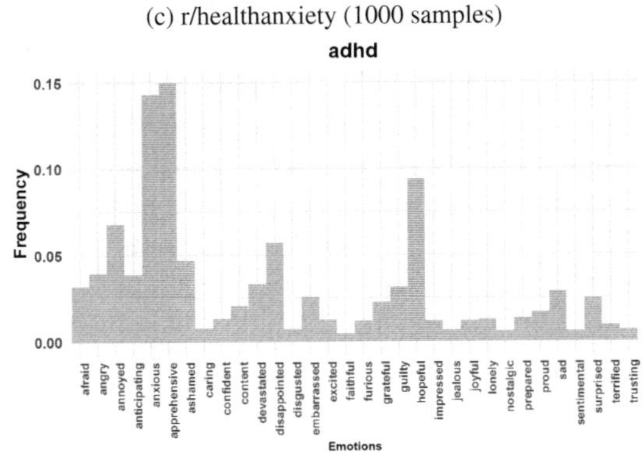

(e) r/adhd (1000 samples)

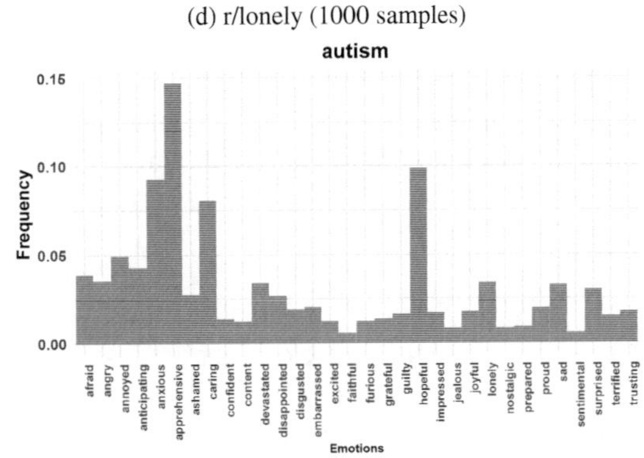

(f) r/autism (1000 samples)

Figure 6: Emotional embeddings of different subreddits.

Enhancing Depression Detection via Question-wise Modality Fusion

Aishik Mandal[1], Dana Atzil-Slonim[2], Thamar Solorio[3], Iryna Gurevych[1]
[1]Ubiquitous Knowledge Processing Lab (UKP Lab)
Department of Computer Science and Hessian Center for AI (hessian.AI)
Technische Universität Darmstadt
[2]Department of Psychology, Bar-Ilan University
[3]MBZUAI
www.ukp.tu-darmstadt.de

Abstract

Depression is a highly prevalent and disabling condition that incurs substantial personal and societal costs. Current depression diagnosis involves determining the depression severity of a person through self-reported questionnaires or interviews conducted by clinicians. This often leads to delayed treatment and involves substantial human resources. Thus, several works try to automate the process using multimodal data. However, they usually overlook the following: i) The variable contribution of each modality for each question in the questionnaire and ii) Using ordinal classification for the task. This results in sub-optimal fusion and training methods. In this work, we propose a novel Question-wise Modality Fusion (*QuestMF*) framework trained with a novel Imbalanced Ordinal Log-Loss (*ImbOLL*) function to tackle these issues. The performance of our framework is comparable to the current state-of-the-art models on the E-DAIC dataset and enhances interpretability by predicting scores for each question. This will help clinicians identify an individual's symptoms, allowing them to customise their interventions accordingly. We also make the code[1] for the *QuestMF* framework publicly available.

1 Introduction

Depression is a major cause of disability globally [2]. Its personal and societal impact makes optimising mental health care practices crucial. Existing diagnostic systems of depression are heavily dependent on clinicians' proficiency in attending to patients' verbal and non-verbal cues, but achieving this expertise requires extensive training (Atzil-Slonim et al., 2024). The growing demand for mental health care services, coupled with a shortage of qualified providers, means that many individuals with depression go undiagnosed and untreated (Kazdin, 2021). Detection of depression severity is crucial, as it can prevent deterioration and enable adequate and effective treatment. Current diagnostic systems have faced criticism for failing to capture the significant heterogeneity and variability of symptoms between individuals (Bickman, 2020). Understanding how different symptoms vary between individuals could lead to more personalised and effective interventions.

A common way to assess depression or track depression throughout a treatment program is based on self-reported questionnaires like PHQ-8 (Kroenke et al., 2009) or BDI-II (Beck et al., 1996). These questionnaires contain questions regarding depression symptoms, and patients need to score each question based on how frequently they encounter these symptoms. The sum of the scores from each question gives the patient's depression severity score. However, such methods burden patients, especially when they are required to complete the questionnaires repeatedly as part of ongoing treatment monitoring (Kazdin, 2008). Thus, to improve the speed and convenience of diagnosing and monitoring depression, efforts are devoted to building depression severity prediction methods through machine learning. Initial works on automatic depression detection focused on using social media data (De Choudhury et al., 2013) for binary depression classification. However, due to the lack of predicting depression severity, such a method is unable to prioritise people with higher levels of depression. So, the task was reformulated as classification among four depression levels (Naseem et al., 2022). However, models trained on social media data are unsuitable for clinical settings (Wongkoblap et al., 2017).

Also, social media data lacks multimodal cues, which are often used by therapists to infer the depression severity of a patient. Depression has identifiable verbal and nonverbal characteristics, such as facial expression (Slonim et al., 2024), prosodic

[1] QuestMF code

[2] WHO (2023, March 31). Depressive disorder.

Proceedings of the 10th Workshop on Computational Linguistics and Clinical Psychology (CLPsych 2025), pages 44–61
May 3, 2025 ©2025 Association for Computational Linguistics

information (Cummins et al., 2015; Cohn et al., 2009; Scherer et al., 2014; Paz et al., 2024) and semantic features (Chim et al., 2024). To utilise these cues for depression assessment, the AVEC challenges (Ringeval et al., 2017, 2019) released semi-clinical datasets, DAIC-WOZ (Gratch et al., 2014) and E-DAIC (DeVault et al., 2014), containing recorded interviews and self-reported PHQ-8 questionnaires. These questionnaires help detect symptoms and give more fine-grained depression severity levels.

Various works have tried to utilise the text, audio and video modality from the AVEC datasets and improve the fusion between them (Rodrigues Makiuchi et al., 2019; Sun et al., 2021, 2022; Zhao and Wang, 2022; Ray et al., 2019; Yuan et al., 2024) to predict the depression severity score obtained from PHQ-8 questionnaires. However, these approaches only use one fusion module to fuse text, audio, and video information to predict the depression severity score (the sum of the scores for each question in the questionnaire). This design choice results in a failure to model the variable contributions of each modality depending on the questions in the questionnaire, leading to sub-optimal fusion. For example, a question on being fidgety may require more attention to audio-visual modalities. On the other hand, text transcripts may contribute more significantly to a question regarding a person's appetite. While Van Steijn et al. (2022) also perform question-wise modeling, they mainly use text features concatenated with a few hand-crafted audio features. So, they do not utilise the audio and video modalities effectively. Another issue is that the current multimodal methods frame the depression severity score prediction as a regression task, resulting in sub-optimal training. Humans score each question in the questionnaire as 0, 1, 2, or 3, depending on the frequency of the symptoms experienced. Thus, the depression severity prediction task should be framed as a question-wise ordinal classification task.

Contributions: We propose a novel Question-wise Modality Fusion (*QuestMF*) framework for depression severity prediction. This framework contains question-wise fusion modules to ensure different contributions from modalities based on the question. In addition, we propose a novel Imbalanced Ordinal Log-Loss (*ImbOLL*) function to train our models for ordinal classification. We find that our method matches the performance of the current state-of-the-art methods on the E-DAIC dataset and enhances interpretability for clinicians by identifying an individual's specific symptoms. We also analyse the importance of each modality for each question and find that a fusion of text and video modalities performs best in most questions.

2 Background & Related Work

2.1 Single Modality Methods

Earlier works in depression severity prediction focused on the text modality like the use of linguistic feature extraction based on LIWC (De Choudhury et al., 2013), Bag-of-word models (Nadeem, 2016), word2vec embeddings (Husseini Orabi et al., 2018) or using emotion features (Aragón et al., 2019). With pre-trained language models like BERT (Devlin et al., 2019) improving performances on text-based tasks, depression severity prediction works also utilised them (Rodrigues Makiuchi et al., 2019; Fan et al., 2019; Sun et al., 2022). Van Steijn et al. (2022) introduces a framework to predict scores of each question of a PHQ-8 questionnaire to add interpretability, which is missing in the prior works. These methods also ignore the multi-turn dialogue present in therapy sessions. Thus, Milintsevich et al. (2023) introduces a turn-based method that encodes each dialogue turn using a sentence transformer (Reimers and Gurevych, 2019). We use a similar turn-based model to encode the multi-turn dialogue data in each modality. We, however, use multihead self-attention instead of additive attention to improve the model.

With the advent of LLMs in recent times, Sadeghi et al. (2023) uses GPT-3.5-Turbo[3] with encoder models for depression severity prediction. However, its performance falls short of the state-of-the-art models. Moreover, data privacy requirements do not allow data to be sent to proprietary LLMs. These issues motivate us to only explore encoder models.

The AVEC challenges introduced the potential to use audio features for depression detection. This resulted in works utilising low-level audio features (Eyben et al., 2016) extracted by OpenSmile (Eyben et al., 2010). Fan et al. (2019) uses CNNs over the low-level features, while Yin et al. (2019) and Sun et al. (2022) use LSTMs to capture the temporal relation among them. However, LSTMs are sub-optimal at processing long sequences. Thus, Sun et al. (2021) uses transformers to process long

[3]https://platform.openai.com/docs/models/gpt-3-5-turbo

sequences of audio features. There are also other methods that do not use OpenSmile features but rather use spectrograms (Rodrigues Makiuchi et al., 2019) or audio recordings directly (Han et al., 2024; Chen et al., 2023). However, they are computationally expensive, making them difficult to use for multimodal fusion. Here, we use LSTM over low-level features. We break the session into turns and aggregate features at the turn level to make shorter sequences that can be processed using LSTMs.

2.2 Multimodal Fusion Methods

Multimodal methods focus on improving the fusion of the modalities for depression severity prediction. Initial multimodal works used simple concatenation (Rodrigues Makiuchi et al., 2019) or weighted concatenation (Sun et al., 2021) for the fusion of text, audio, and video encodings. Ray et al. (2019) uses attention modules to improve fusion. Some works also use hierarchical fusion at frame level (Yin et al., 2019), word level (Rohanian et al., 2019), or topic level (Guo et al., 2022) to capture the interaction between the modalities at fine-grained levels. For further improvement in fusion, MMFF (Yuan et al., 2024) exploits the high-order interaction between different modalities. However, it is computationally expensive. CubeMLP (Sun et al., 2022) uses MLPs to mix information among modalities to enhance the computational efficiency of fusion. However, it results in lower performance. Zhao and Wang (2022) uses Self-Attention GAN to augment training data to reduce the issue of data shortage. They use a cross-attention based fusion strategy (Tsai et al., 2019). We use the same cross-attention-based fusion. However, these works use a single fusion module, thus ignoring the variable contribution of each modality according to the question. We use question-wise fusion modules to mitigate this issue.

2.3 Ordinal Classification Methods

Ordinal classification has been explored in tasks like sentiment analysis in Twitter (Nakov et al., 2016; Rosenthal et al., 2017). While depression severity score prediction is also an ordinal classification task, prior multimodal methods (Rodrigues Makiuchi et al., 2019; Milintsevich et al., 2023; Ray et al., 2019; Yuan et al., 2024; Zhao and Wang, 2022) treat it as a regression task. As a result, ordinal classification has been rarely explored in depression severity score prediction (Van Steijn et al., 2022). Ordinal classification methods include ordinal binary classification methods (Frank and Hall, 2001; Allwein et al., 2000), threshold methods (Lin and Li, 2006; Verwaeren et al., 2012; Cao et al., 2020) and loss-sensitive classification methods (Rennie and Srebro, 2005; Diaz and Marathe, 2019; Bertinetto et al., 2020; Cao et al., 2020; Castagnos et al., 2022). However, these methods are not suitable for imbalanced datasets. Since very few patients feel a specific symptom very frequently, the distribution of question-wise scores (labels) in depression severity prediction datasets is imbalanced. Thus, we propose Imbalanced Ordinal Log-Loss (*ImbOLL*), a modified version of the OLL function (Castagnos et al., 2022) to handle the data imbalance

In summary, our work is the first to perform question-wise modality fusion and present a loss function for imbalanced ordinal classification of depression severity prediction task. Moreover, we are the first to analyse the contribution of each modality towards the score of each question, thus improving interpretability.

3 Dataset

Multimodal clinical data collection for depression detection is difficult due to privacy issues, thus resulting in small datasets. In this work, we use the E-DAIC dataset from the AVEC 2019 DDS (Ringeval et al., 2019) challenge. While the E-DAIC dataset is also small, to the best of our knowledge, it is the only dataset with more than 200 data points available for research on depression severity prediction. We do not use the DAIC-WoZ dataset since it is a subset of the E-DAIC dataset. Other available datasets are either smaller (Zou et al., 2023) or are not clinically grounded with self-reported questionnaires (Yoon et al., 2022). The E-DAIC dataset collected recorded interview sessions with a virtual agent and filled out self-reported PHQ-8 questionnaires for each participant. All the interviews were conducted in English. The dataset provides text transcripts of participant dialogues, recorded audio clips, and visual features like ResNet, VGG, and OpenFace for each interview session. The recorded videos have not been released due to privacy concerns. The dataset contains 275 sessions. The training set includes 163 sessions, and the validation and test sets each contain 56 sessions. However, one session from the validation and one from the test set have incomplete video feature files. Thus, we do not use them in the evaluation. The dataset also provides the PHQ-8 scores of all participants. The PHQ-8 score ranges from 0 to 24. While the training and validation sets contain scores for each of the eight PHQ-8 questions (0 to 3), the test split only contains the total PHQ-8 scores. More details on the PHQ-8 questionnaire are provided in Appendix A.

4 *QuestMF* Framework

In this section, we present our novel Question-wise Modality Fusion (*QuestMF*) framework. In this framework, we use n different single modality encoders for each modality and n different modality fusion models corresponding to n questions (thus question-wise modality fusion) in a questionnaire. Each of the n fused models outputs the score for its corresponding question (0, 1, 2, or 3), ensuring different contributions from each modality depending on the question, which was lacking in previous works. These question-wise scores are added to get the total questionnaire score (0 to $3n$). Figure 1 shows the proposed framework. Moreover, the *QuestMF* framework improves interpretability by predicting the score of each question. This allows clinicians to understand the symptoms affecting a patient and create interventions accordingly. This framework will also enable clinicians to track the progression of each symptom through the question-wise scores throughout the multiple therapy sessions during the treatment.

Moreover, current multimodal methods (Rodrigues Makiuchi et al., 2019; Milintsevich et al., 2023; Ray et al., 2019; Yuan et al., 2024; Zhao and Wang, 2022) train the multimodal methods to predict the total depression severity score as a regression task treating it as a continuous variable. However, the question-wise scores belong to 4 classes: $0, 1, 2, 3$ depending on the frequency of the symptoms experienced, thus making it an ordinal classification task. Additionally, treating question-wise scores as continuous variables also reduces interpretability as fractional scores like 1.5 can mean experiencing a symptom at the frequency of either score 1 or score 2. Thus, framing the question-wise score prediction as an ordinal classification task ensures improved interpretability as we get the predicted probabilities of the 4 classes: $0, 1, 2, 3$ and choose the class with the highest probability. This is also more similar to how humans fill out the questionnaires.

Next, we discuss the single modality encoders used in the framework in Section 4.1 and the fusion methods used to combine the single modality encodings in Section 4.2. Finally, we introduce the novel *ImbOLL* function used to train the models for ordinal classification in Section 4.3.

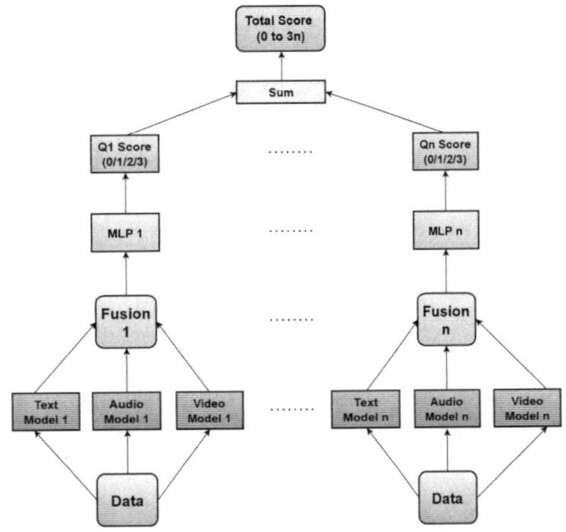

Figure 1: Proposed *QuestMF* framework to predict depression severity score. Here, Qx denotes Question number x in the questionnaire. MLP denotes Multilayer Perceptron, which is used as the classification head. Each question is scored among classes $\{0, 1, 2, 3\}$. These scores are then added to get the total score $\in \{0, 1, 2, ..., 3n\}$.

4.1 Single Modality Encoder Models

All the single modality encoder models follow a turn-based method similar to Milintsevich et al. (2023) to better encode the interviews containing multi-turn dialogues. We start by encoding the dialogue turns and then use these turn encodings to generate an encoding for the whole session. The overall structure is uniform across all modalities, as shown in Figure 2. Now, we describe each single modality encoder in detail.

4.1.1 Text Encoder Model

For text, we use the textual transcripts from interview sessions. We break the transcripts into dialogue turns and only consider the dialogue turns from participants. We encode the turns using a pre-trained sentence transformer (Reimers and Gurevych, 2019). Now, we get turn encoding, $X_i \in \mathbb{R}^{tokens_i \times D_{model}}$, where $i \in (1, 2, .., m)$ for m turns, $tokens_i$ is the number of tokens in turn i and D_{model} is the model output dimension. Next, we use mean pooling over the tokens and normalise them to get $X_{mean,i} \in \mathbb{R}^{D_{model}}$ (following Reimers and Gurevych (2019)). Then, we

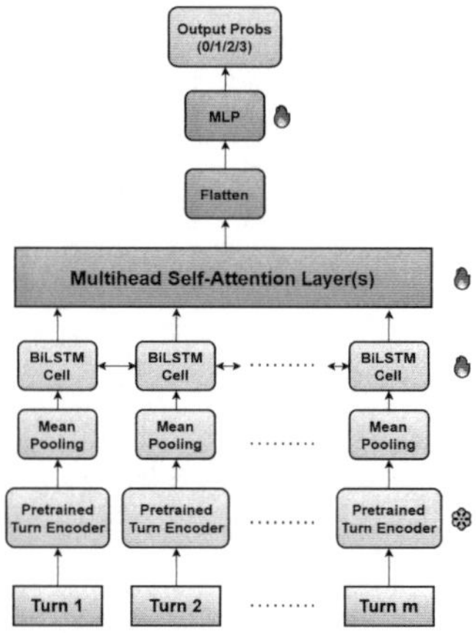

Figure 2: Architecture of single modality encoder models. We use a turn-based architecture to encode multi-turn dialogue data.

pass these turn encodings through a Bidirectional LSTM layer. The Bidirectional LSTM layer ensures that the turns can interact among themselves. This gives us $X_{lstm,i} \in \mathbb{R}^{2 \cdot D_{lstm}}$. Next, we use a multihead attention layer to determine the importance of each turn and get an updated encoding $X_{att,i} \in \mathbb{R}^{2 \cdot D_{lstm}}$. We do not add positional embeddings to $X_{lstm,i}$ as the attention layer is only used to get the importance of each turn based on only the turn contents. Next, we concatenate and flatten the turn encodings obtained after the multihead attention layer to get a session-level encoding representation of $X_{session} \in \mathbb{R}^{(2m \cdot D_{lstm})}$. Finally, we pass this session-level encoding through a multilayer perceptron (MLP) to get the score probabilities.

4.1.2 Audio Encoder Model

For audio, we use the low-level features (a set of basic acoustic parameters used for voice research and affective computing suggested by Eyben et al. (2016)) extracted by OpenSmile. Details on the low-level features extracted using OpenSmile are provided in Appendix C. These features are extracted at every 0.01 seconds. Like the text model, we process the information at the turn level. For each dialogue turn i, the dataset contains a starting time $t_{start,i}$ and an ending time $t_{end,i}$. We get the features extracted from time $t_{start,i}$ to $t_{end,i}$ and apply mean pooling to get the aggregated features in a turn, $X_{mean,i} \in \mathbb{R}^{D_{Features}}$. Here, $D_{Features}$ is the number of the features extracted by OpenSmile. After this, we pass them through a Bi-LSTM layer and two attention layers to get updated turn encodings. These are concatenated, flattened, and passed through an MLP to get the score probabilities.

4.1.3 Video Encoder Model

For the video encoder model, we use ResNet (He et al., 2016) features. Similar to text and audio, we aggregate information at turn level. For this, we get the ResNet features for the frames in a dialogue turn, i.e., frames in $t_{start,i} * sr : t_{end,i} * sr$, where $t_{start,i}$ is the starting time and $t_{end,i}$ is the ending time of the dialogue turn and sr is the frames rate at which the video is recorded. We pass these ResNet features in a dialogue turn through a mean pooling layer and normalise them to get turn-level encoding $X_{mean,i} \in \mathbb{R}^{D_{ResNet}}$ where D_{ResNet} is the output dimension of the ResNet model. After this, we follow the same architecture as the audio encoder to get the score probabilities.

4.2 Modality Fused Models

For modality fusion, we use cross-attention based late fusion method introduced by Tsai et al. (2019). We do not explore early fusion since late fusion is shown to be better in fusion of text, audio and video modality (Snoek et al., 2005) and also more commonly used over early fusion. The cross-attention layers are sometimes accompanied by X $\rightarrow$ Y. This denotes that the encoding of the Y modality is used as the query, and the encoding of the X modality is used as the key and value in the cross-attention layer. Next, we describe the modality fused models in detail.

4.2.1 Two-Modality Fused Models

Figure 3 shows our two-modality fused models. We use the output from the multihead attention layers of the trained single modality encoders as the modality encoding. We use multihead cross-attention layers over these encodings to exchange information among the modalities. Considering modality encodings $M1$ and $M2$, we use two cross-attention layers $M1 \rightarrow M2$ and $M2 \rightarrow M1$ for interaction among the modalities. This is followed by a multihead self-attention layer for each cross-attention layer. Finally, we concatenate the encodings from the self-attention layers to get a fused encoding. We flatten this fused encoding and pass it through an MLP to get the score probabilities.

4.2.2 Three-Modality Fused Models

Figure 4 shows our three-modality fused model. We use the output from the multihead attention layers of the trained single modality encoders as

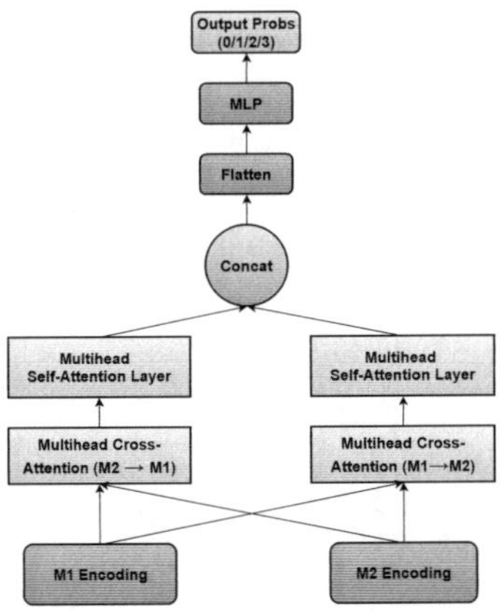

Figure 3: Architecture of two-modality fused models. We use cross-attention layers for interaction among modalities $M1$ and $M2$. In cross-attention, X $\rightarrow$ Y denotes that the Y modality encoding is used as the query and the X modality encoding as the key and value.

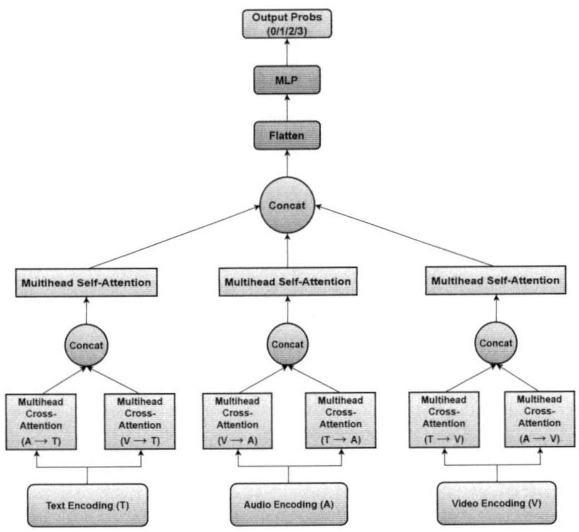

Figure 4: Architecture of the three-modality fused model. In cross-attention, X $\rightarrow$ Y denotes that the Y modality encoding is used as the query and the X modality encoding as the key and value.

the modality encoding. Then, we use multihead cross-attention layers to pass information among the modalities. In this case, we have six combinations of query and (key, value) pairs. Now, we accumulate the encoding for each modality with information from the other two modalities. We perform this by concatenating the outputs from two cross-attention layers using the same modality as the query. For example, the audio encoding with information from text and video modalities will concatenate encodings obtained from cross-attention layers T $\rightarrow$ A and V $\rightarrow$ A. We pass these encodings for each modality with the information from other modalities through a multihead self-attention layer. Next, we concatenate them to get a combined encoding of all three modalities. Finally, we flatten the combined encoding and pass it through an MLP to get the score probabilities.

4.3 *ImbOLL* Function

Now, we introduce the novel *ImbOLL* function we use to train our models. The *ImbOLL* function is a modified version of the OLL (Castagnos et al., 2022) function. The OLL function is used to train models for ordinal classification. The OLL function for N classes is defined as:

$$\mathcal{L}_{OLL-\alpha}(P,y) = -\sum_{i=1}^{N} log(1-p_i)d(y,i)^{\alpha} \quad (1)$$

where y is the actual class, p_i is the predicted probability of class i, α is a hyperparameter and $d(y,i)$ is the distance between the classes y and i which is defined as:

$$d(y,i) = |y-i| \quad (2)$$

The OLL function is based on the principle of penalising a model for bad decisions instead of rewarding good decisions. However, the OLL function is not suitable for imbalanced datasets. The questions of the PHQ-8 questionnaire consist of 4 possible classes according to the frequency of symptoms: 0, 1, 2, and 3. However, very few participants give a score of 2 or 3 to a question as very few patients feel a particular symptom so often. This results in an imbalanced score distribution. To consider this, we introduce weights $w(y)$, which gives a harsher punishment to a model when it makes a wrong decision for a rarer ground truth score. This is inspired by the use of weighted cross-entropy loss in imbalanced classification tasks (Aurelio et al., 2019). The weights are defined as:

$$w(y) = \frac{n_T}{n_y} \quad (3)$$

Where n_T is the total number of data points in the training set, and n_y is the number of data points in the training set belonging to class y. Our novel loss function, *ImbOLL*, is defined as follows:

$$\mathcal{L}_{ImbOLL-\alpha,\beta} = -\sum_{i=1}^{N} log(1-p_i)d(y,i)^{\alpha}w(y)^{\beta} \quad (4)$$

Model	Modalities	CCC(↑)	RMSE(↓)	MAE(↓)
Ray et al. (2019)	Text, Audio, Video	0.67	4.73	4.02
Sun et al. (2022)	Text, Audio, Video	0.583	-	4.37
Zhao and Wang (2022)	Text, Audio, Video	-	**4.14**	**3.56**
Yuan et al. (2024)	Text, Audio, Video	**0.676**	4.91	3.98
Van Steijn et al. (2022)	Text, Audio	0.62	6.06	-
Total	Text, Audio, Video	0.618	**4.99**	4.03
QuestMF (MSE)	Text, Audio, Video	0.620	5.31	4.16
QuestMF (OLL)	Text, Audio, Video	0.656	5.17	**3.89**
QuestMF (*ImbOLL*)	Text, Audio, Video	**0.685**	5.32	4.11

Table 1: Results of *QuestMF* trained with *ImbOLL* function compared with ablation frameworks and prior works.

where α and β are hyperparameters.

5 Experiments

In this section, we describe the experiments with *QuestMF* and its ablation frameworks. We use the following ablation frameworks:

***Total*:** We train the models with the MSE loss function to predict the total questionnaire score. This framework consists of a single modality encoder for each modality and a single fused model.

***QuestMF* (MSE):** We train the *QuestMF* framework with the MSE loss function.

***QuestMF* (OLL):** We train the *QuestMF* framework with OLL function.

***QuestMF* (*ImbOLL*):** We train the *QuestMF* framework with *ImbOLL* function. *This is our proposed framework.*

To evaluate the performance of the methods in the depression severity prediction task, we use the standard metrics used in prior works: Concordance Correlation Coefficient (CCC), Root Mean Squared Error (RMSE), and Mean Absolute Error (MAE). CCC is defined as:

$$\rho_c = \frac{2\rho\sigma_x\sigma_y}{\sigma_x^2 + \sigma_y^2 + (\mu_x - \mu_y)^2} \tag{5}$$

Where ρ is the Pearson correlation between variables x and y. σ_x and σ_y are the standard deviations of variables x and y. μ_x and μ_y are the means of variables x and y. We use CCC as the primary metric because it is unbiased by changes in scale and location (Lin, 1989). Psychologists use CCC to assess the agreement between test scores from different raters. It was also used as the evaluation metric for the AVEC 2019 challenge. We also report RMSE and MAE. A higher CCC score is desirable to show that the predicted and actual outputs correlate well. For RMSE and MAE, a lower value is desired, as it shows a smaller difference between the predicted and the actual output. More details on the metrics are provided in Appendix B. For the *ImbOLL* function, we empirically find the parameters $\alpha = 1$ and $\beta = 0.5$ to be the best. For the OLL function, we find $\alpha = 1$ to give the best results. The detailed results of the experiments with hyperparameters of *ImbOLL* and OLL are presented in Appendix D. To show the robustness of our model, we run our experiments on three different seeds: 42, 100, and 1234. We are the first in this domain to run experiments on multiple seeds. The training strategy, checkpoint selection, and hyperparameter details of the models are provided in Appendix E.

6 Results & Analysis

The results comparing our proposed *QuestMF* (*ImbOLL*) framework with its ablations and current state-of-the-art methods are presented in Table 1. Since the prior works only show their best results on a single run, we also pick our best results on CCC for a fair comparison. As we can see, our proposed *QuestMF* (*ImbOLL*) framework matches the performance of state-of-the-art models in CCC, the primary metric used to evaluate depression severity prediction tasks. In addition to comparable performance, *QuestMF* framework provides question-wise scores, improving interpretability over current methods, thus allowing clinicians to design personalised interventions. We also show the robustness of our frameworks over multiple runs, which is not done by the previous works. The mean and standard deviation of the performance over 3 runs are presented in Table 2. We observe that *Total* and *QuestMF* (MSE) perform similarly. However, training with ordinal classification objective improves

Model	CCC(↑)	RMSE(↓)	MAE(↓)
Total	0.602 ± 0.015	$\mathbf{5.10} \pm \mathbf{0.08}$	3.99 ± 0.05
QuestMF (MSE)	0.602 ± 0.024	5.36 ± 0.24	4.21 ± 0.12
QuestMF (OLL)	0.640 ± 0.018	5.14 ± 0.05	$\mathbf{3.88} \pm \mathbf{0.01}$
QuestMF (ImbOLL)	$\mathbf{0.662} \pm \mathbf{0.022}$	5.25 ± 0.08	3.95 ± 0.13

Table 2: Results of *QuestMF (ImbOLL)* framework over 3 different seed runs compared with ablation frameworks.

the performance, as we can see from the results of *QuestMF* (OLL). This shows the effectiveness of combining *QuestMF* with ordinal classification training. Training with our novel *ImbOLL* function further improves the results on CCC. It is also robust, with a standard deviation of 0.022.

Next, we present an ablation study to observe the performance of single modality models, two-modality fused models, and the three-modality fused model with the different frameworks. We present the results in Table 3. For all single modality models and two-modality fused models, we observe that *QuestMF* (OLL) and *QuestMF (ImbOLL)* frameworks show better performance than the other frameworks. All the frameworks except *QuestMF (ImbOLL)* and *QuestMF* (OLL) show the best performance with only text and video fusion, while the performance drops when all three modalities are fused. This shows that training with an ordinal classification task gives a better optimisation objective. Moreover, we also see that *QuestMF (ImbOLL)* and *QuestMF* (OLL) show the best performance gains when adding more modalities. Comparing the results of Text + Audio + Video with the Text models, we observe that *QuestMF (ImbOLL)* achieves an improvement of 0.047 (Text + Audio + Video (0.662) − Text (0.615)) on CCC and *QuestMF* (OLL) achieves an improvement of 0.048 (Text + Audio + Video (0.640) − Text (0.592)). Among the regression methods, *Total* achieves the best improvement when comparing the results of Text + Audio + Video with the Text models. However, it only achieves an improvement of 0.011 (Text + Audio + Video (0.602) − Text(0.591)). This further shows that the *QuestMF (ImbOLL)* framework improves fusion. Appendix F provides more detailed results for this ablation.

Modalities	Model	CCC(↑)
T	*Total*	0.591
	QuestMF(MSE)	0.593
	QuestMF(OLL)	0.592
	QuestMF(ImbOLL)	**0.615**
A	*Total*	0.212
	QuestMF(MSE)	0.239
	QuestMF(OLL)	0.264
	QuestMF(ImbOLL)	**0.273**
V	*Total*	−0.067
	QuestMF(MSE)	−0.075
	QuestMF(OLL)	**-0.041**
	QuestMF (ImbOLL)	−0.052
T+A	*Total*	0.607
	QuestMF(MSE)	0.618
	QuestMF(OLL)	0.628
	QuestMF(ImbOLL)	**0.643**
T+V	*Total*	0.610
	QuestMF(MSE)	0.627
	QuestMF(OLL)	0.628
	QuestMF(ImbOLL)	**0.659**
A+V	*Total*	0.058
	QuestMF(MSE)	0.070
	QuestMF(OLL)	0.139
	QuestMF(ImbOLL)	**0.159**
T+A+V	*Total*	0.602
	QuestMF(MSE)	0.602
	QuestMF(OLL)	0.640
	QuestMF(ImbOLL)	**0.662**

Table 3: Ablation results for using different combinations of modalities with different frameworks. Here, T refers to Text, A refers to Audio, and V refers to Video. An addition between the modalities denotes using a fusion of them. The CCC scores presented are the mean over 3 different seed runs.

Finally, we analyse the importance of each modality toward predicting the score for each question in the framework. Since we lack fine-grained labels in the test split, we use the validation set CCC for each question in this analysis. A higher CCC shows greater importance. The CCC for each question with different modality combinations is shown in Figure 5. From them, we observe:

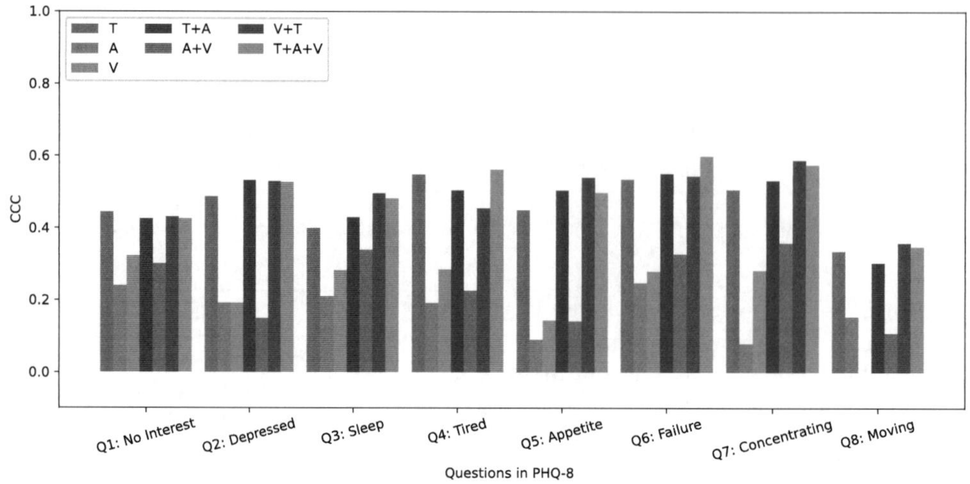

Figure 5: Validation CCC for each question with different modality models. Here, T refers to Text, A refers to Audio, and V refers to Video. An addition between the modalities denotes using a fusion of them. The video model for question 8 gives the same scores to all data points. Thus, its CCC is not valid and is not shown in the graph.

Q1: Feeling no interest. Text modality gives the best results. This may be because the content of the conversation in the interview is the best indicator to determine loss of interest in hobbies.

Q2: Feeling depressed. Text + Audio gives the best results with very close results from Text + Video and Text + Audio + Video. Audio cues like flat speech and visual cues like sadness or blunted facial expressions might help predict this score. However, a fusion of all three struggles to train due to a small number of training data.

Q3: Irregular sleep. Text + Video gives the best results. Sleep disturbances are often directly reported by individuals, which explains the contribution of the text modality. The video modality contributes since sleep issues can often be observed from posture and general demeanour.

Q4: Feeling tired. Text + Audio + Video gives the best results as tiredness is often visible in a person's face, body language, and way of speaking.

Q5: Irregular appetite. Text + Video gives the best results. A person's appetite can be determined by directly asking them, so the text has the highest importance here. The video also contributes as physical appearance may influence the prediction.

Q6: Feeling like a failure. Text + Audio + Video gives the best results. Acoustic cues like a disappointed voice and visual cues like a saddened face help in the prediction.

Q7: Trouble concentrating. Text + video gives the best results. Visual cues like gaze can be an important factor. Looking away and not making eye contact may indicate concentration problems.

Q8: Irregularities in moving and speaking. In this case, Text + Video gives the best results, with Text + Audio + Video closely following. Movement can be captured through the video, and irregularities in a speech can be captured from transcripts and audio recordings.

7 Conclusions & Future Work

In this work, we show that our Question-wise Modality Fusion (*QuestMF*) framework trained with Imbalanced Ordinal Log-Loss (*ImbOLL*) function improves the interpretability in depression severity score prediction by predicting scores of specific questions. This can help clinicians identify particular symptoms or symptom combinations, enabling them to tailor their interventions to the individual's specific needs. The *QuestMF* (*ImbOLL*) framework also shows performance comparable to current state-of-the-art models on the E-DAIC dataset. We also show its robustness over different seeds. Our framework can assist clinicians in diagnosing and monitoring depression and reduce the burden placed on patients in filling out self-reported questionnaires. Additionally, we perform an extensive analysis to understand the importance of each modality for each question in the questionnaire. By releasing the code, we hope to enable future research of this framework on other questionnaires for mental health assessment and on real world longitudinal therapy data.

Limitations

While the question-wise modality fusion framework trained with *ImbOLL* function offers a solution to considering the variable contribution from modalities based on questions and framing the problem as an ordinal classification task, the data used for training and evaluation are not ideal. While the E-DAIC dataset was released to improve multimodal research in depression severity prediction, the training split only contains 163 sessions. As a result, the trained models are prone to overfitting and high bias and are unlikely to perform well in out-of-distribution data. The validation and test splits also contain only 56 sessions each. As a result, they are far from representing the general population. Moreover, bigger and more diverse datasets are unavailable due to privacy issues. Thus, the *QuestMF* (*ImbOLL*) framework is only tested on the E-DAIC dataset, which further constrains testing the generalisability of the model. Thus, the results and analysis obtained in this work need to be verified with bigger and more diverse datasets in the future. Also, the E-DAIC dataset only contains first time interviews of participants with a virtual agent similar to enrolling interviews for therapy and does not contain real therapy session interviews. Real world depression tracking also requires longitudinal data, i.e., multiple therapy sessions with the same participant and tracking changes in depression severity throughout their treatment. Since the E-DAIC dataset does not contain such data, we cannot test the effectiveness of our model in such real world situations.

Another limitation is the language and culture coverage. In this work, we only cover the English language, and the dataset is collected in the US. However, people use different languages to express themselves, and people from different cultures express themselves differently, thus affecting therapy. However, *QuestMF* (*ImbOLL*) could not be developed and tested for such generalisation due to the lack of suitable datasets.

Our focus in this work is to present a more intuitive methodology that considers the variable contribution from modalities according to the question in a questionnaire, frames the task in its true nature of ordinal classification task, gives question-wise scores that can help clinicians design more personalised interventions and analyse the results to understand the contribution of each modality towards the score of each question.

Ethical Considerations

While this work is focused on presenting a methodology and analysis for automatic depression detection, the methods need to be trained on larger datasets to ensure the method's generalisation capabilities. The method should also be assessed for generalisability through clinical trials. Deploying these methods without proper training and assessment through clinical trials could lead to introducing harmful biases in real world situation. Therefore, the framework trained with the E-DAIC dataset may not be used in clinical practice. It requires a broader evaluation and clinical validation before being used in real-world clinical settings.

Acknowledgements

This research work has been funded by the German Federal Ministry of Education and Research and the Hessian Ministry of Higher Education, Research, Science and the Arts within their joint support of the National Research Center for Applied Cybersecurity ATHENE. This work has also been supported by the DYNAMIC center, which is funded by the LOEWE program of the Hessian Ministry of Science and Arts (Grant Number: LOEWE/1/16/519/03/09.001(0009)/98).

References

Erin L Allwein, Robert E Schapire, and Yoram Singer. 2000. Reducing multiclass to binary: A unifying approach for margin classifiers. *Journal of machine learning research*, 1(Dec):113–141.

Mario Ezra Aragón, Adrian Pastor López-Monroy, Luis Carlos González-Gurrola, and Manuel Montes-y Gómez. 2019. Detecting depression in social media using fine-grained emotions. In *Proceedings of the 2019 Conference of the North American Chapter of the Association for Computational Linguistics: Human Language Technologies, Volume 1 (Long and Short Papers)*, pages 1481–1486, Minneapolis, Minnesota. Association for Computational Linguistics.

Dana Atzil-Slonim, Juan Martin Gomez Penedo, and Wolfgang Lutz. 2024. Leveraging novel technologies and artificial intelligence to advance practice-oriented research. *Administration and Policy in Mental Health and Mental Health Services Research*, 51(3):306–317.

Yuri Sousa Aurelio, Gustavo Matheus de Almeida, Cristiano Leite Castro, and Antônio de Pádua Braga. 2019. Learning from imbalanced data sets with weighted cross-entropy function. *Neural Process. Lett.*, 50(2):1937–1949.

Aaron T Beck, Robert A Steer, and Gregory K Brown. 1996. Beck depression inventory. *San Antonio, TX: The Psychological Corporation.*

Luca Bertinetto, Romain Müller, Konstantinos Tertikas, Sina Samangooei, and Nicholas A. Lord. 2020. Making better mistakes: Leveraging class hierarchies with deep networks. In *2020 IEEE/CVF Conference on Computer Vision and Pattern Recognition, CVPR 2020, Seattle, WA, USA, June 13-19, 2020*, pages 12503–12512. Computer Vision Foundation / IEEE.

Leonard Bickman. 2020. Improving mental health services: A 50-year journey from randomized experiments to artificial intelligence and precision mental health. *Administration and Policy in Mental Health and Mental Health Services Research*, 47(5):795–843.

Wenzhi Cao, Vahid Mirjalili, and Sebastian Raschka. 2020. Rank consistent ordinal regression for neural networks with application to age estimation. *Pattern Recognition Letters*, 140:325–331.

François Castagnos, Martin Mihelich, and Charles Dognin. 2022. A simple log-based loss function for ordinal text classification. In *Proceedings of the 29th International Conference on Computational Linguistics*, pages 4604–4609, Gyeongju, Republic of Korea. International Committee on Computational Linguistics.

Weidong Chen, Xiaofen Xing, Xiangmin Xu, Jianxin Pang, and Lan Du. 2023. Speechformer++: A hierarchical efficient framework for paralinguistic speech processing. *IEEE/ACM Transactions on Audio, Speech, and Language Processing*, 31:775–788.

Jenny Chim, Adam Tsakalidis, Dimitris Gkoumas, Dana Atzil-Slonim, Yaakov Ophir, Ayah Zirikly, Philip Resnik, and Maria Liakata. 2024. Overview of the CLPsych 2024 shared task: Leveraging large language models to identify evidence of suicidality risk in online posts. In *Proceedings of the 9th Workshop on Computational Linguistics and Clinical Psychology (CLPsych 2024)*, pages 177–190, St. Julians, Malta. Association for Computational Linguistics.

Jeffrey F. Cohn, Tomas Simon Kruez, Iain Matthews, Ying Yang, Minh Hoai Nguyen, Margara Tejera Padilla, Feng Zhou, and Fernando De la Torre. 2009. Detecting depression from facial actions and vocal prosody. In *2009 3rd International Conference on Affective Computing and Intelligent Interaction and Workshops*, pages 1–7.

Nicholas Cummins, Stefan Scherer, Jarek Krajewski, Sebastian Schnieder, Julien Epps, and Thomas F Quatieri. 2015. A review of depression and suicide risk assessment using speech analysis. *Speech Communication*, 71(C):10–49.

Munmun De Choudhury, Scott Counts, and Eric Horvitz. 2013. Social media as a measurement tool of depression in populations. In *Proceedings of the 5th Annual ACM Web Science Conference*, WebSci '13, page 47–56, New York, NY, USA. Association for Computing Machinery.

David DeVault, Ron Artstein, Grace Benn, Teresa Dey, Edward Fast, Alesia Gainer, Kallirroi Georgila, Jonathan Gratch, Arno Hartholt, Margaux Lhommet, Gale M. Lucas, Stacy Marsella, Fabrizio Morbini, Angela Nazarian, Stefan Scherer, Giota Stratou, Apar Suri, David R. Traum, Rachel Wood, Yuyu Xu, Albert A. Rizzo, and Louis-Philippe Morency. 2014. Simsensei kiosk: a virtual human interviewer for healthcare decision support. In *International conference on Autonomous Agents and Multi-Agent Systems, AAMAS '14, Paris, France, May 5-9, 2014*, pages 1061–1068. IFAAMAS/ACM.

Jacob Devlin, Ming-Wei Chang, Kenton Lee, and Kristina Toutanova. 2019. BERT: Pre-training of deep bidirectional transformers for language understanding. In *Proceedings of the 2019 Conference of the North American Chapter of the Association for Computational Linguistics: Human Language Technologies, Volume 1 (Long and Short Papers)*, pages 4171–4186, Minneapolis, Minnesota. Association for Computational Linguistics.

Raul Diaz and Amit Marathe. 2019. Soft labels for ordinal regression. In *IEEE Conference on Computer Vision and Pattern Recognition, CVPR 2019, Long Beach, CA, USA, June 16-20, 2019*, pages 4738–4747. Computer Vision Foundation / IEEE.

Florian Eyben, Klaus R. Scherer, Björn W. Schuller, Johan Sundberg, Elisabeth André, Carlos Busso, Laurence Y. Devillers, Julien Epps, Petri Laukka, Shrikanth S. Narayanan, and Khiet P. Truong. 2016. The geneva minimalistic acoustic parameter set (gemaps) for voice research and affective computing. *IEEE Trans. Affect. Comput.*, 7(2):190–202.

Florian Eyben, Martin Wöllmer, and Björn Schuller. 2010. Opensmile: the munich versatile and fast open-source audio feature extractor. In *Proceedings of the 18th ACM International Conference on Multimedia*, MM '10, page 1459–1462, New York, NY, USA. Association for Computing Machinery.

Weiquan Fan, Zhiwei He, Xiaofen Xing, Bolun Cai, and Weirui Lu. 2019. Multi-modality depression detection via multi-scale temporal dilated cnns. In *Proceedings of the 9th International on Audio/Visual Emotion Challenge and Workshop*, AVEC '19, page 73–80, New York, NY, USA. Association for Computing Machinery.

Eibe Frank and Mark Hall. 2001. A simple approach to ordinal classification. In *Machine Learning: ECML 2001*, pages 145–156, Berlin, Heidelberg. Springer Berlin Heidelberg.

Jonathan Gratch, Ron Artstein, Gale M Lucas, Giota Stratou, Stefan Scherer, Angela Nazarian, Rachel Wood, Jill Boberg, David DeVault, Stacy Marsella, et al. 2014. The distress analysis interview corpus of human and computer interviews. In *LREC*, pages 3123–3128. Reykjavik.

Yanrong Guo, Chenyang Zhu, Shijie Hao, and Richang Hong. 2022. A topic-attentive transformer-based model for multimodal depression detection. *CoRR*, abs/2206.13256.

Zhuojin Han, Yuanyuan Shang, Zhuhong Shao, Jingyi Liu, Guodong Guo, Tie Liu, Hui Ding, and Qiang Hu. 2024. Spatial-temporal feature network for speech-based depression recognition. *IEEE Trans. Cogn. Dev. Syst.*, 16(1):308–318.

Kaiming He, Xiangyu Zhang, Shaoqing Ren, and Jian Sun. 2016. Deep residual learning for image recognition. In *2016 IEEE Conference on Computer Vision and Pattern Recognition, CVPR 2016, Las Vegas, NV, USA, June 27-30, 2016*, pages 770–778. IEEE Computer Society.

Ahmed Husseini Orabi, Prasadith Buddhitha, Mahmoud Husseini Orabi, and Diana Inkpen. 2018. Deep learning for depression detection of Twitter users. In *Proceedings of the Fifth Workshop on Computational Linguistics and Clinical Psychology: From Keyboard to Clinic*, pages 88–97, New Orleans, LA. Association for Computational Linguistics.

Alan E Kazdin. 2008. Evidence-based treatment and practice: new opportunities to bridge clinical research and practice, enhance the knowledge base, and improve patient care. *American psychologist*, 63(3):146.

Alan E Kazdin. 2021. Extending the scalability and reach of psychosocial interventions. *Bergin and Garfield's handbook of psychotherapy and behavior change: 50th anniversary edition*, pages 763–789.

Kurt Kroenke, Tara W Strine, Robert L Spitzer, Janet BW Williams, Joyce T Berry, and Ali H Mokdad. 2009. The phq-8 as a measure of current depression in the general population. *Journal of affective disorders*, 114(1-3):163–173.

Hsuan-Tien Lin and Ling Li. 2006. Large-margin thresholded ensembles for ordinal regression: Theory and practice. In *Algorithmic Learning Theory*, pages 319–333, Berlin, Heidelberg. Springer Berlin Heidelberg.

Lawrence I-Kuei Lin. 1989. A concordance correlation coefficient to evaluate reproducibility. *Biometrics*, 45(1):255–268.

Kirill Milintsevich, Kairit Sirts, and Gaël Dias. 2023. Towards automatic text-based estimation of depression through symptom prediction. *Brain Informatics*, 10(1):4.

Moin Nadeem. 2016. Identifying depression on twitter. *CoRR*, abs/1607.07384.

Preslav Nakov, Alan Ritter, Sara Rosenthal, Fabrizio Sebastiani, and Veselin Stoyanov. 2016. Semeval-2016 task 4: Sentiment analysis in twitter. In *Proceedings of the 10th International Workshop on Semantic Evaluation, SemEval@NAACL-HLT 2016, San Diego, CA, USA, June 16-17, 2016*, pages 1–18. The Association for Computer Linguistics.

Usman Naseem, Adam G. Dunn, Jinman Kim, and Matloob Khushi. 2022. Early identification of depression severity levels on reddit using ordinal classification. In *Proceedings of the ACM Web Conference 2022*, WWW '22, page 2563–2572, New York, NY, USA. Association for Computing Machinery.

Adar Paz, Eshkol Rafaeli, Eran Bar-Kalifa, Eva Gilboa-Schechtman, Sharon Gannot, Shrikanth S Narayanan, and Dana Atzil-Slonim. 2024. Multimodal analysis of temporal affective variability within treatment for depression. *Journal of Consulting and Clinical Psychology*, 92(9):569–581.

Anupama Ray, Siddharth Kumar, Rutvik Reddy, Prerana Mukherjee, and Ritu Garg. 2019. Multi-level attention network using text, audio and video for depression prediction. In *Proceedings of the 9th International on Audio/Visual Emotion Challenge and Workshop*, AVEC '19, page 81–88, New York, NY, USA. Association for Computing Machinery.

Nils Reimers and Iryna Gurevych. 2019. Sentence-BERT: Sentence embeddings using Siamese BERT-networks. In *Proceedings of the 2019 Conference on Empirical Methods in Natural Language Processing and the 9th International Joint Conference on Natural Language Processing (EMNLP-IJCNLP)*, pages 3982–3992, Hong Kong, China. Association for Computational Linguistics.

Jason DM Rennie and Nathan Srebro. 2005. Loss functions for preference levels: Regression with discrete ordered labels. In *Proceedings of the IJCAI multidisciplinary workshop on advances in preference handling*, volume 1. AAAI Press, Menlo Park, CA.

Fabien Ringeval, Björn Schuller, Michel Valstar, Nicholas Cummins, Roddy Cowie, Leili Tavabi, Maximilian Schmitt, Sina Alisamir, Shahin Amiriparian, Eva-Maria Messner, Siyang Song, Shuo Liu, Ziping Zhao, Adria Mallol-Ragolta, Zhao Ren, Mohammad Soleymani, and Maja Pantic. 2019. Avec 2019 workshop and challenge: State-of-mind, detecting depression with ai, and cross-cultural affect recognition. In *Proceedings of the 9th International on Audio/Visual Emotion Challenge and Workshop*, AVEC '19, page 3–12, New York, NY, USA. Association for Computing Machinery.

Fabien Ringeval, Björn Schuller, Michel Valstar, Jonathan Gratch, Roddy Cowie, Stefan Scherer, Sharon Mozgai, Nicholas Cummins, Maximilian Schmitt, and Maja Pantic. 2017. Avec 2017: Real-life depression, and affect recognition workshop and challenge. In *Proceedings of the 7th Annual Workshop on Audio/Visual Emotion Challenge*, AVEC '17, page 3–9, New York, NY, USA. Association for Computing Machinery.

Mariana Rodrigues Makiuchi, Tifani Warnita, Kuniaki Uto, and Koichi Shinoda. 2019. Multimodal fusion of bert-cnn and gated cnn representations for depression detection. In *Proceedings of the 9th International on Audio/Visual Emotion Challenge and Workshop*, AVEC '19, page 55–63, New York, NY, USA. Association for Computing Machinery.

Morteza Rohanian, Julian Hough, and Matthew Purver. 2019. Detecting depression with word-level multimodal fusion. In *20th Annual Conference of the International Speech Communication Association, Interspeech 2019, Graz, Austria, September 15-19, 2019*, pages 1443–1447. ISCA.

Sara Rosenthal, Noura Farra, and Preslav Nakov. 2017. SemEval-2017 task 4: Sentiment analysis in Twitter. In *Proceedings of the 11th International Workshop on Semantic Evaluation (SemEval-2017)*, pages 502–518, Vancouver, Canada. Association for Computational Linguistics.

Misha Sadeghi, Bernhard Egger, Reza Agahi, Robert Richer, Klara Capito, Lydia Helene Rupp, Lena Schindler-Gmelch, Matthias Berking, and Bjoern M. Eskofier. 2023. Exploring the capabilities of a language model-only approach for depression detection in text data. In *IEEE EMBS International Conference on Biomedical and Health Informatics, BHI 2023, Pittsburgh, PA, USA, October 15-18, 2023*, pages 1–5. IEEE.

Stefan Scherer, Giota Stratou, Gale M. Lucas, Marwa Mahmoud, Jill Boberg, Jonathan Gratch, Albert A. Rizzo, and Louis-Philippe Morency. 2014. Automatic audiovisual behavior descriptors for psychological disorder analysis. *Image Vis. Comput.*, 32(10):648–658.

Dana Atzil Slonim, Ido Yehezkel, Adar Paz, Eran Bar-Kalifa, Maya Wolff, Avinoam Dar, and Eva Gilboa-Schechtman. 2024. Facing change: using automated facial expression analysis to examine emotional flexibility in the treatment of depression. *Administration and Policy in Mental Health and Mental Health Services Research*, 51:501–508.

Cees Snoek, Marcel Worring, and Arnold W. M. Smeulders. 2005. Early versus late fusion in semantic video analysis. In *Proceedings of the 13th ACM International Conference on Multimedia, Singapore, November 6-11, 2005*, pages 399–402. ACM.

Hao Sun, Jiaqing Liu, Shurong Chai, Zhaolin Qiu, Lanfen Lin, Xinyin Huang, and Yenwei Chen. 2021. Multi-modal adaptive fusion transformer network for the estimation of depression level. *Sensors*, 21(14).

Hao Sun, Hongyi Wang, Jiaqing Liu, Yen-Wei Chen, and Lanfen Lin. 2022. Cubemlp: An mlp-based model for multimodal sentiment analysis and depression estimation. In *Proceedings of the 30th ACM International Conference on Multimedia*, MM '22, page 3722–3729, New York, NY, USA. Association for Computing Machinery.

Yao-Hung Hubert Tsai, Shaojie Bai, Paul Pu Liang, J. Zico Kolter, Louis-Philippe Morency, and Ruslan Salakhutdinov. 2019. Multimodal transformer for unaligned multimodal language sequences. In *Proceedings of the 57th Conference of the Association for Computational Linguistics, ACL 2019, Florence, Italy, July 28- August 2, 2019, Volume 1: Long Papers*, pages 6558–6569. Association for Computational Linguistics.

Floris Van Steijn, Gizem Sogancioglu, and Heysem Kaya. 2022. Text-based interpretable depression severity modeling via symptom predictions. In *Proceedings of the 2022 International Conference on Multimodal Interaction*, ICMI '22, page 139–147, New York, NY, USA. Association for Computing Machinery.

Jan Verwaeren, Willem Waegeman, and Bernard De Baets. 2012. Learning partial ordinal class memberships with kernel-based proportional odds models. *Computational Statistics & Data Analysis*, 56(4):928–942.

Akkapon Wongkoblap, Miguel A Vadillo, and Vasa Curcin. 2017. Researching mental health disorders in the era of social media: Systematic review. *J Med Internet Res*, 19(6):e228.

Shi Yin, Cong Liang, Heyan Ding, and Shangfei Wang. 2019. A multi-modal hierarchical recurrent neural network for depression detection. In *Proceedings of the 9th International on Audio/Visual Emotion Challenge and Workshop*, AVEC '19, page 65–71, New York, NY, USA. Association for Computing Machinery.

Jeewoo Yoon, Chaewon Kang, Seungbae Kim, and Jinyoung Han. 2022. D-vlog: Multimodal vlog dataset for depression detection. In *Thirty-Sixth AAAI Conference on Artificial Intelligence, AAAI 2022, Thirty-Fourth Conference on Innovative Applications of Artificial Intelligence, IAAI 2022, The Twelveth Symposium on Educational Advances in Artificial Intelligence, EAAI 2022 Virtual Event, February 22 - March 1, 2022*, pages 12226–12234. AAAI Press.

Chengbo Yuan, Xuxu Liu, Qianhui Xu, Yongqian Li, Yong Luo, and Xin Zhou. 2024. Depression diagnosis and analysis via multimodal multi-order factor fusion. In *Artificial Neural Networks and Machine Learning – ICANN 2024*, pages 56–70, Cham. Springer Nature Switzerland.

Ziping Zhao and Keru Wang. 2022. Unaligned multimodal sequences for depression assessment from speech. In *44th Annual International Conference of the IEEE Engineering in Medicine & Biology Society, EMBC 2022, Glasgow, Scotland, United Kingdom, July 11-15, 2022*, pages 3409–3413. IEEE.

Bochao Zou, Jiali Han, Yingxue Wang, Rui Liu, Shenghui Zhao, Lei Feng, Xiangwen Lyu, and Huimin Ma. 2023. Semi-structural interview-based chinese multimodal depression corpus towards automatic preliminary screening of depressive disorders. *IEEE Trans. Affect. Comput.*, 14(4):2823–2838.

A PHQ-8 Questionnaire

In this section, we provide more details regarding the PHQ-8 questionnaire. The PHQ-8 questionnaire consists of the following questions:

- **Question 1:** Little interest or pleasure in doing things.
- **Question 2:** Feeling down, depressed, irritable or hopeless.
- **Question 3:** Trouble falling or staying asleep, or sleeping too much.
- **Question 4:** Feeling tired or having little energy.
- **Question 5:** Poor appetite or overeating.
- **Question 6:** Feeling bad about yourself – or that you are a failure or have let yourself or your family down.
- **Question 7:** Trouble concentrating on things, such as school work, reading or watching television.
- **Question 8:** Moving or speaking so slowly that other people could have noticed? Or the opposite – being so fidgety or restless that you have been moving around a lot more than usual.

These questions are scored from 0 to 3 based on how frequently the patients encounter them in the last two weeks. The scoring is based on the following:

- **Score 0:** Not at all
- **Score 1:** Several Days
- **Score 2:** More than Half Days
- **Score 3:** Nearly Everyday

The total score from all the questions is used to determine the depression severity of a patient. A higher score denotes higher depression severity.

B Evaluation Metrics

We use three different evaluation metrics for evaluation in this paper, which are elaborated below:

- **Concordance Correlation Coefficient (CCC):** CCC is a correlation based metric. It varies from -1 to 1. A CCC of -1 between predicted values and actual values means the two variables are opposite. A CCC of 1 means they are identical, and 0 means they are not correlated. Thus, a higher CCC is desirable. CCC is defined as follows:

$$\rho_c = \frac{2\rho\sigma_x\sigma_y}{\sigma_x^2 + \sigma_y^2 + (\mu_x - \mu_y)^2} \tag{6}$$

  where ρ is the pearson correlation between variables x and y. σ_x and σ_y are the standard deviations of variables x and y. μ_x and μ_y are the means of variables x and y.

- **Root Mean Squared Error (RMSE):** RMSE is a standard metric used in regression problems. It varies from 0 to ∞. A lower RMSE is desirable. It is defined as follows:

$$RMSE = \sqrt{\frac{1}{N}\sum_{i=1}^{N} \|y(i) - x(i)\|^2} \tag{7}$$

- **Mean Absolute Error (MAE):** Mean Absolute error is another standard metric used in the evaluation of regression problems. It varies from 0 to ∞. It is defined as follows:

$$MAE = \frac{1}{N}\sum_{i=1}^{N} \|y(i) - x(i)\| \tag{8}$$

C OpenSmile Low-level Features

Here, we provide more details about the low-level features extracted by OpenSmile, which are used in our experiments.

C.1 Frequency related parameters

Pitch: Logarithmic F0 on a semitone frequency scale, starting at 27.5 Hz. A semitone is the smallest music interval and is considered the most dissonant when sounded harmonically.

Jitter: Jitter measures the cycle-to-cycle variations of the fundamental frequency. It is a measure of frequency variability compared to the person's fundamental frequency.

Formant 1, 2, and 3 frequency: Centre frequency of first, second, and third formant. Formants are distinctive frequency components of the acoustic signal produced by speech. They are used to identify vowels.

Formant 1 Bandwidth: Bandwidth of the first formant. The formant bandwidth affects the identification of vowels in competition with other vowels.

C.2 Energy/Amplitude related parameters

Shimmer: Shimmer measures the cycle-to-cycle variations of fundamental amplitude. The shimmer changes with the reduction of glottal resistance and mass lesions on the vocal cords and is correlated with the presence of noise emission and breathiness.

Loudness: an estimate of perceived signal intensity from an auditory spectrum.

Harmonics-to-noise ratio: Relation of energy in harmonic components to energy in noise-like components. HNR quantifies the relative amount of additive noise.

C.3 Spectral Parameters

Alpha Ratio: Ratio of the summed energy from 50-1000 Hz and 1-5 kHz.

Hammarberg Index: Ratio of the strongest energy peak in the 0-2 kHz region to the strongest peak in the 2–5 kHz region.

Spectral Slope 0-500 Hz and 500-1500 Hz: Linear regression slope of the logarithmic power spectrum within the two given bands.

Formant 1, 2, and 3 relative energy: The ratio of the energy of the spectral harmonic peak at the first, second, and third formant's centre frequency to the energy of the spectral peak at F0.

MFCC 1-4: Mel-Frequency Cepstral Coefficients 1-4.

Spectral flux: Difference of the spectra of two consecutive frames.

D *ImbOLL* and OLL parameters

Here, we show the experiments conducted to determine the optimal values for α and β for our *ImbOLL* function presented in equation 4. We experiment with $\alpha \in \{1, 1.5, 2\}$ and $\beta \in \{0.5, 1\}$. The results are shown in Table 4.

α	β	Validation CCC(↑)
1	0.5	**0.654 ± 0.014**
	1	0.653 ± 0.022
1.5	0.5	0.639 ± 0.026
	1	0.516 ± 0.024
2	0.5	0.610 ± 0.038
	1	0.422 ± 0.024

Table 4: *ImbOLL* experiments

α	Validation CCC(↑)
1	**0.659 ± 0.024**
1.5	**0.655 ± 0.012**
2	0.645 ± 0.014

Table 5: OLL experiments

From Table 4, we see that $\alpha = 1$ gives the best results. Both $\beta = 0.5$ and $\beta = 1$ give similar results on mean performance. However, $\beta = 0.5$ gives a lower standard deviation. Thus, we choose the value of $\alpha = 1$ and $\beta = 0.5$ for training *QuestMF*.

We also experiment with the hyperparameters of the OLL function presented in equation 1. We experiment with $\alpha \in \{1, 1.5, 2\}$, and the results are presented in Table 5. From the table, we see that $\alpha = 1$ and $\alpha = 1.5$ give good results. While $\alpha = 1$ gives slightly better mean performance, $\alpha = 1.5$ gives a lower standard deviation. So, we train *QuestMF* with both $\alpha = 1$ and $\alpha = 1.5$. The results are presented in Appendix F. From that, we observe that OLL with $\alpha = 1$ gives the best result on the three-modality fused model. Thus, only this result is presented in Section 6.

E Hyperparameter Details and Training Setup

This section presents the hyperparameters used in the single modality and fused two-modality and three-modality models. We use a maximum of 120 participant dialogue turns for all our experiments.

Single Modality Encoder Models: For the single modality models, we first experiment to find the best Bi-LSTM output dimensions. We experiment with hidden dimensions of $d \in 30, 50, 100$. Due to the computational expenses, we only experimented with the text encoder model and extended the same output dimensions to audio and video encoder models. The results of this experiment are presented in Table 6 for the *QuestMF* framework trained with the *ImbOLL* function. The results of *QuestMF* framework trained with OLL with $\alpha = 1$ are presented in Table 7 and with OLL with $\alpha = 1.5$ are shown in Table 8. The results of *QuestMF* framework trained with the MSE loss function are presented in Table 9 and results of the *Total* framework are presented in Table 10. The tables show that a Bi-LSTM hidden dimension of 50 works best for all frameworks. For the multi-

Output Dimension	Validation CCC(↑)
30	0.639 ± 0.027
50	$\mathbf{0.654} \pm \mathbf{0.014}$
100	0.647 ± 0.004

Table 6: Results of experiments with LSTM output dimension for *QuestMF* Framework trained with *ImbOLL* function

Output Dimension	Validation CCC(↑)
30	0.622 ± 0.047
50	$\mathbf{0.659} \pm \mathbf{0.024}$
100	0.639 ± 0.038

Table 7: Results of experiments with LSTM output dimension for *QuestMF* Framework trained with OLL function with $\alpha = 1$

Output Dimension	Validation CCC(↑)
30	0.609 ± 0.028
50	$\mathbf{0.655} \pm \mathbf{0.012}$
100	0.616 ± 0.015

Table 8: Results of experiments with LSTM output dimension for *QuestMF* Framework trained with OLL function with $\alpha = 1.5$

Output Dimension	Validation CCC(↑)
30	0.554 ± 0.022
50	$\mathbf{0.632} \pm \mathbf{0.024}$
100	0.602 ± 0.008

Table 9: Results of experiments with LSTM output dimension for *QuestMF* Framework trained with MSE loss function

Output Dimension	Validation CCC(↑)
30	0.588 ± 0.015
50	$\mathbf{0.614} \pm \mathbf{0.010}$
100	0.610 ± 0.012

Table 10: Results of experiments with LSTM output dimension for *Total* Framework

head attention layer, we use 4 attention heads and a dropout of 0.5 for the text encoder model for all frameworks. For the audio and video encoder models, we use two multihead attention layers with 4 attention heads and a dropout of 0.2 in all frameworks. The MLP in all single modality encoder models and frameworks consists of two linear layers with a hidden dimension of 256, and the ReLU activation function connects the linear layers. A dropout of 0.2 is applied before each linear layer.

For the training of the single modality encoders, we use a learning rate of 5×10^{-4} with AdamW optimiser and a batch size of 10 for all modalities. During training, we freeze the turn encoders and only train the Bi-LSTM layer, attention layer, and MLP. Since the text models fit faster, we train them for 20 epochs. Meanwhile, we train the audio and video models for 50 epochs. We select the model checkpoint with the lowest validation loss for further modality fusion training. To evaluate the models on the depression severity score prediction task, we select the checkpoint with the best validation CCC.

Two-Modality Fused Models: We follow the architecture shown in Figure 3 for the fusion of two modalities. We use 4 heads and a dropout of 0.8 for multihead cross-attention and self-attention layers. We use the very high dropout to reduce overfitting due to the small size of the training dataset. The MLP consists of two linear layers with a hidden dimension of 256. The linear layers are connected through the ReLU activation function. We apply a dropout of 0.8 before the first linear layer and a dropout of 0.5 before the last linear layer. We use a smaller dropout before the last linear layer to avoid underfitting.

For the training of a two-modality fusion encoder, we use a learning rate of 5×10^{-4} with AdamW optimiser and a batch size of 10. If the text modality is involved in the two-modality fusion model, we freeze the weights from the text encoder model while we train the weights in the audio or video encoder models during the fusion. This is because the text model fits the data quickly, so training the weights of the other model with the frozen text model helps information alignment across the modalities and improves their encodings. Another reason is that training the parameters of all single modality encoder models with the small training set would increase the chances of overfitting. In addition to this, we train the cross-attention layers and self-attention layers used for fusion and the MLP. We train the models for 20 epochs. We select the model checkpoint with the best validation CCC.

Three-Modality Fused Models: We follow the architecture shown in Figure 4 for the fusion of three modalities. We use 4 heads and a dropout of 0.8 in multihead cross-attention and self-attention layers. We use an MLP of two linear layers with a hidden dimension of 256. The linear layers are connected through the ReLU activation function. We apply a dropout of 0.8 before the first linear layer and a dropout of 0.5 before the last linear layer.

For the training of the three-modality fused model, we use a learning rate of 5×10^{-4} with AdamW optimiser and a batch size of 10. While training three-modality fused models, we freeze the weights from the text model and train the weights of the audio and video models along with the cross-attention and self-attention layers used for fusion, and the MLP. We train the fusion for 20 epochs. We select the model checkpoint with the best validation CCC.

F Ablation Details

Here, we show more detailed results of our ablation study to observe the performance of single modality models, two-modality fused models, and the three-modality fused model with the frameworks. Here, we show the RMSE and MAE along with CCC results. We also show the standard deviation along with the mean for the three different seed runs. For *QuestMF* (OLL), we have two different frameworks here:

QuestMF (OLL-1): We train the *QuestMF* framework with OLL function with $\alpha = 1$.

QuestMF (OLL-1.5): We train the *QuestMF* framework with OLL function with $\alpha = 1.5$.

We present the results in Table 11.

Modalities	Framework	CCC(↑)	RMSE(↓)	MAE(↓)
Text	*Total*	0.591 ± 0.031	5.51 ± 0.38	4.37 ± 0.30
	QuestMF (MSE)	0.593 ± 0.020	5.52 ± 0.09	4.33 ± 0.06
	QuestMF (OLL-1)	0.592 ± 0.025	5.77 ± 0.31	4.53 ± 0.22
	QuestMF (OLL-1.5)	**0.616 ± 0.019**	**5.22 ± 0.07**	**4.02 ± 0.07**
	QuestMF (ImbOLL)	0.615 ± 0.031	5.71 ± 0.25	4.36 ± 0.26
Audio	*Total*	0.212 ± 0.017	6.41 ± 0.15	5.27 ± 0.09
	QuestMF (MSE)	0.239 ± 0.012	**6.35 ± 0.09**	**5.14 ± 0.03**
	QuestMF (OLL-1)	0.264 ± 0.008	6.96 ± 0.46	5.42 ± 0.20
	QuestMF (OLL-1.5)	0.256 ± 0.023	6.90 ± 0.21	5.36 ± 0.14
	QuestMF (ImbOLL)	**0.273 ± 0.021**	6.67 ± 0.11	5.32 ± 0.05
Video	*Total*	-0.067 ± 0.009	8.21 ± 0.07	6.63 ± 0.04
	QuestMF (MSE)	-0.075 ± 0.007	7.97 ± 0.05	6.46 ± 0.04
	QuestMF (OLL-1)	**-0.041 ± 0.015**	7.91 ± 0.20	6.44 ± 0.18
	QuestMF (OLL-1.5)	-0.044 ± 0.026	**7.79 ± 0.19**	**6.33 ± 0.16**
	QuestMF (ImbOLL)	-0.052 ± 0.028	7.89 ± 0.12	6.44 ± 0.11
Text + Audio	*Total*	0.607 ± 0.020	**5.27 ± 0.28**	**4.12 ± 0.23**
	QuestMF (MSE)	0.618 ± 0.017	5.61 ± 0.31	4.42 ± 0.18
	QuestMF (OLL-1)	0.628 ± 0.013	5.44 ± 0.04	4.17 ± 0.04
	QuestMF (OLL-1.5)	0.622 ± 0.004	5.53 ± 0.07	4.27 ± 0.06
	QuestMF (ImbOLL)	**0.643 ± 0.024**	5.48 ± 0.27	4.21 ± 0.14
Text + Video	*Total*	0.610 ± 0.008	**5.13 ± 0.19**	4.02 ± 0.17
	QuestMF (MSE)	0.627 ± 0.021	5.19 ± 0.16	3.99 ± 0.05
	QuestMF (OLL-1)	0.628 ± 0.006	5.34 ± 0.07	4.04 ± 0.20
	QuestMF (OLL-1.5)	0.630 ± 0.039	5.23 ± 0.31	4.05 ± 0.30
	QuestMF (ImbOLL)	**0.659 ± 0.018**	5.22 ± 0.09	**3.92 ± 0.08**
Audio + Video	*Total*	0.058 ± 0.030	7.35 ± 0.10	5.86 ± 0.11
	QuestMF (MSE)	0.070 ± 0.011	7.35 ± 0.21	5.81 ± 0.16
	QuestMF (OLL-1)	0.139 ± 0.016	7.07 ± 0.11	5.69 ± 0.08
	QuestMF (OLL-1.5)	0.080 ± 0.047	7.42 ± 0.19	6.02 ± 0.20
	QuestMF (ImbOLL)	**0.159 ± 0.039**	**7.03 ± 0.18**	**5.67 ± 0.25**
Text + Audio + Video	*Total*	0.602 ± 0.015	**5.10 ± 0.08**	3.99 ± 0.05
	QuestMF (MSE)	0.602 ± 0.025	5.36 ± 0.24	4.21 ± 0.12
	QuestMF (OLL-1)	0.640 ± 0.018	5.14 ± 0.05	**3.88 ± 0.01**
	QuestMF (OLL-1.5)	0.599 ± 0.023	5.30 ± 0.22	4.06 ± 0.25
	QuestMF (ImbOLL)	**0.662 ± 0.022**	5.25 ± 0.08	3.95 ± 0.13

Table 11: Ablation results for using different combinations of modalities with different frameworks. The CCC, RMSE and MAE scores presented are the mean and standard deviation over 3 different seed runs.

Linking Language-based Distortion Detection to Mental Health Outcomes

Vasudha Varadarajan[1], **Allison Lahnala**[2], **Sujeeth Vankudari**[1] **Akshay Raghavan**[1], **Scott Feltman**[1], **Syeda Mahwish**[1], **Camilo Ruggero**[1], **Roman Kotov**[1], **H. Andrew Schwartz**[1]
[1]Stony Brook University, [2]McMaster University
{vvaradarajan,has}@cs.stonybrook.edu

Abstract

Recent work has suggested detection of cognitive distortions as an impactful task for NLP in the clinical space, but the connection between language-detected distortions and validated mental health outcomes has been elusive. In this work, we evaluate the co-occurrence of (a) 10 distortions derived from language-based detectors trained over two common distortion datasets with (b) 12 mental health outcomes contained within two new language-to-mental-health datasets: DS4UD and iHiTOP. We find higher rates of distortions for those with greater mental health condition severity (ranging from $r = 0.16$ for thought disorders to $r = 0.46$ for depressed mood), and that the specific distortions of *should statements* and *fortune telling* were associated with a depressed mood and being emotionally drained, respectively. This suggested that language-based assessments of cognitive distortion could play a significant role in detection and monitoring of mental health conditions.

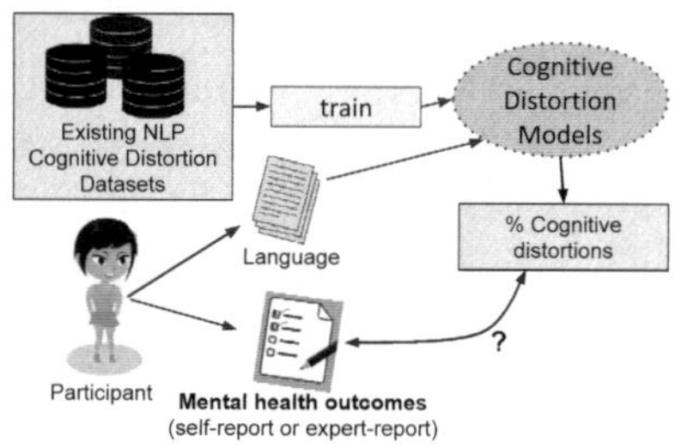

Figure 1: We train distortion detection models on existing cognitive distortion datasets, apply them to identify cognitive distortions in language, and evaluate their relationship with HiTOP (Kotov et al., 2022) and DS4UD (Nilsson et al., 2024) mental health outcomes over two new datasets.

1 Introduction

Cognitive distortions–systematic thinking patterns that cause inaccurate perceptions of reality–contribute to maintaining or worsening mental health conditions, such as depression and anxiety (Beck, 1963). The practice of recognizing one's own cognitive distortions is a core component of cognitive behavioral therapy (CBT), one of the most effective non-medicinal therapies for depression (Hofmann et al., 2012). Recent advances in natural language processing (NLP) have opened new avenues for automatically detecting distortions as well as generating text to reframe the distortions (de Toledo Rodriguez et al., 2021; Lim et al., 2024), potentially extending accessibility to therapeutic practices like CBT. Reliable detection of cognitive distortions in text holds promise for scalable mental health assessments to increase their efficacy and adds a layer of explainability. However, a key step in this vision is to validate that language-detected distortions do in fact have associations with validated mental health outcomes.

This study seeks to empirically evaluate whether language-detected distortions do in fact show connections to expected mental health outcomes over both clinical interviews as well as standard self-report assessments. Our paper highlights two key findings: (1) our analyses validate the co-occurrence of cognitive distortions with mental health conditions, demonstrating that higher rates of distorted thinking patterns generally correspond to greater severity of mental health symptoms; (2) we identify specific distortion types that exhibit stronger correlations with certain mental health indicators, suggesting they may be useful language markers of particular health indicators. We also identify where better detection performance and connections to mental health outcomes could be stronger, motivating directions for future work.

The established links between distortions and mental health conditions have motivated language analysis on social networks for early detection of depression markers of depression in social media posts (Ophir et al., 2017; Bathina et al., 2021;

Proceedings of the 10th Workshop on Computational Linguistics and Clinical Psychology (CLPsych 2025), pages 62–68
May 3, 2025

A. Rutter et al., 2025). Our study underscores that NLP models of cognitive distortions effectively align language with actual mental health conditions, and contributes to real-world monitoring or intervention strategies through advanced detection capabilities.

2 Background

Cognitive distortions are systematic patterns of biased thinking and false self-beliefs that can lead to negative moods and behaviors, playing a role mental health conditions like depression (Beck, 1963). Therapies like cognitive behavioral therapy (CBT) involve the practice of identifying and reframing distortions to support individuals in adopting healthier thinking patterns (Rupke et al., 2006). There is strong evidence that this form of therapy is effective for managing conditions like anxiety and depression (Hofmann et al., 2012). Since the COVID-19 pandemic, therapy has increasingly transitioned into the telehealth space (Leroy et al., 2025), highlighting a need for automated detection tools in conversations which would allow therapists in recognizing distorted thinking within the vast amount of information they process during a therapy session. This shift has encouraged the exploration of various ways in which agents within telehealth sessions can assist therapists in recognizing patterns, creating opportunities for the application of distortion detection tools. Tools such as this can enable timely interventions, helping patients recognize and work through cognitive distortions. Additionally, distortion detection tools can be integrated into developing assistive agents for therapy homework after CBT, bringing a more patient-facing support by flagging cognitive distortions and prompting the need for reappraisal (Stade et al., 2024).

As distortion reframing occurs through language, recent research has explored NLP-based approaches for cognitive distortion detection, reframing, and positive reformulation. Various efforts have been dedicated toward cognitive distortion detection and classification models (Shreevastava and Foltz, 2021; Chen et al., 2023; Lim et al., 2024). Datasets of situations, thoughts, and reframes have been created to train generative models (Sharma et al., 2023; Maddela et al., 2023). This research has focused on models that perform positive reformulation of distorted thoughts to more constructive ones (de Toledo Rodriguez et al., 2021), by adopting, for instance, strategies from positive psychology (Ziems et al., 2022). Others have also aimed to build chat systems that guide users through cognitive restructuring (Sharma et al., 2024). These studies highlight the potential of NLP-driven interventions in fostering cognitive shifts and improving mental health, yet there is limited work into how automatically detected distortions correspond to existing mental health conditions.

3 Data

3.1 Mental Health Outcomes Datasets

iHiTOP The iHiTOP dataset contains transcribed clinical interviews with psychiatric outpatients, aligned with the HiTOP taxonomy—a modern mental health taxonomy mappable to DSM-V (Kotov et al., 2022; Regier et al., 2009). These semi-structured interviews, lasting 45-90 minutes, were diarized and transcribed using NVIDIA NeMo and openai/whisper-large-v2.[1] We use the transcribed text of the interviewee in our analysis.

Dataset	Num participants	Num messages	Num spans	Mental Health Outcomes	Report
iHiTOP	536	536	568989	Internalizing Mania Anankastia Thought disorder Detachment Disinhibition Antagonism	Expert
DS4UD	587	32773	58103	Depressed Mood Daily Stress Daily Drain Wave Anxiety Wave Depression	Self

Table 1: Descriptions of Mental Health datasets.

The dataset includes patient scores across seven “spectra”: *internalizing*, *mania*, *anankastia*, *thought disorder*, *detachment*, *disinhibition*, and *antagonism*. After filtering segments shorter than 4 words, the average segment length was 12 words, with interviews averaging 8,217 words per patient.

DS4UD The Data Science for Unhealthy Drinking Study (DS4UD) dataset (Nilsson et al., 2024) comprises mental health assessments and language data collected from U.S. service industry workers over two years. We focus on daily diary language from Ecological Momentary Assessments (EMAs). Participants provided three daily EMA responses across six 14-day waves, responding to: *"Please describe in 2 to 3 sentences how you are currently*

[1]‘iHiTOP’ is also the name of the instrument used to assess HiTOP mental health scores. This is the first dataset to use it so it is named the same.

feeling." With responses averaging 50 words (average 11 words per sentence), each participant could contribute up to 252 responses. The dataset includes daily metrics (affect, stress, alcohol consumption, and cravings) and WAVE measurements of anxiety and depression.

3.2 Cognitive Distortions Training Data

Patient Queries dataset (PQ) Shreevastava and Foltz (2021) contains patient queries to therapists, which include questions, concerns, descriptions of circumstances, and symptoms, among other topics. Each example is labeled with 1-2 dominant cognitive distortions, from 10 common types – All-or-Nothing Thinking, Overgeneralizing, Labeling, Fortune Telling, Mind Reading, Emotional Reasoning, Should Statements, Personalization, Mental Filter, and Magnification. There are 1597 instances of distorted spans (average length: 36 words) annotated with one of the ten types, from a total of 2530 messages (average length: 166 words).

Thinking Traps dataset (TT) Sharma et al. (2023) covers a set of 13 cognitive distortions: *All-or-Nothing Thinking, Overgeneralizing, Labeling, Fortune Telling, Mind Reading, Emotional Reasoning, Should Statements, Personalization, Disqualifying the Positive (Mental Filter), Catastrophizing (Magnification), Comparing and Despairing, Blaming, Negative Feeling or Emotion.* We drop the classes *Blaming, Comparing* and *Negative Emotion* due to the lack of enough examples in the dataset, and to maintain the same set of distortions in both the datasets. Our final dataset contains 1011 spans (average length: 21 words) that describe a situation and lead to a distorted thought leading from the situation.

4 Methods

We develop models to detect cognitive distortions in text as a means to study their relationship with mental health outcomes. Following established approaches in mental health NLP (Ganesan et al., 2021), we utilize transformer-based language models (LMs) and their contextual embeddings rather than pursuing incremental architectural improvements. These models enable us to quantify distortion rates per participant and examine their associations with mental health measures, addressing our primary research question.

Task 1: Distortion Detection This binary classification task assessed the models' ability to distinguish between messages containing cognitive distortions and those without. The objective was to make a fundamental present/absent determination for distorted thinking patterns.

	Detection		Classification	
Model	**F1**	**AUC**	**F1**	**AUC**
TT	.597	.813	.276	.755
PQ (span)	.823	.917	**.369**	**.876**
PQ (full)	.693	.766	-	-
TT + PQ (span)	**.833**	**.921**	.366	.847

Table 2: Cross-validation metrics for distortion detection and 11-way classification models. Note that PQ (full) contains full passages and could contain many distortions, so it wasn't used for the classification task.

Model	F1	AUC
All-or-Nothing Thinking	.506	.768
Overgeneralizing	.581	.735
Labeling	**.607**	**.853**
Fortune Telling	**.612**	**.878**
Mind Reading	**.675**	**.871**
Emotional Reasoning	.525	.753
Should Statements	**.696**	**.874**
Personalization	.554	.797
Mental Filter	.526	.783
Catastrophizing	.522	.706

Table 3: Cross-validation metrics for one-vs-rest distortion classification models. We pick the models with F1 > 0.6 (bolded) for validation on the mental health datasets.

Task 2: Distortion Classification We formulate this in two ways: a multi-class task (Table 2) and a one-vs-rest task (Table 3). The multi-class classification task required models to categorize messages according to specific distortion types identified in the training data. Notably, we included "No Distortion" as a distinct category by augmenting with sentences from the PQ dataset that were not annotated with a distortion, representing the absence of any recognized cognitive distortion patterns. This approach allowed for a more nuanced analysis of distinct cognitive distortion types in relation to mental health outcomes. Further, one-vs-rest task was used to build distortion type-specific classifiers with positive class being a distortion type and the rest of the examples from all the other classes (including No Distortion).

We report the F1 and AUC scores for comparing the performance of the models [2].

[2]F1 is a metric that is calculated as the harmonic mean

Model	iHiTOP								DS4UD					
									EMA			Wave		
	Interna-lizing	Mania	Anank-astia	Thought Disorder	Detach-ment	Disihi-bition	Antago-nism	Distort Rate (%)	Depressed Mood	Emotionally Drained	Nervous Stress	Anxiety	Depression	Distort Rate (%)
TT + PQ	0.24	0.18	0.16	0.16	0.11	0.29	0.19	9.78	**0.42**	**0.29**	**0.17**	**0.30**	**0.27**	8.47
PQ	**0.34**	**0.21**	**0.22**	**0.16**	**0.24**	**0.35**	**0.21**	11.36	**0.46**	**0.29**	**0.25**	**0.32**	**0.28**	15.03

Table 4: **Between-user correlations** (Pearson r) between overall percentage of distortions and mental health assessment scores in the iHiTOP and DS4UD dataset. Bold indicates statistically significant (p-values < 0.05). Correlations between a behavior (distortion mention) and psychological variables have a modal correlation between 0.1 to 0.4 and those above are considered very large (Roberts et al., 2007).

4.1 Modeling

We implemented two distinct approaches for fine-tuning encoder models:

Span-only (span) In this approach, we utilized only the text spans explicitly annotated as cognitive distortions. These spans were processed through a RoBERTa-base model, from which we derived averaged embeddings to train task-specific linear classifiers. This methodology was employed for both the therapist QA dataset and the thinking traps corpus (the latter consisting exclusively of short, distortion-containing sentences).

Full message (full) We expanded the input to encompass the complete message context from which the distortion spans were originally annotated in the therapist QA dataset. The processing pipeline remained consistent with the span-only approach, utilizing the same model architecture and classification framework. Both approaches leveraged the DistilRoBERTa-base architecture as our foundation, with subsequent linear classification layers optimized for each specific task.

We trained and validated our models on stratified train-test splits of the distortion-labeled datasets. We selected the **PQ (span)** and **TT + PQ (span)** to apply to our mental health outcomes data as they were the top performers by F1 score (Table 2) for detection. Some of the one-vs-rest models perform better than the others (Table 3, this could be attributed to PQ dataset: it has 73% examples with more than one distortion type annotated, the models might not pick up on the signals for certain types effectively. We select four one-vs-rest classification models for distortion classification due to their superior performance compared to the other models.

of precision and recall for a class. AUC is short for AUC-ROC, which stands for Area under the Receiving Operating Characteristic curve, a measure of binary classification models' ability to distinguish two classes.

4.2 Predictions on Mental Health Outcomes Dataset

We applied our trained detection and classification models to the DS4UD and iHiTOP texts to quantify the presence of distortions within users' language to analyze in relation to their mental health scores. We then compute the percentage of sentences that contain a detected distortion. For each of the distortion classes, we likewise compute the percent of sentences where the distortion was detected.

5 Results

We examine correlations between detected cognitive distortions and mental health outcomes at two levels: between users (in both DS4UD and iHiTOP datasets) and within users over time (in DS4UD). Using Pearson's r, we analyze how distortion rates correlate with mental health indicators across users, as well as how individual-level fluctuations in distortion rates relate to changes in mental health states.

Finding: Cognitive distortions detected in language are linked to mental health outcomes. The results of the between-user correlation analysis are shown in Table 4. Overall, rates of detected distortions are positively correlated with the mental health outcome scores, reflecting increased severity of mental health conditions as associated with elevated patterns of distorted thinking.

In the iHiTOP dataset, the overall rate of detected distortions by the PQ model correlates significantly positively with the spectra. Significant correlations were observed between all indicators and the rate of distortions detected by the PQ model.

In the DS4UD data, both models identified distortions at rates that correlate significantly with levels of user depression, being emotionally drained, and having nervous stress in the EMAs, and anxiety and depression scores measured by WAVE.

Model	DS4UD				
	EMA			Wave	
	Depressed Mood	Emot Drained	Nervous Stress	Anx	Dep
Should Statements	**0.32**	**0.12**	0.04	0.10	**0.18**
Fortune Telling	0.09	**0.16**	0.09	0.12	**0.13**
Mind Reading	-0.07	-0.04	-0.03	-0.06	-0.07
Labeling	0.26	-0.07	-0.10	-0.10	-0.05

Table 5: **Between-user correlations** (Pearson r) between overall percentage of distortions and mental health assessment scores in the DS4UD dataset. Bold indicates statistically significant (p-values < 0.05).

Distortion Type	Depressed	Emotionally Drained	Nervous Stress
TT + PQ	0.12	0.12	0.13
PQ	0.25	0.13	0.16
Should Statements	-0.02	0.03	0.00
Fortune Telling	0.14	0.08	0.07
Mind Reading	0.00	0.03	0.00
Labeling	-0.23	0.05	0.00

Table 6: **Within user correlations** (mean of Pearson r across users) of the overall percentage of distortions with Psychological State Indicators aligned in time. A higher value means that as the distortion increases so too does the reported condition severity where as a negative correlation indicates the severity decreases as the condition increases.

Considering specific classes of distortions, weak but significant positive correlations were observed between the presence of *should statements* with depressed and emotionally drained states from the EMAs, and depression from WAVE, and likewise for *fortune telling* except which does not have a significant relationship with the depressed EMA state. The difference in the rates of various types of distortions has also been observed in other studies with respect to emotional stress, depressive symptoms and anxiety (Jha et al., 2022; Wang et al., 2025), which could indicate distinct thinking patterns for specific mental health conditions. However, the low degree of associations observed should not necessarily mean that some of the cognitive distortion types could be disregarded.

The within-person analysis for DS4UD dataset is discussed in Table 6. Significance is not reported for this analysis since it captures the average Pearson correlations across a user timeline, and the maximum number of user EMAs is 252 (See §3.1). However, we still observe positive r values for the EMA-level outcomes, which indicates that increase in cognitive distortions expressed in language is weakly positively correlated to worsening mental health scores, even at a user-level. We note that we have a small number of repeated measures for these users (six), which limits the scope of observing within-person patterns in the WAVE outcomes. Future research can explore these relationships with more longitudinal data to assess whether models would detect within-user fluctuations in mental health states and thinking patterns.

6 Conclusion

We evaluated the link between distortion models and mental health outcomes for the authors of the language across two language-to-mental-health datasets: DS4UD and iHiTOP. We found automatically detected distortions in language correlated in general with higher anxiety and depression-related outcomes. In particular, we found that *should statements* and *fortune telling* associated with depressed states. Other types of distortions were not as easy to detect, suggesting further development may unlock additional benefits of NLP-based distortion detectors. Our findings establish language-based distortion detection as a promising tool for mental health professionals, offering empirically-validated support for identifying and addressing cognitive distortions in clinical settings. Our work contributes to advancing methods for early detection of mental health conditions like depression that can be integrated in real-world monitoring and intervention strategies.

Limitations

Our study faces methodological and data constraints that warrant consideration. The classification models show variable performance across different types of cognitive distortions, with some categories like mind reading and personalization showing particularly weak correlations with mental health outcomes. We also drop three classes of cognitive distortions from TT in our analyses for ease of combining the datasets. This suggests room for improvement in capturing more nuanced forms of distorted thinking and exploring more complex frameworks. We have limited our analysis to two datasets that could have potential sampling biases. We limit the analysis to English-language. Further, our computational approach faces several challenges. The reliance on automated detection methods may miss contextual nuances that human clinicians typically observe. These methods should be used as an assistance rather than a replacement for human clinicians.

Ethics Statement

As NLP continues to advance in enhancing human-centered applications such as improving mental health assessments, striking a balance between respecting human privacy and promoting open data sharing becomes increasingly important. In this case, the data was shared with consent solely for academic research and was anonymized. Open sharing would breach the trust with participants and violate agreements with ethical review boards. Ideally, all data should be released while maintaining privacy, however, the limited availability of data underscores the need for those with access to share their work openly within established ethical guidelines, such as the training datasets used in this work.

All data collection, storage, and secondary analyses procedures were approved by an academic ethics institutional review board.

7 Acknowledgements

This work was supported in part by a grant from the NIH-NIAAA (R01 AA028032) awarded to H. Andrew Schwartz at Stony Brook University. The conclusions contained herein are those of the authors and should not be interpreted as necessarily representing the official policies, either expressed or implied, NIH, any other government organization, or the U.S. Government.

References

Lauren A. Rutter, Andy Edinger, Lorenzo Lorenzo-Luaces, Marijn ten Thij, Danny Valdez, and Johan Bollen. 2025. Anxiety and depression are associated with more distorted thinking on social media: A longitudinal multi-method study. *Cognitive Therapy and Research*, pages 1–9.

Krishna C Bathina, Marijn Ten Thij, Lorenzo Lorenzo-Luaces, Lauren A Rutter, and Johan Bollen. 2021. Individuals with depression express more distorted thinking on social media. *Nature human behaviour*, 5(4):458–466.

Aaron T Beck. 1963. Thinking and depression: I. idiosyncratic content and cognitive distortions. *Archives of general psychiatry*, 9(4):324–333.

Zhiyu Chen, Yujie Lu, and William Wang. 2023. Empowering psychotherapy with large language models: Cognitive distortion detection through diagnosis of thought prompting. In *Findings of the Association for Computational Linguistics: EMNLP 2023*, pages 4295–4304, Singapore. Association for Computational Linguistics.

Ignacio de Toledo Rodriguez, Giancarlo Salton, and Robert Ross. 2021. Formulating automated responses to cognitive distortions for CBT interactions. In *Proceedings of the 4th International Conference on Natural Language and Speech Processing (IC-NLSP 2021)*, pages 108–116, Trento, Italy. Association for Computational Linguistics.

Adithya V Ganesan, Matthew Matero, Aravind Reddy Ravula, Huy Vu, and H Andrew Schwartz. 2021. Empirical evaluation of pre-trained transformers for human-level nlp: The role of sample size and dimensionality. In *Proceedings of the conference. Association for Computational Linguistics. North American Chapter. Meeting*, volume 2021, page 4515. NIH Public Access.

Stefan G Hofmann, Anu Asnaani, Imke JJ Vonk, Alice T Sawyer, and Angela Fang. 2012. The efficacy of cognitive behavioral therapy: A review of meta-analyses. *Cognitive therapy and research*, 36:427–440.

Ajeya Jha, Akash Kumar Bhoi, Saibal Kumar Saha, Ankit Singh, Samrat Mukherjee, Bibeth Sharma, and Jayarani. 2022. Impact of select cognitive distortions on emotional stress. *Cognitive Computing for Risk Management*, pages 31–44.

Roman Kotov, David C Cicero, Christopher C Conway, Colin G DeYoung, Alexandre Dombrovski, Nicholas R Eaton, Michael B First, Miriam K Forbes, Steven E Hyman, Katherine G Jonas, et al. 2022. The hierarchical taxonomy of psychopathology (hitop) in psychiatric practice and research. *Psychological medicine*, 52(9):1666–1678.

Tatiana Leroy, David Nicholas Top Jr., Russell J. Bailey, Christina Bartholomew, Julia Toomey, Taylor Baker, Josephine Schwalbe, and Kirra D. Jensen and. 2025. Accessing care: qualitative analysis of counseling center therapists,Äô experiences of transitioning to telehealth. *Cogent Mental Health*, 4(1):1–46.

Sehee Lim, Yejin Kim, Chi-Hyun Choi, Jy-yong Sohn, and Byung-Hoon Kim. 2024. ERD: A framework for improving LLM reasoning for cognitive distortion classification. In *Proceedings of the 6th Clinical Natural Language Processing Workshop*, pages 292–300, Mexico City, Mexico. Association for Computational Linguistics.

Mounica Maddela, Megan Ung, Jing Xu, Andrea Madotto, Heather Foran, and Y-Lan Boureau. 2023. Training models to generate, recognize, and reframe unhelpful thoughts. In *Proceedings of the 61st Annual Meeting of the Association for Computational Linguistics (Volume 1: Long Papers)*, pages 13641–13660, Toronto, Canada. Association for Computational Linguistics.

August Håkan Nilsson, Hansen Andrew Schwartz, Richard N Rosenthal, James R McKay, Huy Vu, Young-Min Cho, Syeda Mahwish, Adithya V Ganesan, and Lyle Ungar. 2024. Language-based ema assessments help understand problematic alcohol consumption. *Plos one*, 19(3):e0298300.

Yaakov Ophir, Christa SC Asterhan, and Baruch B Schwarz. 2017. Unfolding the notes from the walls: Adolescents' depression manifestations on facebook. *Computers in Human Behavior*, 72:96–107.

Darrel A Regier, William E Narrow, Emily A Kuhl, and David J Kupfer. 2009. The conceptual development of dsm-v. *American Journal of Psychiatry*, 166(6):645–650.

Brent W Roberts, Nathan R Kuncel, Rebecca Shiner, Avshalom Caspi, and Lewis R Goldberg. 2007. The power of personality: The comparative validity of personality traits, socioeconomic status, and cognitive ability for predicting important life outcomes. *Perspectives on Psychological science*, 2(4):313–345.

Stuart J Rupke, David Blecke, and Marjorie Renfrow. 2006. Cognitive therapy for depression. *American family physician*, 73(1):83–86.

Ashish Sharma, Kevin Rushton, Inna Lin, David Wadden, Khendra Lucas, Adam Miner, Theresa Nguyen, and Tim Althoff. 2023. Cognitive reframing of negative thoughts through human-language model interaction. In *Proceedings of the 61st Annual Meeting of the Association for Computational Linguistics (Volume 1: Long Papers)*, pages 9977–10000, Toronto, Canada. Association for Computational Linguistics.

Ashish Sharma, Kevin Rushton, Inna Wanyin Lin, Theresa Nguyen, and Tim Althoff. 2024. Facilitating self-guided mental health interventions through human-language model interaction: A case study of cognitive restructuring. In *Proceedings of the CHI Conference on Human Factors in Computing Systems*, pages 1–29.

Sagarika Shreevastava and Peter Foltz. 2021. Detecting cognitive distortions from patient-therapist interactions. In *Proceedings of the Seventh Workshop on Computational Linguistics and Clinical Psychology: Improving Access*, pages 151–158, Online. Association for Computational Linguistics.

Elizabeth C Stade, Shannon Wiltsey Stirman, Lyle H Ungar, Cody L Boland, H Andrew Schwartz, David B Yaden, João Sedoc, Robert J DeRubeis, Robb Willer, and Johannes C Eichstaedt. 2024. Large language models could change the future of behavioral healthcare: a proposal for responsible development and evaluation. *NPJ Mental Health Research*, 3(1):12.

Xi Wang, Yujia Zhou, and Guangyu Zhou. 2025. Unveiling the cognitive burden: The impact of stigma on distorted thinking among individuals living with hepatitis b. *International Journal of Clinical and Health Psychology*, 25(1):100556.

Caleb Ziems, Minzhi Li, Anthony Zhang, and Diyi Yang. 2022. Inducing positive perspectives with text reframing. In *Proceedings of the 60th Annual Meeting of the Association for Computational Linguistics (Volume 1: Long Papers)*, pages 3682–3700, Dublin, Ireland. Association for Computational Linguistics.

A Within-person Analysis

Significance is not reported for this analysis since it captures the average Pearson correlations across a user timeline, and the maximum number of user EMAs is 252 (See §3.1). However, we still observe positive r values for the EMA-level outcomes, which indicates that increase in cognitive distortions expressed in language is weakly positively correlated to worsening mental health scores, even at a user-level. We note that we have a small number of repeated measures for these users (six), which limits the scope of observing within-person patterns in the WAVE outcomes. Future research can explore these relationships with more longitudinal data to assess whether models would detect within-user fluctuations in mental health states and thinking patterns.

Measuring Mental Health Variables in Computational Research: Toward Validated, Dimensional, and Transdiagnostic Approaches

Chen Shani
Computer Science
Stanford University
cshani@stanford.edu

Elizabeth C. Stade
Human-Centered Artificial Intelligence
Stanford University
ecs@stanford.edu

Abstract

Computational mental health research develops models to predict and understand psychological phenomena, but often relies on inappropriate measures of psychopathology constructs, undermining validity. We identify three key issues: (1) reliance on unvalidated measures (e.g., self-declared diagnosis) over validated ones (e.g., diagnosis by clinician); (2) treating mental health constructs as categorical rather than dimensional; and (3) focusing on disorder-specific constructs instead of transdiagnostic ones. We outline the benefits of using validated, dimensional, and transdiagnostic measures and offer practical recommendations for practitioners. Using valid measures that reflect the nature and structure of psychopathology is essential for computational mental health research.

1 Introduction

In computational mental health research, significant effort is invested in designing models to predict and understand psychological phenomena. Yet the validity and utility of these models can be undermined when they rely on flawed or inappropriate representations of psychopathology (i.e., mental disorder). For example, a classifier cannot validly predict depression if its training data is based on an invalid measure of depression.

Measuring psychopathology *constructs* – representations of psychological states or processes like "depression" or "neuroticism" (Fried, 2017b) – is challenging. Constructs are abstract and cannot be directly observed (e.g., there is no single biological indicator for depression) and diagnostic systems like *DSM-5* (American Psychiatric Association, 2013) and *ICD-10* (World Health Organization, 2004) have slight variations in how they define syndromes.

Yet clinical psychology has for decades invested in validating measures of psychopathology (Cronbach and Meehl, 1955) and advancing measurement techniques that reflect developments in psychological science (Stanton et al., 2020). Recommended measurement techniques include clinician-administered interviews, self-report questionnaires, and informant reports (Stanton et al., 2020). While no measurement scheme is without error, techniques exist to help ensure that a measure actually taps the construct it purports to (a process known as *construct validity* (Campbell and Fiske (1959)).

However, computational psychopathology research has slow to adopt measurement techniques from clinical psychological science, hindering progress. Here, we highlight three key concepts from clinical science—measurement validity, dimensional measurement, and transdiagnostic measurement. Acknowledging that computational research often turns to inappropriate measures of psychopathology constructs due to the constraints of computational research – collecting large-scale clinician-assessed data is expensive and time-consuming, and existing or archival data can be difficult or impossible to access due to privacy concerns – we offer practical recommendations for improved assessment of psychological phenomena in computational research.

2 Measurement Validity

Computational research often infers mental health conditions using methods with poor or unknown validity. For example, some studies assume a diagnosis based on forum membership (e.g., `r/Depression`) or self-declarations (e.g., "I have depression") on social media (Guntuku et al., 2017), "proxy diagnostic signals" which have been shown to have poor external validity (Ernala et al., 2019). In other cases, computational researchers write their own single-item measure of a psychological construct, rather than selecting an existing measure with good validity (Allen et al., 2022). The clinical science literature provides justifica-

Proceedings of the 10th Workshop on Computational Linguistics and Clinical Psychology (CLPsych 2025), pages 69–78
May 3, 2025 ©2025 Association for Computational Linguistics

tion that all measurement methods are not equal vis-a-vis validity: For instance, self-reported history of depression diagnosis only modestly agrees with semi-structured diagnostic interview findings (Stuart et al., 2014; Sanchez-Villegas et al., 2008).

To mitigate this, computational researchers should use measures of psychological constructs with good validity. Gold standard psychopathology assessment typically involves a clinician-administered structured or semi-structured interview (e.g., Structured Clinical Interview for *DSM-5*; First et al. (2016) based on an established psychopathology classification system (e.g., *DSM-5*. As an alternative, self- or informant-report measures that have undergone rigorous psychometric evaluation, such as the PHQ-9 (Kroenke et al., 2001), can be used (Stanton et al., 2020).

2.1 Limitations of Self-Report Questionnaires

Self-report methodology, while offering high levels of convenience, has meaningful limitations of which computational researchers should be aware:

Bias. Self-report measures are vulnerable to biased patterns of responding such as participant lack of insight or yay- or nay-saying biases (Hunt et al., 2003), which can introduce systematic errors into computational models.

Specificity of Constructs. Self-report tools may be imprecise measures of psychological constructs. Evidence exists that putative self-report measures of depression may in fact capture general distress (Coyne, 1994; Kendall et al., 1987) or anxiety (Breslau, 1985) rather than just depression. This lack of precision can weaken model predictions and blur construct boundaries.

Using Tools Outside Their Intended Setting. Self-report measures may be less effective when used outside their original context. For example, the PHQ-9, designed for primary care, has low specificity and PPV in specialty mental health settings (Inoue et al., 2012), limiting its validity in clinical samples. Developed as a screening tool, it maps perfectly onto *DSM-5* criteria, but it does not assess other symptoms associated with depression, like self-dislike or low libido, meaning it has a narrower possible range than measures that capture a wide range of symptoms. Using tools beyond their intended purpose may reduce model accuracy, and range restriction can attenuate effects (see below).

3 Dimensional Measurement

Mental health constructs can be assessed dimensionally (e.g., on a scale from 0 [*no depression*] to 10 [*extreme depression*]) or using categorical labels (*not depressed* vs. *depressed*). Despite computational research tending to employ categorical measurement schemes, most psychopathology constructs are inherently dimensional, as evidenced by the following:

At the **manifest (observable) level**, symptoms show a monotonic relationship with functional outcomes (Kessler et al. (2006); Ruscio et al. (2007, 2008); Cuijpers et al. (2004); Judd et al. (1997)). Even mild or infrequent depression symptoms that fall below the *DSM-5* criteria for major depressive disorder are associated with impairment. However, categorical representations of psychopathology group all subthreshold symptom presentations together, obscuring mild yet clinically meaningful dysfunction (Ruscio, 2019).

Furthermore, longitudinal research reveals that individuals frequently fluctuate between levels of severity of symptoms over time, including crossing thresholds above and below thresholds week-by-week (Chen et al., 2000; Judd et al., 1997), making diagnoses somewhat arbitrary depending on the time of evaluation.

At the **latent level**, taxometric analysis of psychometric variables, which compares categorical and dimensional models (Ruscio and Ruscio, 2000), typically yields dimensional solutions. This indicates that constructs like depression and anxiety have a natural, underlying structure that is dimensional, not categorical (Haslam et al., 2012).

To reflect the true nature and structure of psychopathology, computational researchers should treat most mental health variables as continuous. Using dimensional measures of psychopathology can improve granularity of computational models (e.g., allowing models to differentiate between moderate and severe symptom presentations), improving accuracy and clinical utility.

4 Transdiagnostic Measurement

While diagnostic systems like the *DSM-5* dominate clinical practice and research, there is growing recognition of *transdiagnostic processes* that cut across the mental disorders (Harvey et al., 2004). For example, *avoidance* is shared by many different types of anxiety disorders, including panic disorder, specific phobia, and social anxiety disorder. Growing evidence supports transdiagnostic conceptualizations of psychopathology, including:

High Comorbidity. Mental disorders frequently co-occur at rates far exceeding chance. For example, people diagnosed with one anxiety disorder are six times more likely to have another (Kessler, 1997), suggesting shared underlying mechanisms.

High Diagnostic Crossover. Many patients transition between diagnoses over time. For example, 20-50% of individuals diagnosed with anorexia nervosa later develop bulimia (Eddy et al., 2008), suggesting that existing diagnostic categories may sub-optimally represent psychopathology.

Non-Specific Treatment Effects. Treatments targeting one disorder often alleviate symptoms of co-occurring conditions—for example, PTSD treatments frequently reduce depression symptoms (Barlow et al., 2014). This suggests that interventions may be acting on transdiagnostic mechanisms rather than disorder-specific factors.

In response, the field has introduced transdiagnostic classification models, such as **NIMH's Research Domain Criteria (RDoC)** (Insel et al., 2010), which defines cross-cutting dimensions like "reward responsiveness" and "potential threat," and the **Hierarchical Taxonomy of Psychopathology (HiTOP)** (Kotov et al., 2017), which groups together frequently co-occurring symptoms.

Transdiagnostic measures offer increased parsimony and can better reflect the psychopathology vis-à-vis syndrome-specific alternatives (e.g., Stade et al. (2023a); See Stanton et al. (2020) for guidance on selecting transdiagnostic measures).

5 Impact of Poor Measurement

Using measures with poor validity, or using categorical or syndrome-specific measures, can negatively impact computational research. Key implications include:

Reduced Resolution. Using categorical labels for psychopathology oversimplifies complex constructs, discarding valuable information about symptom severity, which is important for model accuracy and has clinical utility (e.g., the difference between moderate and severe depression is meaningful to clinicians).

Mis-Classifying Boundary Cases. Relatedly, categorical representation of psychopathology risks misclassifying individuals who fall close to the diagnostic boundary.

Risk of Type II Errors. Dichotomizing variables that are continuous in nature sacrifices statistical power (Cohen, 1983) and reduces reliability (Markon et al., 2011), increasing the risk of Type II errors.

Overfitting and Poor Generalization. Noisy measurements cause models to learn spurious patterns, reducing reliability and real-world applicability.

Misleading Interpretations. Poor measurements can cause misleading conclusions about mental health constructs, such as conflating depression overlapping yet different constructs, like anxiety or negative emotionality.

Erosion of Clinical Utility. For computational models to have practical relevance in mental health care, they must provide insights or predictions that clinicians can act upon. Models based on bad measurements often lack this clinical utility.

Bias Amplification and Inequities. Inaccurate measurement can amplify bias, reinforcing disparities and inequities in mental health care.

Missed Opportunities for Scientific Progress. Bad measurements limit scientific progress, preventing meaningful contributions and advancements in understanding mental health.

6 The Path Forward

By addressing the limitations of current measurement practices in computational mental health research, we hope to create more accurate, robust, and impactful models. To strengthen the scientific rigor and relevance of research, we offer the following recommendations:

Consult Experts. Computational researchers can refer to established guidelines and evidence-based recommendations for the assessment of specific constructs or disorders (e.g., Klein et al. (2005); Antony and Rowa (2005); Shear and Maser (1994)). Collaborating with clinical science colleagues across departments can help guide appropriate measure selection. For projects aiming at immediate clinical application, working closely with a clinician is essential. Clinicians can offer expertise on the tool's clinical utility, including its relevance to real-world practice, ease of use in clinical settings, and alignment with existing diagnostic and treatment workflows. They can also provide feedback on whether the tool offers actionable insights for patient care, supports case conceptualization and treatment planning, and meets the practical needs of diverse clinical populations.

Improve Methodological Rigor When Using Proxies. As previously described, proxy diagnostic signals, such as self-identified diagnoses on social media, have poor validity (Ernala et al., 2019). Suggestions for improving the methodological rigor of research using proxies include paring proxy diagnostic signals with offline clinical datasets (Inkster et al., 2016) – a strong correlation between the proxy and an established clinical outcome, even in just a subsample of participants, could serve to demonstrate the validity of the proxy variable – and combining multiple proxies to improve reliability (Ernala et al., 2019). At the very least, researchers using proxies should clearly state their limitations, e.g., a Twitter-based depression variable should be distinguished from a clinically validated diagnosis, with a note that future research using higher validity measures is needed.

Adopt Dimensional Measurement. Avoid measuring mental health constructs into binary categories (e.g., "depressed" vs. "not depressed"). Choose dimensional measures that capture severity gradients and avoid dichotomizing continuous variables to form diagnostic categories. Researchers interested in diagnostic status could test this variable in secondary analyses (e.g., (Stade et al., 2023b)).

Critically Evaluate Disorder-Specific Measures. Before selecting a construct of interest and its corresponding measure, carefully evaluate whether a disorder-specific approach is necessary. For example, many researchers express interest in indexing anxiety, yet do so using the GAD-7 (Spitzer et al., 2006), which measures the symptoms of generalized anxiety disorder, a disorder of frequent and uncontrollable worry. Yet uncontrollable worry only represents one form of anxiety pathology. Measures that are not disorder-specific (e.g., MASQ Anxious Symptoms subscale (Watson et al., 1995)) better capture features and processes that cut across the anxiety disorders. Researchers should use syndrome-specific measures only when this truly aligns with their research goals.

Adopt a Process-Oriented Approach. Instead of focusing solely on specific disorders or syndromes, consider examining transdiagnostic processes. For example, studying constructs that encompass multiple diagnostic categories, such as "internalizing psychopathology" or "fear," can offer more generalizable and integrative insights than research limited to a single diagnosis (e.g., major depression, specific phobia).

Think Beyond Psychopathology. There are many non *DSM-5* constructs that are important for health and well-being, especially those that confer risk or protection for psychopathology, such as neuroticism, perfectionism, resilience, and disinhibition. The HiTOP "components/traits" level of analysis (e.g., DeYoung et al. (2022) offers a starting place for exploring non *DSM-5* constructs.

Maximize Range on Variables of Interest. Since many psychopathology constructs are dimensional, researchers should recruit participants with varying levels of the construct. For example, when studying depression, aim to include the widest possible range of depression severity scores, including subthreshold presentations (e.g., individuals who have symptoms of depression that do not meet *DSM-5* major depressive disorder criteria). Maximizing range on variables of interest should also yield greater effect sizes, since range restriction attenuates effects (Linn, 1968).

Attend to Reliability. Reliability sets the upper limit of validity and is crucial for research. Assess reliability using metrics like Cronbach's alpha for self-report (Cronbach, 1951), the intraclass correlation coefficient, for dimensional observer ratings, or Cohen's Kappa, for categorical observer ratings (Hallgren, 2012). Training raters thoroughly and enhancing rater competency can ensure good reliability (Reichelt et al., 2003; Creed et al., 2016).

Consider Condition Heterogeneity. Mental health conditions are highly heterogeneous, with significant variability in symptom presentation and individual experiences. Two patients with a *DSM-5* diagnosis of major depressive disorder may not share a single symptom (Fried, 2017a). Study designs should account for this variability – including by analyzing individual symptoms (Fried and Nesse, 2015) – to avoid oversimplification.

Address Comorbidity. Mental health conditions often co-occur and share overlapping symptoms, hinting that effects thought to be driven by one disorder could be driven by co-occurring conditions. Researchers can account for comorbidity using statistical controls (e.g., (Stade et al., 2023b)) or measures that disentangle overlapping conditions (e.g., (Watson et al., 1995)), increasing confidence that effects are unique to a given condition.

Adopt Longitudinal Measurement. Psychopathology dynamically evolves over time, both in terms of severity and diagnostic label. Longitudinal methods of data collection, and analyses using temporally-aware models or time-series analyses, could help address this reality. Although not historically accounted for in computational research, recent work has begun to examine the relationship between symptoms and language features over time (e.g., Nook et al. (2022)).

We acknowledge that some of the proposed suggestions are challenging for NLP researchers to implement: Computational researcher may require sample sizes prohibitively large for conducting semi-structured diagnostic interviewing (prohibitive from a resource perspective, and even the interaction required to collect self-reported scores, as opposed to social-media based proxy measurement, which may require no interaction between researchers and participants, may be more resource-intensive or involved than is actually feasible. Especially if doing social-media based research requires no interaction with participants whatsoever. Therefore, while advocating for clinician-assessed, dimensional psychopathology measurement as the gold-standard, we suggest that researchers seeking to strengthen measurement approaches can adopt an "n+1" approach, where they seek to take one step towards improved measurement. For example, researchers planning to administer a single-item measure can weigh the pros and cons of this approach (Allen et al., 2022) and select a measure with demonstrated validity in their population of interest (e.g., Joiner et al. (2025)) rather than writing their own item from scratch. Researchers can follow the guidelines for selecting measures in line with transdiagnostic frameworks (e.g. Stanton et al. (2020)) rather than using disorder-specific measures. To demonstrate what different measurement strategies can look like, we present in Table 1 a matrix demonstrating measures of social anxiety that systematically vary on the categories we have highlighted in this paper (validated vs. unvalidated, categorical vs. dimensional, and disorder-specific vs. transdiagnostic). Even incremental improvements can significantly improve validity and utility.

Beyond this, the field sorely needs large, publicly available datasets that include natural language from well-characterized clinical samples, perhaps created leveraging something like a practice-research network (Parry et al., 2010). Given that many academic mental health clinics routinely administer the same semi-structured interviews, the aggregation of such recordings could be utilized. Diagnostic interviews can be a particularly efficient source of data, because they yield language as well as measurement of psychopathology constructs, and they are often audio recorded. It is possible to use to predict diagnostic severity scores obtained from a separate section of the interview (e.g., Stade et al. (2023b).

Accruing this type of large, shared dataset is not without challenges, one of which is the issue of confidentiality. It is difficult to acquire natural language data that are not identifiable or semi-identifiable in some way; and to conduct computational research, this dataset would need clinical variables; risking the disclosure of PHI. However, a potential workaround is not making the raw language public but instead extracting a range of linguistic features (including basic, dictionary-based features as well as more sophisticated, transformer/embedding based features) available.

7 Discussion and Conclusions

We highlight challenges in measuring mental health constructs in computational research and propose ways to improve validity. Key issues include overreliance on categorical frameworks, neglect of condition heterogeneity, and inadequate transdiagnostic measures.

Categorical frameworks like the *DSM-5* oversimplify constructs, while dimensional approaches—capturing severity and shared symptoms—enhance model accuracy. Focusing on transdiagnostic constructs, like "negative affect," provides a holistic understanding of mental health.

Condition heterogeneity complicates analysis, but transdiagnostic approaches can address comorbidity and overlapping symptoms. Poor measurement practices introduce errors and biases, so researchers should prioritize validated instruments and diverse datasets.

We argue that researchers should adopt dimensional measures, assess disorder-specific metrics critically, and ensure sample diversity. Implementing transdiagnostic approaches, rater calibration, and reliability checks will further enhance validity.

In conclusion, improving measurement practices is crucial for advancing computational models and mental health care, capturing the complexity of psychopathology, and driving progress in the field.

8 Limitations

While this work highlights critical issues in the measurement of mental health constructs and provides practical recommendations, it is important to acknowledge its limitations.

First, although we emphasize the importance of validated measures, we recognize that resource constraints and practical barriers may prevent many researchers from employing clinician-rated assessments or developing fully validated instruments. These barriers underscore the need for scalable, cost-effective alternatives that balance feasibility and validity.

Second, while we center our discussion on practices and pitfalls in the computational research community, we do not mean to imply that computational researchers are the only ones making these mistakes. These measurement issues we outline here are common in behavioral health research more broadly, including psychiatric and psychological research. Addressing these challenges comprehensively will require a broader interdisciplinary effort that includes collaboration across fields.

Third, while we focus on dimensional and transdiagnostic measurement approaches, we acknowledge that these may not be universally applicable. Certain clinical scenarios may necessitate categorical diagnoses for treatment decisions, and some researchers may have justifiable reasons for focusing on specific disorders. Future work should aim to provide clearer guidance on when categorical, dimensional, or transdiagnostic approaches are best.

Fourth, this work does not claim to exhaustively address all the challenges in the measurement of mental health constructs. Other significant issues, such as the influence of cultural biases, ethical considerations in mental health data collection, and the challenges of interpreting results from large-scale datasets, also warrant attention but fall outside the scope of this discussion.

Finally, while we provide practical recommendations, the field still lacks consensus on "best practices" for measuring mental health constructs in computational research. More empirical studies are needed to evaluate the relative merits of different measurement approaches and their impacts on model performance and real-world applications.

Despite these limitations, we hope that this work stimulates critical reflection and contributes to advancing the validity and utility of mental health research in computational contexts.

9 Acknowledgements

Dan Jurafsky provided critical feedback on an earlier version of this manuscript, and a conversation with Ayelet M. Ruscio shaped our thinking about the different measurement strategies appearing in Table 1. This work was supported in part by the Koret Foundation grant for Smart Cities and Digital Living.

References

Mark S Allen, Dragos Iliescu, and Samuel Greiff. 2022. Single item measures in psychological science: A call to action. *European journal of psychological assessment*, 38(1):1–5.

American Psychiatric Association. 2013. *Diagnostic and Statistical Manual of Mental Disorders: DSM-5*, volume 5. American psychiatric association Washington, DC.

Martin M Antony and Karen Rowa. 2005. Evidence-based assessment of anxiety disorders in adults. *Psychological Assessment*, 17(3):256–266.

David H Barlow, Shannon Sauer-Zavala, Jenna R Carl, Jacqueline R Bullis, and Kristen K Ellard. 2014. The nature, diagnosis, and treatment of neuroticism: Back to the future. *Clinical Psychological Science*, 2(3):344–365.

Naomi Breslau. 1985. Depressive symptoms, major depression, and generalized anxiety: A comparison of self-reports on ces-d and results from diagnostic interviews. *Psychiatry Research*, 15(3):219–229.

Timothy A Brown and David H Barlow. 2014. *Anxiety and Related Disorders Interview Schedule for DSM-5 (ADIS-5L): Lifetime Version. Client Interview Schedule.* Oxford University Press.

Donald T Campbell and Donald W Fiske. 1959. Convergent and discriminant validation by the multitrait-multimethod matrix. *Psychological Bulletin*, 56(2):81–105.

Li-Shiun Chen, William W Eaton, Joseph J Gallo, Gerald Nestadt, and Rosa M Crum. 2000. Empirical examination of current depression categories in a population-based study: symptoms, course, and risk factors. *American Journal of Psychiatry*, 157(4):573–580.

Jacob Cohen. 1983. The cost of dichotomization. *Applied Psychological Measurement*, 7(3):249–253.

Kathryn M Connor, Jonathan RT Davidson, L Erik Churchill, Andrew Sherwood, Richard H Weisler, and Edna Foa. 2000. Psychometric properties of the social phobia inventory (SPIN): New self-rating scale. *The British Journal of Psychiatry*, 176(4):379–386.

James C Coyne. 1994. Self-reported distress: Analog or ersatz depression? *Psychological Bulletin*, 116(1):29–45.

Torrey A Creed, Courtney Benjamin Wolk, Betsy Feinberg, Arthur C Evans, and Aaron T Beck. 2016. Beyond the label: Relationship between community therapists' self-report of a cognitive behavioral therapy orientation and observed skills. *Administration and Policy in Mental Health and Mental Health Services Research*, 43:36–43.

Lee J Cronbach. 1951. Coefficient alpha and the internal structure of tests. *Psychometrika*, 16(3):297–334.

Lee J Cronbach and Paul E. Meehl. 1955. Construct validity in psychological tests. *Psychological Bulletin*, 52(4):281–302.

Pim Cuijpers, Ron de Graaf, and Saskia van Dorsselaer. 2004. Minor depression: risk profiles, functional disability, health care use and risk of developing major depression. *Journal of Affective Disorders*, 79(1-3):71–79.

Colin G DeYoung, Michael Chmielewski, Lee Anna Clark, David M Condon, Roman Kotov, Robert F Krueger, Donald R Lynam, Kristian E Markon, Joshua D Miller, Stephanie N Mullins-Sweatt, et al. 2022. The distinction between symptoms and traits in the hierarchical taxonomy of psychopathology (hitop). *Journal of Personality*, 90(1):20–33.

Kamryn T Eddy, David J Dorer, Debra L Franko, Kavita Tahilani, Heather Thompson-Brenner, and David B Herzog. 2008. Diagnostic crossover in anorexia nervosa and bulimia nervosa: Implications for DSM-V. *American Journal of Psychiatry*, 165(2):245–250.

Sindhu Kiranmai Ernala, Michael L Birnbaum, Kristin A Candan, Asra F Rizvi, William A Sterling, John M Kane, and Munmun De Choudhury. 2019. Methodological gaps in predicting mental health states from social media: Triangulating diagnostic signals. In *Proceedings of the 2019 CHI Conference on Human Factors in Computing Systems*, pages 1–16.

Michael B. First, Janet B. W. Williams, Rhonda S. Karg, and Robert L. Spitzer. 2016. *SCID-5-CV: Structured Clinical Interview for DSM-5 Disorders: Clinician Version*. American Psychiatric Association Publishing, Arlington, VA.

Eiko I Fried. 2017a. Moving forward: How depression heterogeneity hinders progress in treatment and research. *Expert Review of Neurotherapeutics*, 17(5):423–425.

Eiko I. Fried. 2017b. What are psychological constructs? On the nature and statistical modelling of emotions, intelligence, personality traits and mental disorders. *Health Psychology Review*, 11(2):130–134.

Eiko I Fried and Randolph M Nesse. 2015. Depression is not a consistent syndrome: An investigation of unique symptom patterns in the STAR*D study. *Journal of Affective Disorders*, 172:96–102.

Sharath Chandra Guntuku, David B Yaden, Margaret L Kern, Lyle H Ungar, and Johannes C Eichstaedt. 2017. Detecting depression and mental illness on social media: An integrative review. *Current Opinion in Behavioral Sciences*, 18:43–49.

Kevin A Hallgren. 2012. Computing inter-rater reliability for observational data: an overview and tutorial. *Tutorials in Quantitative Methods for Psychology*, 8(1):23.

Allison G Harvey, Edward Watkins, and Warren Mansell. 2004. *Cognitive behavioural processes across psychological disorders: A transdiagnostic approach to research and treatment*. Oxford University Press, USA.

N. Haslam, E. Holland, and P. Kuppens. 2012. Categories versus dimensions in personality and psychopathology: A quantitative review of taxometric research. *Psychological Medicine*, 42(5):903–920.

Melissa Hunt, Joseph Auriemma, and Ashara CA Cashaw. 2003. Self-report bias and underreporting of depression on the BDI-II. *Journal of Personality Assessment*, 80(1):26–30.

Becky Inkster, David Stillwell, Michal Kosinski, and Peter Jones. 2016. A decade into facebook: Where is psychiatry in the digital age? *The Lancet Psychiatry*, 3(11):1087–1090.

Takeshi Inoue, Teruaki Tanaka, Shin Nakagawa, Yasuya Nakato, Rie Kameyama, Shuken Boku, Hiroyuki Toda, Tsugiko Kurita, and Tsukasa Koyama. 2012. Utility and limitations of PHQ-9 in a clinic specializing in psychiatric care. *BMC Psychiatry*, 12:1–6.

Thomas Insel, Bruce Cuthbert, Marjorie Garvey, Robert Heinssen, Daniel S Pine, Kevin Quinn, Charles Sanislow, and Philip Wang. 2010. Research domain criteria (RDoC): Toward a new classification framework for research on mental disorders. *American Journal of Psychiatry*, 167(7):748–751.

Thomas E Joiner, Morgan Robison, Jamie Manwaring, Renee D Rienecke, Daniel Le Grange, Alan Duffy, Philip S Mehler, and Dan V Blalock. 2025. The reliability and construct validity of a single-item assessment of suicidal ideation among patients with mood and anxiety disorders. *Journal of Clinical Psychology*. Advanced online publication.

Lewis L Judd, Hagop S Akiskal, and Martin P Paulus. 1997. The role and clinical significance of subsyndromal depressive symptoms (SSD) in unipolar major depressive disorder. *Journal of Affective Disorders*, 45(1-2):5–18.

Philip C Kendall, Steven D Hollon, Aaron T Beck, Constance L Hammen, and Rick E Ingram. 1987. Issues and recommendations regarding use of the Beck Depression Inventory. *Cognitive Therapy and Research*, 11:289–299.

Ronald C. Kessler. 1997. The prevalence of psychiatric comorbidity. In Scott Ed Wetzler and William C Sanderson, editors, *Treatment strategies for patients with psychiatric comorbidity.*, pages 23–48. John Wiley & Sons Inc.

Ronald C. Kessler, Wai Tat Chiu, Robert Jin, Ayelet Meron Ruscio, Katherine Shear, and Ellen E. Walters. 2006. The epidemiology of panic attacks, panic disorder, and agoraphobia in the National Comorbidity Survey Replication. *Archives of General Psychiatry*, 63(4):415–424.

Daniel N Klein, Lea R Dougherty, and Thomas M Olino. 2005. Toward guidelines for evidence-based assessment of depression in children and adolescents. *Journal of Clinical Child and Adolescent Psychology*, 34(3):412–432.

R. Kotov, G. Perlman, W. Gámez, and D. Watson. 2015. The structure and short-term stability of the emotional disorders: a dimensional approach. *Psychological Medicine*, 45(8):1687–1698.

Roman Kotov, Robert F Krueger, David Watson, Thomas M Achenbach, Robert R Althoff, R Michael Bagby, Timothy A Brown, William T Carpenter, Avshalom Caspi, Lee Anna Clark, et al. 2017. The hierarchical taxonomy of psychopathology (hitop): A dimensional alternative to traditional nosologies. *Journal of Abnormal Psychology*, 126(4):454.

K. Kroenke, R. L. Spitzer, and J. B. Williams. 2001. The phq-9: validity of a brief depression severity measure. *Journal of General Internal Medicine*, 16:606–613.

Robert L Linn. 1968. Range restriction problems in the use of self-selected groups for test validation. *Psychological Bulletin*, 69(1):69–73.

Kristian E Markon, Michael Chmielewski, and Christopher J Miller. 2011. The reliability and validity of discrete and continuous measures of psychopathology: a quantitative review. *Psychological Bulletin*, 137(5):856–879.

Erik C Nook, Thomas D Hull, Matthew K Nock, and Leah H Somerville. 2022. Linguistic measures of psychological distance track symptom levels and treatment outcomes in a large set of psychotherapy transcripts. *Proceedings of the National Academy of Sciences*, 119(13):e2114737119.

Glenys Parry, Louis G Castonguay, Tom D Borkovec, and Abraham W Wolf. 2010. Practice research networks and psychological services research in the uk and usa. *Developing and delivering practice-based evidence: A guide for the psychological therapies*, pages 311–325.

F Katharina Reichelt, Ian A James, and Ivy-Marie Blackburn. 2003. Impact of training on rating competence in cognitive therapy. *Journal of Behavior Therapy and Experimental Psychiatry*, 34(2):87–99.

Ayelet Meron Ruscio. 2019. Normal versus pathological mood: Implications for diagnosis. *Annual Review of Clinical Psychology*, 15(1):179–205.

Ayelet Meron Ruscio, Timothy A Brown, Wai Tat Chiu, Jitender Sareen, and Ronald C Kessler. 2008. Social fears and social phobia in the United States: Results from the National Comorbidity Survey Replication. *Psychological Medicine*, 38(1):15–28.

Ayelet Meron Ruscio, Wai Tat Chiu, Peter Roy-Byrne, Paul E. Stang, Dan J. Stein, Hans-Ulrich Wittchen, and Ronald C. Kessler. 2007. Broadening the definition of generalized anxiety disorder: Effects on prevalence and associations with other disorders in the National Comorbidity Survey Replication. *Journal of Anxiety Disorders*, 21(5):662–676.

John Ruscio and Ayelet Meron Ruscio. 2000. Informing the continuity controversy: A taxometric analysis of depression. *Journal of Abnormal Psychology*, 109(3):473–487.

Almudena Sanchez-Villegas, Javier Schlatter, Felipe Ortuno, Francisca Lahortiga, Jorge Pla, Silvia Benito, and Miguel A Martinez-Gonzalez. 2008. Validity of a self-reported diagnosis of depression among participants in a cohort study using the Structured Clinical Interview for DSM-IV (SCID-I). *BMC Psychiatry*, 8:1–8.

M Katherine Shear and Jack D Maser. 1994. Standardized assessment for panic disorder research: A conference report. *Archives of General Psychiatry*, 51(5):346–354.

Robert L Spitzer, Kurt Kroenke, Janet BW Williams, and Bernd Löwe. 2006. A brief measure for assessing generalized anxiety disorder: The GAD-7. *Archives of Internal Medicine*, 166(10):1092–1097.

Elizabeth C Stade, Robert J DeRubeis, Lyle Ungar, and Ayelet Meron Ruscio. 2023a. A transdiagnostic, dimensional classification of anxiety shows improved parsimony and predictive noninferiority to DSM. *Journal of Psychopathology and Clinical Science*, 132(8):937–948.

Elizabeth C Stade, Lyle Ungar, Johannes C Eichstaedt, Garrick Sherman, and Ayelet Meron Ruscio. 2023b. Depression and anxiety have distinct and overlapping language patterns: Results from a clinical interview. *Journal of Psychopathology and Clinical Science*, 132(8):972–983.

Kasey Stanton, Christina G McDonnell, Elizabeth P Hayden, and David Watson. 2020. Transdiagnostic approaches to psychopathology measurement: Recommendations for measure selection, data analysis, and participant recruitment. *Journal of Abnormal Psychology*, 129(1):21.

Amanda L Stuart, Julie A Pasco, Felice N Jacka, Sharon L Brennan, Michael Berk, and Lana J Williams. 2014. Comparison of self-report and structured clinical interview in the identification of depression. *Comprehensive Psychiatry*, 55(4):866–869.

David Watson, Michael W O'Hara, Kristin Naragon-Gainey, Erin Koffel, Michael Chmielewski, Roman Kotov, Sara M Stasik, and Camilo J Ruggero. 2012. Development and validation of new anxiety and bipolar symptom scales for an expanded version of the IDAS (the IDAS-II). *Assessment*, 19(4):399–420.

David Watson, Kris Weber, Jana Smith Assenheimer, Lee Anna Clark, Milton E Strauss, and Richard A McCormick. 1995. Testing a tripartite model: I. evaluating the convergent and discriminant validity of anxiety and depression symptom scales. *Journal of Abnormal Psychology*, 104(1):3–14.

World Health Organization. 2004. *ICD-10 : International Statistical Classification of Diseases and Related Health Problems: Tenth Revision, 2nd Edition*. World Health Organization.

10 Appendix

	Not dimensional		Dimensional	
	Syndrome-specific	**Transdiagnostic**	**Syndrome-specific**	**Transdiagnostic**
Not validated	r/SocialAnxiety membership status (member vs. not member) Twitter-disclosed social anxiety disorder diagnosis (present vs. absent) Self-reported history of social anxiety disorder diagnosis	r/Anxiety membership status (member vs. not member) Twitter-disclosed anxiety (present vs. absent)	Self-reported social anxiety disorder severity (e.g., "How much have you been bothered by social anxiety disorder symptoms in the past month, on a 0-10 scale?")	Self-reported anxiety severity (e.g., "How much have you been bothered by anxiety symptoms in the past month, on a 0-10 scale?")
Validated	Clinician-rated SCID-5 social anxiety disorder status (present vs. absent) Self-reported SPIN score, converted to dichotomous variable reflecting the presence (score > 19) or absence (score ≤ 18) of social anxiety disorder	Self-reported MASQ Anxious Symptoms subscale, dichotomized ("high anxious symptoms" vs. "low anxious symptoms") Clinician-rated IMAS fear subfactor score, converted to dichotomous variable reflecting high fear or low fear	Clinician-rated ADIS-5L social anxiety disorder severity on 0 (*none*) to 8 (*very severe*) scale. Self-reported SPIN (0-68 score) Self-reported IDAS-II Social Anxiety Scale	Self-reported MASQ Anxious Symptoms subscale Clinician-rated IMAS fear subfactor

Table 1: ADIS-5L = Anxiety and Related Disorders Interview Schedule for *DSM-5* (Brown and Barlow, 2014); IDAS-II = Inventory of Depression and Anxiety Symptoms (Watson et al., 2012); IMAS = Interview for Mood and Anxiety Symptoms (Kotov et al., 2015); MASQ = Mood and Anxiety Symptom Questionnaire (Watson et al., 1995); SCID-5 = Structured Clinical Interview for *DSM-5* (First et al., 2016); SPIN = Social Phobia Inventory (Connor et al., 2000).

Automatic Scoring of an Open-Response Measure of Advanced Mind-Reading Using Large Language Models

Yixiao Wang[1], Russel Dsouza[1], Robert Lee[2], Ian Apperly[2], Rory T. Devine[2], Sanne W. van der Kleij[2], Mark Lee[1]
[1]School of Computer Science, University of Birmingham, UK
[2]School of Psychology, University of Birmingham, UK
{y.wang.37, r.s.dsouza, r.lee.5, i.a.apperly, r.t.devine, s.w.vanderkleij, m.g.lee}@bham.ac.uk

Abstract

A rigorous psychometric approach is crucial for the accurate measurement of mind-reading abilities. Traditional scoring methods for such tests, which involve lengthy free-text responses, require considerable time and human effort. This study investigates the use of large language models (LLMs) to automate the scoring of psychometric tests. Data were collected from participants aged 13 to 30 years and scored by trained human coders to establish a benchmark. We evaluated multiple LLMs against human assessments, exploring various prompting strategies to optimize performance and fine-tuning the models using a subset of the collected data to enhance accuracy. Our results demonstrate that LLMs can assess advanced mind-reading abilities with over 90% accuracy on average. Notably, in most test items, the LLMs achieved higher Kappa agreement with the lead coder than two trained human coders, highlighting their potential to reliably score open-response psychometric tests.

1 Introduction

Theory of Mind (ToM), commonly referred to as mind-reading, is a crucial social cognitive skill that enables individuals to understand, analyze, and use mental states to predict and explain the behavior of others (Apperly, 2010). Researchers have extensively studied the emergence and development of mind-reading abilities in young children, focusing on how they begin to grasp concepts such as perspective-taking and intention recognition (Perner et al., 1987; Wimmer and Perner, 1983; Gopnik and Astington, 1988). There is growing evidence (Apperly et al., 2011; Devine, 2021) to suggest that ToM continues to develop throughout middle childhood and adolescence and that there are individual differences in mind-reading across this age range.

Individual differences in a child's ability to understand others' perspectives remain stable over time, are frequently disrupted in clinical and mental health conditions, and have a significant impact on long-term outcomes. (Hughes and Devine, 2015). These outcomes include the quality of peer relationships, experiences of loneliness, mental health, overall well-being, and success in educational settings. Given its importance in mental health, individual differences in mind-reading offer a target for intervention. Such interventions can be tailored for individuals in therapeutic settings or applied broadly to larger populations by improving social environments. It is plausible that mind-reading will be equally important to the mental health and well-being of older adolescents and adults. However, researchers currently lack reliable and valid tools to study individual differences in mind reading beyond middle adolescence to adulthood (Yeung et al., 2024). This work addresses the significant challenges of creating sufficiently difficult mind-reading tasks that are scalable to large samples.

To create a sufficiently difficult task we reasoned that a core challenge for performing advanced mindreading is to apply mindreading abilities across a variety of people and contexts. Building on established theoretical frameworks, as outlined in previous research (Dziobek et al., 2006), we collected authentic social narratives from a demographically diverse group of individuals aged 17-18 to serve as test items, ensuring that the assessment effectively measures mind-reading ability. Story authors' interpretation of the mental states of characters in their story became the ground-truth against which mindreading accuracy was assessed. To maximize the potential for individual differences in performance, participants were asked to provide open-ended responses explaining their reasoning. This approach generated rich qualitative data that were graded by trained human coders who evaluated answers based on predefined rubrics. While this approach ensures a nuanced understanding of participants' mental state inferences, it is labour-intensive, time-

Proceedings of the 10th Workshop on Computational Linguistics and Clinical Psychology (CLPsych 2025), pages 79–89
May 3, 2025 ©2025 Association for Computational Linguistics

consuming, and prone to variation due to subjective interpretation (Devine et al., 2023).

Automation to overcome the need for human coding is needed for employing the new task at scale. However, automated coding of such responses poses challenges because, by design, the mindreading involved is highly sensitive to the story context, and the expression of correct and incorrect answers is highly variable. Recent advancements in natural language processing, particularly the development of large language models (LLMs), present a promising solution to automate this process. LLMs have demonstrated impressive capabilities in understanding, generating, and evaluating human language (Achiam et al., 2023; Dubey et al., 2024). They have been successfully used to grade free-text responses in educational settings (Xiao et al., 2024; Nilsson and Tuvstedt, 2023), making them strong candidates for evaluating individual differences in advanced mind-reading ability. However, unlike standard text classification, scoring advanced mind-reading responses is particularly challenging due to the complexity of following and applying the coding scheme consistently. Even for human coders, extensive training is required to achieve reliable scoring.

In this study, we explore the potential of LLMs to address these challenges and improve the automation of mind-reading assessment. Specifically, we investigate the following key questions:

1. How well do state-of-the-art LLMs measure advanced mind-reading ability compared to human coders?

2. What prompting strategies optimize the grading performance of these models?

3. To what extent does fine-tuning improve LLMs' grading accuracy?

To address these questions, we designed a set of mind-reading tests based on 10 selected social narratives, collected and coded responses from 1733 participants aged 13-30 before benchmarking several LLMs against human-coded scores. In particular, we assessed the impact of various prompting techniques and fine-tuning strategies on model performance. To further enhance the models, we applied data augmentation to expand the dataset, improving the effectiveness of fine-tuning. Our results show that LLMs, particularly those fine-tuned on the augmented dataset, achieve high accuracy and consistency, significantly reducing the effort required for human grading while maintaining reliability. This automated scoring approach provides clinicians with a fast, scalable, and reliable tool for assessing mind-reading ability. By addressing the scalability limitations of human-coded evaluations, it improves screening for conditions such as autism spectrum disorder and social communication disorders, where difficulties in mind-reading are prevalent (Dziobek et al., 2008; Happé, 2015). Our contributions can be summarized as follows

- We designed and implemented innovative psychological tests to measure advanced mind-reading abilities, addressing a critical need for robust and scalable assessment tools in psychometrics.

- We collected a unique dataset from participants aged 13 to 30 years and will publicly release this dataset, along with our code and fine-tuned models[1] .

- We systematically optimized the performance of LLMs through various prompting strategies and fine-tuning based on data augmentation, achieving over 90% accuracy in scoring psychometric tests.

2 Related Work

Automated grading of psychometric tests with open-ended responses has attracted significant interest in recent years. Early efforts focused on rule-based systems, which relied on manually defined patterns and logic to assess responses (Williamson et al., 2006). While these systems provided consistency in scoring, they struggled to handle the variability and nuance of open-ended responses (Burrows et al., 2015).

Over time, machine learning techniques have gained prominence as a more versatile and adaptable solution to address the problem (Mohler et al., 2011). Machine learning models frequently employ supervised learning methods, which rely on annotated datasets containing labeled examples (Bailey and Meurers, 2008; Nielsen et al., 2008; Madnani et al., 2013). These datasets enable the models to train classifiers that learn patterns and relationships between input features and their corresponding outcomes, allowing them to predict scores or make

[1]All code and data to replicate our experiments is available at `https://github.com/YixiaoWang/ToM-automatic-scoring-using-LLMs/`.

informed decisions when presented with unseen data. Additionally, machine learning also incorporates unsupervised learning approaches (Alfonseca and Pérez, 2004; Pérez et al., 2005; Mohler and Mihalcea, 2009). These methods identify hidden patterns, groupings, or structures within the data itself, such as clustering similar items or detecting anomalies. However, the performance of these machine learning models remained constrained by the quality and size of the training data, as well as their limited ability to capture deeper semantic understanding.

The advent of pre-trained language models marked a significant leap forward in automating text-based assessments. The model DistilBERT (Sanh, 2019), have been applied to scoring the open-response for mind-reading, where they have shown promise in scoring standardized tests of children's mind-reading (Devine et al., 2023). To further enhance the effectiveness of fine-tuning in these language models, data augmentation techniques can be employed to artificially expand the training dataset, thereby improving model generalization and robustness (Kovatchev et al., 2021). Methods such as synonym replacement, back-translation, and paraphrasing introduce variability in training samples, reducing the risk of overfitting to limited datasets.

The emergence of foundation models (Bommasani et al., 2021) trained on larger datasets with substantially more parameters to capture deeper contextual relationships has significantly enhanced performance in text-based tasks. Research efforts have successfully used LLMs to develop automatic grading systems in education setting (Xiao et al., 2024; Nilsson and Tuvstedt, 2023), enabling accurate evaluation of student writing and essay grading, often matching human evaluators in accuracy.

Assessing mind-reading ability poses significant challenges, as it requires the interpretation of nuanced psychological cues that are often deeply context-dependent, extending beyond surface-level or factual knowledge. Recent studies (Strachan et al., 2024; Kosinski, 2023; He et al., 2023) have demonstrated that LLMs are capable of making mental inferences, highlighting their suitability for this task. Although the application of LLMs to evaluate advanced mind-reading assessments remains underexplored in the broader literature, prior work by Devine et al. (2023) has made notable progress by automating the scoring of mind-reading ability using DistilBERT. This study builds on that foundation while advancing it in two key directions: (1) introducing a novel open-ended test designed for adults, which requires inferring more subtle and context-dependent mental states, and (2) leveraging LLMs instead of lightweight models to enable more sophisticated evaluations. By applying LLMs to assess responses in advanced mind-reading tests, this study seeks to further explore their potential in assessing complex human cognition.

3 Methodology

3.1 Data

The mind-reading test included 10 social narratives, each followed by a question that asked participants to interpret the mental states of the characters. An example of one such narrative, along with the corresponding mind-reading question and coding scheme, is presented in the table 1. A total of 1,733 participants aged 13–30 provided free-text responses after completing these psychometric test either in schools or online via Prolific.co. The labeling process was conducted by one lead coder and four trained coders. After an initial training phase, during which coders achieved inter-rater reliability (Cohen's Kappa > 0.7) with the lead coder, they independently coded different portions of the dataset. To ensure consistency and accuracy, each coder periodically re-coded responses from another coder. Discrepancies were resolved by the lead coder, ensuring high reliability throughout the process.

The final labeled data confirmed that the designed task was sufficiently challenging. The table 2 below shows the percentages of participants who successfully completed the mind-reading test for each of the 10 social narratives. This rigorous, multi-step process generated a high-quality, gold-standard dataset for training and evaluating LLMs. We assessed LLMs by comparing their predictions with labels of the dataset, using accuracy as the evaluation metric. Through systematic benchmarking, we aim to identify the most effective LLM for automated grading.

3.2 Model Selection

To assess the suitability of different LLMs for the mind-reading evaluation, we select a diverse set of state-of-the-art models as shown in Table 3. By comparing these models, we aim to analyze the trade-offs between model size, computational cost, and task performance for automating the mind-reading evaluation process.

Story	It was October last year, and I went to a theme park that had extra attractions for Halloween. One highlight was the "Dungeon Experience". This had actors playing characters who interact with you as you pass through it. I went in. It was really fun, but I have sensory needs and I couldn't believe how loud it was. For the first half of the experience, I had to keep my fingers in my ears, and I felt really self-conscious. I got to the bit of the experience where you get to ride on a boat through the 'Black River'. A Ferryman was wearing dark robes, limping, and carrying a lantern. He greeted us in a raspy voice then started warning us about the journey to come. He saw how I looked and put his finger up for us to wait. He hobbled off to one side, then returned a moment later and pressed a small package into my hand. It was a pack of earplugs. I put them in my ears and the ferryman caught my eye and raised an eyebrow. I gave him a thumbs up back and he grinned then returned to warning us about the journey in his raspy voice, before giving my sister two riddles to solve. He didn't even really break character!
Question	Why did she appreciate that the actor stayed in character?
Partial Coding Scheme	Correct Responses: 1 part required for a 'correct response': She appreciated that the actor was able to help her without drawing excessive attention to her sensory needs/differences (about which she felt self-conscious) (the Ferryman drew less unwanted attention by staying in character, but it was the not drawing unnecessary attention rather than specifically staying in character per se) This may be phrased in a number of different ways, for example: • He did not make her feel 'different', 'strange', or 'weird' through his actions did not make a big deal of it did not make it into an emergency (whilst also meeting her needs) • The Ferryman was able to help discretely helped without making an unnecessary fuss scene • He did not make her feel embarrassed/awkward/ self-conscious, • He did not make her feel like an inconvenience or 'a nuisance', • He did not treat her differently (other than by supporting her needs) • It meant that the actor helped her discretely (without drawing unwanted attention). • It did not make her feel more conspicuous (and therefore more self-conscious). • ... Incorrect or Incomplete Mindreading responses: Fail to mention or indicate that she was glad that the actor was able to help in a way that did not draw unwanted attention to her needs/differences • For example, responses that just mention that it didn't break the immersion for herself/others (without considering the context that made this important) [e.g. this may be expressed as 'it didn't ruin the magic'] would be incomplete mindreading responses.(The actor staying in character was his way of not drawing unnecessary attention, and not making her feel embarrassed but the not breaking immersion for others was only an add-on, not the key reason that the author appreciated him staying in character) • "It made her feel included/not left out"/"it included her in the experience" are incomplete responses, since the actor giving her the earplugs would help with this, regardless of whether he stayed in character. • Responses that focus on how helping her 'didn't ruin the experience/immersion for others' • ... Non-mindreading responses: Express an opinion on the situation, rather than trying to take the author's perspective. Or just describes the general situation without linking this to the author's experience, e.g. 'it kept it fun', 'it didn't break immersion' (for whom?)
Correct Sample	She appreciated that the actor was able to help her without drawing excessive attention to her sensory needs.
Incorrect Sample	Because the immersion wasn't totally ruined for the author and the other people in the experience.

Table 1: One Example from the 10 Test Items

	Test 1	Test 2	Test 3	Test 4	Test 5	Test 6	Test 7	Test 8	Test 9	Test 10
Proportion of Correct Answers (%)	56.7	58.3	56.8	42.1	16.1	51.4	30.7	35.7	20.8	42.2

Table 2: Percentages of participants (out of 1,733 total) accurately performing mind-reading tasks across 10 distinct test items

Model Name	Reference
allenai/longformer-base-4096	Beltagy et al. (2020)
meta/llama-3.2-3B-instruct	Dubey et al. (2024)
microsoft/phi-3.5-mini-instruct	Abdin et al. (2024a)
mistralai/mistral-7B-v0.3-Instruct	Jiang et al. (2023)
microsoft/phi-4	Abdin et al. (2024b)
openai/gpt-4o-2024-08-06	Achiam et al. (2023)
openai/gpt-4o-mini-2024-07-18	Achiam et al. (2023)
BERT	Devlin (2018)
RoBERTa	Liu (2019)

Table 3: List of Models and References

3.3 Prompt Strategies

These experiments explore the influence of various prompting strategies on the performance of LLMs in the task. We conducted a series of experiments to assess how different input formats, grading schemes, and prompting techniques impact a LLM performance. First, we compared the effect of different input formats, including plain text, XML, and JSON, on the task results. The goal was to determine whether more structured formats, such as XML and JSON, yield better results than plain text input. Next, we evaluated the impact of different grading schemes included in the prompts. The original grading scheme, which is highly detailed but often difficult for humans to interpret, was compared to two alternative formulations: a rephrased version and a summarized version generated by GPT-4o. This comparison aimed to identify which grading scheme provided the clearest and most effective guidance for mind-reading responses evaluation.

Based on the findings from the input format and grading scheme experiments, we selected the most effective combination of syntax format and grading scheme for the remaining experiments. Following the tradition in prompt engineering (Ouyang et al., 2022), the LLM to be tested was given two prompts as components: a system prompt and a user prompt. The system prompt provided the LLM with basic instructions for the task, while the user prompt contained the specific mind-reading context, including the narrative, question, corresponding grading schemes, and the participants' responses to be graded. The LLM was expected to provide a binary response (0 or 1), indicating the true value of the given response. This process is visualized in Figure 1.

Finally, we compared the performance of LLMs under zero-shot and few-shot prompting conditions. In the few-shot condition, the user prompt included a small number of labeled responses for each test. In contrast, no labeled responses from the dataset were included in the user prompt under the zero-shot condition. This comparison aimed to assess whether providing labeled responses within the prompt improves the model's performance on the mind-reading evaluation task.

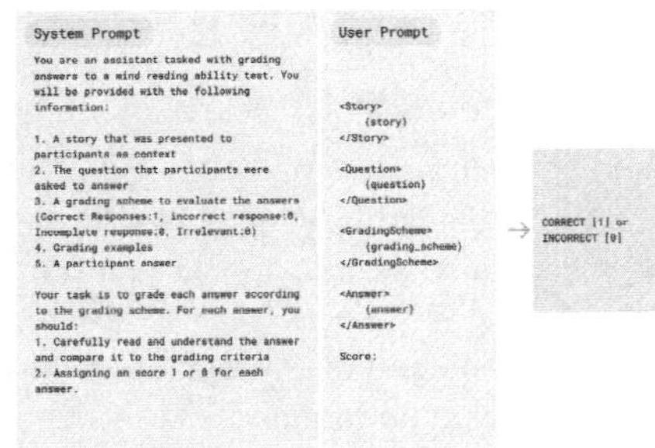

Figure 1: Prompt structure

3.4 Fine-Tuning

To enhance the performance of LLMs in grading free-text responses, we fine-tuned selected models using a labeled dataset. The dataset was split into 80% for training, 10% for validation and 10% for testing. The models to be fine-tuned are listed in Table 3. Fine-tuning aimed to align these models with the grading rubric, improving their ability to interpret and assess responses accurately. Since different models have distinct architectures and constraints, we adopted three fine-tuning strategies:

- Propreitary Models (GPT-4o, GPT-4o-mini): Fine-tuning was conducted using OpenAI's API service. Due to token limitations, we could include a maximum of 50 labeled responses per test, totaling 500 responses across 10 test items.

- Open-source Models (Llama, Mistral, and Phi): Given our computational constraints, we employed LoRA (Low-Rank Adaptation) (Hu et al., 2021) instead of full-parameter fine-tuning. LoRA performs on par with full fine-tuning, but requires significantly less memory. However, LoRA requires careful hyperparameter tuning, which we select using Bayesian search to achieve the best performance.

- BERT and RoBERTa: Fine-tuning these models differs significantly from other LLMs. Unlike other models, it involves training a binary classifier for each test item. Each binary classifier receive individual response from its corresponding test and predicts its truth values without considering contextual elements like the question, narrative, or grading rubric.

After fine-tuning, we evaluated the models on the test set using accuracy as the primary metric.

3.5 Data Augmentation

A key challenge in fine-tuning LLMs is the limited availability of labelled training data. To address this, we investigated the role of data augmentation in enhancing fine-tuning performance. Specifically, we used GPT-4o to generate paraphrased versions of all responses in the training split of our gold-standard dataset. These paraphrased response preserved the original meaning while varying in vocabulary and sentence structure. To ensure labelling consistency, a human coder randomly selected 50 paraphrased responses per story test item for labeling. The coder's labels were then compared to those of the original responses, achieving an agreement rate of over 90%, indicating a high level of consistency. After generating the paraphrased responses, we incorporated them into the training split of the original dataset, effectively doubling the size of the available training data. This augmented dataset was then used to fine-tune the LLMs.

4 Results and Analysis

Syntax Format	Accuracy
Plain	0.82
XML	**0.84**
JSON	0.83

Table 4: Results testing the effect of syntax format in prompting GPT-4o in terms of grading accuracy.

Scheme	Accuracy
Original Grading Scheme	**0.88**
Grading Scheme Summary	0.86
Paraphrase Summary Scheme	0.85

Table 5: Results testing the effect of grading scheme on GPT-4o prompt in terms of grading accuracy.

4.1 Result of Prompt Engineering

We first analyze how different input formats affected the performance of the LLM (GPT-4o) in the mind-reading evaluation task. As shown in the table 4, structured formats, XML and JSON, slightly outperform plain text in terms of accuracy. Then, we compared the effect of different grading schemes incorporated into the prompts. As shown in the Table 5, the original grading scheme, although highly detailed and challenging for human coders to employ consistently, surprisingly produced the better results, outperforming both the rephrased version and summarized version generated by GPT-4o. This finding suggests that, despite the complexity of the original scheme, LLMs are capable of capturing the relevant information embedded in highly detailed text. Based on these results, we use XML as prompt syntax and original grading scheme as default coding rubric to prompt all LLMs, both in zero-shot setting and few-shot setting. The detailed performance of zero-shot prompting and few-shot prompting are included in the Table 6 and Table 7.

In the zero-shot condition, each model's performance was assessed by comparing the results to those assigned by trained human coders and calculating its accuracy in scoring answers for each test. Overall performance was determined by averaging accuracy rates across all 10 test items. Among the models tested, GPT-4o achieved the highest accuracy at 89.4 %, significantly outperforming the others. Phi-4 followed with a strong 81.5%, while Mistral-7B and Phi-3.5 scored 77.1% and 73.5%, respectively. Llama-3.2 trailed at 64.3%, and Longformer, the smallest model in the table, lagged further at just 50%—likely due to its limited capacity to process complex information. These results indicate that larger language models tend to perform better on mind-reading ability scoring task.

Building on the observation from zero-shot results, we evaluated the performance of the best-performing model, GPT-4o, along with GPT-4o-mini, under few-shot conditions. These models were selected due to their outstanding performance in the zero-shot evaluation and their larger capacity to handle more complex prompts. In the first few-shot test, where 10 labelled answers from the dataset were provided for each test, we observed a slight improvement in performance for both models. GPT-4o achieved an accuracy rate of 89.5%, marginally outperforming its zero-shot result of 89.4%. Similarly, GPT-4o-mini saw an increase, with its accuracy rate rising to 81.4% from 79.7% in the zero-shot condition. However, when the number of labelled answer was increased to 50 for each test, the results shifted, GPT-4o's accuracy rate decreased to 88.1%, and GPT-4o-mini's accuracy rate dropped to 80.1%. These results highlight an important insight in few-shot prompting: While providing a certain number of examples can enhance model performance, increasing this number beyond a certain threshold does not always lead to improved outcomes.

4.2 Result of LLMs fine-tuning

As is shown in the Table 8, all models in the evaluation show significant improvements after fine-tuning, highlighting the effectiveness of this approach for the task of psychometric scoring. GPT-4o achieves the best results, with its accuracy increasing from 89.4% to 92.8%. Notably, its performance is further supported by a kappa value of 0.83, indicating strong agreement that far exceeds what would be expected by chance. GPT-4o-mini benefits greatly from fine-tuning, rising from 79.7% to 90.5%. This success is particularly remarkable considering that GPT-4o-mini was fine-tuned on only 50 examples per test. Longformer, initially starting at 50.0%, shows a remarkable jump to 86.7%, and Llama moves from 64.3% to 91.1%. Models like Mistral and Phi-4, which started with strong zero-shot accuracy, also see significant improvements. These results underscore the substantial benefits of fine-tuning in improving model accuracy.

Notably, the BERT family of models has demonstrated impressive performance despite their smaller sizes. BERT-base and BERT-large achieved accuracies of 90.2% and 90.5%, respectively, matching or even surpassing larger models like GPT-4o-mini. This is particularly remarkable given BERT's more compact architecture, highlighting its competitive edge when fine-tuned for specific test items. However, fine-tuning BERT models differs significantly from that of other LLMs. Unlike LLMs, which are fine-tuned as single scoring systems to handle all test items, BERT and RoBERTa are trained into 10 distinct classifiers, each dedicated to a specific test item. These classifiers are test-specific and cannot be transferred to other test items, so while their specialization enhances performance on individual test, it limits their flexibility across a range of tests. Additionally, BERT and RoBERTa fall short of LLMs in providing explanations or feedback to justify the scores they assign, making their high performance both impressive and somewhat constrained in comparison.

4.3 Effect of Data Augmentation

Data augmentation has a positive effect on performance for most models, although the improvements are not consistent across all of them. Longformer sees a notable gain, increasing from 86.7% to 91.6%, demonstrating the clear benefit of augmented data. Mistral and Phi-3.5 also benefit, with Mistral rising from 88.7% to 91.6% and Phi-3.5 improving from 83.8% to 90.1%. However, Llama experiences a slight drop, from 91.1% to 90.5%, and Phi-4 shows only a small increase, from 87.5% to 87.6%. These results indicate that while data augmentation often enhances model accuracy, its impact can vary depending on the model and test item.

4.4 Comparison between human coders and LLMs

Building on the previous findings that LLMs can grade psychometric tests with high accuracy, we now compare their performance to human coders in both accuracy and efficiency. Initially, all four trained human coders demonstrated adequate Inter-Rater Reliability (Cohen's Kappa > .7) with the lead coder across 10 test items before being assigned different portions of the main dataset to code. However, the following spot checks revealed that two trained coders drifted in their application of the marking criteria for certain test items. To address this, the fine-tuned GPT-4o was used to reassess all participant responses for those cases. Whenever the LLM and the trained coder disagreed, the lead coder made the final decision. The table below summarizes the relative accuracy of human coders and the fine-tuned GPT-4o under this procedure. Importantly, the Kappa agreement score was calculated only for cases where the LLM and the human coder initially disagreed. The results indicate a clear trend: except for test item 5, fine-tuned GPT-4o consistently showed higher agreement with the lead coder than the trained coders did. This suggests that, for the majority of test items (1, 2, 3, 6, 7, 8, and 10), the LLM provided more reliable coding across many cases.

In terms of time efficiency, training a single human coder requires at least 14 hours before they can pass the Inter-Rater Reliability check. With four human coders trained, this amounts to a total of 56 hours of training time. After passing the check, each coder takes an average of 33 seconds to grade a single response. Given 10 test items and 1,733

	Test 1	Test 2	Test 3	Test 4	Test 5	Test 6	Test 7	Test 8	Test 9	Test 10	AVG
gpt-4o	**88.5**	**86.7**	**90.2**	**91.3**	**87.3**	**91.9**	**94.2**	78.1	**95.9**	**90.2**	**89.4**
gpt-4o-mini	80.4	83.3	83.9	82.1	80.1	83.9	81	73.5	65.5	83.3	79.7
longformer-4096-base	68.5	63.3	66.5	49.7	22.3	60.2	42.6	47.1	28.5	48	50
Llama-3.2-3B	72.7	78.6	71.5	66.2	43.9	70.7	65.5	61.4	48.8	61.7	64.3
mistral-7b-v0.3-instruct	81.1	78.6	77.8	74.5	82	78.2	87.1	70	65.8	74.7	77.1
phi-3.5-mini-instruct	81.8	79.4	72.1	65.6	81.3	73.7	75	72.8	72.3	63.6	73.5
phi-4	86	78.6	79.1	83.4	87.1	82.7	71	**80**	84.5	83.1	81.5

Table 6: Evaluation results of LLMs on 10 psychometric tests using zero-shot prompting. Results are reported in terms of accuracy (%)

	Test 1	Test 2	Test 3	Test 4	Test 5	Test 6	Test 7	Test 8	Test 9	Test 10	AVG
gpt-4o (10 shots)	83.9	87.3	**90.2**	91.3	**89.6**	**91.9**	**90.8**	**85**	**96.5**	**89**	**89.5**
gpt-4o-mini (10 shots)	78.7	85.6	82.7	86.2	81	84.4	83.4	77.5	68.9	85.6	81.4
gpt-4o (50 shots)	**86.2**	**89.6**	86.2	**93.6**	89	86.7	89	82.7	94.8	83.9	88.17
gpt-4o-mini (50 shots)	80.4	87.3	85	81	80.2	83.9	79.8	77.5	64.3	82.1	80.1

Table 7: Grading results of LLMs on 10 psychometric tests using few-shot prompting. Results are reported in terms of accuracy rate (%)

	Test 1	Test 2	Test 3	Test 4	Test 5	Test 6	Test 7	Test 8	Test 9	Test 10	AVG
BERT-base	86.2	83.3	90.2	93.1	94.8	91.9	89	87.3	94.2	90.2	90.2
BERT-large	**89.6**	83.3	90.8	93.6	93.1	88.5	91.9	88.5	92.5	92.5	90.4
RoBERTa-base	88.5	86.7	93.6	91.9	91.9	91.9	91.3	83.3	94.2	90.8	90.4
RoBERTa-large	89.6	87.9	**94.8**	92.5	94.8	91.3	92.5	87.3	95.4	94.2	92
gpt-4o	89	**91.3**	93.1	94.2	**94.8**	**93.6**	**94.2**	85	**97.1**	96.5	**92.8**
gpt-4o-mini	86.7	83.3	91.9	94.8	93.1	92.5	90.8	79.3	97.1	95.9	90.5
longformer-4096-base	87.0	80.9	91.7	94.9	77.7	89.5	87.1	89.3	81.3	87.6	86.7
Llama-3.2-3B	86.7	85.5	91.1	**96.8**	90.6	90.2	87.8	**90**	95.1	**97.4**	91.1
mistral-7b-v0.3-instruct	86.7	80.9	89.2	94.9	84.9	91.7	89.2	82.1	91.9	95.4	88.7
phi-3.5-mini-instruct	76.2	86.3	89.9	69.3	87.1	80.4	85.8	88.6	90.2	85.7	83.8
phi-4	87.4	84	84.8	90.4	89.9	83.5	85.8	86.4	91.9	90.9	87.5

Table 8: Results of evaluation of fine-tuned LLMs in the 10 psychometric tests. Results are reported in terms of accuracy (%)

	Test 1	Test 2	Test 3	Test 4	Test 5	Test 6	Test 7	Test 8	Test 9	Test 10	AVG
longformer-4096-base	**91.6**	83.9	**94.3**	94.9	89.9	91	**90.5**	90.7	93.5	96.1	91.6
Llama-3.2-3B	85.3	86.2	93	**94.9**	89.2	89.4	89.8	89.3	93.5	94.1	90.5
mistral-7b-v0.3-instruct	90.9	**87**	90.5	94.3	91.4	90.2	89.2	**91.4**	94.3	**97.4**	**91.6**
phi-3.5-mini-instruct	88.8	81.7	87.3	93.6	**92.8**	90.2	87.8	87.9	**96.7**	94.2	90.1
phi-4	86	85.5	84.8	91.7	89.2	**91**	88.5	81.4	88.6	88.9	87.6

Table 9: Results of evaluation of fine-tuned LLMs on augmented train split in the 10 psychometric tests. Results are reported in terms of accuracy (%)

Participant Numbers	ID number assigned to human coder	Item (Story and Question)	Kappa agreement of human coder with lead coder	Kappa agreement of GPT-4o with lead coder
1325–2013	1	2	0.791	0.906
		5	0.873	0.786
		6	0.758	0.837
2014–2157	3	1	0.929	0.971
		2	0.756	0.878
		3	0.889	0.845
		5	0.974	0.638
		8	0.580	0.928
		10	0.833	0.889

Table 10: Kappa agreement of trained human coder and GPT-4o with lead coder.

participants, the entire dataset requires approximately 158 hours to label. In contrast, fine-tuning LLMs (e.g., Llama-3.2-3B) takes approximately 16-24 hours, including 8-16 hours for hyperparameter tuning. Once fine-tuned, the LLM can score each response in milliseconds, a dramatic reduction compared to the time required by human coders. This highlights the LLM's exceptional efficiency in processing speed.

5 Discussion

Our findings highlight the transformative potential of LLMs in automating the scoring of open-ended responses in complex mind-reading tests. Fine-tuning, particularly when paired with augmented training data, enables LLMs to better grasp the test-specific nuances of intricate coding manuals, resulting in more accurate evaluation. Despite the inherent complexity of the task, LLMs demonstrated an impressive ability to interpret and apply these detailed coding guidelines effectively. This adaptability suggests that LLMs could be valuable tools for automating the scoring of other psychometric tests, particularly those that involve open-ended responses. Such applications could help overcome the ceiling effect often seen in closed-ended questions, making it possible to quantify reliably the abilities of more developmentally advanced participants (i.e. older adolescents and adults) than has previously been possible.

Our exploration of prompt strategies further revealed that a relatively small number of examples led to noticeable improvements in performance. However, increasing the number of examples beyond a certain point did not produce gains. As our results show, fine-tuning is a more effective strategy than prompting, particularly when leveraging a larger set of examples to enhance model performance. This highlights that fine-tuning, rather than prompting, is the more powerful tool for maximizing LLM capabilities in psychometric task scoring.

Furthermore, the BERT family of models continues to be a highly effective and practical choice for scoring open-response psychometric tasks. While a BERT classifier trained on one test may not directly transfer to others due to the distinct nature of each test, its strength lies in its simplicity and computational efficiency. BERT models are relatively easy to implement and require fewer computational resources compared to other LLMs, making them an ideal option for users with limited computational resources or specific task requirements.

6 Conclusion

This paper demonstrates the effectiveness of LLMs in scoring psychometric tests designed to assess advanced mind-reading ability. By optimizing prompting strategies and fine-tuning models, we achieve results that not only align closely with human evaluations but also surpass the performance of some trained human coders on most of test items. This highlights LLMs' potential to reliably assess complex cognitive processes, offering a scalable, efficient, and consistent approach to psychometric testing. While current methods use LLMs to evaluate responses against pre-defined answers, LLMs also excel at analyzing patterns in mind-reading responses. This goes beyond identifying performance gaps in individuals with neurodevelopmental or psychiatric conditions, allowing researchers to explore whether they mind-read in systematically different ways. Such insights could transform our understanding of individual differences in mind-read processes. Future work should explore these applications, further expanding the utility of LLMs in psychometric research.

7 Limitations

While the performance of LLMs in scoring mind-reading responses is impressive, it raises the question of what enables them to excel in this task. Are LLMs inherently skilled at mind-reading, allowing them to assess responses reliably, or do they simply follow the complex coding manual with high accuracy? This study does not provide a definitive answer, and further research is needed to explore the underlying mechanisms of LLM judgment.

8 Ethical Considerations

The project gained ethical review and approval from the Science Technology Engineering and Mathematics ethical review panel at the University of Birmingham UK, project approval ID: ERN_2311-Jun2024.

References

Marah Abdin, Jyoti Aneja, Hany Awadalla, Ahmed Awadallah, Ammar Ahmad Awan, Nguyen Bach, Amit Bahree, Arash Bakhtiari, Jianmin Bao, Harkirat Behl, et al. 2024a. Phi-3 technical report: A highly capable language model locally on your phone. *arXiv preprint arXiv:2404.14219.*

Marah Abdin, Jyoti Aneja, Harkirat Behl, Sébastien Bubeck, Ronen Eldan, Suriya Gunasekar, Michael Harrison, Russell J Hewett, Mojan Javaheripi, Piero Kauffmann, et al. 2024b. Phi-4 technical report. *arXiv preprint arXiv:2412.08905.*

Josh Achiam, Steven Adler, Sandhini Agarwal, Lama Ahmad, Ilge Akkaya, Florencia Leoni Aleman, Diogo Almeida, Janko Altenschmidt, Sam Altman, Shyamal Anadkat, et al. 2023. Gpt-4 technical report. *arXiv preprint arXiv:2303.08774.*

Enrique Alfonseca and Diana Pérez. 2004. Automatic assessment of open ended questions with a bleu-inspired algorithm and shallow nlp. In *Advances in Natural Language Processing: 4th International Conference, EsTAL 2004, Alicante, Spain, October 20-22, 2004. Proceedings 4*, pages 25–35. Springer.

Ian Apperly. 2010. *Mindreaders: the cognitive basis of" theory of mind".* Psychology Press.

Ian A Apperly, Frances Warren, Benjamin J Andrews, Jay Grant, and Sophie Todd. 2011. Developmental continuity in theory of mind: Speed and accuracy of belief–desire reasoning in children and adults. *Child development*, 82(5):1691–1703.

Stacey Bailey and Detmar Meurers. 2008. Diagnosing meaning errors in short answers to reading comprehension questions. In *Proceedings of the third workshop on innovative use of NLP for building educational applications*, pages 107–115.

Iz Beltagy, Matthew E Peters, and Arman Cohan. 2020. Longformer: The long-document transformer. *arXiv preprint arXiv:2004.05150.*

Rishi Bommasani, Drew A Hudson, Ehsan Adeli, Russ Altman, Simran Arora, Sydney von Arx, Michael S Bernstein, Jeannette Bohg, Antoine Bosselut, Emma Brunskill, et al. 2021. On the opportunities and risks of foundation models. *arXiv preprint arXiv:2108.07258.*

Steven Burrows, Iryna Gurevych, and Benno Stein. 2015. The eras and trends of automatic short answer grading. *International journal of artificial intelligence in education*, 25:60–117.

Rory T Devine. 2021. Individual differences in theory of mind in middle childhood and adolescence. In *Theory of mind in middle childhood and adolescence*, pages 55–76. Routledge.

Rory T Devine, Venelin Kovatchev, Imogen Grumley Traynor, Phillip Smith, and Mark Lee. 2023. Machine learning and deep learning systems for automated measurement of "advanced" theory of mind: Reliability and validity in children and adolescents. *Psychological Assessment*, 35(2):165.

Jacob Devlin. 2018. Bert: Pre-training of deep bidirectional transformers for language understanding. *arXiv preprint arXiv:1810.04805.*

Abhimanyu Dubey, Abhinav Jauhri, Abhinav Pandey, Abhishek Kadian, Ahmad Al-Dahle, Aiesha Letman, Akhil Mathur, Alan Schelten, Amy Yang, Angela Fan, et al. 2024. The llama 3 herd of models. *arXiv preprint arXiv:2407.21783.*

Isabel Dziobek, Stefan Fleck, Elke Kalbe, Kimberley Rogers, Jason Hassenstab, Matthias Brand, Josef Kessler, Jan K Woike, Oliver T Wolf, and Antonio Convit. 2006. Introducing masc: a movie for the assessment of social cognition. *Journal of autism and developmental disorders*, 36:623–636.

Isabel Dziobek, Kimberley Rogers, Stefan Fleck, Markus Bahnemann, Hauke R Heekeren, Oliver T Wolf, and Antonio Convit. 2008. Dissociation of cognitive and emotional empathy in adults with asperger syndrome using the multifaceted empathy test (met). *Journal of autism and developmental disorders*, 38:464–473.

Alison Gopnik and Janet W Astington. 1988. Children's understanding of representational change and its relation to the understanding of false belief and the appearance-reality distinction. *Child development*, pages 26–37.

FRANCESCA Happé. 2015. Autism as a neurodevelopmental disorder of mind-reading. *Journal of the British Academy*, 3(1):197–209.

Yinghui He, Yufan Wu, Yilin Jia, Rada Mihalcea, Yulong Chen, and Naihao Deng. 2023. Hi-tom: A benchmark for evaluating higher-order theory of mind reasoning in large language models. *arXiv preprint arXiv:2310.16755.*

Edward J Hu, Yelong Shen, Phillip Wallis, Zeyuan Allen-Zhu, Yuanzhi Li, Shean Wang, Lu Wang, and Weizhu Chen. 2021. Lora: Low-rank adaptation of large language models. *arXiv preprint arXiv:2106.09685.*

Claire Hughes and Rory T Devine. 2015. Individual differences in theory of mind from preschool to adolescence: Achievements and directions. *Child development perspectives*, 9(3):149–153.

Albert Q Jiang, Alexandre Sablayrolles, Arthur Mensch, Chris Bamford, Devendra Singh Chaplot, Diego de las Casas, Florian Bressand, Gianna Lengyel, Guillaume Lample, Lucile Saulnier, et al. 2023. Mistral 7b. *arXiv preprint arXiv:2310.06825*.

Michal Kosinski. 2023. Theory of mind may have spontaneously emerged in large language models. *arXiv preprint arXiv:2302.02083*, 4:169.

Venelin Kovatchev, Phillip Smith, Mark Lee, and Rory Devine. 2021. Can vectors read minds better than experts? comparing data augmentation strategies for the automated scoring of children's mindreading ability. *arXiv preprint arXiv:2106.01635*.

Yinhan Liu. 2019. Roberta: A robustly optimized bert pretraining approach. *arXiv preprint arXiv:1907.11692*, 364.

Nitin Madnani, Jill Burstein, John Sabatini, and Tenaha O'Reilly. 2013. Automated scoring of summary-writing tasks designed to measure reading comprehension. *Grantee Submission*.

Michael Mohler, Razvan Bunescu, and Rada Mihalcea. 2011. Learning to grade short answer questions using semantic similarity measures and dependency graph alignments. In *Proceedings of the 49th annual meeting of the association for computational linguistics: Human language technologies*, pages 752–762.

Michael Mohler and Rada Mihalcea. 2009. Text-to-text semantic similarity for automatic short answer grading. In *Proceedings of the 12th Conference of the European Chapter of the ACL (EACL 2009)*, pages 567–575.

Rodney D Nielsen, Wayne H Ward, and James H Martin. 2008. Learning to assess low-level conceptual understanding. In *FLAIRS*, pages 427–432.

Filippa Nilsson and Jonatan Tuvstedt. 2023. Gpt-4 as an automatic grader: The accuracy of grades set by gpt-4 on introductory programming assignments.

Long Ouyang, Jeffrey Wu, Xu Jiang, Diogo Almeida, Carroll Wainwright, Pamela Mishkin, Chong Zhang, Sandhini Agarwal, Katarina Slama, Alex Ray, et al. 2022. Training language models to follow instructions with human feedback. *Advances in neural information processing systems*, 35:27730–27744.

Diana Pérez, Enrique Alfonseca, Pilar Rodríguez, Alfio Gliozzo, Carlo Strapparava, and Bernardo Magnini. 2005. About the effects of combining latent semantic analysis with natural language processing techniques for free-text assessment. *Revista signos*, 38(59):325–343.

Josef Perner, Susan R Leekam, and Heinz Wimmer. 1987. Three-year-olds' difficulty with false belief: The case for a conceptual deficit. *British journal of developmental psychology*, 5(2):125–137.

V Sanh. 2019. Distilbert, a distilled version of bert: smaller, faster, cheaper and lighter. *arXiv preprint arXiv:1910.01108*.

James WA Strachan, Dalila Albergo, Giulia Borghini, Oriana Pansardi, Eugenio Scaliti, Saurabh Gupta, Krati Saxena, Alessandro Rufo, Stefano Panzeri, Guido Manzi, et al. 2024. Testing theory of mind in large language models and humans. *Nature Human Behaviour*, pages 1–11.

David M Williamson, Robert J Mislevy, and Isaac I Bejar. 2006. *Automated scoring of complex tasks in computer-based testing*. Lawrence Erlbaum Associates Mahwah, NJ.

Heinz Wimmer and Josef Perner. 1983. Beliefs about beliefs: Representation and constraining function of wrong beliefs in young children's understanding of deception. *Cognition*, 13(1):103–128.

Changrong Xiao, Wenxing Ma, Sean Xin Xu, Kunpeng Zhang, Yufang Wang, and Qi Fu. 2024. From automation to augmentation: Large language models elevating essay scoring landscape. *arXiv preprint arXiv:2401.06431*.

Elaine Kit Ling Yeung, Ian A Apperly, and Rory T Devine. 2024. Measures of individual differences in adult theory of mind: A systematic review. *Neuroscience & Biobehavioral Reviews*, 157:105481.

Bigger But Not Better: Small Neural Language Models Outperform Large Language Models in Detection of Thought Disorder

Changye Li[1*] Weizhe Xu[1*] Serguei Pakhomov[2]

Ellen Bradley[3] Dror Ben-Zeev[1] Trevor Cohen[1]

[1]University of Washington
[2]University of Minnesota
[3]University of California, San Francisco
{changyel, xuweizhe}@uw.edu

Abstract

Disorganized thinking is a key diagnostic indicator of schizophrenia-spectrum disorders. Recently, clinical estimates of the severity of disorganized thinking have been shown to correlate with measures of how difficult speech transcripts would be for large language models (LLMs) to predict. However, LLMs' deployment challenges – including privacy concerns, computational and financial costs, and lack of transparency of training data – limit their clinical utility. We investigate whether smaller neural language models can serve as effective alternatives for detecting positive formal thought disorder, using the same sliding window based perplexity measurements that proved effective with larger models. Surprisingly, our results show that smaller models are more sensitive to linguistic differences associated with formal thought disorder than their larger counterparts. Detection capability declines beyond a certain model size and context length, challenging the common assumption of "bigger is better" for LLM-based applications. Our findings generalize across audio diaries and clinical interview speech samples from individuals with psychotic symptoms, suggesting a promising direction for developing efficient, cost-effective, and privacy-preserving screening tools that can be deployed in both clinical and naturalistic settings.

1 Introduction

With an estimated prevalence of 15.2 in 100,000 persons (McGrath et al., 2008), schizophrenia-spectrum disorders (SSDs) are debilitating conditions that can lead to impaired social and occupational functioning, and poor healthcare outcomes including early mortality (Laursen et al., 2014). Formal Thought Disorder (FTD) – a breakdown in the structure of an individual's thinking – is a diagnostic feature of schizophrenia (Kircher et al., 2018), and is recognized by observing speech that appears incoherent. Traditional evaluation of FTD relies on clinical interviews and standardized rating scales, which require extensive training and can be time-consuming. Natural language processing (NLP) methods have emerged as promising computational tools for automated evaluation of FTD. These data-driven approaches can systematically analyze linguistic patterns and discourse structure in patients' speech, offering objective quantitative measures of semantic coherence that correspond with clinical estimates of FTD severity (Elvevåg et al., 2007; Corcoran et al., 2020; Xu et al., 2021; Sarzynska-Wawer et al., 2021; Xu et al., 2022, *inter alia*).

Among the language impairments detectable by NLP methods, FTD represents a particularly complex set of disruptions in thought and speech organization. Patients with FTD exhibit distinct thinking patterns including tangentiality (gradual topic drift) and derailment (completely/partially unrelated thoughts), which may indicate a relative insensitivity to global discourse context (Kuperberg, 2010a,b). Previous psycholinguistic and neurolinguistic studies in *language comprehension* have shown an impaired ability to use global linguistic context (e.g., information from early in longer sentences) and relatively intact ability for local linguistic context (e.g., information from shorter sentences, local priming) in SSDs (Sitnikova et al., 2002; Swaab et al., 2013). With recent advances in neural language models (LMs) in particular, it is now possible to measure dependence upon global and local context in *language production* in SSDs. Recent work (Sharpe et al., 2024) suggests that disorganized speech can be characterized by comparing the global (probabilities estimated for observed text when including proximal and distal context) and local (probabilities estimated for observed speech when including proximal context only) lexical probabilities retrieved from GPT-3 (Brown

*Equal contribution

Proceedings of the 10th Workshop on Computational Linguistics and Clinical Psychology (CLPsych 2025), pages 90–105
May 3, 2025

et al., 2020) when applied to speech samples from SSDs patients and neurotypical demographically-matched controls. A key finding from this research is that models that include longer context appear better-equipped to recognize language from participants with SSDs, in accordance with prior work showing SSDs patients are less influenced by overall sentence context during text comprehension (Sitnikova et al., 2002).

Commercially-developed large language models (LLMs) such as GPT-3 are pre-trained on large text corpora to enhance their linguistic capabilities. However, their usage and deployment raise concerns in clinical settings. Healthcare applications require stringent privacy protections, yet commercial LLMs mostly operate through cloud-based APIs, requiring the sharing of patient data with third-party commercial services. Even services compliant with the Health Information Portability and Accountability Act (HIPAA) or European counterpart General Data Protection Regulation (GDPR), LLMs can present certain risks to patient privacy especially when used for health-related purposes outside of a healthcare environment (Marks and Haupt, 2023). In addition to privacy concerns, computational requirements and associated costs of accessing commercial LLMs may also restrict their application in clinical settings. Lastly, the proprietary nature of commercial LLMs and a lack of transparency regarding their training data make it difficult to investigate or mitigate sources of bias.

While these limitations present challenges, there is a pressing need to explore the potential of LLMs in healthcare settings. Healthcare systems continuously face significant workforce shortages, particularly in specialized areas requiring extensive training (Thomas et al., 2009; Butryn et al., 2017). Current diagnostic and monitoring approaches rely heavily on in-person evaluations, creating bottlenecks in patient care and limiting access to specialized services, especially in resource-constrained settings. LLMs, if properly implemented with appropriate privacy safeguards, could help address these challenges by facilitating the development of technology-assisted diagnostic and monitoring tools, potentially improving the efficiency and accessibility of healthcare services.

Beyond these obstacles, we posit that commercial LLMs' extensive exposure to diverse linguistic patterns drawn from the internet and other sources – while beneficial for their remarkable text generation capabilities – may paradoxically reduce their sensitivity to subtle linguistic differences. This hypothesis is supported by recent works suggesting that broad exposure to diverse linguistic data leads LLMs to prioritize general patterns over fine-grained linguistic sensitivity (Lee et al., 2024a; Cong, 2024; Wilson et al., 2023), potentially diminishing their sensitivity to subtle deviations characteristic of FTD. We hypothesize that smaller LMs, such as those in the Pythia suite from EleutherAI (Biderman et al., 2023), may exhibit enhanced sensitivity to these linguistic phenomena. These models, ranging from 70M to 12B parameters, are trained on identical public datasets in the same order. As they differ in size only, it is possible to assess the extent to which they respond to nuanced linguistic patterns differently with the constraints in their capacity. In contrast, larger models' potentially excessive capacity to model complex textual relationships may obscure these subtle linguistic markers beneath layers of broader contextual understanding learned from vast amounts of data.

LMs' sensitivity to linguistic manifestations can be measured with perplexity (PPL). PPL is an intrinsic measure used to evaluate the performance of language models on unseen data. The more different the input is from a LM's training data, the "harder" it is for the model to predict the next word, resulting in higher PPL. Therefore, it is reasonable to hypothesize that PPL may have some degree of diagnostic utility, as has been documented by prior work using PPL to evaluate cognitive impairment in Dementia of the Alzheimer's Type (Orimaye et al., 2018; Fritsch et al., 2019; Cohen and Pakhomov, 2020; Li et al., 2022, 2024, *inter alia*) and psychosis (Colla et al., 2022; He et al., 2024).

Building upon Sharpe et al. (2024)'s findings with LLMs, our study seeks to assess smaller LMs' sensitivity to linguistic patterns associated with *positive* FTD by analyzing PPLs derived from Pythia models (exemplifying smaller LMs) and LLaMA (Dubey et al., 2024) (exemplifying a LLM) across both monologue and conversational speech samples from individuals with psychotic symptoms and clinically diagnosed SSDs respectively, and evaluating their correlation with the corresponding clinical ratings.

The contributions of this work can be summarized as follows: a) we provide empirical evidence suggesting that smaller LMs are more sensitive to linguistic patterns associated with FTD; b) we demonstrate that the degree of sensitivity starts to decline after models exceed a threshold of a certain number of parameters, suggesting a diminishing relationship between model size and detection capability; and c) the sliding window PPL approach generalizes to both monologue and conversational speech samples of individuals with psychotic symptoms and clinically diagnosed SSDs respectively, suggesting its potential utility for screening and monitoring of SSDs in diverse clinical settings.[1]

2 Related Work

2.1 FTD in schizophrenia

Traditionally, FTD is evaluated through clinically administered rating scales such as the thought and language index (Liddle et al., 2002) or thought and language disorder (TALD) scale (Kircher et al., 2014) in research settings, which captures the full variety of FTD phenomenology including subjective experiences. It can also be more conveniently evaluated through self-reporting scales (Barrera et al., 2008). However, there are inherent problems associated with each approach: administering the clinical scales is time-consuming and requires specific training and expertise. Additionally, even when the required expertise is readily available, the clinical assessment only provides intermittent measures during office visits, making it difficult to paint a continuous picture in a more ecological setting. On the other hand, the self-reporting scale lacks objectivity because each patient may have subjective views on the scale's severity settings, and the ability to self-appraise may be impaired in psychosis. These inherent problems and the advancement of computational technologies have inspired the usage of NLP methods to evaluate and quantify the severity of FTD.

2.2 Assessing FTD and SSDs with NLP methods

Advancements in computational systems have introduced innovative methods for automated FTD assessment. A seminal approach by Foltz et al. (1998) utilizes distributional similarity, specifically measuring semantic relatedness between consecutive text segments using latent semantic analysis (LSA) (Landauer and Dumais, 1997) to measure coherence, providing a proxy that was later used to quantify the degree FTD. This method's diagnostic utility was demonstrated by Elvevåg et al. (2007), who found significant differences in automated coherence metrics when comparing individuals with schizophrenia to healthy controls, as well as among patients with varying levels of thought disorder.

Building on this work, a subsequent study integrated LSA-based coherence metrics into a machine learning classifier that accurately predicted psychosis onset in a small sample of at-risk youth, achieving perfect leave-one-out cross-validation accuracy (Bedi et al., 2015). An adapted version maintained 83% accuracy in predicting psychosis onset in a larger, independent dataset (Corcoran et al., 2018). More recently, neural word embeddings (Mikolov et al., 2013), which represent words as vectors derived from neural networks trained to predict nearby words, have been explored as an alternative to LSA for coherence analysis. Similarity metrics from these embeddings showed promising results in aligning with clinical assessments of thought disorder (Just et al., 2019, 2020).

With advances in NLP methods, recent studies have used sentence embeddings from BERT (Bidirectional Encoder Representations from Transformers) (Devlin et al., 2019) to identify coherence differences between transcripts from individuals with SSDs and those from healthy controls (Tang et al., 2021). As a transformer-based model, BERT generates context-specific representations of tokens by dynamically incorporating information from surrounding words, unlike LSA or neural word embeddings, which rely on static word representations derived from all of the contexts a word is observed in during training. Prior research also introduced methods for assessing global coherence – estimating the relationship between a unit of text and the overarching theme of a text – using these methods to improve coherence evaluation in automatic speech recognition by extracting time series features for machine learning (Xu et al., 2021, 2022).

With the emergence of autoregressive LMs, some recent studies (Palaniyappan et al., 2023; Fradkin et al., 2023; Sharpe et al., 2024) have examined the assessment of psychosis using such models to demonstrate that LMs can be utilized with in-silico experimental research to gain better understanding of the linguistic manifestation of FTD. In contrast to BERT, which is a bidirectional model that utilizes tokens on both sides of a target token for prediction, autoregressive LMs are designed to predict only the next token in a sequence. While BERT-derived representations are highly effective for estimating semantic relatedness, autoregressive LMs are specifically optimized for generating co-

[1] Our code is publicly available on `https://github.com/LinguisticAnomalies/small-lm-sliding-windows`

herent and fluent sequences of text, offering potential for developing alternative approaches to FTD evaluation. However, these approaches have primarily relied on such models without exploring how model size and granular context windows affect sensitivity to linguistic manifestations of FTD.

3 Methods

3.1 Data

AVH Dataset Speech monologue samples from native English speaking participants who experienced auditory verbal hallucinations (AVH) using a smartphone application were collected during the course of a previous study (Ben-Zeev et al., 2020). Participants experiencing AVH were recruited in-person and online, and prompted to describe their experiences of AVH and anything else they would like to share or think would be helpful for the research team to know. Informed consent from participants was obtained through a rigorous procedure involving triple confirmations from a screening questionnaire. The study was approved by the Institutional Review Boards (IRB) of the University of Washington and Dartmouth College. Two annotators labeled the manual transcripts of the audio recordings for their degree of incoherence based on the TALD scale, using the construct of derailment. The TALD score ranges from 0-4 and represents greater incoherence as the score increases. The inter-rater agreement between annotators was 0.71, as measured by weighted Kappa. This set contained samples with a mean TALD score of 1.18 and a standard deviation of 0.83. We select 310 recordings that: a) have manual transcriptions; and b) are annotated with TALD. The transcript-level demographic information for this dataset is summarized in Table A.1 in the Appendix.

Clinical Interview Dataset This set contains semi-structured clinical interviews of San Francisco Bay Area male outpatients diagnosed with SSDs participating in a study of oxytocin conducted independently at University of California, San Francisco (UCSF) (Bradley et al., 2024). All participants are provided with written informed consent and study protocols were approved by the IRB at the UCSF. Following a prior work (Poole et al., 2000), the clinical assessments were conducted by trained raters in the form of a composite score combining the conceptual disorganization item (ranging from 1-7 with increasing severity) (Kay et al., 1987) from the Positive and Negative Syndrome Scale (PANSS) and the incoherent speech item (ranging from 0-5 with increasing severity) from the Comprehensive Assessment of Symptom and History (CASH) (Andreasen et al., 1992) to supplement the disorganized symptom subscale and the measure of suicidality. To avoid any potential confusion between these terms referring to different types of ratings, in the remainder of this paper we will refer this score as composite PANSS. We use manually transcribed interviews from 39 participants with corresponding composite PANSS between 2 and 8 (in the range of 0-12), with a mean of 3.36 and a standard deviation of 1.80. The transcript-level demographic information for this dataset is summarized in Table A.2 in Appendix.

3.2 Language models

Pythia is the first LLM suite deliberately designed to enable scientific research on LLMs. The Pythia suite offers pre-trained decoder-only autoregressive LMs ranging from 70M to 12B parameters. The Pythia suite is trained on the Pile corpus (Gao et al., 2020), which is a publicly available and curated collection of English language. In particular, we select Pythia checkpoints (70m, 160m, 410m, 1b, 1.4b, 2.8b, 6.9b, and 12b in parameter size) that are pre-trained on a *deduplicated* Pile corpus containing approximately 207B tokens. We select these checkpoints as deduplication has demonstrated its benefits in LLM training process (Lee et al., 2022). The Pythia suite largely follows the architecture and hyperparameters of GPT-3, but differs in several aspects: a) it uses fully dense attention layers; b) it is pre-trained using Flash Attention (Dao et al., 2022) for improved device throughput; and c) it uses rotary positional embeddings (Su et al., 2024) for a flexible mechanism to encode positional information.

We also compare Pythia suite with locally-hosted LLaMA-3.1-405b (Dubey et al., 2024) model, which is quantized with 4-bit precision using ExLlamaV2[2] (prior work indicates that quantization does not significantly degrade model performance (Lee et al., 2024b)). As initial experimentation in previous work showed comparable results to those obtained with the base model (without instruction tuning), we use an instruction-tuned model, hosted locally on a secure server.

[2] `https://github.com/turboderp-org/exllamav2`

3.3 Global and sliding window PPLs

We compute PPL for a transcript using two approaches: a) a global PPL that evaluates the full transcript as a single sequence; and b) a local PPL using sliding windows of 8, 16, 32, 64, and 128 for the Pythia suite, and a sliding window of 64 for the LLaMA model, as prior work indicates that restricting to a short input (e.g., context length of 128) can substantially improve the performance of LMs (Press et al., 2021). The sliding window is defined as a window of a corresponding number of tokens moved sequentially through the transcript. PPL is calculated for each window position as it shifts one token at a time until reaching the end of the transcript. If the transcript is shorter than the designated sliding window, then we calculate the global PPL for the transcript instead. As window size increases, the sliding window PPL approach allows the model to have more dynamic context when making each prediction, resulting in a more accurate approximation of the fully-factorized PPL (i.e., the global PPL). This can be particularly useful for evaluating spontaneous speech where the context is more fragmented than with read speech (Auer, 2009; Shriberg, 2001; Agmon et al., 2023; Wang et al., 2010). To generate a transcript-level measure, we use the maximum and the averaged PPL across the estimated sliding window PPLs, in addition to the global PPL for each transcript. For each measure, we compare the Spearman ρ with the TALD and composite PANSS scores for the AVH and clinical interview datasets respectively.

We opt to use maximum sliding window PPL as our primary transcript-level metric for detecting incoherent language. The rationale for this choice is evident in the distinct separation between transcripts that exhibit *mild* derailment (with TALD derailment < 3, labeled as 0) and those that exhibit *severe* derailment (with TALD derailment ≥ 3, labeled as 1) in the AVH dataset (Figure A.1 in Appendix). Transcripts rated with TALD ≥ 3 consistently exhibit higher maximum sliding window PPL spikes across different model sizes (particularly visible with the sliding window length of 64), while transcripts rated below this threshold maintain relatively stable, lower PPL patterns. We observed a similar trend in Figure A.2 in Appendix, the variation of PPL spikes across different severities of composite PANSS, suggesting that maximum sliding window PPL reflects disorganized speech.[3]

4 Results

4.1 Global PPL as a proxy for FTD-related linguistic patterns

As illustrated in Figure 1, smaller Pythia models (Pythia-70m and Pythia-160m) consistently exhibited higher global PPLs compared to their larger counterparts across both the AVH and clinical interview datasets. Larger models (6.9b and 12b parameters) tended to cluster together at lower PPLs, suggesting diminishing effects on PPL estimation as model size increases. We observed minimal correlation ($\rho < 0.01$) between global PPL and TALD (in the AVH set), and this was statistically insignificant across all model sizes. While correlations between global PPL and composite PANSS (in the clinical interview set) were present (Spearman's ρ between 0.20 and 0.39), statistical significance was achieved only for the larger models, including Pythia-2.8b, Pythia-6.9b, and Pythia-12b, at $\alpha = 0.1$.

4.2 Sliding window PPL performance

4.2.1 The AVH dataset

Model	Sliding windows				
	8	16	32	64	128
70m	0.366***	0.375***	0.427***	**0.440*** **	0.370***
160m	0.369***	0.360***	0.426***	**0.451*** **	0.378***
410m	0.347***	0.336***	0.430***	**0.458*** **	0.378***
1b	0.329***	0.328***	0.431***	**0.458*** **	0.367***
1.4b	0.331***	0.305***	0.423***	**0.486*** **	0.388***
2.8b	0.315***	0.316***	0.435***	**0.464*** **	0.365***
6.9b	0.319***	0.310***	0.420***	**0.475*** **	0.370***
12b	0.317***	0.307***	0.421***	**0.470*** **	0.372***
LLaMA	–	–	–	0.457***	–

$^{***}p < 0.01$, $^{**}p < 0.05$, $^{*}p < 0.1$

Table 1: The AVH dataset Spearman's ρ between the *maximum* sliding window PPL and TALD across model size. **Bold** indicates the highest ρ for a model.

As shown in Table 1, all correlations between *maximum* sliding window PPL and TALD scores were statistically significant (p-value < 0.01) across all model and sliding window sizes. The strongest correlations consistently occurred with a 64-token sliding window, with coefficients peaking at 0.486 for the 1.4b model, and remaining moderately correlated with TALD scores for all

[3]Our experiments are conducted on 3 H100 GPUs.

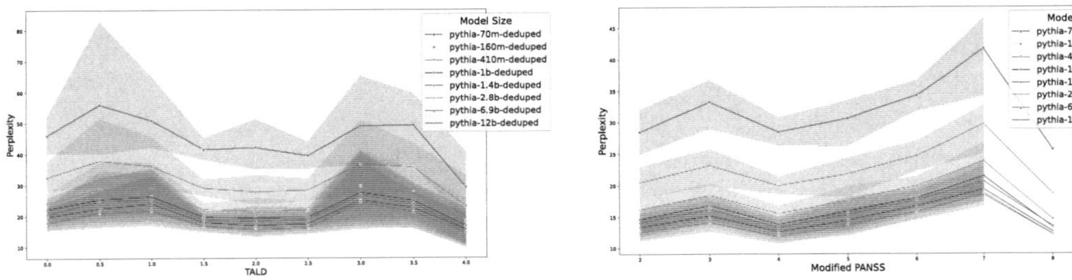

(a) Global PPLs of transcripts in the AVH dataset

(b) Global PPLs of transcripts in the clinical interview dataset

Figure 1: Global PPLs estimated by the Pythia suite. The shaded area represents 95% confidence intervals of the global PPLs estimated from a pre-trained Pythia model.

model variants. Interestingly, the 4-bit quantized LLaMA 405b model did not outperform the Pythia suite, attaining a lower Spearman ρ of 0.457 on the 64-token sliding window.

Table A.3 in Appendix shows similar patterns of correlation between *averaged* sliding window PPL for a transcript and TALD scores. While all correlations remain significant (p-value < 0.01) across all model sizes and sliding windows, the correlation coefficients are generally lower compared to those for maximum sliding window PPLs. The 64-token sliding window again emerged as the optimal configuration, with correlation coefficients ranging from 0.202 to 0.251. The 1.4b model achieved the strongest correlation ($\rho = 0.251$), followed closely by the 6.9b model ($\rho = 0.249$). The LLaMA model achieved the highest correlation ($\rho = 0.371$) for the sliding window of 64 tokens.

4.2.2 The clinical interview dataset

Model	Sliding windows				
	8	16	32	64	128
70m	0.265	**0.482*****	0.338**	0.356**	0.344**
160m	0.414***	0.433***	0.316*	0.354***	0.325**
410m	0.385*	0.433***	0.316*	0.354**	0.325**
1b	0.415***	0.352**	0.380**	0.410***	0.313*
1.4b	0.458***	0.370***	0.382**	0.418***	0.326**
2.8b	0.425***	0.385**	0.369**	**0.428*****	0.348***
6.9b	**0.478*****	0.352**	0.394**	0.414***	**0.368****
12b	0.441***	0.313*	**0.404****	0.412***	0.357*
LLaMA	–	–	–	0.249	–

$^{***}p < 0.01$, $^{**}p < 0.05$, $^{*}p < 0.1$

Table 2: The clinical interview dataset Spearman's ρ between the *maximum* sliding window PPL and modified PANSS across model size. **Bold** indicates the highest ρ for a model.

The patterns observed in Table 2 of the *maximum* sliding window PPLs on clinical interview dataset vary compared to those from audio diary data shown in Table 1. Pythia-70m with a sliding window of 16 has the highest correlation ($\rho = 0.482$, p-value < 0.01), while the Pythia-6.9b model shows comparable performance with a sliding window of 8 ($\rho = 0.478$, p-value < 0.01). Unlike the patterns with TALD, the sliding window size of 64 tokens was not optimal across all models, though it did yield strong correlations for several model sizes, particularly with the Pythia-2.8b model ($\rho = 0.428$, p-value < 0.01). Similarly, we also observed that LLaMA achieved the lowest and insignificant Spearman ρ with a 64-token sliding window.

Table A.4 in Appendix shows more moderate relationships for *averaged* sliding window PPLs in the clinical interview data, with correlation coefficients ranging from 0.248 to 0.360. The Pythia-1.4b model demonstrated consistently strong performance across all window sizes, achieving the highest correlation coefficients of all models, with its peak at a sliding window of 64 ($\rho = 0.360$, p-value < 0.05). LLaMA achieved the lowest Spearman ρ in the sliding window of 64, despite this being moderately significant.

4.3 The comparison of model sizes and sliding windows

For TALD correlations (Table 1 and Table A.3 in Appendix), larger sliding window sizes (32 and 64) consistently showed stronger correlation as model size increased. However, this trend was less evident with smaller sliding window sizes (8 and 16), where the correlation coefficients remained relatively stable across model sizes. In contrast, the composite PANSS correlations (Table 2 and Table A.4 in Appendix) exhibited a different pattern:

smaller sliding window sizes (8 and 16) showed more variation across model sizes with maximum PPLs, with correlation coefficients fluctuating. For example, with the sliding window of 8, the correlation coefficients ranged from $\rho = 0.265$ to $\rho = 0.478$. The correlation coefficients with averaged PPLs show more consistent behavior, as they gradually increase up to the 1.4b model across all sliding window sizes before they plateau or slightly decline with in larger models.

As can be observed in Table 1, which shows the correlations between maximum PPL for a transcript and the TALD on the AVH dataset, all models exhibit a consistent pattern where correlation coefficients generally increase with a sliding window of 8 and 16, peak at a sliding window of 64, and then decrease with a sliding window of 128. There is a similar trend in Table A.3, but with more moderate increases and decreases. In contrast, there are more variable patterns with the clinical interview results shown in Table 2. Smaller models (70m-410m) tended to achieve peak correlations at smaller window sizes, while larger models show more distributed peaks across different window sizes. The correlation between the averaged sliding window PPL and composite PANSS shows the most consistent pattern across window sizes, as is particularly evident with the 1.4b model, which maintains relatively stable correlations ranging from $\rho = 0.272$ to $\rho = 0.360$ across all window sizes. Notably, a sliding window of 128 consistently produced the weakest correlations in both datasets, suggesting that larger window size may dampen the PPL response to local patterns as compared with medium-sized windows. Interestingly, there is also a general trend of diminishing effects of sliding window size with both datasets, with the correlation coefficients declining with larger models (e.g., billions of parameter size) at the same sliding window size.

5 Discussion

Our key findings are as follows. First, our results suggest LM PPL can potentially serve as an objective computational marker for capturing subtle linguistic patterns associated with FTD. This aligns with previous studies indicating that such abnormal linguistic patterns manifest in ways that can be quantitatively measured (Colla et al., 2022; He et al., 2024; Xu et al., 2021, 2022; Sharpe et al., 2024). Second, our results extend Sharpe et al. (2024)'s work by examining fine-grained sliding window PPLs to capture semantic variations across longer sequences using models ranging from 70m to 405b parameters for two model families. Our results indicate that model size does not necessarily correlate linearly with its capability for detecting FTD-related linguistic manifestations. Furthermore, our findings suggest that small/medium-sized sliding windows consistently demonstrate optimal performance across different model sizes, indicating an effective balance point between clinical utility and computational efficiency. That performance declines with larger windows may suggest the approximate range within which contextual inconsistencies manifest in FTD. These findings collectively suggest that calibrated smaller LMs can be at least as effective as their larger counterparts, offering practical advantages for real-world deployment while maintaining clinically-validated and robust performance in detecting FTD-related language differences.

Our work also demonstrates a more nuanced relationship between window size and the linguistic manifestations of FTD. While prior work (Sharpe et al., 2024) using GPT-3 indicates that differences in lexical probability (i.e., the intermediate products of PPLs) differ more between cases and controls with larger context windows (i.e., up to 50 tokens) for FTD, our work provides a more granular characterization of optimal sliding window sizes for alignment with human ratings in both monologue and conversational speech samples. This is particularly important for detecting linguistic inconsistencies that span across longer text segments, an aspect of comprehension that prior research suggests may be selectively impaired in people with SSDs (Kuperberg, 2010a,b; Sharpe et al., 2024). In addition, it reveals diminishing correlation coefficients at larger model and sliding window size, suggesting an upper bound to the utility of increasing both context window size and model size. Interestingly, the clinical interview dataset shows more variable optimal sliding window sizes across different model scales, with smaller models performing best at shorter windows and larger models showing distributed peaks across different window sizes. These varied patterns suggest that the manifestation of FTD – particularly in conversational language – may operate at multiple scales, rather than simply becoming more apparent with larger contexts. This in turn suggests that such LM-based methods may require calibrated combinations of both model size and context window, rather than simply maximizing either dimension.

Our findings demonstrate the generalizability of PPL-based computational and automated assessment across both monologue (AVH) and conversational data (clinical interview), suggesting that changes in language associated with FTD can be effectively captured regardless of the communicative setting. Spontaneous speech presents unique complexities due to its impromptu nature, where speakers have minimal time to organize their thoughts. The challenges include a lack of clear syntactic boundaries (Auer, 2009; Shriberg, 2001; Agmon et al., 2023; Wang et al., 2010), complex interaction of linguistic demands due to mental states (Menn and Obler, 1989; Caplan and Hanna, 1998), and context sensitivity. They make it particularly challenging for generalizable computational analysis. However, our results show that PPL-based measures can effectively operate within these complexities, yielding statistically significant correlations with human ratings across both monologue and conversational datasets. This capability to perform consistently across different communication contexts is important for clinical applications, where assessment tools may need to maintain reliability across various real-world scenarios. The consistency of our results across both data sources indicates the potential of sliding-window based LM perplexity as an automated and computational assessment tool.

With respect to model size, 4-bit quantized LLaMA 405b, despite its significantly larger scale and strong performance on open-domain tasks (Lee et al., 2024b), consistently underperformed compared to smaller Pythia models. This finding supports our hypothesis regarding the potential advantages of smaller LMs in detecting subtle linguistic patterns associated with FTD, though it remains to be determined whether this advantage is attributable to model size or a relatively constrained amount of training data. Larger LMs, with their extensive pre-training on vast corpora of text of uncertain provenance, may find the subtle linguistic patterns that characterize FTD more predictable. In contrast, smaller LMs' more limited exposure to coherent language patterns and/or constrained capacity (as proxied by parameter size) may paradoxically enhance their sensitivity to linguistic patterns associated with FTD. This suggests that the relationship between model scale and clinical assessment capability is not strictly linear (i.e., bigger is not necessarily better), and that optimal performance may be achieved by models that maintain adequate linguistic competence while remaining sensitive to deviations from typical language patterns. A further advantage of the Pythia suite is that their training data are publicly available, and therefore amenable to analyses to identify sources of biased assessment, such as the absence of training data reflecting dialectical variation characteristic of particular population groups. These findings and observations collectively suggest that pursuing ever-larger models may not necessarily yield better clinical assessment capabilities and utilities.

Our analysis across multiple model sizes provides empirical guidance for sliding window size selection in clinical practice. The finding that small- to medium-sized sliding windows (typically 16 to 64 tokens) consistently demonstrate optimal performance across different model sizes suggests an effective range for practical implementation. This observation is consistent with previous studies demonstrating that linguistic inconsistencies manifest as local coherence disruptions (Sitnikova et al., 2002; Swaab et al., 2013; Kuperberg, 2010a,b). The observed performance decline with larger windows (>128 tokens) further supports this understanding. Notably, the optimal sliding window sizes remain consistent across both shorter monologues (≈ 100 tokens) and longer clinical interviews (>1000 tokens), suggesting that the linguistic manifestation of FTD operates at a fragmented level independent of overall discourse length or interaction type. This pattern suggests that aspects of FTD may be best characterized in intermittent steps rather than as global narrative incoherence. Additionally, the sliding window size sensitivity remains remarkably consistent across model scales from 70m to 12b, suggesting that PPL, as a computational marker, can effectively capture such linguistic manifestations, providing a compelling evidence for the context-sensitive nature of FTD and its variable expression across different communicative demands.

Our findings suggest that PPL derived from smaller LMs with granular sliding windows offer promising clinical utility in addition to existing assessment methods. Furthermore, The reduced computational requirements of our approach also makes it particularly suitable for resource-constrained settings, potentially enabling automated FTD screening in underserved communities. These models' ability to detect subtle linguistic manifestations of FTD opens several promising application pathways in clinical practice. Most notably, the efficiency of smaller LMs (70m-410m parameters) enables privacy-preserving, on-device processing that could streamline the mental health monitoring and early intervention. For example, these lightweight models could be integrated into telehealth platforms to analyze discourse during remote psychiatric consultations in real-time, providing clinicians with immediate linguistic computational markers while ensuring all patient data remains on local devices. In ambient monitoring scenarios, these models could be deployed on smartphones to periodically assess everyday conversation with participants' prior consent, creating longitudinal datasets that track subtle changes in FTD, enabling the comparison of cross-sectional linguistic patterns to identify preliminary warning signs that might surf unnoticed. The resulting data could also used for near-real-time flagging of warning signs, which could be shared with their clinical teams to enable time-sensitive interventions that may help prevent further deterioration (Ben-Zeev et al., 2017).

While our token-level sliding window PPL method demonstrates promising results, we acknowledge that certain sentence-level proximity-based methods (Xu et al., 2022) achieved comparable or higher correlations. However, the sliding window PPL method is responsive to linguistic inconsistencies at varying granularities that complement existing methods, potentially capturing dynamic aspects of FTD that may be missed by static sentence-level measures.

6 Conclusion

We presented experimental findings consistent with prior work suggesting sliding-window PPL as an efficient measurement for linguistic patterns associated with FTD. Surprisingly, our findings suggest that smaller LMs with calibrated sliding window sizes, are more sensitive to such linguistic manifestations. The comparable effectiveness of smaller models opens new possibilities for implementing automated and computational language assessment tools in resource-constrained clinical settings while remaining cost-efficient and privacy-preserving.

Limitations

The work presented here has several limitations. All participants represented in both data sets are English speakers, and it remains unclear the extent to which our findings apply to other languages. Our analysis relied on transcribed speech data, which may not fully capture the nuances of spoken language, including prosody, pauses, and other paralinguistic features that could be clinically relevant (for a related review, see Ehlen et al. (2023)). While our findings demonstrate correlations between PPL and human ratings, these ratings do not constitute clinical diagnoses, which would be needed for case/control comparisons. Furthermore our analysis does not account for potential confounding variables – such as age, gender, origin and socioeconomic status – which may influence language patterns. While smaller LMs show promising results, we have not yet established clear clinical thresholds that would be necessary for diagnosis, or assessed the utility of measurements over time as indicators of change in clinical status. We also note that the severity scores for both datasets are relatively low on average, and that datasets with more representation of severe FTD may be needed to establish optimal parameter settings in this context. Our study focused specifically on positive FTD, a key diagnostic feature for SSDs. Therefore, the extent to which sliding window PPL is responsive to linguistic manifestations of other conditions remains to be established. Future work to address these limitations will be required to reach the potential of these methods for clinical deployment. Finally, while we included a larger language model with a 64-token sliding window, including additional sliding window sizes with LLaMA would make for a more comprehensive analysis that is left for future work.

Acknowledgment

This work was supported by National Institute of Mental Health under award number R01MH112641, U01MH13590, and the Department of Veterans Affairs under award number IK1 CX002092-02. We also want to thank the reviewers for their insightful critiques and suggestions that helped improve this manuscript.

References

Galit Agmon, Manuela Jaeger, Reut Tsarfaty, Martin G Bleichner, and Elana Zion Golumbic. 2023. "um…, it's really difficult to… um… speak fluently": Neural tracking of spontaneous speech. *Neurobiology of Language*, 4(3):435–454.

Nancy C. Andreasen, Michael Flaum, and Stephan Arndt. 1992. The comprehensive assessment of symptoms and history (cash): An instrument for assessing diagnosis and psychopathology. *Archives of General Psychiatry*, 49.

Peter Auer. 2009. On-line syntax: Thoughts on the temporality of spoken language. *Language Sciences*, 31(1):1–13.

Alvaro Barrera, Peter J. McKenna, and German E. Berrios. 2008. Two new scales of formal thought disorder in schizophrenia. *Psychiatry Research*, 157.

Gillinder Bedi, Facundo Carrillo, Guillermo A. Cecchi, Diego Fernández Slezak, Mariano Sigman, Natália B. Mota, Sidarta Ribeiro, Daniel C. Javitt, Mauro Copelli, and Cheryl M. Corcoran. 2015. Automated analysis of free speech predicts psychosis onset in high-risk youths. *npj Schizophrenia*.

Dror Ben-Zeev, Rachel Brian, Rui Wang, Weichen Wang, Andrew T Campbell, Min SH Aung, Michael Merrill, Vincent WS Tseng, Tanzeem Choudhury, Marta Hauser, et al. 2017. Crosscheck: Integrating self-report, behavioral sensing, and smartphone use to identify digital indicators of psychotic relapse. *Psychiatric rehabilitation journal*, 40(3):266.

Dror Ben-Zeev, Benjamin Buck, Ayesha Chander, Rachel Brian, Weichen Wang, David Atkins, Carolyn J Brenner, Trevor Cohen, Andrew Campbell, and Jeffrey Munson. 2020. Mobile rdoc: Using smartphones to understand the relationship between auditory verbal hallucinations and need for care. *Schizophrenia Bulletin Open*.

Stella Biderman, Hailey Schoelkopf, Quentin Gregory Anthony, Herbie Bradley, Kyle O'Brien, Eric Hallahan, Mohammad Aflah Khan, Shivanshu Purohit, USVSN Sai Prashanth, Edward Raff, et al. 2023. Pythia: A suite for analyzing large language models across training and scaling. In *International Conference on Machine Learning*, pages 2397–2430. PMLR.

Ellen R. Bradley, Jake Portanova, Josh D. Woolley, Benjamin Buck, Ian S. Painter, Michael Hankin, Weizhe Xu, and Trevor Cohen. 2024. Quantifying abnormal emotion processing: A novel computational assessment method and application in schizophrenia. *Psychiatry Research*, 336:115893.

Tom Brown, Benjamin Mann, Nick Ryder, Melanie Subbiah, Jared D Kaplan, Prafulla Dhariwal, Arvind Neelakantan, Pranav Shyam, Girish Sastry, Amanda Askell, et al. 2020. Language models are few-shot learners. *Advances in neural information processing systems*, 33:1877–1901.

Tracy Butryn, Leah Bryant, Christine Marchionni, and Farhad Sholevar. 2017. The shortage of psychiatrists and other mental health providers: causes, current state, and potential solutions. *International Journal of Academic Medicine*, 3(1):5–9.

David Caplan and Joy E. Hanna. 1998. Sentence production by aphasic patients in a constrained task. *Brain and Language*, 63(2):184–218.

Trevor Cohen and Serguei Pakhomov. 2020. A tale of two perplexities: Sensitivity of neural language models to lexical retrieval deficits in dementia of the Alzheimer's type. In *Proceedings of the 58th Annual Meeting of the Association for Computational Linguistics*, pages 1946–1957, Online. Association for Computational Linguistics.

Davide Colla, Matteo Delsanto, Marco Agosto, Benedetto Vitiello, and Daniele P Radicioni. 2022. Semantic coherence markers: The contribution of perplexity metrics. *Artificial Intelligence in Medicine*, 134:102393.

Yan Cong. 2024. Manner implicatures in large language models. *Scientific Reports*, 14(1):29113.

Cheryl M. Corcoran, Facundo Carrillo, Diego Fernández-Slezak, Gillinder Bedi, Casimir Klim, Daniel C. Javitt, Carrie E. Bearden, and Guillermo A. Cecchi. 2018. Prediction of psychosis across protocols and risk cohorts using automated language analysis. *World Psychiatry*.

Cheryl M. Corcoran, Vijay A. Mittal, Carrie E. Bearden, Raquel E. Gur, Kasia Hitczenko, Zarina Bilgrami, Aleksandar Savic, Guillermo A. Cecchi, and Phillip Wolff. 2020. Language as a biomarker for psychosis: A natural language processing approach. *Schizophrenia Research*, 226:158–166.

Tri Dao, Dan Fu, Stefano Ermon, Atri Rudra, and Christopher Ré. 2022. Flashattention: Fast and memory-efficient exact attention with io-awareness. *Advances in Neural Information Processing Systems*, 35:16344–16359.

Jacob Devlin, Ming-Wei Chang, Kenton Lee, and Kristina Toutanova. 2019. BERT: Pre-training of deep bidirectional transformers for language understanding. In *Proceedings of the 2019 Conference of the North American Chapter of the Association for Computational Linguistics: Human Language Technologies, Volume 1 (Long and Short Papers)*, pages 4171–4186, Minneapolis, Minnesota. Association for Computational Linguistics.

Abhimanyu Dubey, Abhinav Jauhri, Abhinav Pandey, Abhishek Kadian, Ahmad Al-Dahle, Aiesha Letman, Akhil Mathur, Alan Schelten, Amy Yang, Angela Fan, et al. 2024. The llama 3 herd of models. *arXiv preprint arXiv:2407.21783*.

Felicitas Ehlen, Christiane Montag, Karolina Leopold, and Andreas Heinz. 2023. Linguistic findings in persons with schizophrenia—a review of the current literature. *Frontiers in Psychology*, 14.

Brita Elvevåg, Peter W. Foltz, Daniel R. Weinberger, and Terry E. Goldberg. 2007. Quantifying incoherence in speech: An automated methodology and novel application to schizophrenia. *Schizophrenia Research*.

Peter W. Foltz, Walter Kintsch, and Thomas K Landauer. 1998. The measurement of textual coherence with latent semantic analysis. *Discourse Processes*.

Isaac Fradkin, Matthew M. Nour, and Raymond J. Dolan. 2023. Theory-driven analysis of natural language processing measures of thought disorder using generative language modeling. *Biological Psychiatry: Cognitive Neuroscience and Neuroimaging*, 8(10):1013–1023. Natural Language Processing in Psychiatry and Clinical Neuroscience Research.

Julian Fritsch, Sebastian Wankerl, and Elmar Nöth. 2019. Automatic diagnosis of alzheimer's disease using neural network language models. In *ICASSP 2019-2019 IEEE International Conference on Acoustics, Speech and Signal Processing (ICASSP)*, pages 5841–5845. IEEE.

Leo Gao, Stella Biderman, Sid Black, Laurence Golding, Travis Hoppe, Charles Foster, Jason Phang, Horace He, Anish Thite, Noa Nabeshima, et al. 2020. The pile: An 800gb dataset of diverse text for language modeling. *arXiv preprint arXiv:2101.00027*.

Rui He, Claudio Palominos, Han Zhang, Maria Francisca Alonso-Sánchez, Lena Palaniyappan, and Wolfram Hinzen. 2024. Navigating the semantic space: Unraveling the structure of meaning in psychosis using different computational language models. *Psychiatry Research*, 333:115752.

Sandra Just, Erik Haegert, Nora Kořánová, Anna-Lena Bröcker, Ivan Nenchev, Jakob Funcke, Christiane Montag, and Manfred Stede. 2019. Coherence models in schizophrenia. In *Proceedings of the Sixth Workshop on Computational Linguistics and Clinical Psychology*, pages 126–136, Minneapolis, Minnesota. Association for Computational Linguistics.

Sandra A. Just, Erik Haegert, Nora Kořánová, Anna Lena Bröcker, Ivan Nenchev, Jakob Funcke, Andreas Heinz, Felix Bermpohl, Manfred Stede, and Christiane Montag. 2020. Modeling incoherent discourse in non-affective psychosis. *Frontiers in Psychiatry*.

S. R. Kay, A. Fiszbein, and L. A. Opler. 1987. The positive and negative syndrome scale (panss) for schizophrenia. *Schizophrenia Bulletin*, 13.

Tilo Kircher, Henrike Bröhl, Felicitas Meier, and Jennifer Engelen. 2018. Formal thought disorders: from phenomenology to neurobiology. *The Lancet Psychiatry*, 5(6):515–526.

Tilo Kircher, Axel Krug, Mirjam Stratmann, Sayed Ghazi, Christian Schales, Michael Frauenheim, Lena Turner, Paul Fährmann, Tobias Hornig, Michael Katzev, Michael Grosvald, Rüdiger Müller-Isberner, and Arne Nagels. 2014. A rating scale for the assessment of objective and subjective formal thought and language disorder (tald). *Schizophrenia Research*.

Gina R Kuperberg. 2010a. Language in schizophrenia part 1: an introduction. *Language and linguistics compass*, 4(8):576–589.

Gina R Kuperberg. 2010b. Language in schizophrenia part 2: What can psycholinguistics bring to the study of schizophrenia... and vice versa? *Language and linguistics compass*, 4(8):590–604.

Thomas K Landauer and Susan T Dumais. 1997. A solution to plato's problem: The latent semantic analysis theory of acquisition, induction, and representation of knowledge. *Psychological review*, 104(2):211.

Thomas Munk Laursen, Merete Nordentoft, and Preben Bo Mortensen. 2014. Excess early mortality in schizophrenia. *Annual review of clinical psychology*, 10(1):425–448.

Eun-Kyoung Rosa Lee, Sathvik Nair, and Naomi Feldman. 2024a. A psycholinguistic evaluation of language models' sensitivity to argument roles. In *Findings of the Association for Computational Linguistics: EMNLP 2024*, pages 3262–3274, Miami, Florida, USA. Association for Computational Linguistics.

Jemin Lee, Sihyeong Park, Jinse Kwon, Jihun Oh, and Yongin Kwon. 2024b. A comprehensive evaluation of quantized instruction-tuned large language models: An experimental analysis up to 405b. *arXiv preprint arXiv:2409.11055*.

Katherine Lee, Daphne Ippolito, Andrew Nystrom, Chiyuan Zhang, Douglas Eck, Chris Callison-Burch, and Nicholas Carlini. 2022. Deduplicating training data makes language models better. In *Proceedings of the 60th Annual Meeting of the Association for Computational Linguistics (Volume 1: Long Papers)*, pages 8424–8445, Dublin, Ireland. Association for Computational Linguistics.

Changye Li, David Knopman, Weizhe Xu, Trevor Cohen, and Serguei Pakhomov. 2022. GPT-D: Inducing dementia-related linguistic anomalies by deliberate degradation of artificial neural language models. In *Proceedings of the 60th Annual Meeting of the Association for Computational Linguistics (Volume 1: Long Papers)*, pages 1866–1877, Dublin, Ireland. Association for Computational Linguistics.

Changye Li, Zhecheng Sheng, Trevor Cohen, and Serguei Pakhomov. 2024. Too big to fail: Larger language models are disproportionately resilient to induction of dementia-related linguistic anomalies. In *Findings of the Association for Computational Linguistics: ACL 2024*, pages 6363–6377, Bangkok, Thailand. Association for Computational Linguistics.

Peter F Liddle, Elton TC Ngan, Stephanie L Caissie, Cameron M Anderson, Alan T Bates, Digby J Quested, Richard White, and Rowena Weg. 2002.

Thought and language index: an instrument for assessing thought and language in schizophrenia. *The British Journal of Psychiatry*, 181(4):326–330.

Mason Marks and Claudia E Haupt. 2023. Ai chatbots, health privacy, and challenges to hipaa compliance. *Jama*.

John McGrath, Sukanta Saha, David Chant, Joy Welham, et al. 2008. Schizophrenia: a concise overview of incidence, prevalence, and mortality. *Epidemiologic reviews*, 30(1):67–76.

Lise Menn and Loraine K Obler. 1989. Cross-language data and theories of agrammatism. In *Agrammatic aphasia*, pages 1369–1389. John Benjamins.

Tomás Mikolov, Kai Chen, Greg Corrado, and Jeffrey Dean. 2013. Efficient estimation of word representations in vector space. In *1st International Conference on Learning Representations, ICLR 2013, Scottsdale, Arizona, USA, May 2-4, 2013, Workshop Track Proceedings*.

Sylvester Olubolu Orimaye, Jojo Sze-Meng Wong, and Chee Piau Wong. 2018. Deep language space neural network for classifying mild cognitive impairment and alzheimer-type dementia. *PloS one*, 13(11):e0205636.

Lena Palaniyappan, David Benrimoh, Alban Voppel, and Roberta Rocca. 2023. Studying Psychosis Using Natural Language Generation: A Review of Emerging Opportunities. *Biological Psychiatry: Cognitive Neuroscience and Neuroimaging*, 8(10):994–1004.

John H Poole, Faith C Tobias, and Sophia Vinogradov. 2000. The functional relevance of affect recognition errors in schizophrenia. *Journal of the International Neuropsychological Society*, 6(6):649–658.

Ofir Press, Noah A. Smith, and Mike Lewis. 2021. Shortformer: Better language modeling using shorter inputs. In *Proceedings of the 59th Annual Meeting of the Association for Computational Linguistics and the 11th International Joint Conference on Natural Language Processing (Volume 1: Long Papers)*, pages 5493–5505, Online. Association for Computational Linguistics.

Justyna Sarzynska-Wawer, Aleksander Wawer, Aleksandra Pawlak, Julia Szymanowska, Izabela Stefaniak, Michal Jarkiewicz, and Lukasz Okruszek. 2021. Detecting formal thought disorder by deep contextualized word representations. *Psychiatry Research*, 304:114135.

Victoria Sharpe, Michael MacKinley, Samer Nour Eddine, Lin Wang, Lena Palaniyappan, and Gina R Kuperberg. 2024. Gpt-3 reveals selective insensitivity to global vs. local linguistic context in speech produced by treatment-naive patients with positive thought disorder. *bioRxiv*, pages 2024–07.

Elizabeth Shriberg. 2001. To 'errrr'is human: ecology and acoustics of speech disfluencies. *Journal of the international phonetic association*, 31(1):153–169.

Tatiana Sitnikova, Dean F Salisbury, Gina Kuperberg, and Phillip J Holcomb. 2002. Electrophysiological insights into language processing in schizophrenia. *Psychophysiology*, 39(6):851–860.

Jianlin Su, Murtadha Ahmed, Yu Lu, Shengfeng Pan, Wen Bo, and Yunfeng Liu. 2024. Roformer: Enhanced transformer with rotary position embedding. *Neurocomputing*, 568:127063.

Tamara Y Swaab, Megan A Boudewyn, Debra L Long, Steve J Luck, Ann M Kring, J Daniel Ragland, Charan Ranganath, Tyler Lesh, Tara Niendam, Marjorie Solomon, et al. 2013. Spared and impaired spoken discourse processing in schizophrenia: effects of local and global language context. *Journal of Neuroscience*, 33(39):15578–15587.

Sunny X. Tang, Reno Kriz, Sunghye Cho, Suh Jung Park, Jenna Harowitz, Raquel E. Gur, Mahendra T. Bhati, Daniel H. Wolf, João Sedoc, and Mark Y. Liberman. 2021. Natural language processing methods are sensitive to sub-clinical linguistic differences in schizophrenia spectrum disorders. *npj Schizophrenia*.

Kathleen C Thomas, Alan R Ellis, Thomas R Konrad, Charles E Holzer, and Joseph P Morrissey. 2009. County-level estimates of mental health professional shortage in the united states. *Psychiatric services*, 60(10):1323–1328.

Yu-Tsai Wang, Jordan R Green, Ignatius SB Nip, Ray D Kent, and Jane Finley Kent. 2010. Breath group analysis for reading and spontaneous speech in healthy adults. *Folia Phoniatrica et Logopaedica*, 62(6):297–302.

Michael Wilson, Jackson Petty, and Robert Frank. 2023. How abstract is linguistic generalization in large language models? experiments with argument structure. *Transactions of the Association for Computational Linguistics*, 11:1377–1395.

Weizhe Xu, Jake Portanova, Ayesha Chander, Dror Ben-Zeev, and Trevor Cohen. 2021. The centroid cannot hold: comparing sequential and global estimates of coherence as indicators of formal thought disorder. In *AMIA Annual Symposium Proceedings*, volume 2020, page 1315.

Weizhe Xu, Weichen Wang, Jake Portanova, Ayesha Chander, Andrew Campbell, Serguei Pakhomov, Dror Ben-Zeev, and Trevor Cohen. 2022. Fully automated detection of formal thought disorder with time-series augmented representations for detection of incoherent speech (TARDIS). *Journal of Biomedical Informatics*, 126:103998.

A Appendix

	Level	Mild	Severe
# of transcripts		292	18
Age (mean (SD))		40.45 (10.71)	38.33 (8.97)
Gender (%)	Female	162 (55.5)	10 (55.6)
	Male	118 (40.4)	8 (44.4)
	Transgendered: FTM	3 (1.0)	0 (0.0)
	Transgendered: MTF	9 (3.1)	0 (0.0)
Education (%)		1 (0.3)	0 (0.0)
	Associates Degree	51 (17.5)	3 (16.7)
	Bachelors Degree	25 (8.6)	0 (0.0)
	Doctorate Degree	3 (1.0)	0 (0.0)
	Grade school	4 (1.4)	4 (22.2)
	High School Diploma /GED	171 (58.6)	9 (50.0)
	Junior High	22 (7.5)	2 (11.1)
	Masters Degree	15 (5.1)	0 (0.0)
Race (%)		1 (0.3)	0 (0.0)
	American Indian or Alaskan Native	6 (2.1)	0 (0.0)
	Asian	6 (2.1)	0 (0.0)
	Black or African American	61 (20.9)	8 (44.4)
	More than one race	35 (12.0)	2 (11.1)
	White	183 (62.7)	8 (44.4)
# of words per transcript (mean (SD))		182.76 (139.17)	300.50 (170.42)
TALD (mean (SD))		1.08 (0.70)	3.33 (0.34)

Table A.1: Basic transcript-level demographic information for the AVH dataset. `Mild` denotes as mild symptoms of positive FTD where TALD score < 3, and `Severe` denotes severe symptoms of positive FTD, where TALD score ≥ 3.

	level	2	3	4	5	6	7	8
# of transcripts		20	5	5	3	2	3	1
Gender (%)	Male	20 (100.0)	5 (100.0)	5 (100.0)	3 (100.0)	2 (100.0)	3 (100.0)	1 (100.0)
Age (mean (SD))		32.35 (10.84)	28.20 (6.26)	35.20 (13.77)	41.00 (3.61)	24.50 (3.54)	36.00 (23.39)	58.00 (NA)
Race (%)	African American	4 (20.0)	0 (0.0)	2 (40.0)	0 (0.0)	0 (0.0)	1 (33.3)	0 (0.0)
	Asian	2 (10.0)	1 (20.0)	0 (0.0)	0 (0.0)	0 (0.0)	0 (0.0)	1 (100.0)
	Latino	1 (5.0)	1 (20.0)	0 (0.0)	0 (0.0)	1 (50.0)	0 (0.0)	0 (0.0)
	Other (including mixed)	2 (10.0)	0 (0.0)	1 (20.0)	1 (33.3)	1 (50.0)	1 (33.3)	0 (0.0)
	White	11 (55.0)	3 (60.0)	2 (40.0)	2 (66.7)	0 (0.0)	1 (33.3)	0 (0.0)
Education (%)	Associates Degree	2 (10.0)	0 (0.0)	0 (0.0)	0 (0.0)	0 (0.0)	1 (33.3)	0 (0.0)
	Bachelors Degree	3 (15.0)	2 (40.0)	2 (40.0)	1 (33.3)	0 (0.0)	0 (0.0)	0 (0.0)
	GED	1 (5.0)	1 (20.0)	0 (0.0)	1 (33.3)	0 (0.0)	0 (0.0)	0 (0.0)
	H.S. Diploma	13 (65.0)	1 (20.0)	3 (60.0)	1 (33.3)	2 (100.0)	2 (66.7)	1 (100.0)
	Other	0 (0.0)	1 (20.0)	0 (0.0)	0 (0.0)	0 (0.0)	0 (0.0)	0 (0.0)
	Vocational Certification	1 (5.0)	0 (0.0)	0 (0.0)	0 (0.0)	0 (0.0)	0 (0.0)	0 (0.0)
# of words per transcript (mean (SD))		1494.20 (1053.27)	1178.40 (536.35)	1822.00 (1774.69)	1478.67 (601.23)	2004.50 (340.12)	3023.67 (2197.36)	4339.00 (NA)

Table A.2: Basic transcript-level demographic information for the clinical interview dataset. The top row represents the value of the composite PANSS.

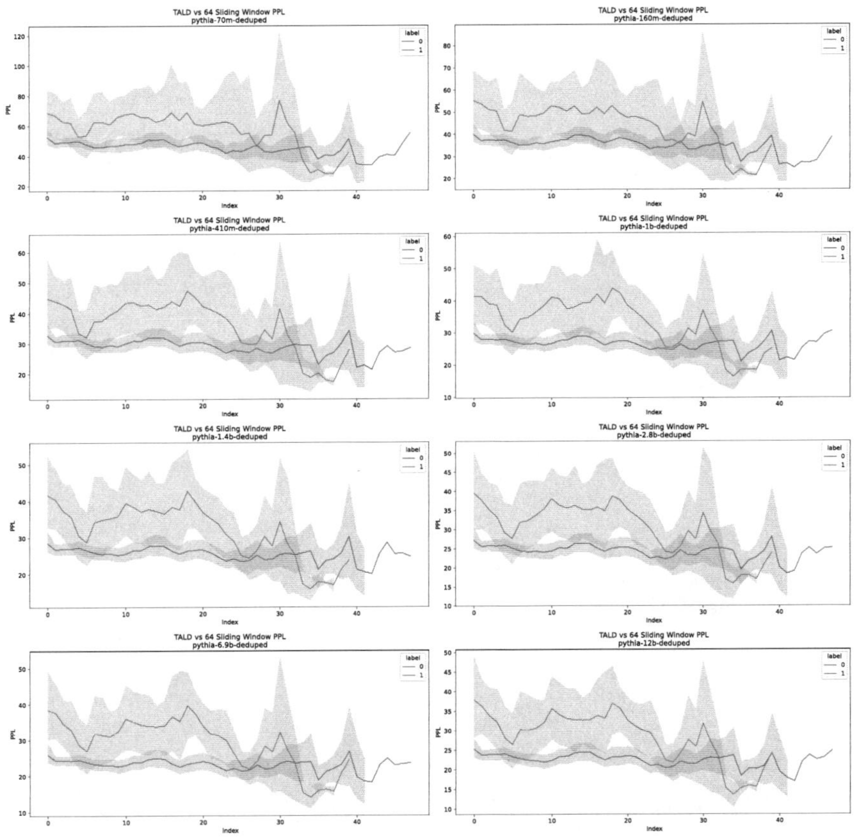

Figure A.1: The distribution of sliding window PPLs using a sliding window of 64 tokens on the AVH dataset, where the x-axis represents the index of sliding window PPLs in transcripts. The shaded area represents the 95% confidence interval of the estimated sliding window PPL on a given index. The label of 0 is defined where TALD < 3, serving as a proxy label for cognitively healthy individuals, whereas the label of 1 serves as the proxy label of FTD individuals.

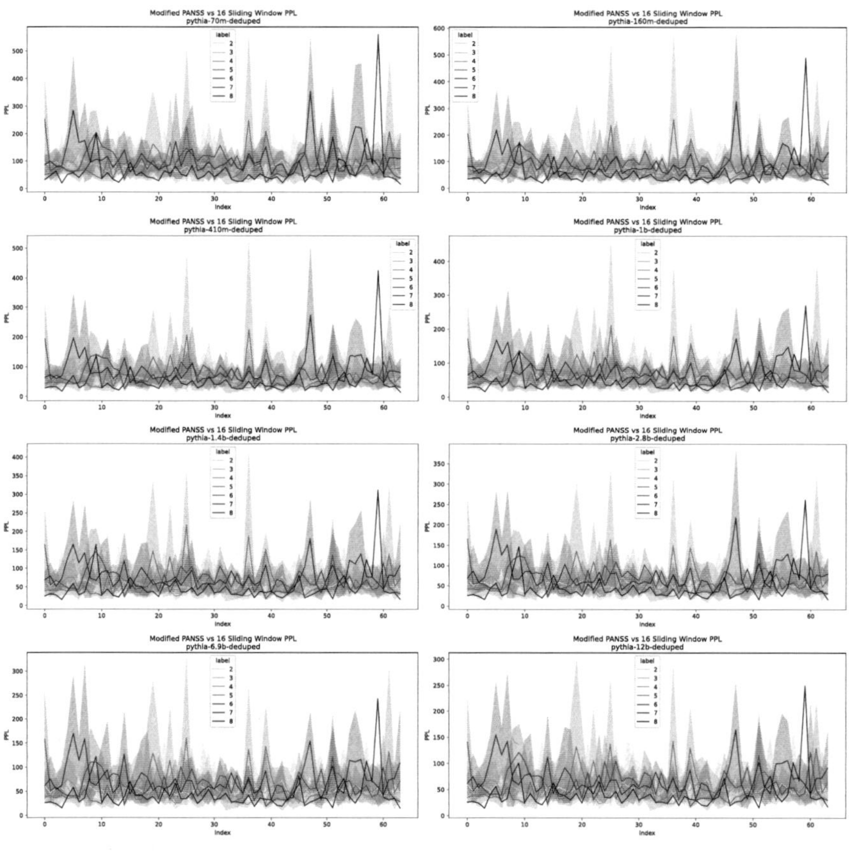

Figure A.2: The distribution of sliding window PPLs using the sliding window of 16 tokens on the clinical interview dataset, where the x-axis represents the index of sliding window PPLs in transcripts. The shaded area represents the 95% confidence interval of the estimated sliding window PPL on a given index.

Model	Sliding windows				
	8	16	32	64	128
70m	0.158***	0.156***	0.176***	**0.202*****	0.180***
160m	0.177***	0.170***	0.177***	**0.206*****	0.183***
410m	0.175***	0.179***	0.196***	**0.225*****	0.189***
1b	0.178***	0.176***	0.205***	**0.230*****	0.180***
1.4b	0.181***	0.179***	0.208***	**0.251*****	0.201***
2.8b	0.168***	0.165***	0.204***	**0.240*****	0.194***
6.9b	0.172***	0.169***	0.206***	**0.249*****	0.194***
12b	0.175***	0.174***	0.204***	**0.245*****	0.195***
LLaMA	–	–	–	0.371***	–

***$p < 0.01$, **$p < 0.05$, *$p < 0.1$

Table A.3: The AVH dataset Spearman's ρ between the *averaged* sliding window PPL and TALD across model size. **Bold** indicates the highest ρ for a model.

Model	Sliding windows				
	8	16	32	64	128
70m	0.258	0.248	0.274*	0.276*	0.276*
160m	0.264	0.278*	0.296*	0.313*	0.294*
410m	0.263	0.276*	0.324**	0.318**	0.301*
1b	0.266	0.292*	0.318**	0.330***	0.305*
1.4b	**0.272***	**0.334****	**0.355****	**0.360****	**0.342****
2.8b	0.261	0.324**	0.344**	0.343**	0.325**
6.9b	0.269*	0.315*	0.342**	0.315*	0.310*
12b	0.270*	0.302*	0.338**	0.334**	0.326**
LLaMA	–	–	–	0.200*	–

***$p < 0.01$, **$p < 0.05$, *$p < 0.1$

Table A.4: The clinical interview dataset Spearman's ρ between the *averaged* sliding window PPL and composite PANSS across model size. **Bold** indicates the highest ρ for a model.

CFiCS: Graph-Based Classification of Common Factors and Microcounseling Skills

Fabian Schmidt[1], Karin Hammerfald[2], Henrik Haaland Jahren[3], Vladimir Vlassov[1]

[1]Department of Computer Science, KTH Royal Institute of Technology, Stockholm, Sweden
[2]Department of Psychology, University of Oslo, Oslo, Norway
[3]Braive AS, Oslo, Norway
Correspondence: fschm@kth.se

Abstract

Common factors and microcounseling skills are critical to the effectiveness of psychotherapy. Understanding and measuring these elements provides valuable insights into therapeutic processes and outcomes. However, automatic identification of these change principles from textual data remains challenging due to the nuanced and context-dependent nature of therapeutic dialogue. This paper introduces CFiCS, a hierarchical classification framework integrating graph machine learning with pretrained contextual embeddings. We represent common factors, intervention concepts, and microcounseling skills as a heterogeneous graph, where textual information from ClinicalBERT enriches each node. This structure captures both the hierarchical relationships (e.g., skill-level nodes linking to broad factors) and the semantic properties of therapeutic concepts. By leveraging graph neural networks, CFiCS learns inductive node embeddings that generalize to unseen text samples lacking explicit connections. Our results demonstrate that integrating ClinicalBERT node features and graph structure significantly improves classification performance, especially in fine-grained skill prediction. CFiCS achieves substantial gains in both micro and macro F1 scores across all tasks compared to baselines, including random forests, BERT-based multi-task models, and graph-based methods.

1 Introduction

Psychotherapy is a complex process that, despite diverse theories and techniques—from cognitive-behavioral and psychodynamic methods to humanistic approaches—shares common change principles that reinforce its effectiveness. These universal elements, known as common factors (CFs), include the therapeutic relationship, expectancy factors, corrective experiencing, insight, and self-efficacy (Bailey and Ogles, 2023) and account for around 30% of therapy outcomes (Lambert, 1992), with the therapeutic relationship being a particularly strong predictor of positive change (Nahum et al., 2019). Intertwined with these CFs are microcounseling skills—discrete, teachable behaviors introduced by Ivey et al. (1968), such as reflective listening and the strategic use of open-ended questions. These skills enable therapists to evoke change principles in practice. For example, fostering the therapeutic bond (an element of the CF therapeutic relationship) by using reflective listening (a microcounseling skill) to convey an empathic and validating attitude (an intervention concept) will likely improve client involvement and therapeutic effectiveness.

Monitoring therapists' behaviors in relation to therapeutic processes can provide deeper insights into how these processes, in turn, contribute to improved treatment outcomes. Moreover, automating skill and CF identification in place of human-based coding or post-session questionnaires improves scalability, lowers costs, and enables the analysis of within-session micro-processes (Falkenström and Larsson, 2017). This approach also offers targeted, session-by-session feedback, enabling clinicians to refine their techniques and adapt interventions to individual client needs.

One effective way to structure change principles systematically is by modeling CFs and microcounseling skills as a graph-based taxonomy. In this taxonomy, CF elements serve as higher-level categories, while microcounseling skills act as specific subcategories or methods used to elicit these factors. The hierarchical relationships within the graph illustrate how particular skills are applied in the context of broader factors. Moreover, this graph can be further enriched by incorporating node attributes, such as detailed descriptions and contextual examples, that clarify how each skill functions within its corresponding CF.

Building on this structured representation, we can leverage graph machine learning (ML) models to classify text by encoding these relationships.

Proceedings of the 10th Workshop on Computational Linguistics and Clinical Psychology (CLPsych 2025), pages 106–115
May 3, 2025 ©2025 Association for Computational Linguistics

These models can learn embeddings for each node, effectively capturing both the structural and feature-based information embedded in the taxonomy. For example, when analyzing a therapeutic interaction, the model can identify relevant microcounseling skills (like reflective listening or validation) and link them to higher-level CFs (such as the therapeutic bond). Based on this framework, we propose a classification method *CFiCS* [1] that employs graph ML to aggregate information from the interconnected network of **c**ommon **f**actors, **i**ntervention **c**oncepts, **s**kills, and examples. This approach allows us to inductively predict associations between previously unseen text and the corresponding CFs (e.g., therapeutic relationship), CF elements (e.g., therapeutic bond), intervention categories (e.g., collaboration and partnership), and skills (e.g., reflective listening), ultimately enhancing our ability to assess and improve therapeutic interactions. We demonstrate through experiments that integrating the graph outperforms baselines. The most accurate configuration combines ClinicalBERT embeddings with GraphSage.

2 Background and Related Work

2.1 Automatic Detection of Therapeutic Elements in Clinical Text

The growing use of technology in psychotherapy has expanded the possibilities of automating text data collection, such as therapy transcripts. This has increased interest in using natural language processing and ML to automatically detect, classify, or score therapist behaviors and client responses. For instance, recent research has attempted to identify empathy-related cues in counseling dialogue (Tao et al., 2024; Tavabi et al., 2023). Other studies have focused on classifying types of reflections or questions posed by therapists (Can et al., 2016; Pérez-Rosas et al., 2017). Despite this progress, several significant challenges remain. First, therapy transcripts inherently contain sensitive information, restricting the available data for model training. Second, counselor behaviors are highly context-dependent; the same microcounseling skill may evoke different CFs. For example, respect for the client's autonomy generally conveys an attitude of collaboration and partnership but is also an inherent part of goal alignment. Third, most existing studies focus on one specific behavioral construct (e.g., identifying therapist empathy alone) rather than a broad taxonomy encompassing multiple change principles and a wide range of microcounseling skills. Recent work explored fine-grained analysis of psychotherapy sessions. Mayer et al. (2024) developed models that predict client emotions and therapist interventions at the utterance level. Similarly, Gibson et al. (2019) introduced multi-label, multi-task deep learning methods that simultaneously predict multiple behavioral codes within therapy dialogues. Despite these advancements, most existing studies still emphasize isolated behavioral constructs (e.g., identifying therapist empathy alone) rather than addressing a broader taxonomy of multiple change principles and diverse microcounseling skills. These limitations motivate the need for more holistic, theory-driven computational approaches that can parse complex therapeutic interactions at multiple levels of granularity.

2.2 Taxonomies and Graph-Based Modeling Approaches

Researchers have explored structured representations like taxonomies or ontologies to capture the hierarchical and interconnected nature of therapeutic elements, such as broad CFs and more granular microcounseling skills, and have applied these frameworks to classify symptoms, diagnoses, and interventions in mental health research (Evans et al., 2021). However, few existing taxonomies systematically link higher-level CFs (e.g., the therapeutic relationship) to actionable skills (e.g., reflective listening, validating) that instantiate those factors in practice. Graph-based ML offers a robust avenue for modeling these relationships. With methods such as Graph Convolutional Networks (GCNs) (Kipf and Welling, 2017) or GraphSAGE (Hamilton et al., 2017), one can encode both the textual features of nodes (e.g., descriptions of skills) and the relational structure (e.g., which skills evoke which CFs) into a unified embedding space. Graph ML has shown promise in diverse classification tasks—among others, in detecting suicidality (Lee et al., 2022)—suggesting that a similar strategy could be applied to psychotherapy discourse.

3 The CFiCS Graph

We construct a structured knowledge graph of therapeutic practices centered around five main types of nodes: the root of our graph, the CF therapeutic alliance; the *CF elements* (i.e., Bond, Goal Align-

[1] Code available on GitHub

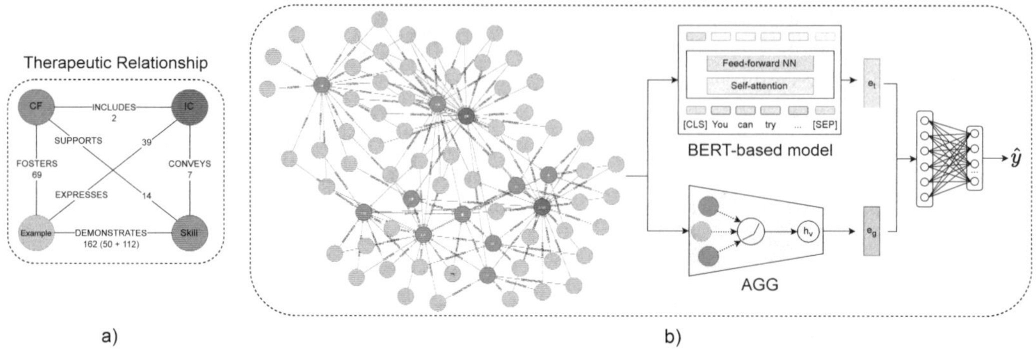

Figure 1: (a) Edge types and connection patterns between the node types for the common factor Therapeutic Relationship and (b) Common factor and microcounseling skill prediction with our CFiCS classification model.

ment, and Task Agreement), which define integral components of the CF therapeutic alliance; *intervention concepts (ICs)* (i.e., Empathy, Acceptance, and Positive Regard; Collaboration and Partnership), which express specific approaches to fostering the CF elements; *therapeutic skills* (i.e., Open-Ended Questions, Reflective Listening, Affirmation, Validation, Genuineness, Respect for Autonomy, Asking for Permission), which are practical techniques therapists use; and *examples*, which are therapist statements that illustrate how specific skills and concepts are applied. Figure 1a visualizes the connection patterns between the node types. The graph is heterogeneous, incorporating different node types and relationships, and sparse. Most examples link to one CF or one IC and one or two skills. Hence, the average path length is short due to the triadic pattern: example $\rightarrow$ skill $\rightarrow$ CF or IC. The structure is hierarchical, with CFs at the top, ICs as an intermediate layer, and therapeutic skills and examples forming the practical, grounded components. Clusters naturally form around specific CFs and ICs, creating thematic groupings. The graph's relationships are multi-relational and include edges like *fosters*, linking examples to the CFs they develop; *expresses*, connecting skills to ICs or examples to concepts; and *demonstrates*, linking examples to the therapeutic skills they showcase. This semantic structure provides a foundation for our classification approach.

4 The CFiCS Classification Model

We propose a model shown in Figure 1b to classify CFs, ICs, and *associated* therapeutic skills for textual input. We leverage graph ML to exploit the topology of the nodes in the CFiCS graph and combine it with the textual embeddings produced by a pretrained language model. CFiCS enables inductive classification of previously unseen nodes, which do not have explicit edges but can still leverage the structural patterns learned from the graph during training in addition to the textual features.

Input graph structure Let $\mathcal{G} = (\mathcal{V}, \mathcal{E})$ be a heterogeneous graph where: $\mathcal{V} = \mathcal{V}_r \cup \mathcal{V}_f \cup \mathcal{V}_c \cup \mathcal{V}_s \cup \mathcal{V}_e$ represents the set of all nodes, where $\mathcal{V}_r$ the root node, $\mathcal{V}_f$ the CF nodes, $\mathcal{V}_c$ the IC nodes, $\mathcal{V}_s$ the skill nodes, and $\mathcal{V}_e$ the example nodes. $\mathcal{E} \subseteq (v_i, v_j) | v_i, v_j \in \mathcal{V}$ represents the bidirectional edges between nodes in the graph. There exist six distinct edge types, also visualized in Figure 1b:

1. **Fosters relation**: (v_e, v_f) denotes a relationship between an example node and a CF node, indicating that the example fosters the development of a specific CF in therapy.

2. **Expresses relation**: (v_e, v_c) connects an example node to an IC node, signifying that the example expresses the therapeutic intention of an IC (e.g., reflective listening expresses empathy, acceptance, and positive regard).

3. **Demonstrates relation**: (v_e, v_s) links an example node to a skill node, showing that the example demonstrates the practical application of a specific therapeutic skill.

4. **Includes relation**: (v_f, v_c) links a CF node to an IC node, indicating that the IC is a specific approach to operationalizing the broader CF.

5. **Conveys relation**: (v_s, v_c) connects a skill node to an IC node, signifying that the skill conveys the therapeutic intention of an IC.

6. **Supports relation**: (v_s, v_f) connects a skill node to a CF node, highlighting that a micro-counseling skill supports a CF element.

Node features For each node $v \in \mathcal{V}$, let τ_v denote the associated textual input. For nodes representing CFs, ICs, or skills, τ_v consists of the node's name and detailed description. In contrast, for example nodes, τ_v comprises solely the example text. A pretrained language model $\mathcal{M}$ (e.g., BERT (Devlin et al., 2019) or ClinicalBERT (Alsentzer et al., 2019)) computes the feature vector for each node.

$$\mathbf{x}_v = \mathcal{M}(\tau_v)$$

where $\mathbf{x}_v \in \mathbb{R}^d$ and d is the embedding dimension of the chosen model (e.g., $d = 768$ for BERT).

Learning task We seek to learn node embeddings via a *message passing* framework inspired by GraphSAGE (Hamilton et al., 2017). Given training node representations $\{\mathbf{x}_v \mid v \in \mathcal{V}\}$ and graph structure $\mathcal{E}$, we iteratively update each node v's representation as

$$\mathbf{h}_v^{(l)} = \gamma^{(l)}\Big(\mathbf{h}_v^{(l-1)}, \text{AGG}\big\{\mathbf{h}_u^{(l-1)} \mid u \in \mathcal{N}(v)\big\}\Big)$$

where $\mathbf{h}_v^{(l)}$ is v's embedding at layer l, $\gamma^{(l)}$ is a learnable transformation (often a nonlinear MLP), AGG is a neighborhood aggregation function (e.g., mean or pooling aggregator), and $\mathcal{N}(v)$ denotes v's neighbors. Although we present this in a GraphSAGE-oriented formulation, *the same learning task is fully model-agnostic*: by substituting different forms of AGG (such as attention-based aggregation in GAT (Veličković et al., 2018) or weighted normalized sums in GCN (Kipf and Welling, 2017)) and choosing a suitable $\gamma^{(l)}$, one can instantiate a variety of GNN variants without altering the underlying message passing framework.

Classification tasks Given an example node $v \in \mathcal{V}_e$ with embedding $\mathbf{h}_v$, we define a single linear classification layer that encompasses all labels (i.e., for CFs, ICs, and skills). Let $\mathcal{T} = \{\text{CF}, \text{IC}, \text{Skill}\}$ denote the task types, and let $\mathcal{V}_t$ be the set of labels for each task $t \in \mathcal{T}$. We construct a single parameter matrix W and bias vector $\mathbf{b}$ such that the row segments of W and the corresponding portions of $\mathbf{b}$ map to the different tasks. Formally, the probability of assigning label y from $\mathcal{V}_t$ to node $v \in \mathcal{V}_e$ is computed by slicing the relevant portion of the linear output and applying a softmax.

$$p_t(y \mid v) = \text{softmax}\Big(\big(W_{\text{slice}(t)}\, \mathbf{h}_v + \mathbf{b}_{\text{slice}(t)}\big)\Big)$$

where $y \in \{1, \ldots, |\mathcal{V}_t|\}$, $\mathbf{h}_v$ is the shared node embedding (e.g., obtained from a graph neural network), $W_{\text{slice}(t)}$ and $\mathbf{b}_{\text{slice}(t)}$ refer to the rows in W and $\mathbf{b}$ that correspond to the label set $\mathcal{V}_t$, and $\mathrm{softmax}(\cdot)$ is applied to the sliced logits to form a probability distribution specific to the task t.

Optimization objective We treat each task $t \in \mathcal{T}$ as a separate multi-class classification problem and define a cross-entropy loss $\mathcal{L}_t$ on the predictions $p_t(y_v \mid v)$. Formally,

$$\mathcal{L}_t = -\sum_{v \in \mathcal{V}_e} \log\, p_t(y_v \mid v),$$

where $y_v \in \mathcal{V}_t$ denotes the ground truth label for node v in task t. Our overall multi-task objective is a linear combination of these losses

$$\mathcal{L} = \sum_{t \in \mathcal{T}} \lambda_t\, \mathcal{L}_t,$$

with weights $\{\lambda_t\}$ controlling the relative importance of each task. Intuitively, each $\mathcal{L}_t$ measures how well the model performs on task t, and the hyperparameters λ_t balance their contributions to the total loss.

Inference A new node is isolated during inference, meaning it has no edges and lacks direct neighbors in the graph. When the set of neighbors $\mathcal{N}(v)$ is empty, the aggregator defaults to relying solely on $h_v^{(l-1)}$. However, the model's learned weights still capture global patterns from the training graph. The aggregator, which processes node features, has learned the overall graph structure and label signals, allowing it to embed the new node in a graph-aware feature space. Even without access to neighbors, the aggregator's learned MLP transforms the new node's features with knowledge learned during training on graph edges.

5 Experiments

5.1 Implementation Details

We use either *bert-base-uncased* or *ClinicalBERT* from Huggingface Transformers (Wolf et al., 2020) to encode the node name and description by average-pooling the last hidden state. We implement the model in Python using PyTorch Geometric. The model processes 2×768 input channels, 768 for the text embedding and 768 for the graph embedding, through hidden layers of 64 channels in a three-layer architecture with a 0.5 dropout rate.

Table 1: Three-fold cross-validation micro and macro F1 scores.

Model	Micro F1			Macro F1		
	CF	IC	Skill	CF	IC	Skill
RF (TF-IDF multi-task)	52.50 ± 3.11	74.59 ± 1.48	53.02 ± 8.39	20.43 ± 3.82	38.04 ± 1.65	49.77 ± 4.24
BERT (multi-task)	59.69 ± 4.95	79.02 ± 1.39	59.60 ± 8.97	34.34 ± 11.0	46.29 ± 6.84	59.23 ± 10.3
GCN without BERT	55.70 ± 5.77	70.06 ± 4.38	19.37 ± 4.63	17.86 ± 1.19	27.45 ± 1.00	4.04 ± 0.81
GAT without BERT	56.89 ± 1.16	71.27 ± 2.42	23.24 ± 8.39	18.13 ± 0.24	27.74 ± 0.55	4.65 ± 1.34
GraphSage without BERT	56.86 ± 2.08	70.04 ± 2.63	14.39 ± 1.27	20.19 ± 4.00	27.45 ± 0.60	3.14 ± 0.24
CFiCS GCN with ClinicalBERT	74.53 ± 16.62	86.30 ± 9.05	91.36 ± 14.97	66.06 ± 21.02	74.79 ± 17.69	88.63 ± 19.68
CFiCS GAT with ClinicalBERT	91.98 ± 12.33	93.21 ± 10.2	93.83 ± 10.69	82.39 ± 27.45	89.35 ± 14.99	92.18 ± 13.55
CFiCS with ClinicalBERT (**ours**)	**95.04 ± 7.00**	**100.0 ± 0.00**	96.30 ± 6.42	**88.95 ± 16.14**	**100.0 ± 0.00**	96.09 ± 6.78
CFiCS with BERT (**ours**)	**95.04 ± 5.33**	97.53 ± 4.28	**96.89 ± 3.84**	84.24 ± 16.43	95.88 ± 7.13	**97.14 ± 3.44**

It handles three task components: three CFs, two ICs, seven skills, and a *neutral* class for each task. During training, a forward pass computes logits for these tasks, followed by slicing the output into separate components and computing the task-specific losses weighted by predefined task weights (defaulting to uniform). The Adam optimizer (Kingma and Ba, 2017) updates the parameters, with training configured for up to 400 epochs, a learning rate of 1e-3, and a weight decay of 1e-4. The model performs validation at each epoch, tracks the lowest loss, and stops early if it detects no improvement for 50 epochs.

5.2 Dataset

Our dataset consists solely of manually created and curated examples drawn from established psychotherapy literature, rather than real patient conversations. The dataset is structured as a heterogeneous, undirected, and multi-relational CFiCS graph modeling therapeutic practices. It includes three CF elements (e.g., Bond), two ICs (e.g., Collaboration and Partnership), and seven therapeutic skills (e.g., Reflective Listening). The dataset contains 69 fully annotated examples, including CF, IC, and skill annotations, and an additional 112 examples annotated only for therapeutic skills, illustrating their application contextually. An expert selected and curated these examples from respected psychotherapy literature on therapeutic alliance. Specifically, we identified reference samples directly from Fuertes (2019); Miller and Moyers (2021); Bailey and Ogles (2023) as representative instances of therapeutic interaction for each class. These original excerpts served as a reference for generating new synthetic samples using ChatGPT, ensuring alignment with the themes, styles, and therapeutic concepts illustrated in the literature. Examples not assigned to any class are designated as *neutral*. The dataset is split into training and testing subsets (80/20), with k-fold cross-validation applied to the training data for model evaluation.

5.3 Baselines

We compare our approach against two baselines: a Random Forest (RF) (Breiman, 2001) and a BERT-based architecture. Additionally, we evaluate and compare several graph ML methods, including GAT, GCN, and GraphSage, to assess their effectiveness in modeling the relationships within the CFiCS graph. For the RF baseline, we convert each utterance into TF-IDF features and create a multi-output target vector where CFs, ICs, and skills are multi-class tasks. We then train a `MultiOutputClassifier`, effectively training one RF per output dimension. For the BERT-based model, we finetune a pretrained encoder that extracts a pooled `[CLS]` representation and optimizes three classification heads using cross-entropy loss for the CF, IC, and skill prediction.

5.4 Metrics

We report macro- and micro-averaged F1 scores for the multi-class tasks. The macro F1 treats each class equally, computing the mean F1 over classes, whereas the micro F1 aggregates contributions from all classes to compute precision and recall overall. The micro F1 tallies the total number of correctly predicted skill labels versus all predictions, while the macro F1 averages the F1 values per skill category. In addition, we use Precision@k and Recall@k as indicators of cluster quality to

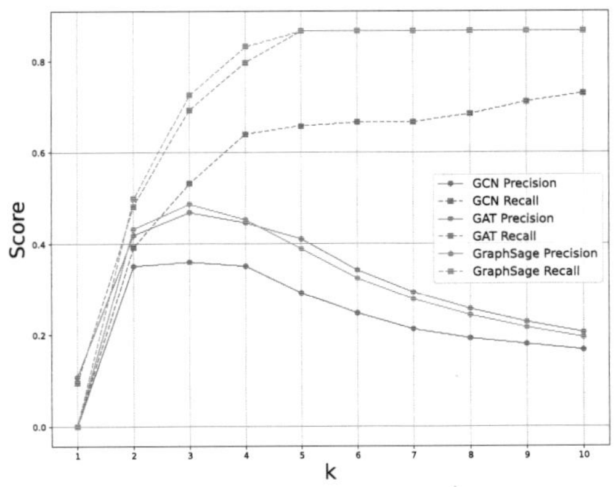

Figure 2: Precision and recall at different values of k (ranging from 1 to 10) for graph-based models, including GCN, GAT, and GraphSage.

evaluate how well the model ranks and groups related concepts. Precision@k measures the proportion of relevant samples among the top-k similar nodes, reflecting how accurately the model embedds input features. Recall@k quantifies the proportion of relevant samples retrieved within the top-k similar samples, indicating how well the model captures relevant clusters. Higher Precision@k and Recall@k scores suggest that similar concepts are embedded closely together, providing additional validation of the quality of learned representations.

6 Results

6.1 Quantitative Results

Performance of non-graph baselines Table 1 reports the micro and macro F1 scores for various models on the CF, IC, and skill classification. We compare several baselines, including traditional TF-IDF and BERT multi-task classifiers, graph-based models without BERT, and our proposed CFiCS variants that integrate clinical BERT features. The TF-IDF multi-task Random Forest baseline achieves relatively modest performance, with micro F1 scores of 52.50, 74.59, and 53.02 for CF, IC, and Skill, respectively, and corresponding macro F1 scores of 20.43, 38.04, and 49.77. The BERT multi-task classifier improves these numbers considerably (e.g., obtaining 59.69 and 59.60 micro F1 for CF and Skill, respectively), indicating the benefit of richer contextual representations.

Effect of graph structure on performance Graph-based models without BERT features exhibit mixed results. For example, the GCN and GAT models without BERT yield micro F1 scores in the range of 55.70–56.89 for CF and around 70 for IC. However, their performance on the skill task is markedly lower (with micro F1 scores of 19.37, 23.24, and 14.39 for GCN, GAT, and GraphSage, respectively). This suggests that relying solely on graph structure without contextual text representations can be detrimental, particularly for the more nuanced skill classification. In contrast, our proposed methods that incorporate clinical BERT features within the CFiCS framework demonstrate substantial improvements. Both the CFiCS GCN and CFiCS GAT variants with ClinicalBERT improve performance across all tasks. In particular, the CFiCS GAT with ClinicalBERT variant achieves micro F1 scores of 91.98, 93.21, and 93.83 for CF, IC, and Skill, respectively, with corresponding macro F1 scores of 82.39, 89.35, and 92.18. The CFiCS models that integrate BERT features achieve the best results. The model labeled as *CFiCS with ClinicalBERT (ours)* achieves a micro F1 of 95.04 on CF and 96.30 on Skill, with perfect performance on the IC task (100.00 in both micro and macro F1). Similarly, *CFiCS with BERT (ours)* shows competitive performance with micro F1 scores exceeding 95% for CF, IC, and Skill, and macro F1 scores that are consistently high.

Precision and recall at different k We evaluate the embedding quality by comparing the precision and recall at varying k ranging from one to ten. Figure 2 presents the aggregated precision and recall for values of k ranging from 1 to 10. As expected, recall increases as k grows since more relevant items are retrieved, while precision declines due to the broader set of top-k retrievals. Table 2 provides a more detailed breakdown of precision and recall across different class types. GraphSage demonstrates superior recall and precision at higher k, suggesting that its embeddings create more cohesive clusters of relevant nodes, making it more effective for retrieving multiple correct labels. GCN lags behind both models in rank-1 precision but improves recall at k=10. While its embeddings do not strongly differentiate the best match, they still capture helpful information for broader retrieval. Overall, GAT is best for fine-grained differentiation, GraphSage generates well-structured clusters that enhance overall representation quality, and GCN provides moderate performance with embeddings that favor broader contextual generalization.

Table 2: Precision@ k and Recall@ k for GAT, GCN, and GraphSage with ClinicalBERT ($k \in \{1, 5, 10\}$).

Class	Model	P@1	R@1	P@5	R@5	P@10	R@10
Common Factors	GAT	**10.81**	**1.03**	57.84	26.46	**63.78**	50.46
	GCN	0.00	0.00	45.41	24.04	46.22	41.80
	GraphSage	0.00	0.00	**58.92**	**33.47**	59.73	**57.96**
Intervention Concepts	GAT	**13.51**	**1.29**	71.35	28.83	74.32	45.85
	GCN	0.00	0.00	62.16	20.47	64.05	35.44
	GraphSage	0.00	0.00	**76.76**	**32.82**	**77.03**	**46.62**
Skills	GAT	**13.51**	**3.60**	55.14	**71.17**	29.19	76.13
	GCN	0.00	0.00	48.65	64.11	32.97	**77.18**
	GraphSage	0.00	0.00	**57.30**	55.08	**37.84**	71.98

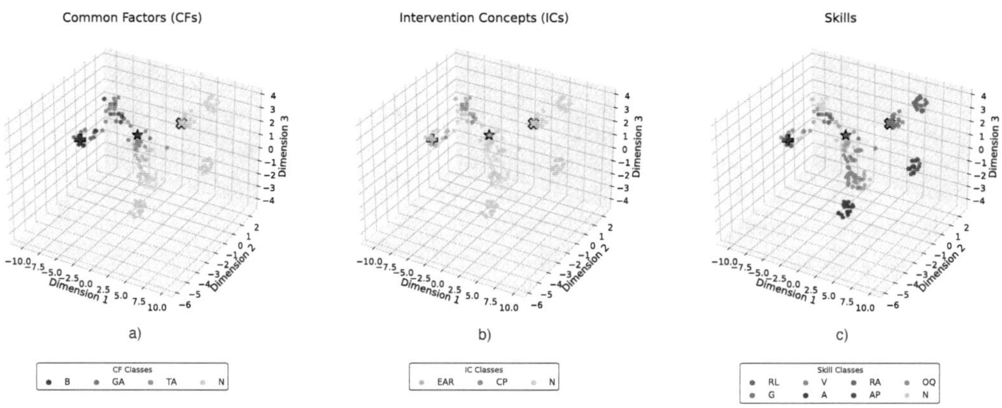

Figure 3: Visualization of the 197 learned embeddings after aggregating contextual information from the graph. TSNE reduces the 832-dimensional vectors (768 text features + 64 graph features) to two dimensions. Each point represents a node embedding, with colors indicating the three different class types. The embeddings are extracted from the final hidden layer during the forward pass.

6.2 Qualitative Results

In Figure 3, we visualize the learned embeddings in 3-dimensional space using the dimension reduction method t-SNE (van der Maaten and Hinton, 2008). We color the samples based on their respective class for all three class types (CF, IC, and skill). We observe that skills form well-separated clusters, indicating that the model effectively distinguishes between different microcounseling techniques. Additionally, higher-level CF influence representation structure, as utterances containing a skill alone (e.g., Genuineness, ✖ in Figure 3c) are embedded separately from those where the skill co-occurs with broader therapeutic elements (e.g., Bond (B) or Empathy, Acceptance, and Positive Regard (EAR) (see the ☆ and ✚ sign in Figure 3a and b). This suggests that the model captures fine-grained skill differentiation and hierarchical relationships between skills and ICs. Moreover, higher-level features such as the CF and IC cluster more closely together, showing less separation. The context of the skills separates them at the lower, more fine-grained skill level. This finding indicates that while the model captures distinct skill representations, the broader context in which practitioners apply these skills is crucial for differentiation.

7 Ethical and Impact Considerations

Psychotherapy transcripts contain highly sensitive and personal information, and patients are particularly vulnerable data subjects. However, our dataset consists solely of manually created and curated examples from the literature, not real patient conversations, ensuring no private or identifiable data is used. Using this method with real-world therapy data would require strict attention to privacy and confidentiality, ensuring compliance with data protection regulations. ML models trained on limited or biased datasets can inadvertently learn and propagate biases in the data. Since our dataset is relatively small and manually curated, there is a risk that certain features are over- or underrepresented, potentially impacting the generalizability of our results. Furthermore, automated classification of psychotherapy content could be misused if applied without proper oversight. For instance, misclassification of therapeutic interactions could lead to inaccurate feedback for therapists, and reliance on imperfect AI-driven assessments might undermine professional judgment. Therefore, the model should be deployed as an assistive tool rather than replacing human expertise.

Our work has implications for both psychotherapy research and practice. Firstly, traditional research methods often fail to capture the complexity of the patient-therapist interaction process (Lundh and Falkenström, 2019). For example, studies on therapeutic alliance patterns typically rely on post-session evaluations, which may oversimplify evolving patient-therapist dynamics (Falkenström and Larsson, 2017). Automatic assessment of in-session "microprocesses" (Lundh and Falkenström, 2019) could offer a more precise understanding of common factors development, identifying key therapist skills linked to treatment success across modalities and client profiles.

Secondly, psychotherapy quality in practice depends on research-driven training and performance-based feedback (Baldwin and Imel, 2013). Yet, many clinicians receive little feedback after initial training (Moyers et al., 2005). A system providing session-by-session feedback on common factor usage on various levels of granularity could help therapists set incremental improvement goals and track progress in real time (Rousmaniere, 2016). Thirdly, automating common factor feedback would enable integration with digital health tools, linking therapist skill use to broader treatment data, including symptom levels and session attendance.

8 Conclusion

We presented a graph ML classification method *CFiCS* to classify common factors, intervention concepts, and associated skill usage. Overall, the results demonstrate that combining textual features from ClinicalBERT with graph-based ML in the CFiCS framework significantly enhances classification performance, particularly for the challenging skill prediction task, and outperforms conventional TF-IDF, BERT, and pure graph-based baselines.

9 Limitations

One primary limitation is the dataset size. We evaluate our method using a manually curated dataset alongside examples from the literature. While the model demonstrates promising performance, the relatively small sample size may limit generalizability and increase the risk of overfitting to specific linguistic patterns or annotation biases. A second limitation is language dependence. Our study focuses exclusively on English-language data, and we do not assess whether the method generalizes to other languages or multilingual settings. Given that therapeutic discourse varies linguistically and culturally, future work should explore cross-lingual adaptations and assess whether pretrained multilingual models (e.g., XLM-R, mBERT) can extend classification performance to other languages. An additional challenge is linguistic ambiguity per se. Identical statements can have different meanings depending on the context. Prosodic features play a key role in language comprehension (Dahan, 2015), and models trained on spoken language outperform text-based approaches (Singla et al., 2020). Thus, CFiCS classification could benefit from incorporating auditory and visual cues. Another limitation is the lack of external validation on out-of-distribution datasets. Our dataset consists solely of manually curated literature examples and synthetically generated examples, rather than real therapist-patient interactions. While this approach has ethical advantages by avoiding privacy concerns, it limits the clinical relevance of the dataset. Additionally, the effectiveness of common factor usage depends on their thoughtful application rather than mere frequency. Therapist responsiveness, seen as a "metacompetency" integrating skills like executive functioning and reflection (Hatcher, 2015), is more valuable than rigid technique use (Stiles and Horvath, 2017).

10 Future Work

Future work should expand the dataset and use real therapy interactions in different settings and with therapists using different approaches. Additionally, it may be beneficial to explore multilingual extensions, expand the CFiCS graph structure, validate on external corpora, and consider integrating prosodic features.

Acknowledgments

This work is funded by the Research Council of Norway (grant 321561) through the Adaptive Level of Effective and Continuous Care (ALEC2) project and was partially supported by the Wallenberg AI, Autonomous Systems and Software Program (WASP), funded by the Knut and Alice Wallenberg Foundation. The model training was enabled by the Berzelius resource provided by the Knut and Alice Wallenberg Foundation at the National Supercomputer Centre.

References

Emily Alsentzer, John Murphy, William Boag, Wei-Hung Weng, Di Jindi, Tristan Naumann, and Matthew McDermott. 2019. Publicly available clinical BERT embeddings. In *Proceedings of the 2nd Clinical Natural Language Processing Workshop*, pages 72–78, Minneapolis, Minnesota, USA. Association for Computational Linguistics.

Russell J Bailey and Benjamin M Ogles. 2023. *Common factors therapy: A principle-based treatment framework*. American Psychological Association.

Scott A Baldwin and Zack E Imel. 2013. Therapist effects: Findings and methods. *Bergin and Garfield's handbook of psychotherapy and behavior change*, 6:258–297.

Leo Breiman. 2001. Random forests. *Machine learning*, 45:5–32.

Doğan Can, Rebeca A Marín, Panayiotis G Georgiou, Zac E Imel, David C Atkins, and Shrikanth S Narayanan. 2016. "it sounds like...": A natural language processing approach to detecting counselor reflections in motivational interviewing. *Journal of counseling psychology*, 63(3):343.

Delphine Dahan. 2015. Prosody and language comprehension. *Wiley Interdisciplinary Reviews: Cognitive Science*, 6(5):441–452.

Jacob Devlin, Ming-Wei Chang, Kenton Lee, and Kristina Toutanova. 2019. BERT: Pre-training of deep bidirectional transformers for language understanding. In *Proceedings of the 2019 Conference of the North American Chapter of the Association for Computational Linguistics: Human Language Technologies, Volume 1 (Long and Short Papers)*, pages 4171–4186, Minneapolis, Minnesota. Association for Computational Linguistics.

Spencer C Evans, Michael C Roberts, Jessy Guler, Jared W Keeley, and Geoffrey M Reed. 2021. Taxonomy and utility in the diagnostic classification of mental disorders. *Journal of clinical psychology*, 77(9):1921–1936.

Fredrik Falkenström and Mattias Holmqvist Larsson. 2017. The working alliance: From global outcome prediction to micro-analyses of within-session fluctuations. *Psychoanalytic Inquiry*, 37(3):167–178.

Jairo N Fuertes. 2019. *Working alliance skills for mental health professionals*. Oxford University Press, USA.

James Gibson, David C Atkins, Torrey A Creed, Zac Imel, Panayiotis Georgiou, and Shrikanth Narayanan. 2019. Multi-label multi-task deep learning for behavioral coding. *IEEE Transactions on Affective Computing*, 13(1):508–518.

Will Hamilton, Zhitao Ying, and Jure Leskovec. 2017. Inductive representation learning on large graphs. *Advances in neural information processing systems*, 30.

Robert L Hatcher. 2015. Interpersonal competencies: Responsiveness, technique, and training in psychotherapy. *American Psychologist*, 70(8):747.

Allen E Ivey, Cheryl J Normington, C Dean Miller, Weston H Morrill, and Richard F Haase. 1968. Microcounseling and attending behavior: An approach to prepracticum counselor training. *Journal of Counseling Psychology*, 15(5p2):1.

Diederik P. Kingma and Jimmy Ba. 2017. Adam: A method for stochastic optimization. *Preprint*, arXiv:1412.6980.

Thomas N. Kipf and Max Welling. 2017. Semi-supervised classification with graph convolutional networks. In *International Conference on Learning Representations*.

Michael J. Lambert. 1992. Psychotherapy outcome research: Implications for integrative and eclectic therapists. In John C. Norcross and Marvin R. Goldfried, editors, *Handbook of Psychotherapy Integration*, pages 94–129. Basic Books.

Daeun Lee, Migyeong Kang, Minji Kim, and Jinyoung Han. 2022. Detecting suicidality with a contextual graph neural network. In *Proceedings of the Eighth Workshop on Computational Linguistics and Clinical Psychology*, pages 116–125, Seattle, USA. Association for Computational Linguistics.

Lars-Gunnar Lundh and Fredrik Falkenström. 2019. Towards a person-oriented approach to psychotherapy research. *Journal for person-oriented research*, 5(2):65.

Tobias Mayer, Neha Warikoo, Amir Eliassaf, Dana Atzil-Slonim, and Iryna Gurevych. 2024. Predicting client emotions and therapist interventions in psychotherapy dialogues. In *Proceedings of the 18th*

Conference of the European Chapter of the Association for Computational Linguistics (Volume 1: Long Papers), pages 1463–1477, St. Julian's, Malta. Association for Computational Linguistics.

William R Miller and Theresa B Moyers. 2021. *Effective psychotherapists*. Guilford Publications.

Theresa B Moyers, Tim Martin, Jennifer K Manuel, Stacey ML Hendrickson, and William R Miller. 2005. Assessing competence in the use of motivational interviewing. *Journal of substance abuse treatment*, 28(1):19–26.

Daniel Nahum, César A Alfonso, and Ekin Sönmez. 2019. Common factors in psychotherapy. *Advances in psychiatry*, pages 471–481.

Verónica Pérez-Rosas, Rada Mihalcea, Kenneth Resnicow, Satinder Singh, Lawrence An, Kathy J Goggin, and Delwyn Catley. 2017. Predicting counselor behaviors in motivational interviewing encounters. In *Proceedings of the 15th Conference of the European Chapter of the Association for Computational Linguistics: Volume 1, Long Papers*, pages 1128–1137.

Tony Rousmaniere. 2016. *Deliberate practice for psychotherapists: A guide to improving clinical effectiveness*. Routledge.

Karan Singla, Zhuohao Chen, David C Atkins, and Shrikanth Narayanan. 2020. Towards end-2-end learning for predicting behavior codes from spoken utterances in psychotherapy conversations. In *Proceedings of the conference. Association for Computational Linguistics. Meeting*, volume 2020, page 3797. NIH Public Access.

William B Stiles and Adam O Horvath. 2017. *Appropriate responsiveness as a contribution to therapist effects*. American Psychological Association.

Dehua Tao, Tan Lee, Harold Chui, and Sarah Luk. 2024. Learning representation of therapist empathy in counseling conversation using siamese hierarchical attention network. In *Interspeech 2024*, pages 1085–1089.

Leili Tavabi, Trang Tran, Brian Borsari, Joannalyn Delacruz, Joshua D Woolley, Stefan Scherer, and Mohammad Soleymani. 2023. Therapist empathy assessment in motivational interviews. In *2023 11th International Conference on Affective Computing and Intelligent Interaction (ACII)*, pages 1–8. IEEE.

Laurens van der Maaten and Geoffrey Hinton. 2008. Visualizing data using t-sne. *Journal of Machine Learning Research*, 9(86):2579–2605.

Petar Veličković, Guillem Cucurull, Arantxa Casanova, Adriana Romero, Pietro Liò, and Yoshua Bengio. 2018. Graph attention networks. In *International Conference on Learning Representations*.

Thomas Wolf, Lysandre Debut, Victor Sanh, Julien Chaumond, Clement Delangue, Anthony Moi, Pierric Cistac, Tim Rault, Remi Louf, Morgan Funtowicz, Joe Davison, Sam Shleifer, Patrick von Platen, Clara Ma, Yacine Jernite, Julien Plu, Canwen Xu, Teven Le Scao, Sylvain Gugger, Mariama Drame, Quentin Lhoest, and Alexander Rush. 2020. Transformers: State-of-the-art natural language processing. In *Proceedings of the 2020 Conference on Empirical Methods in Natural Language Processing: System Demonstrations*, pages 38–45, Online. Association for Computational Linguistics.

Datasets for Depression Modeling in Social Media: An Overview

Ana-Maria Bucur[1,2], Andreea-Codrina Moldovan[1], Krutika Parvatikar[3]
Marcos Zampieri[4], Ashiqur R. KhudaBukhsh[3], Liviu P Dinu[5,6]

[1]Interdisciplinary School of Doctoral Studies, University of Bucharest, Romania
[2]PRHLT Research Center, Universitat Politècnica de València, Spain
[3]Rochester Institute of Technology, USA, [4]George Mason University, USA
[5]Faculty of Mathematics and Computer Science, [6]HLT Research Center,
University of Bucharest, Romania

ana-maria.bucur@drd.unibuc.ro

Abstract

Depression is the most common mental health disorder, and its prevalence increased during the COVID-19 pandemic. As one of the most extensively researched psychological conditions, recent research has increasingly focused on leveraging social media data to enhance traditional methods of depression screening. This paper addresses the growing interest in interdisciplinary research on depression, and aims to support early-career researchers by providing a comprehensive and up-to-date list of datasets for analyzing and predicting depression through social media data. We present an overview of datasets published between 2019 and 2024. We also make the comprehensive list of datasets available online as a continuously updated resource, with the hope that it will facilitate further interdisciplinary research into the linguistic expressions of depression on social media.

1 Introduction

Depression is the most common mental health disorder, and its prevalence has increased further during the COVID-19 pandemic (Wolohan, 2020; Kaseb et al., 2022; Bucur et al., 2025). Depression is also one of the most extensively researched mental health disorders in the field of psychology (Xu et al., 2021). Since the past decade, interdisciplinary researchers have explored this widespread mental disorder using data from social media (De Choudhury et al., 2013; Yates et al., 2017; Orabi et al., 2018; Aragón et al., 2019; Fine et al., 2020; Uban et al., 2021; Nguyen et al., 2022; Wang et al., 2024; Raihan et al., 2024; Abdelkadir et al., 2024). The language used on social media has been shown to predict future depression diagnoses recorded in medical files, suggesting that social media data could be a valuable supplement to traditional depression screening methods (Eichstaedt et al., 2018).

Interdisciplinary research has gained popularity through workshops and shared tasks focused on computational approaches for analyzing mental disorders, including CLPsych (Chim et al., 2024), LT-EDI (Kayalvizhi et al., 2023), eRisk (Parapar et al., 2024), and MentalRiskES (Mármol-Romero et al., 2023). As the research community shows increasing interest in examining how depression is expressed in social media language, we aim to support early-career researchers and anyone interested in this field by providing a comprehensive list of datasets for analyzing or predicting depression using social media data. Our motivation stems from recent changes in the terms of service and API rate limits for popular social media platforms, such as Twitter and Reddit, which have been the primary sources for data collection (Harrigian et al., 2021). These changes have made it more challenging and costly to gather new data. Therefore, we focus on the availability of the datasets in this overview.

The most recent review of social media data for mental health research was conducted by Harrigian et al. (2021), which covered datasets published between 2014 and 2019. Our current work aims to provide an updated overview of social media datasets specifically related to depression research. Since the latest dataset included by Harrigian et al. (2021) is from 2019, our focus will be on datasets published between 2019 and 2024.

This paper contributes to the computational research in depression by providing a meticulously curated, up-to-date, and continuously updated list of data collections.[1] We hope that the resources presented in this overview will further contribute to the interdisciplinary research on depression manifestations in social media language and aid in developing effective interventions for those affected by depression.

[1]We make the list available online at https://github.com/bucuram/depression-datasets-nlp.

Proceedings of the 10th Workshop on Computational Linguistics and Clinical Psychology (CLPsych 2025), pages 116–126
May 3, 2025

2 Methodology

We have conducted a comprehensive literature search on the major publication databases, including ACL Anthology, IEEE Xplore, Scopus, ACM Digital Library, Springer Nature Link, ScienceDirect, and Google Scholar to search for papers using NLP models for depression modeling or papers presenting novel depression-related data collections from social media. We formulated the following search query to retrieve relevant papers:
("depression" OR "depression detection" OR "depression prediction" OR "depression monitoring" OR "depression analysis") AND ("social media" OR "online" OR "Twitter" OR "Reddit" OR "Facebook")

For this overview, we selected papers published between 2019 and 2024 that specifically analyze depression using social media data. We excluded any papers not written in English. To determine if the retrieved papers included analyses related to depression based on social media data or described new data collections, we manually inspected the full texts. We focused on data in the English language. In total, we identified 310 relevant papers, of which 59 proposed new data collections for depression-related research using social media data.

3 Datasets

In Figure 1, we show the number of papers on depression modeling from social media data published each year.

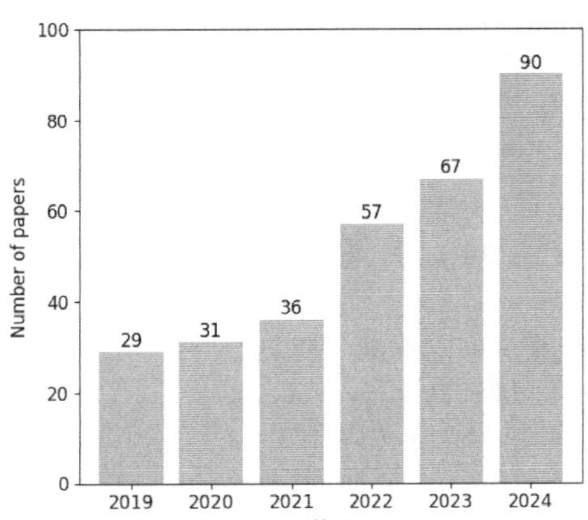

Figure 1: Number of papers on depression modeling published each year in peer-reviewed conferences or journals.

We observe a growing trend in interdisciplinary research on depression, which may have been partly influenced by the COVID-19 pandemic, as there has been an increase in depression rates during this time (Wolohan, 2020; Kaseb et al., 2022). In addition, there has been more research focused on using NLP models for mental health surveillance on social media platforms to assess the pandemic's impact on the population (Dhelim et al., 2023).

In Figure 2, we present the most used datasets in the 310 papers found through our search. Most of the papers have used the datasets from the LT-EDI Workshop (DepSign dataset (Sampath and Durairaj, 2022)), the eRisk Lab (Losada et al., 2017, 2018, 2019, 2020; Parapar et al., 2021; Crestani et al., 2022), or the CLPsych 2015 Shared Task dataset (Coppersmith et al., 2015). All the aforementioned datasets were released as part of shared tasks or competitions, and the data was a valuable resource that was further used after the end of the shared task. Other benchmark datasets are from Shen et al. (2017), Pirina and Çöltekin (2018), or RSDD (Yates et al., 2017).

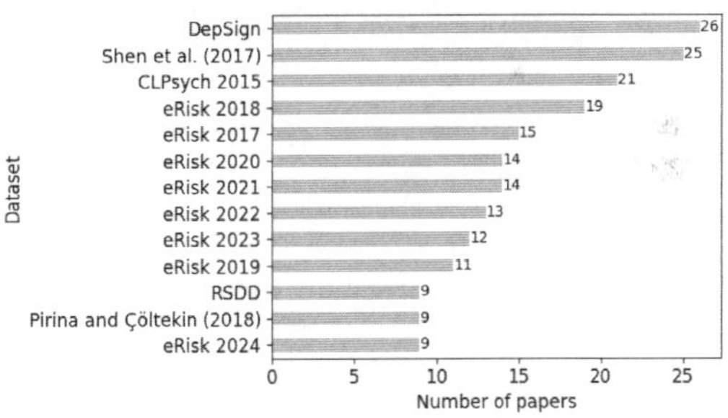

Figure 2: The most used datasets for depression modeling.

The availability of data collections has advanced the development of state-of-the-art depression prediction models. Of the 310 papers published during 2019 and 2024, 59 of them collect and annotate new data from online platforms. In Appendix 6 Table 1, we present detailed information for each of the data collections, such as the platform used for data gathering, the annotation procedure, and the level of annotation (either for each post or user), the labels that are provided for the data, the size of the dataset and its availability.

Platform In Figure 3, we present the social media platforms used for gathering datasets for depression modeling. Reddit and Twitter were the most commonly used platforms for data collection due to easy access to dedicated APIs. However, recent changes in the terms of service and API rate limits for both Twitter / X[2] and Reddit[3] have complicated data collection from these platforms. These updates may hinder the reproduction of datasets where authors only provide Twitter or Reddit IDs instead of the raw text. In addition, these changes make the process of collecting new data more challenging, costly and time-consuming.

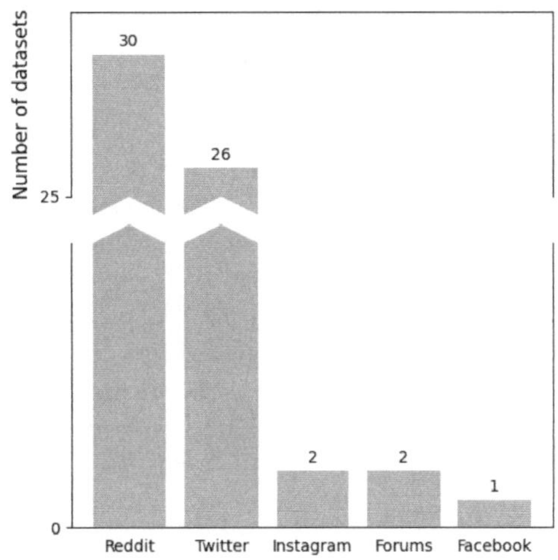

Figure 3: The most used platforms for the data collections presented in this overview.

Annotation procedure and labels For depression detection from social media data, the most common method of annotation from the datasets presented in this work is the annotation based on self-disclosure (Figure 4), labeling users binary, depending on whether they mention online a depression diagnosis or not. In 20 of the data collections, researchers use self-mentions of depression diagnoses (e.g., "I was diagnosed with depression") for their annotation processes. This approach allows for the compilation of large datasets containing hundreds of thousands of users.

Another common annotation procedure is manual annotation, used for 18 of the data collections. These annotations can be performed by mental health experts, graduate students, or laypeople. Most procedures for manual annotations are performed at the post level. Manual annotation is used

[2]https://developer.twitter.com/en/docs/twitter-api/rate-limits

[3]https://support.reddithelp.com/hc/en-us/articles/16160319875092-Reddit-Data-API-Wiki

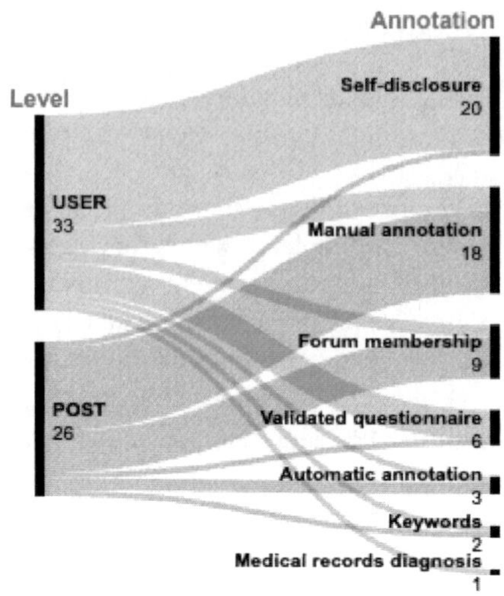

Figure 4: Overview of the annotation levels within each dataset, at either the user or post level, along with the procedures used for annotation.

to label the data binary (depression vs. control), to label data for depression severity (no signs of depression, mild, moderate, severe, etc.), and for symptoms measured by different validated questionnaires, or symptoms from The Diagnostic and Statistical Manual of Mental Disorders, Fifth Edition (DSM-V) (American Psychiatric Association, 2013). Recently, datasets have shifted from binary labeling to labeling based on depression symptoms, leading to the development of explainable methods for depression modeling (Pérez et al., 2023c; Bao et al., 2024).

Data annotation can also be performed by asking social media users to fill in validated self-report questionnaires, such as the Beck's Depression Inventory (BDI) or Patient Health Questionnaire-9 (PHQ-9). However, even if psychometric tools produce a more reliable assessment of depression, fewer people are willing to participate in the data collection, resulting in small sample sizes. Only six datasets rely on self-report questionnaires for the annotation procedure, and one of them relies on the diagnosis from medical records.

Another method for annotation, which is noisier and more prone to errors, is labeling posts by the presence of specific depression-related keywords or automatic annotation performed via an NLP model trained on mental health data. These methods are used less frequently in the data collections included in this overview, with only three data collections being labeled automatically and two datasets being labeled using depression-related keywords.

Availability Due to the sensitive nature of the information in the datasets used for depression modeling, their availability varies. Our exploration of data availability was inspired by the work of Harrigian et al. (2021). However, unlike their study, we have decided not to consider datasets that can be reproduced using APIs from social media platforms as readily available. This decision was influenced by recent changes in the terms of service of platforms such as Reddit and Twitter / X, which have complicated the reproduction of data and made it difficult to retrieve social media posts using the IDs included in the data collections via APIs.

Out of the 59 papers proposing new datasets, 16 are publicly available and hosted online for anyone to use, 15 can be made available after signing a data usage agreement, and 11 collections can be made available by contacting the authors of the dataset. The availability of the rest of the datasets is unknown.

4 Discussion

Data availability One of the primary motivations for this overview were the recent changes in social media platforms, which may hinder the development of new research collections. Our aim was to provide the research community with a comprehensive list of data collections that can be used for interdisciplinary research on the manifestations of depression in social media. We included availability information for each dataset in this overview. We have found that 16 of the datasets are publicly available and free for anyone interested to download and use. As detailed in Section 3, data collections that were part of shared tasks or easily accessible were successfully used by the research community.

Annotation reliability One common method for user-level labeling involves relying on individuals to self-disclose their depression diagnoses. However, this approach is not reliable. Even when annotators manually review posts that contain self-disclosed information, there is no way to verify the authenticity of these disclosures or the accuracy of the users' statements. In addition, for the control group, which includes users who do not mention any depression diagnoses, their actual mental health status remains unknown. We cannot assume that these individuals do not suffer from mental disorders because they have not disclosed this information. It is essential to recognize that relying on self-reported diagnoses for mental health data collection can lead to self-selection bias (Amir et al., 2019). This means that the data obtained may only represent individuals who are willing to openly discuss their mental health issues, which may not accurately reflect the entire population of people with mental disorders.

5 Conclusion and Future Work

We presented a comprehensive and up-to-date overview of datasets used for depression modeling from social media data. We review papers published in international conferences and journals between 2019 and 2024. Due to the research community's efforts to organize shared tasks, the availability of benchmark datasets has increased, offering researchers the resources to build online screening methods for depression and to analyze the depression-related discourse online.

This paper not only aims to offer information about the available datasets for depression manifestation in social media language, but to encourage further interdisciplinary collaboration and exploration. We hope that the comprehensive list of resources provided will inspire researchers, particularly those in the early stages of their careers, to explore this field more deeply. This could lead to a better understanding of depression as expressed in social media and improved interventions.

In this overview, we focused on English datasets, as it is one of the languages that are most used for data collection (Harrigian et al., 2021; Skaik and Inkpen, 2020). However, studying the manifestations of mental health problems in low-resourced languages is an important step toward providing depression screening solutions that can improve the mental health outcomes of people from all around the world (Garg, 2024). In future work, we aim to extend this effort to include social media datasets in languages other than English. Furthermore, we would like to explore the relationship between datasets curated for depression detection and those used in related tasks. This would provide insights on the relationship between depression detection and related social media tasks (Bucur et al., 2021) as well as support multi-task learning efforts (Benton et al., 2017b; Kodati and Tene, 2025).

Limitations

In this paper, we aim to provide a comprehensive overview of the current state of social media data for computational research on depression and present a list of datasets available for researchers in this field. Our study includes 59 data collections, each of which has been carefully reviewed. However, it is possible that we may have overlooked some works that do not explicitly mention depression-related analyses using social media data in their titles or abstracts.

Ethical Considerations

Addressing ethical considerations in mental health research that uses social media data is essential for protecting the privacy, confidentiality, and well-being of individuals whose data is being analyzed (Chancellor and De Choudhury, 2020; Benton et al., 2017a; Chancellor et al., 2019). In this overview, we present the datasets available for studying the manifestations of mental disorders on social media. Although we do not conduct any analyses on the data presented in this work, we want to emphasize that collecting social media data from individuals affected by mental disorders must adhere to ethical research protocols (Benton et al., 2017a). Additionally, researchers who use these datasets should follow the same ethical guidelines and recommendations for health research involving social media.

Acknowledgements

This research was partially supported by the project "Romanian Hub for Artificial Intelligence - HRIA", Smart Growth, Digitization and Financial Instruments Program, 2021-2027, MySMIS no. 334906.

References

Nuredin Ali Abdelkadir, Charles Zhang, Ned Mayo, and Stevie Chancellor. 2024. Diverse perspectives, divergent models: Cross-cultural evaluation of depression detection on twitter. In *Proceedings of the 2024 Conference of the North American Chapter of the Association for Computational Linguistics*, pages 672–680.

V Adarsh, P Arun Kumar, V Lavanya, and G.R. Gangadharan. 2023. Fair and explainable depression detection in social media. *Information Processing & Management*, 60(1):103168.

Soroush Zamani Alavijeh, Fattane Zarrinkalam, Zeinab Noorian, Anahita Mehrpour, and Kobra Etminani. 2023. What users' musical preference on twitter reveals about psychological disorders. *Information Processing & Management*, 60(3):103269.

Falwah Alhamed, Julia Ive, and Lucia Specia. 2024. Classifying social media users before and after depression diagnosis via their language usage: A dataset and study. In *Proceedings of the 2024 Joint International Conference on Computational Linguistics, Language Resources and Evaluation (LREC-COLING 2024)*, pages 3250–3260.

Salma Almouzini, Asem Alageel, et al. 2019. Detecting arabic depressed users from twitter data. *Procedia Computer Science*, 163:257–265.

American Psychiatric Association. 2013. *Diagnostic and statistical manual of mental disorders: DSM-5*, 5th ed. edition. Autor, Washington, DC.

Silvio Amir, Mark Dredze, and John W. Ayers. 2019. Mental health surveillance over social media with digital cohorts. In *Proceedings of the Sixth Workshop on Computational Linguistics and Clinical Psychology*, pages 114–120, Minneapolis, Minnesota. Association for Computational Linguistics.

Ashutosh Anshul, Gumpili Sai Pranav, Mohammad Zia Ur Rehman, and Nagendra Kumar. 2023. A multimodal framework for depression detection during covid-19 via harvesting social media. *IEEE Transactions on Computational Social Systems*.

Mario Ezra Aragón, Adrian Pastor López-Monroy, Luis Carlos González-Gurrola, and Manuel Montes. 2019. Detecting depression in social media using fine-grained emotions. In *Proceedings of NAACL*, pages 1481–1486.

Eliseo Bao, Anxo Pérez, and Javier Parapar. 2024. Explainable depression symptom detection in social media. *Health Information Science and Systems*, 12(1):47.

Krishna C Bathina, Marijn Ten Thij, Lorenzo Lorenzo-Luaces, Lauren A Rutter, and Johan Bollen. 2021. Individuals with depression express more distorted thinking on social media. *Nature human behaviour*, 5(4):458–466.

Rohit Beniwal and Pavi Saraswat. 2024. A hybrid bert-cnn approach for depression detection on social media using multimodal data. *The Computer Journal*, page bxae018.

Adrian Benton, Glen Coppersmith, and Mark Dredze. 2017a. Ethical research protocols for social media health research. In *Proceedings of the EthNLP Workshop*, pages 94–102.

Adrian Benton, Margaret Mitchell, and Dirk Hovy. 2017b. Multitask learning for mental health conditions with limited social media data. In *Proceedings of the 15th Conference of the European Chapter of the Association for Computational Linguistics*, pages 152–162.

Michael Birnbaum, Raquel Norel, Anna Van Meter, Asra Ali, Elizabeth Arenare, Elif Eyigoz, Carla Agurto, Nicole Germano, John Kane, and Guillermo Cecchi. 2020. Identifying signals associated with psychiatric illness utilizing language and images posted to facebook. *npj Schizophrenia*, 6.

Sravani Boinepelli, Tathagata Raha, Harika Abburi, Pulkit Parikh, Niyati Chhaya, and Vasudeva Varma. 2022. Leveraging mental health forums for user-level depression detection on social media. In *Proceedings of LREC*, pages 5418–5427, Marseille, France. European Language Resources Association.

Ana-Maria Bucur, Andreea-Codrina Moldovan, Krutika Parvatikar, Marcos Zampieri, Ashiqur R. KhudaBukhsh, and Liviu P Dinu. 2025. On the state of nlp approaches to modeling depression in social media: A post-covid-19 outlook. *IEEE Journal of Biomedical and Health Informatics*.

Ana-Maria Bucur, Marcos Zampieri, and Liviu P Dinu. 2021. An exploratory analysis of the relation between offensive language and mental health. In *Findings of ACL*.

Junyeop Cha, Seoyun Kim, and Eunil Park. 2022. A lexicon-based approach to examine depression detection in social media: the case of twitter and university community. *Humanities and Social Sciences Communications*, 9(1):1–10.

Stevie Chancellor, Michael L Birnbaum, Eric D Caine, Vincent MB Silenzio, and Munmun De Choudhury. 2019. A taxonomy of ethical tensions in inferring mental health states from social media. In *Proceedings of the conference on fairness, accountability, and transparency*, pages 79–88.

Stevie Chancellor and Munmun De Choudhury. 2020. Methods in predictive techniques for mental health status on social media: a critical review. *NPJ digital medicine*, 3(1):1–11.

Sharath Chandra Guntuku, Daniel Preotiuc-Pietro, Johannes C. Eichstaedt, and Lyle H. Ungar. 2019. What twitter profile and posted images reveal about depression and anxiety. *Proceedings of AAAI ICWSM*, 13(01):236–246.

Jenny Chim, Adam Tsakalidis, Dimitris Gkoumas, Dana Atzil-Slonim, Yaakov Ophir, Ayah Zirikly, Philip Resnik, and Maria Liakata. 2024. Overview of the clpsych 2024 shared task: Leveraging large language models to identify evidence of suicidality risk in online posts. In *Proceedings of the 9th Workshop on Computational Linguistics and Clinical Psychology (CLPsych 2024)*, pages 177–190.

Chun Yueh Chiu, Hsien Yuan Lane, Jia Ling Koh, and Arbee LP Chen. 2021. Multimodal depression detection on instagram considering time interval of posts. *Journal of Intelligent Information Systems*, 56(1):25–47.

Glen Coppersmith, Mark Dredze, Craig Harman, Kristy Hollingshead, and Margaret Mitchell. 2015. Clpsych 2015 shared task: Depression and ptsd on twitter. In *Proceedings of the 2nd Workshop on Computational Linguistics and Clinical Psychology: From Linguistic Signal to Clinical Reality*, pages 31–39.

Fabio Crestani, David E Losada, and Javier Parapar. 2022. *Early Detection of Mental Health Disorders by Social Media Monitoring: The First Five Years of the eRisk Project*, volume 1018. Springer Nature.

Brent Davis, Dawn McKnight, Daniela Teodorescu, Anabel Quan-Haase, Rumi Chunara, Alona Fyshe, and Daniel Lizotte. 2022. Quantifying depression-related language on social media during the covid-19 pandemic. *International Journal of Population Data Science*, 5.

Munmun De Choudhury, Michael Gamon, Scott Counts, and Eric Horvitz. 2013. Predicting depression via social media. In *Proceedings of ICWSM*.

Sahraoui Dhelim, Liming Chen, Sajal K Das, Huansheng Ning, Chris Nugent, Gerard Leavey, Dirk Pesch, Eleanor Bantry-White, and Devin Burns. 2023. Detecting mental distresses using social behavior analysis in the context of covid-19: A survey. *ACM Computing Surveys*.

Johannes C Eichstaedt, Robert J Smith, Raina M Merchant, Lyle H Ungar, Patrick Crutchley, Daniel Preoţiuc-Pietro, David A Asch, and H Andrew Schwartz. 2018. Facebook language predicts depression in medical records. *Proceedings of the National Academy of Sciences*, 115(44):11203–11208.

Ivette Fernández-Barrera, Sebastián Bravo-Bustos, and Mabel Vidal. 2022. Evaluating the social media users' mental health status during covid-19 pandemic using deep learning. In *International Conference on Biomedical and Health Informatics*, volume 14.

Alex Fine, Patrick Crutchley, Jenny Blase, Joshua Carroll, and Glen Coppersmith. 2020. Assessing population-level symptoms of anxiety, depression, and suicide risk in real time using nlp applied to social media data. In *Proceedings of NLP+CSS Workshop*, pages 50–54.

Muskan Garg. 2024. Towards mental health analysis in social media for low-resourced languages. *ACM Transactions on Asian and Low-Resource Language Information Processing*, 23(3):1–22.

Muskan Garg, Chandni Saxena, Sriparna Saha, Veena Krishnan, Ruchi Joshi, and Vijay Mago. 2022. CAMS: An annotated corpus for causal analysis of mental health issues in social media posts. In *Proceedings of LREC*, pages 6387–6396, Marseille, France. European Language Resources Association.

Tao Gui, Liang Zhu, Qi Zhang, Minlong Peng, Xu Zhou, Keyu Ding, and Zhigang Chen. 2019. Cooperative multimodal approach to depression detection in twitter. In *Proc. of AAAI*, volume 33, pages 110–117.

Xiaobo Guo, Yaojia Sun, and Soroush Vosoughi. 2021. Emotion-based modeling of mental disorders on social media. In *Proceedings of IEEE/WIC/ACM WI-IAT*, pages 8–16.

Shrey Gupta, Anmol Agarwal, Manas Gaur, Kaushik Roy, Vignesh Narayanan, Ponnurangam Kumaraguru, and Amit Sheth. 2022. Learning to automate follow-up question generation using process knowledge for depression triage on Reddit posts. In *Proceedings of CLPsych Workshop, NAACL*, pages 137–147, Seattle, USA. ACL.

Ayaan Haque, Viraaj Reddi, and Tyler Giallanza. 2021. Deep learning for suicide and depression identification with unsupervised label correction. In *Artificial Neural Networks and Machine Learning – ICANN 2021*, pages 436–447, Cham. Springer International Publishing.

Keith Harrigian, Carlos Aguirre, and Mark Dredze. 2021. On the state of social media data for mental health research. In *Proceedings of CLPsych Workshop, NAACL*, pages 15–24.

Mohsinul Kabir, Tasnim Ahmed, Md. Bakhtiar Hasan, Md Tahmid Rahman Laskar, Tarun Kumar Joarder, Hasan Mahmud, and Kamrul Hasan. 2023. Deptweet: A typology for social media texts to detect depression severities. *Computers in Human Behavior*, 139:107503.

Abdelrahman Kaseb, Omar Galal, and Dina Elreedy. 2022. Analysis on tweets towards covid-19 pandemic: An application of text-based depression detection. In *Proceedings od NILES*, pages 131–136. IEEE.

S Kayalvizhi, Durairaj Thenmozhi, Bharathi Raja Chakravarthi, SV Kogilavani, Pratik Anil Rahood, et al. 2023. Overview of the shared task on detecting signs of depression from social media text. In *Proceedings of the Third Workshop on Language Technology for Equality, Diversity and Inclusion*, pages 25–30.

Dheeraj Kodati and Ramakrishnudu Tene. 2025. Advancing mental health detection in texts via multi-task learning with soft-parameter sharing transformers. *Neural Computing and Applications*, 37(5):3077–3110.

Daeun Lee, Hyolim Jeon, Sejung Son, Chaewon Park, Ji hyun An, Seungbae Kim, and Jinyoung Han. 2024. Detecting bipolar disorder from misdiagnosed major depressive disorder with mood-aware multi-task learning. In *Proceedings of the 2024 Conference of the North American Chapter of the Association for Computational Linguistics: Human Language Technologies (Volume 1: Long Papers)*, pages 4954–4970.

Yunji Liang, Lei Liu, Yapeng Ji, Luwen Huangfu, and Daniel Dajun Zeng. 2023. Identifying emotional causes of mental disorders from social media for effective intervention. *Information Processing & Management*, 60(4):103407.

Tingting Liu, Devansh Jain, Shivani R Rapole, Brenda Curtis, Johannes C. Eichstaedt, Lyle H. Ungar, and Sharath Chandra Guntuku. 2023a. Detecting symptoms of depression on reddit. In *Proceedings of ACM Web Science Conference*, WebSci '23, page 174–183, New York, NY, USA. ACM.

Yujian Liu, Laura Biester, and Rada Mihalcea. 2023b. Improving mental health classifier generalization with pre-diagnosis data. In *Proceedings of AAAI ICWSM*, volume 17, pages 566–577.

David E Losada, Fabio Crestani, and Javier Parapar. 2017. Clef 2017 erisk overview: Early risk prediction on the internet: Experimental foundations. *CLEF (Working Notes)*, 850.

David E Losada, Fabio Crestani, and Javier Parapar. 2019. Overview of erisk 2019 early risk prediction on the internet. In *Proc. of CLEF*, pages 340–357. Springer.

David E. Losada, Fabio Crestani, and Javier Parapar. 2020. Overview of erisk 2020: Early risk prediction on the internet. In *Experimental IR Meets Multilinguality, Multimodality, and Interaction: 11th International Conference of the CLEF Association, CLEF 2020, Thessaloniki, Greece, September 22–25, 2020, Proceedings*, page 272–287, Berlin, Heidelberg. Springer-Verlag.

David E. Losada, Fabio A. Crestani, and Javier Parapar. 2018. Overview of erisk 2018: Early risk prediction on the internet (extended lab overview). In *CLEF (Working Notes)*.

Ivan Mihov, Haiquan Chen, Xiao Qin, Wei-Shinn Ku, Da Yan, and Yuhong Liu. 2022. Mentalnet: Heterogeneous graph representation for early depression detection. In *Proceedings of ICDM*, pages 1113–1118. IEEE.

Kirill Milintsevich, Kairit Sirts, and Gaël Dias. 2024. Your model is not predicting depression well and that is why: A case study of primate dataset. In *Proceedings of the 9th Workshop on Computational Linguistics and Clinical Psychology (CLPsych 2024)*, pages 166–171.

Anna Monreale, Benedetta Iavarone, Elena Rossetto, and Andrea Beretta. 2022. Detecting addiction, anxiety, and depression by users psychometric profiles. In *Companion Proceedings of the Web Conference 2022*, WWW '22, page 1189–1197, New York, NY, USA. ACM.

Alba María Mármol-Romero, Adrián Moreno-Muñoz, Flor Miriam Plaza-del Arco, María Dolores Molina-González, Maria Teresa Martín-Valdivia, Luis Alfonso Ureña-López, and Arturo Montejo-Raéz. 2023. Overview of mentalriskes at iberlef 2023: Early detection of mental disorders risk in spanish. *Procesamiento del Lenguaje Natural*, 71:329–350.

Isuri Anuradha Nanomi Arachchige, Vihangi Himaya Jayasuriya, and Ruvan Weerasinghe. 2021. A dataset for research on modelling depression severity in online forum data. In *Proceedings of the Student Research Workshop Associated with RANLP 2021*, pages 144–153, Online. INCOMA Ltd.

Usman Naseem, Adam G. Dunn, Jinman Kim, and Matloob Khushi. 2022. Early identification of depression severity levels on reddit using ordinal classification. WWW '22, page 2563–2572, New York, NY, USA. ACM.

Thong Nguyen, Andrew Yates, Ayah Zirikly, Bart Desmet, and Arman Cohan. 2022. Improving the generalizability of depression detection by leveraging clinical questionnaires. In *Proceedings of ACL*, pages 8446–8459.

Ahmed Husseini Orabi, Prasadith Buddhitha, Mahmoud Husseini Orabi, and Diana Inkpen. 2018. Deep learning for depression detection of twitter users. In *Proceedings of the Fifth Workshop on Computational Linguistics and Clinical Psychology: From Keyboard to Clinic*, pages 88–97.

David Owen, Jose Camacho Collados, and Luis Espinosa-Anke. 2020. Towards preemptive detection of depression and anxiety in twitter. *Preprint*, arXiv:2011.05249.

Javier Parapar, Patricia Martin-Rodilla, David E Losada, and Fabio Crestani. 2021. Overview of erisk 2021: Early risk prediction on the internet.

Javier Parapar, Patricia Martín-Rodilla, David E. Losada, and Fabio Crestani. 2024. Overview of erisk 2024: Early risk prediction on the internet. In *Experimental IR Meets Multilinguality, Multimodality, and Interaction*, pages 73–92, Cham. Springer Nature Switzerland.

Jahandad Pirayesh, Haiquan Chen, Xiao Qin, Wei-Shinn Ku, and Da Yan. 2021. Mentalspot: Effective early screening for depression based on social contagion. In *Proceedings of the 30th ACM International Conference on Information & Knowledge Management*, CIKM '21, page 1437–1446, New York, NY, USA. ACM.

Inna Pirina and Çağrı Çöltekin. 2018. Identifying depression on Reddit: The effect of training data. In *Proceedings of the 2018 EMNLP Workshop SMM4H: The 3rd Social Media Mining for Health Applications Workshop & Shared Task*, pages 9–12.

YHPP Priyadarshana, Zilu Liang, and Ian Piumarta. 2023. Heladepdet: A novel multi-class classification model for detecting the severity of human depression. In *International Conference on Collaboration Technologies and Social Computing*, pages 3–18. Springer.

Anxo Pérez, Marcos Fernández-Pichel, Javier Parapar, and David E Losada. 2023a. Depresym: A depression symptom annotated corpus and the role of llms as assessors of psychological markers. *arXiv preprint arXiv:2308.10758*.

Anxo Pérez, Javier Parapar, álvaro Barreiro, and Silvia Lopez-Larrosa. 2023b. Bdi-sen: A sentence dataset for clinical symptoms of depression. In *Proceedings of ACM SIGIR*, pages 2996–3006.

Anxo Pérez, Paloma Piot-Pérez-Abadín, Javier Parapar, and álvaro Barreiro. 2023c. Psyprof: A platform for assisted screening of depression in social media. In *Advances in Information Retrieval*, pages 300–306, Cham. Springer Nature Switzerland.

Nishat Raihan, Sadiya Sayara Chowdhury Puspo, Shafkat Farabi, Ana-Maria Bucur, Tharindu Ranasinghe, and Marcos Zampieri. 2024. Mentalhelp: A multi-task dataset for mental health in social media. In *Proceedings of the 2024 Joint International Conference on Computational Linguistics, Language Resources and Evaluation (LREC-COLING 2024)*, pages 11196–11203.

Esteban A Ríssola, Seyed Ali Bahrainian, and Fabio Crestani. 2020. A dataset for research on depression in social media. In *Proceedings of UMAP*.

Ramin Safa, Peyman Bayat, and Leila Moghtader. 2022. Automatic detection of depression symptoms in twitter using multimodal analysis. *The Journal of Supercomputing*, 78.

Kayalvizhi Sampath and Thenmozhi Durairaj. 2022. Data set creation and empirical analysis for detecting signs of depression from social media postings. In *Computational Intelligence in Data Science*, pages 136–151, Cham. Springer International Publishing.

Guangyao Shen, Jia Jia, Liqiang Nie, Fuli Feng, Cunjun Zhang, Tianrui Hu, Tat-Seng Chua, and Wenwu Zhu. 2017. Depression detection via harvesting social media: A multimodal dictionary learning solution. In *Proc. of IJCAI*, pages 3838–3844.

Eli Sherman, Keith Harrigian, Carlos Aguirre, and Mark Dredze. 2021. Towards understanding the role of gender in deploying social media-based mental health surveillance models. In *Proceedings of CLPsych Workshop, NAACL*, pages 217–223, Online. ACL.

Asmit Kumar Singh, Udit Arora, Somyadeep Shrivastava, Aryaveer Singh, Rajiv Ratn Shah, Ponnurangam Kumaraguru, et al. 2022. Twitter-stmhd: An extensive user-level database of multiple mental health disorders. In *Proceedings of AAAI ICWSM*, volume 16, pages 1182–1191.

Ruba Skaik and Diana Inkpen. 2020. Using twitter social media for depression detection in the canadian population. In *Proc. of AICCC*, pages 109–114.

Hoyun Song, Jisu Shin, Huije Lee, and Jong C. Park. 2023. A simple and flexible modeling for mental disorder detection by learning from clinical questionnaires. *Preprint*, arXiv:2306.02955.

Sajad Sotudeh, Nazli Goharian, and Zachary Young. 2022. Mentsum: A resource for exploring summarization of mental health online posts. *Preprint*, arXiv:2206.00856.

Tom Tabak and Matthew Purver. 2020. Temporal mental health dynamics on social media. In *Workshop on NLP for COVID-19 (Part 2) at EMNLP 2020*.

Faye Beatriz Tumaliuan, Lorelie Grepo, and Eugene Rex Jalao. 2024. Development of depression data sets and a language model for depression detection: mixed methods study. *JMIR Data*, 5:e53365.

Ana-Sabina Uban, Berta Chulvi, and Paolo Rosso. 2021. An emotion and cognitive based analysis of mental health disorders from social media data. *Future Generation Computer Systems*.

Ana-Sabina Uban, Berta Chulvi, and Paolo Rosso. 2022. Explainability of depression detection on social media: From deep learning models to psychological interpretations and multimodality. In *Early Detection of Mental Health Disorders by Social Media Monitoring*, pages 289–320. Springer.

Miryam Elizabeth Villa-Pérez, Luis A. Trejo, Maisha Binte Moin, and Eleni Stroulia. 2023. Extracting mental health indicators from english and spanish social media: A machine learning approach. *IEEE Access*, 11:128135–128152.

Yuxi Wang, Diana Inkpen, and Prasadith Kirinde Gamaarachchige. 2024. Explainable depression detection using large language models on social media data. In *Proceedings of the 9th Workshop on Computational Linguistics and Clinical Psychology (CLPsych 2024)*, pages 108–126.

Anuradha Welivita, Chun-Hung Yeh, and Pearl Pu. 2023. Empathetic response generation for distress support. In *Proceedings of the 24th Annual Meeting of the Special Interest Group on Discourse and Dialogue*, pages 632–644.

JT Wolohan. 2020. Estimating the effect of covid-19 on mental health: Linguistic indicators of depression during a global pandemic. In *Proceedings of NLP-COVID19 Workshop, EMNLP*.

Jiageng Wu, Xian Wu, Yining Hua, Shixu Lin, Yefeng Zheng, and Jie Yang. 2023. Exploring social media for early detection of depression in covid-19 patients. In *Proceedings of ACM Web Conference*, pages 3968–3977.

Dong Xu, Yi-Lun Wang, Kun-Tang Wang, Yue Wang, Xin-Ran Dong, Jie Tang, and Yuan-Lu Cui. 2021. A scientometrics analysis and visualization of depressive disorder. *Current neuropharmacology*, 19(6):766–786.

Shweta Yadav, Cornelia Caragea, Chenye Zhao, Naincy Kumari, Marvin Solberg, and Tanmay Sharma. 2023. Towards identifying fine-grained depression symptoms from memes. In *Proceedings of the 61st Annual Meeting of the Association for Computational Linguistics*, pages 8890–8905.

Shweta Yadav, Jainish Chauhan, Joy Prakash Sain, Krishnaprasad Thirunarayan, Amit P. Sheth, and Jeremiah Schumm. 2020. Identifying depressive symptoms from tweets: Figurative language enabled multitask learning framework. *CoRR*, abs/2011.06149.

Andrew Yates, Arman Cohan, and Nazli Goharian. 2017. Depression and self-harm risk assessment in online forums. In *Proceedings of the 2017 Conference on Empirical Methods in Natural Language Processing*, pages 2968–2978.

Amir Hossein Yazdavar, Mohammad Saeid Mahdavinejad, Goonmeet Bajaj, William Lee Romine, A. Sheth, Amir Hassan Monadjemi, Krishnaprasad Thirunarayan, John M. Meddar, Annie C. Myers, Jyotishman Pathak, and Pascal Hitzler. 2020. Multimodal mental health analysis in social media. *PLoS ONE*, 15.

Yipeng Zhang, Hanjia Lyu, Yubao Liu, Xiyang Zhang, Yu Wang, and Jiebo Luo. 2021. Monitoring depression trends on twitter during the covid-19 pandemic: Observational study. *JMIR infodemiology*, 1(1):e26769.

Zhiling Zhang, Siyuan Chen, Mengyue Wu, and Kenny Zhu. 2022. Symptom identification for interpretable detection of multiple mental disorders on social media. In *Proceedings of the 2022 Conference on Empirical Methods in Natural Language Processing*, pages 9970–9985, Abu Dhabi, United Arab Emirates. ACL.

Jianlong Zhou, Hamad Zogan, Shuiqiao Yang, Shoaib Jameel, Guandong Xu, and Fang Chen. 2021. Detecting community depression dynamics due to covid-19 pandemic in australia. *IEEE Transactions on Computational Social Systems*, 8(4):982–991.

Hamad Zogan, Imran Razzak, Shoaib Jameel, and Guandong Xu. 2023. Hierarchical convolutional attention network for depression detection on social media and its impact during pandemic. *IEEE Journal of Biomedical and Health Informatics*.

6 Appendix

Table 1: List of available datasets for depression modeling using data posted on online platforms. The labels for availability are the following: **FREE** - the dataset is publicly available and hosted online for anyone to access, **AUTH** - the data can be accessed by contacting the paper's authors, **DUA** - the data is available only after a data usage agreement is signed, **UNK** - the dataset availability is unknown; the authors do not mention if the data is available to the research community.

Dataset	Platform	Level	Annotation Procedure	Label	Size	Availab.
Gui et al. (2019)	Twitter	USER	Self-disclosure	Binary	2.8K users	UNK
Chandra Guntuku et al. (2019)	Twitter	USER	BDI	Binary	887 users	UNK
Almouzini et al. (2019)	Twitter	USER, POST	Manual annotation	Binary	89 users	UNK
eRisk2019 (Losada et al., 2019)	Reddit	USER	BDI-II	BDI filled-in	20 users	DUA
Owen et al. (2020)	Twitter	POST	Manual annotation	Binary	1K posts	FREE
Bathina et al. (2021)	Twitter	USER	Self-disclosure	Binary	1.2K users	AUTH
Ríssola et al. (2020)	Reddit	POST	Self-disclosure, heuristics	Binary	14K posts	DUA
Birnbaum et al. (2020)	Facebook	USER	Medical records diagnosis	Binary	223 users	AUTH
D2S (Yadav et al., 2020)	Twitter	POST	PHQ-9	PHQ-9 symptoms	12K posts	AUTH
eRisk 2020 (Losada et al., 2020)	Reddit	USER	BDI-II	BDI filled-in	90 users	DUA
Tabak and Purver (2020)	Twitter	USER	Self-disclosure	Binary	5K users	UNK
Yazdavar et al. (2020)	Twitter	USER	Manual annotation	Binary	8.7K users	DUA
Haque et al. (2021)	Reddit	POST	Subreddit participation	Depression vs. suicide	1.8K posts	FREE
Chiu et al. (2021)	Instagram	USER	Depression-related keywords	Binary	520 users	UNK
Nanomi Arachchige et al. (2021)	Online forums	POST	Manual annotation	Depression severity	2.1K posts	UNK
Sherman et al. (2021)	Reddit	USER	Self-disclosure	Binary	31K users	DUA
eRisk 2021 (Parapar et al., 2021)	Reddit	USER	BDI-II	BDI filled-in	170 users	DUA
Pirayesh et al. (2021)	Twitter	USER	Self-disclosure	Binary	817 users	AUTH
Guo et al. (2021)	Reddit	USER	Self-disclosure	Labels for multiple disorders	7.9 K users	UNK
Zhang et al. (2021)	Twitter	USER	Self-disclosure	Binary	5K users	UNK
Zhou et al. (2021)	Twitter	USER	Self-disclosure	Binary	1.8M posts	UNK
Safa et al. (2022)	Twitter	USER	Self-disclosure	Binary	1.1 K users	AUTH
Naseem et al. (2022)	Reddit	POST	Manual annotation	Depression severity	3.5 K posts	FREE
PsySym (Zhang et al., 2022)	Reddit	USER, POST	Automatic and manual annotation	DSM-5 symptoms for multiple disorders	26K users, 8.5K posts	AUTH
MHB (Boinepelli et al., 2022)	Online forums	USER	Forum participation	Only depression	9.3K users	FREE
CAMS (Garg et al., 2022)	Reddit	POST	Manual annotation	Causes for depression	3.1 K posts	FREE
Sotudeh et al. (2022)	Reddit	POST	Subreddit participation	Summarization	24 k posts	DUA
Sampath and Durairaj (2022)	Reddit	POST	Manual annotation	Depression severity	16K posts	FREE
eRisk2022 (Crestani et al., 2022)	Reddit	USER	Self-disclosure	Binary	3.1K users	DUA
Monreale et al. (2022)	Reddit	POST	Subreddit participation	Labels for multiple disorders	16 K posts	UNK
PRIMATE (Gupta et al., 2022)	Reddit	POST	Manual annotation	PHQ-9 symptoms	2K posts	DUA
PsycheNet-G (Mihov et al., 2022)	Twitter	USER	Self-disclosure	Binary	591 users	UNK
Twitter-STMHD (Singh et al., 2022)	Twitter	USER	Self-disclosure, manual annotation	Labels for multiple disorders	33K users	FREE
multiRedditDep (Uban et al., 2022)	Reddit	USER	Self-disclosure	Binary	3.7K users	AUTH
Davis et al. (2022)	Reddit	USER	Subreddit participation	Binary	81K users	UNK
Fernández-Barrera et al. (2022)	Flickr	POST	Depression tags	Only depression	14.5K posts	UNK
Cha et al. (2022)	Twitter, Everytime	POST	Lexicon-based automatic annotation	Binary	26M posts, 22K posts	AUTH

Dataset	Platform	Level	Annotation Procedure	Label	Size	Availab.
DEPTWEET (Kabir et al., 2023)	Twitter	POST	Manual annotation	Depression severity	40K posts	FREE
Alavijeh et al. (2023)	Twitter	USER	Self-disclosure	Labels for multiple disorders	1.5K users	FREE
Adarsh et al. (2023)	Reddit	POST	Subreddit participation	Binary	60K posts	UNK
Liu et al. (2023a)	Reddit	POST	Subreddit participation	Symptoms	1.3M posts	FREE
BDI-Sen (Pérez et al., 2023b)	Reddit	POST	Manual annotation	BDI-II symptoms	4.9K posts	DUA
Song et al. (2023)	Reddit	POST	Subreddit participation	Labels for multiple disorders	85K posts	UNK
RedditCE (Liang et al., 2023)	Reddit	POST	Manual annotation	Emotion-cause labels	35K posts	FREE
Liu et al. (2023b)	Reddit, Twitter	USER	Self-disclosure	Binary	205K users, 255 users	UNK
RESTORE (Yadav et al., 2023)	Reddit, Twitter, Pinterest	POST	Manual and automatic annotation	PHQ-9 symptoms	9.8K images	AUTH
Zogan et al. (2023)	Twitter	USER	Self-disclosure	Binary	1.4K users	UNK
Wu et al. (2023)	Twitter	USER	Self-disclosure, manual annotation	Binary	10K users	DUA
DepreSym (Pérez et al., 2023a)	Reddit	POST	Manual annotation	BDI-II symptoms	21K posts	DUA
Villa-Pérez et al. (2023)	Twitter	USER	Self-disclosure	Labels for multiple disorders	6K users	DUA
HelaDepDet (Priyadarshana et al., 2023)	Twitter, Reddit	POST	Manual annotation	Depression severity	40K posts	FREE
Anshul et al. (2023)	Twitter	USER	Self-disclosure, Manual annotation	Binary	1.5K users	FREE
RED (Welivita et al., 2023)	Reddit	POST	Subreddit participation	Labels for multiple disorders	1.2M posts	FREE
Alhamed et al. (2024)	Twitter	USER	Manual annotation	Before/After diagnosis	120 users	FREE
Milintsevich et al. (2024)	Reddit	POST	Manual annotation	Anhedonia	167 posts	DUA
MentalHelp (Raihan et al., 2024)	Reddit	POST	Automatic annotation	Binary	14M posts	FREE
Lee et al. (2024)	Reddit	USER	Manual annotation	Binary	1K users	DUA
Beniwal and Saraswat (2024)	Instagram	POST	Manual annotation	Binary	10K posts	AUTH
Tumaliuan et al. (2024)	Twitter	USER	PHQ-9	Binary	72 users	AUTH

Exploratory Study into Relations between Cognitive Distortions and Emotional Appraisals

Navneet Agarwal
Institute of Computer Science
University of Tartu
navneet.agarwal@ut.ee

Kairit Sirts
Institute of Computer Science
University of Tartu
kairit.sirts@ut.ee

Abstract

In recent years, there has been growing interest in studying cognitive distortions and emotional appraisals from both computational and psychological perspectives. Despite considerable similarities between emotional reappraisal and cognitive reframing as emotion regulation techniques, these concepts have largely been examined in isolation. This research explores the relationship between cognitive distortions and emotional appraisal dimensions, examining their potential connections and relevance for future interdisciplinary studies. Under this pretext, we conduct an exploratory computational study, aimed at investigating the relationship between cognitive distortion and emotional appraisals. We show that the patterns of statistically significant relationships between cognitive distortions and appraisal dimensions vary across different distortion categories, giving rise to distinct appraisal profiles for individual distortion classes. Additionally, we analyze the impact of cognitive restructuring on appraisal dimensions, exemplifying the emotion regulation aspect of cognitive restructuring.

1 Introduction

Understanding the intricate relationship between cognition, emotion, and behavior has long been a central focus of neuroscience and cognitive science. The advent of artificial intelligence (AI) and recent advances in natural language processing (NLP) have enabled computational researchers to contribute to this field by developing models capable of analyzing individuals' mental and emotional states from textual data. Within this rapidly evolving domain, the automated extraction of cognitive patterns that shape emotions and behaviors has gained significant traction, bridging the gap between psychological theories and computational innovation.

Emotions are expressed through various modalities, including tone of voice, facial expressions, gestures, and language, particularly in written text. This multifaceted expression of emotions has attracted the interest from NLP and computational researchers in recent years (Wang et al., 2022; Plaza-del Arco et al., 2024). While discrete emotional states such as anger, joy, and fear are deemed universal and thus form the basis for automated emotion recognition research, a smaller number of studies have explored dimensional models, representing discrete emotions in continuous spaces (Plaza-del Arco et al., 2024). Appraisal theories define emotions as responses that arise from an individual's evaluation of and event's significance to their personal goals and well-being, emphasizing that the quality and intensity of emotional responses depend on appraisals, which are the subjective interpretations of the situation (Moors et al., 2013). In contrast to discrete emotional categories, appraisal theories maps an individual's emotional state to a continuous space with each dimension representing an appraisal dimension. This not only provides a more detailed understanding of a person's state, but also allows comparison between emotions.

Negative thoughts are a natural part of human experience; however, they can have a more profound impact on individuals with mental disorders, often becoming entrenched, automatic, and emotionally triggering. Cognitive distortions refer to irrationally exaggerated negative assessments of oneself or situations (Beck, 1963) and they are linked to the states of depression (Joormann and Stanton, 2016) and anxiety (Yazici-Çelebi and Kaya, 2022). Moreover, cognitive distortions have been found to correlate with the use of non-adaptive emotion regulation strategies (Deperrois, 2022). Cognitive restructuring, also known as cognitive reframing, is a therapeutic intervention designed to encourage a more positive outlook towards situations by addressing these negative thought patterns (Clark, 2013). This technique involves replacing negative thoughts with more neutral or hopeful "reframed

Proceedings of the 10th Workshop on Computational Linguistics and Clinical Psychology (CLPsych 2025), pages 127–139
May 3, 2025 ©2025 Association for Computational Linguistics

thoughts", which provide a softer alternative perspective on the situation.

Although cognitive reappraisal and cognitive restructuring focus on different aspects of cognition—reappraisal aims to change the appraisal of specific events, while restructuring addresses broader thinking patterns—they are both emotion regulation methods that target thoughts to influence emotional states. This suggests the potential for a systematic relationship between emotional appraisal dimensions and cognitive distortions. However, to our knowledge, this relationship has not been thoroughly explored. NLP offers a more accessible and efficient means for conducting such exploratory research compared to traditional psychological studies, which requires recruiting human subjects and eliciting relevant information from them.

Our aim in this work is to bring together emotion appraisals and cognitive distortions, to explore the link between these two different but related psychological constructs. We believe that such a relation (if it exists) could be exploited to define more robust systems for automated emotion regulation. For instance, understanding of such relations could help to devise more deliberate and personalized ways for encouraging cognitive reframing and/or emotional reappraisal. We begin by training appraisal prediction models to perform automated appraisal annotations on a dataset annotated with cognitive distortions enabling a combined analysis of both constructs. We analyze the distribution of appraisal values for each distortion-appraisal pair individually, and find statistically significant relations between cognitive distortions and appraisal dimensions, suggesting that different distortion patterns may exhibit distinct appraisal profiles. Finally, when comparing the appraisal profiles between original and reframed texts, we observed a considerable positive shift in several appraisal dimensions, further demonstrating the link between the two constructs and supporting the need for combined study of the two areas.

2 Appraisal Modeling

In our attempt to analyze the relationship between emotional appraisals and cognitive distortions, we require both appraisal and distortion labels for the same text inputs. No such dataset with both labels is currently available. Therefore, we elected to perform automated data annotation in order to generate the desired labels. Although there are datasets that have been annotated for appraisals (Troiano et al., 2023), these are collected from neutral sources (since cognitive appraisal is a normative phenomenon of all emotions, functional and dysfunctional), and thus these texts are not likely to contain too many cognitive distortions. To verify this, we conducted a preliminary experiment and applied a trained cognitive distortion prediction model to the appraisal-annnotated dataset (Troiano et al., 2023), which resulted in approximately 80% data points assigned to the "no distortion" class, confirming our assumptions. Consequently, the alternate approach was adopted. In particular, we train appraisal prediction model on the *crowd-enVent dataset* (Troiano et al., 2023), and apply it to the *thinking trap dataset* (Sharma et al., 2023b). The remainder of this section explains our methodology for training the appraisal prediction model.

2.1 Crowd-enVent Dataset

The *crowd-enVent* is an emotion and appraisal based corpus of event descriptions collected by Troiano et al. (2023) as part of their research on emotional appraisals. During the data collection process, annotators recalled personal events and annotated them based on their recollection of emotions and feelings they experienced at the time of the event. The dataset contains 6600 event descriptions annotated with 21 appraisal dimensions on a 5-point Likert scale.[1] The dataset is available in pre-defined splits of training (4320 entries), validation (1080 entries) and test (1200 entries) sets.[2] Please refer to the original paper by Troiano et al. (2023) for more details on the dataset.

2.2 Model Architecture

We use a multi-regression model to predict the ratings of all appraisal dimensions simultaneously. Specifically, we adopt the multi-regression model by Milintsevich et al. (2023) who used it for predicting the severity of eight depression symptoms. The original model was a hierarchical model implemented with sentence-transformers to encode longer documents. For our sentence-level prediction task, we forgo of the hierarchical definition of the model and directly use the sentence-level embeddings for final predictions. Furthermore, the

[1] Appendix C provides definitions of appraisal dimensions considered in this study.

[2] Available from `https://www.romanklinger.de/data-sets/crowd-enVent2023.zip`.

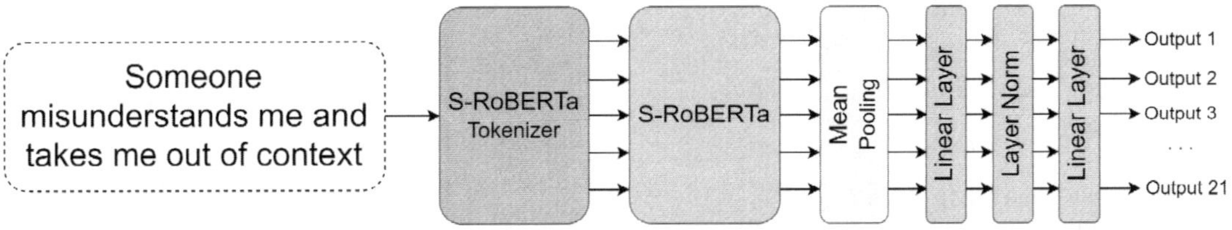

Figure 1: Overview of the appraisal prediction model.

prediction head now produces 21 regression outputs, one for each emotional appraisal dimension considered. Figure 1 provides an overview of the proposed model.

2.3 Experimental Setup

We use the S-RoBERTa Base model for encoding the input text, which is combined with the corresponding S-RoBERTa tokenizer[3]. The model is trained using *AdamW* optimizer with a learning rate of 10^{-5} and *SmoothL1Loss* as the loss function. The network architecture applies a dropout of 0.3, along with layer norm regularization in the regression head. The overall code is a modified version of work by Milintsevich et al. (2023).[4]

2.4 Results

Our model performs on par with the results from Troiano et al. (2023). We use the root-mean-squared error (RMSE) as the evaluation metric and report macro-RMSE averaged over all appraisal dimensions on the test set of 1.36 compared to 1.40 reported in the original paper. Furthermore, Figure 2 compares the performance of the two models for each appraisal dimension, with our model outperforming Troiano et al. (2023) for 13 out of 21 appraisal dimensions. Figure 2 also reports the RMSE of just predicting the median ratings for individual appraisal dimensions, with both trained models performing better than the baseline median predictor in most dimensions. The average RMSE of the median predictor was 1.55, which is clearly worse than the trained models. Thus, we deem our trained model good enough to be used for automated appraisal annotation in the remainder of the paper.

[3]https://huggingface.co

[4]`https://git.unicaen.fr/kirill.milintsevich/hierarchical-depression-symptom-classifier`

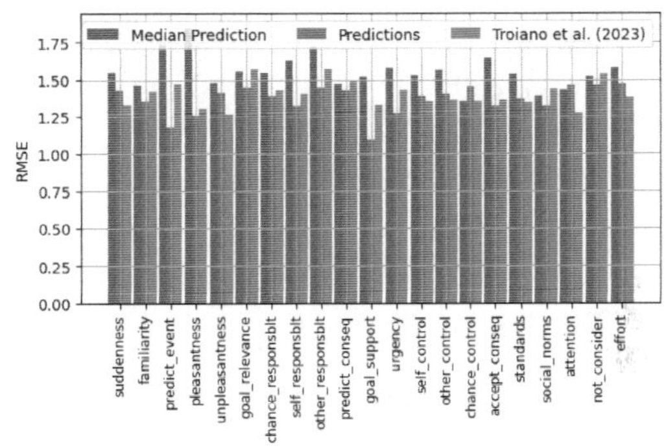

Figure 2: Appraisal ratings prediction accuracy in RMSE for our model, the results reported by Troiano et al. (2023), and those calculated against median predictions for each appraisal dimension as baseline.

3 Cognitive Distortion and Emotional Appraisal

With the aim of analyzing the relationship between appraisals and distortions, we apply the appraisal prediction model trained in the previous section to the *Thinking Trap dataset* (Sharma et al., 2023b), obtaining a dataset with 22 labels per input text (one cognitive distortion label and 21 emotional appraisal ratings). The resulting dataset forms the basis for the analysis discussed in the remainder of the paper. This section provides details on the dataset and explains the statistical methods used to analyze the relationships between cognitive distortions and appraisal dimensions.

3.1 Thinking trap Dataset

The *Thinking Trap* dataset was collected by Sharma et al. (2023b) as part of cognitive reframing research. The dataset contains entries from the existing *Thought Records Dataset* (Burger et al., 2021), along with additional entries collected through an online survey on the Mental Health America website, constituting 300 entries in total. This data was further annotated by mental health practitioners and clinical psychology graduate students, provid-

Distortion Class	# entries
All-or-nothing thinking	99
Blaming	34
Catastrophizing	68
Comparing and Despairing	12
Disqualifying the Positive	40
Emotional reasoning	43
Fortune telling	78
Labeling	102
Magnification	15
Mind reading	71
Negative feeling or emotion	151
Overgeneralization	107
Personalization	98
Should statements	22
Not distorted	96

Table 1: Statistics of the cognitive distortion labels of our version of the Thinking Trap dataset.

ing annotations for 15 cognitive distortion labels (14 distortions and one "no distortion" class) addressed in the text along with corresponding reframed thoughts[5].

The original annotation process of the *Thinking Trap* dataset resulted in a multi-label dataset, with each text associated with one or more distortion labels. Because in this research we are interested in each cognitive distortion category separately, we converted the dataset into a multi-class format by repeating the data points once for each associated distortion label. The resulting dataset contained only 19 data points belonging to the "no distortion" class, amounting to only 1.9% of the total data. Because we wanted to contrast appraisal profiles for texts with and without cognitive distortions, we included 77 data points with the "no distortion" class from an additional dataset also collected by the same authors. The final dataset contains 1036 data points with the class distribution provided in Table 1. For more details on the dataset, please refer to the original work by Sharma et al. (2023b).

3.2 Statistical Analysis

To investigate the relationship between cognitive distortions and appraisal, we analyzed the statistical significance between each distortion category and each appraisal dimension. For each distortion-appraisal pair, we formed two groups of texts: a *positive group (p)*, consisting of texts annotated with the cognitive distortion, and a *negative group (n)*, consisting of texts without the distortion. This grouping allowed us to compare appraisal values in the presence and absence of each cognitive distortion.

We performed an independent statistical analysis for each distortion-appraisal pair to isolate the effect of each distortion on the appraisal dimensions. Specifically, we employed the non-parametric Mann-Whitney U Rank Test (Mann and Whitney, 1947) to assess differences between the positive and negative groups. Under the null hypothesis, we posited that there would be no difference in appraisal values between the two groups ($p=n$, where p represents the positive group and n represents the negative group). To account for multiple comparisons across 14 cognitive distortion classes and 21 appraisal dimensions, we applied a Bonferroni correction (Abdi et al., 2007), setting a base p-value of 0.05, which was divided by the number of comparisons, which is 307 (the product of 14 and 21).

3.3 Negative groups

One major consideration is the definition of the negative group within this analysis. The relationship between cognitive distortions and emotional appraisal dimensions, as inferred from the Mann-Whitney test, is strongly influenced by how the positive and negative groups are defined. Although the definition of the positive group p is fixed, the negative group can have different meanings. Therefore, we consider the following three different definitions of the negative group n:

No distortion: in this case, the negative group only contains entries without any distortion, i.e., those belonging to the "no distortion" class. This group represents the appraisal profile of texts without cognitive distortions and acts as a global baseline against which we can compare individual distortion profiles.

Exclusive: here, the negative group contains entries that do not belong to the given distortion class (defining the positive group) but belong to other cognitive distortion classes (excluding "no distortion"). By utilizing this negative group, we can identify differences in appraisal values between various distortion classes. This approach enables us to analyze the appraisal profile of a specific distortion in relation to other distortions, rather than

[5]Please refer to appendix C for distortion definitions.

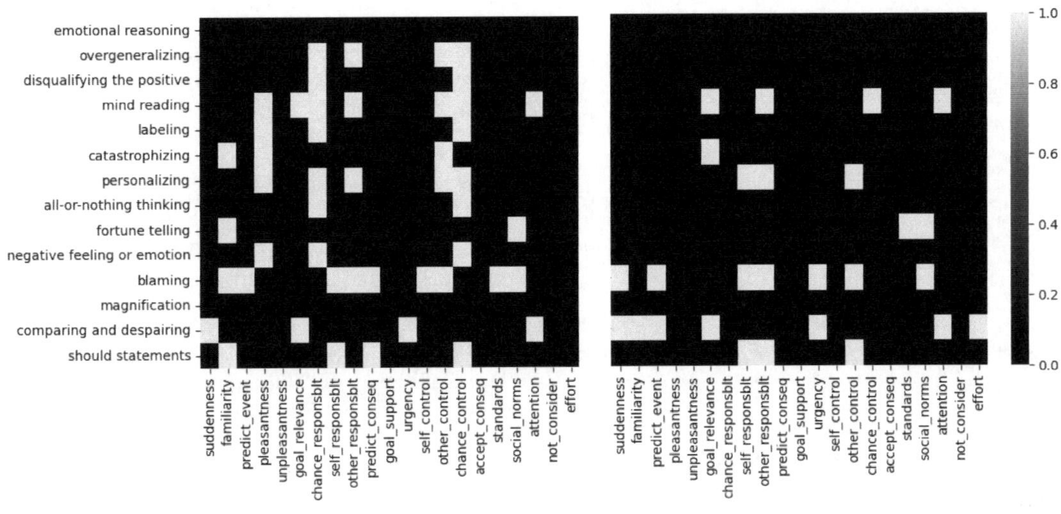

(a) No-distortion (b) Exclusive

Figure 3: Discretized significance plots of distortion-appraisal pairs for two definitions of the negative groups. White cell implies a statistically significant relation between the distortion-appraisal pair, the black cell represents no statistically significant difference.

comparing it to a global baseline.

All others: in this configuration we combine both settings by including in the negative group all entries that do not belong to the specified distortion class (defining the positive group).

The remainder of the paper mostly focuses on the first two categories of negative groups since behavior of *all others* and *exclusive* categories was found to be identical in all our experiments. This can be attributed to the fact that roughly 90% of the data points express a distortion, thereby dominating the behavior of the "no distortion" class.

3.4 Results

Figure 3 plots the discretized significance values for two definitions of the negative groups considered in our study.[6] In these plots, a cell colored white represents a statistically significant difference in appraisal values between the positive and negative groups, while a black cell indicates no statistically significant difference.

First, we observe notable similarities between the two plots. For instance, both plots indicate a lack of statistical significance for *emotional reasoning* and *magnification* across all appraisal dimensions considered. Similarly, *unpleasantness*, *goal support*, and *not consider* dimensions exhibit a lack of statistical significance across all distortion classes. However, we also observe certain differences between the two plots. In the "no distortion" setting (Figure 3(a)), appraisal dimensions like *chance responsibility* and *chance control* show a significant correlation with more than half of the cognitive distortions. However, in the "exclusive" setting (Figure 3(b)), these dimensions lack significant correlation with any of the distortions. While these plots reveal significant correlations between cognitive distortions and appraisal dimensions, they do not indicate the direction of strength of these correlations, thus motivating the further analysis conducted in the next section.

4 Distortions and Corresponding Appraisal Profiles

The previous section showed systematic relations between cognitive distortions and emotional appraisals. In this section, we delve deeper into these correlations, examining their nature, and studying specific appraisal profiles associated with each distortion class.

4.1 Methodology

We begin by defining the "baseline" appraisal profile using the "no distortion" negative group. This choice is motivated by the desire to establish a common baseline that represents the appraisal profile of inputs devoid of any cognitive distortion.

[6]Please refer to Appendix A for the remaining plots.

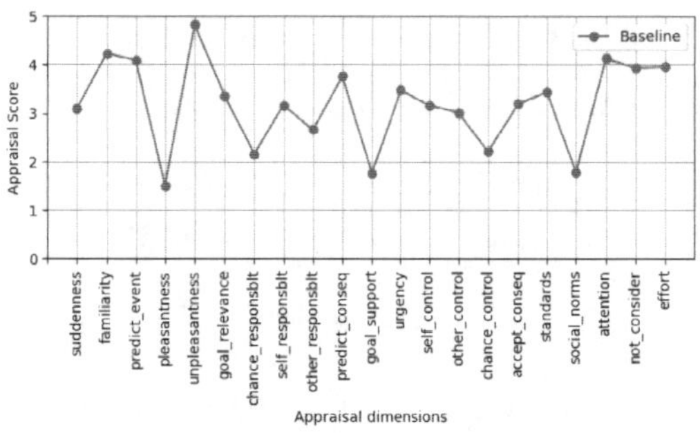

Figure 4: Baseline appraisal profile associated with the "no distortion" class.

Specifically, the baseline profile is determined by calculating the median value for each appraisal dimension using the data from the negative group. Similarly, the appraisal profile for each distortion class is defined by calculating the median value for each appraisal dimension using the corresponding positive group data. Finally, to compare the distortion-specific appraisal profiles with the baseline "no distortion" profile, we subtract the baseline profile from each distortion-specific profile. In this manner, we obtain an appraisal profile for each cognitive distortion that is relative to the baseline profile.

4.2 Results

The baseline profile shown in Figure 4 represents the median appraisal values of event descriptions without any cognitive distortions. Firstly, for most appraisal dimensions the median values cluster around the middle of the scale (between scores 2 and 4). Only four dimensions—*familiarity*, *predictability*, *unpleasantness*, and *attention*—exhibit median values above 4, while three dimensions—*pleasantness*, *goal support*, and *social norms*—have median values below 2. We also notice a high median value for *unpleasantness* stemming from the bias in the *Thinking Trap* dataset, which primarily focused on negative thoughts and situations.

Figure 5 illustrates appraisal profiles associated with individual distortion classes, relative to the baseline profile. The distortion profiles generally exhibit similar patterns, compared to the baseline profile. The two notable exceptions from others are *should statements* and *comparing and despairing*, which display deviations in the dimensions of *familiarity, other's responsibility, and other's control.* Regardless of the similarity of the overall pattern, some cognitive distortions show notable peaks in some appraisal dimensions, such as high *self responsibility* for the *should statements* or high *other's responsibility* and low *self responsibility* for *blaming*.

Note that these plots illustrate the relative differences between the appraisal profiles of cognitive distortions and the baseline profile, but they do not indicate which of the distortion-appraisal relations were statistically significant. The following subsection discusses the appraisal profiles, considering the statistical significance analyses presented in Section 3.

4.3 Discussion

While Figure 3 reveals variations in the statistical significance of appraisal dimension correlations across different settings (presence/absence of distortions), Figure 5 demonstrates that the magnitude and direction of appraisal shifts relative to a non-distorted baseline are broadly similar across most distortion classes. This indicates that the presence of cognitive distortions, rather than their specific type, may be the primary driver of altered emotional experiences.

Furthermore, we observe that the appraisal dimensions of *suddenness*, *unpleasantness*, *goal support*, *accept consequences*, *not consider*, and *effort* exhibit a lack of significant correlation with nearly all distortion classes, as illustrated in Figure 3(a). In Figure 5, these dimensions also demonstrate a relatively balanced distribution of distortion profiles around the baseline. In contrast, the appraisal dimensions of *chance responsibility*, *other's responsibility*, *other's control*, and *chance control* show the highest number of significant correlations with distortion classes, as indicated in Figure 3. Additionally, these dimensions exhibit highly polarized values in their distortion profiles, as seen in Figure 5. This correlation between statistical significance and profile polarization suggests that most cognitive distortions exert similar effects on some appraisal dimensions.

Finally, we illustrate specific distortion-appraisal correlations from Figure 3 that align with established psychological principles. To this end, Figure 6 depicts the appraisal profiles for two distortion classes: *mind reading* and *catastrophizing*. In the case of *mind reading*, the observed appraisal values for responsibility (namely, *self responsibility*, *other's responsibility*, and *chance responsibility*)

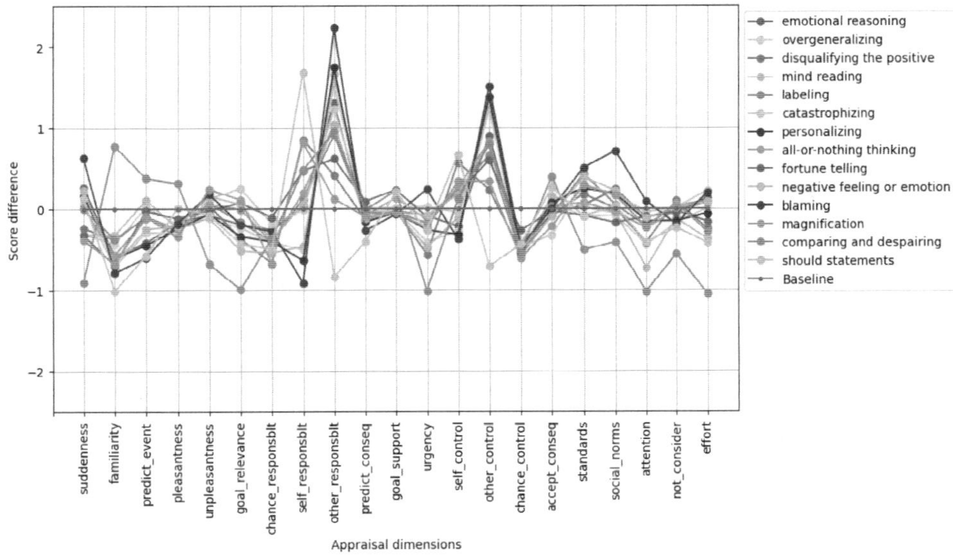

Figure 5: Appraisal profiles (relative to the "no distortion" baseline) for all distortion classes. The x-axis represents different appraisal dimensions considered while the y-axis plots the difference in appraisal scores between individual distortions and the baseline ($score(distortion) - score(baseline)$).

and control (namely, *self control*, *other's control*, and *chance control*) are consistent with the distortion's underlying mechanism. *Mind reading*, by definition, involves assuming knowledge of another person's thoughts and intentions, thereby implicitly attributing greater responsibility and control to that other person rather than to oneself or to chance. Conversely, *catastrophizing* exhibits a negative correlation with the *familiarity* and *accept consequences* dimensions. This is psychologically plausible, as an individual's lack of familiarity with a situation would likely amplify feelings of uncertainty and uncontrollability, making the potential outcomes seem more catastrophic. Furthermore, a reduced ability to accept consequences would logically exacerbate the perceived severity of potential negative outcomes, thus fueling catastrophic thinking.

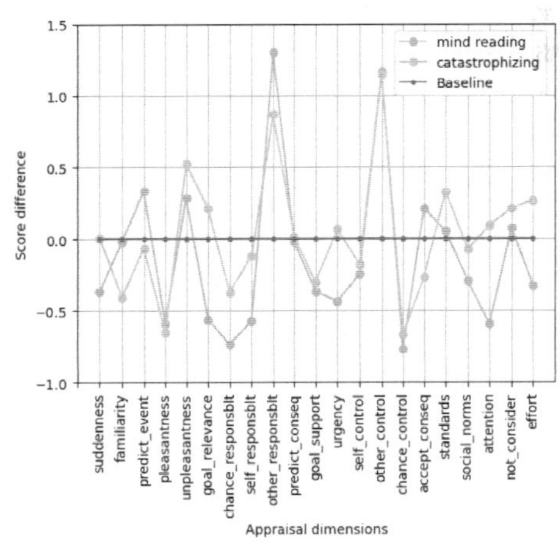

Figure 6: Appraisal profiles (relative to the baseline) for selected distortion classes. The x-axis represents different appraisal dimensions considered while the y-axis plots the difference in appraisal scores between individual distortions and the baseline ($score(distortion) - score(baseline)$).

5 Cognitive Reframing and Emotional Regulation

In the final analysis of this research, we examine the impact of cognitive restructuring on appraisal profiles. Cognitive restructuring aims to regulate an individual's emotional state. Therefore, a significant change in appraisal profiles is expected following the restructuring process. Specifically, we anticipate a positive shift in the appraisal profiles as a result of cognitive restructuring.

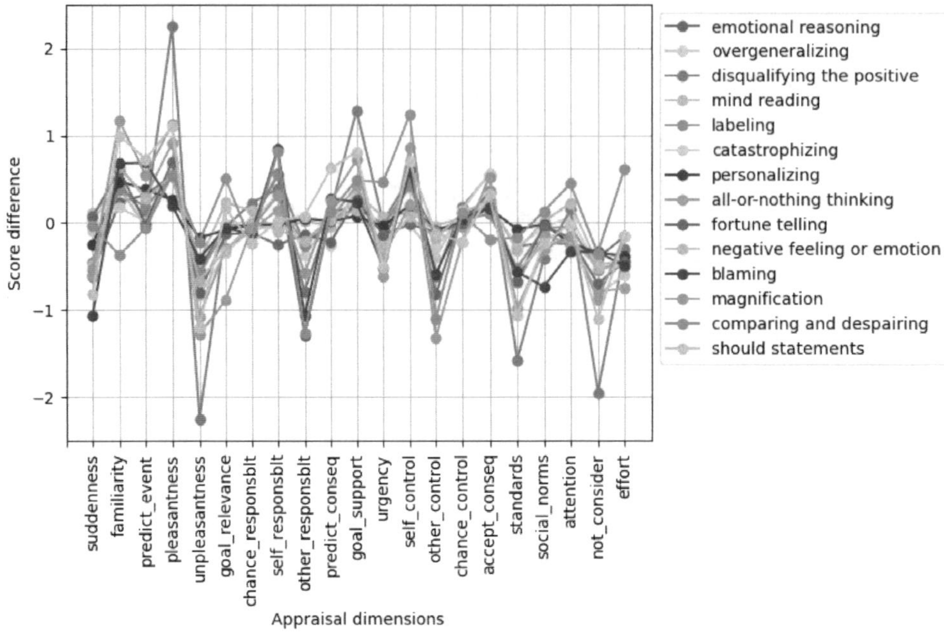

Figure 7: The shift in appraisal profiles after cognitive reframing. Positive values mean that median appraisal value increased after the reframing, while negative values mean indicate a decrease.

5.1 Dataset and Methodology

The *Thinking Trap* dataset, which formed the basis of our analysis thus far, also provides reframings of the input text. Annotators were asked to write reframes that are rational, specific, readable and actionable. Appraisal labels were generated for the reframes using automated annotation, employing the same prediction model trained in Section 2. Then, based on the reframed inputs, appraisal profiles for different distortion classes were generated using the same methodology detailed in Section 4.1.

5.2 Inferences and Interpretations

Figure 7 plots the difference between appraisal profiles generated from the reframed inputs and the original profiles shown in Figure 5. The plots illustrate a considerable increase in values for *pleasantness*, while showcasing a decrease in values for dimensions such as *unpleasantness* and *not consider*. The observed increase in *pleasantness*, along with the decrease in *unpleasantness*, indicates a positive shift in people's perception of the situations. We also observe a decrease in the appraisal dimension *not consider* across all distortion classes, reflecting an increased willingness to engage with, rather than avoid, situations. Overall, the changes in appraisal dimensions plotted in Figure 7 provide evidence that cognitive reframing is associated with a potential positive shift in emotional appraisal. This finding further strengthens the proposed link between cognitive distortions and emotional appraisals, and the necessity of studying these concepts together. Finally, this positive shift in appraisal profiles for reframed texts also supports the validity of the automated appraisal prediction model used in our study.

6 Related Work

Although automatic prediction of emotion appraisals has been less studied than predicting discrete emotions, in recent years, several works have emerged in this direction. Several papers have contributed datasets annotated with different sets of appraisal dimensions (Hofmann et al., 2020; Troiano et al., 2022, 2023; Zhan et al., 2023). Few experiments have been presented to demonstrate the utility of adopting NLP models to predict the appraisal values based on text, training CNN-based neural classifier (Hofmann et al., 2020), fine-tuning RoBERTa-based models (Wegge et al., 2022; Troiano et al., 2023), or prompting large language models (LLMs) (Zhan et al., 2023).

In relation to cognitive distortions, similarly, few datasets annotated with cognitive distortion categories have been published (Shreevastava and Foltz,

2021; Wang et al., 2023; Sharma et al., 2023b) with various classification approaches adopted to predict the cognitive distortions from text, using both fine-tuned BERT-based models (Tauscher et al., 2023; Maddela et al., 2023) and recently also increasingly prompting LLMs (Chen et al., 2023; Lim et al., 2024). Furthermore, NLP researchers have developed methods for a variety of reframing tasks including sentiment and empathy writing (Reif et al., 2022; Sharma et al., 2023a), positive reframing (Ziems et al., 2022; Goel et al., 2024; Jia et al., 2025), and cognitive restructuring (Sharma et al., 2023b; Maddela et al., 2023; Zhan et al., 2024; Xiao et al., 2024).

7 Conclusion

Since both cognitive restructuring and emotional reappraisal serve as emotion regulation strategies, this paper explored the connection between the two constructs from a computational standpoint. As a first step, we automatically annotated appraisal ratings on a dataset of cognitive distortions, producing a new dataset supporting combined analysis. Our analysis at the distortion-appraisal pair level revealed statistically significant relations between cognitive distortions and emotional appraisal dimensions, demonstrating systematic links between the two constructs. By constructing appraisal profiles for individual cognitive distortions and comparing them to a "no distortion" baseline, our analysis showed similar patterns across distortion profiles, indicating a clear distinction between profiles with and without cognitive distortions. Analyzing the impact of cognitive restructuring on the appraisal profiles revealed a shift towards appraisal values indicative of a more positive interpretation of the situations, consistent with the established definition of cognitive reframing. It is our hope that these preliminary results demonstrate the existence of a relationship between cognitive distortions and emotional appraisal dimensions, and illustrate the potential benefits of jointly studying these constructs and motivating further computational research in this area.

Limitations

While we believe this research represents first steps toward understanding correlations between cognitive distortions and emotional appraisals, some concerns remain. A primary issue in computational mental health research is the quality of available data and annotations. The *Thinking Trap* dataset also suffers from the same problem: manual examination reveals instances with incorrect distortion labels. This classification error stems, in part, from the subjective nature of the task, leading to inconsistencies in labeling even among experienced psychologists. Another concern with this dataset is the length of the input texts. Classifying cognitive distortions or any mental health related aspects based on such short texts is unrealistic and contributes to noise in the data. Some examples illustrating this issue are included in Appendix B.

Another major limitation of this work is that it is exploratory research conducted from a computational standpoint, which lacks the in-depth considerations and reasoning from a psychological perspective. This research needs to be complemented by detailed psychological studies, but this is beyond the scope of this paper as well as our knowledge and expertise.

Ethical Considerations

Despite the growing use of NLP (and AI in general) for analyzing the mental and emotional state of individuals based on a variety of input data sources, some ethical considerations need to be taken into account within the research process.

One major area of consideration is the data collection process for such studies. In this work, we use an existing publicly available dataset whose authors reported considering the ethical aspects of their data collection and annotation and also sought approval for their procedures from their institution's review board (see (Sharma et al., 2023b) Section 4.3).

This research direction also comes with significant ethical considerations pertaining to the use of such models. Despite the growing interest in automated systems for mental health analysis and monitoring, improper use of these systems can lead to issues like labeling and stigma. Within our study, we are not developing systems for making predictions about an individual's mental health, but rather studying the general patterns over groups.

References

Hervé Abdi et al. 2007. Bonferroni and šidák corrections for multiple comparisons. *Encyclopedia of measurement and statistics*, 3(01):2007.

Aaron T Beck. 1963. Thinking and depression:

I. idiosyncratic content and cognitive distortions. *Archives of general psychiatry*, 9(4):324–333.

Franziska Burger, Mark A Neerincx, and Willem-Paul Brinkman. 2021. Natural language processing for cognitive therapy: extracting schemas from thought records. *PloS one*, 16(10):e0257832.

Zhiyu Chen, Yujie Lu, and William Wang. 2023. Empowering psychotherapy with large language models: Cognitive distortion detection through diagnosis of thought prompting. In *Findings of the Association for Computational Linguistics: EMNLP 2023*, pages 4295–4304, Singapore. Association for Computational Linguistics.

David A Clark. 2013. Cognitive restructuring. *The Wiley handbook of cognitive behavioral therapy*, pages 1–22.

Romain Deperrois. 2022. Links between cognitive distortions and cognitive emotion regulation strategies in non-clinical young adulthood. *Cognitive Processing*, 23(1):69–77.

Anmol Goel, Nico Daheim, and Iryna Gurevych. 2024. Socratic reasoning improves positive text rewriting. *arXiv preprint arXiv:2403.03029*.

Jan Hofmann, Enrica Troiano, Kai Sassenberg, and Roman Klinger. 2020. Appraisal theories for emotion classification in text. In *Proceedings of the 28th International Conference on Computational Linguistics*, pages 125–138.

Shutong Jia, Biwei Cao, Qingqing Gao, Jiuxin Cao, and Bo Liu. 2025. Positive text reframing under multi-strategy optimization. In *Proceedings of the 31st International Conference on Computational Linguistics*, pages 429–447, Abu Dhabi, UAE. Association for Computational Linguistics.

Jutta Joormann and Colin H Stanton. 2016. Examining emotion regulation in depression: A review and future directions. *Behaviour research and therapy*, 86:35–49.

Sehee Lim, Yejin Kim, Chi-Hyun Choi, Jy-yong Sohn, and Byung-Hoon Kim. 2024. Erd: A framework for improving llm reasoning for cognitive distortion classification. *arXiv preprint arXiv:2403.14255*.

Mounica Maddela, Megan Ung, Jing Xu, Andrea Madotto, Heather Foran, and Y-Lan Boureau. 2023. Training models to generate, recognize, and reframe unhelpful thoughts. In *Proceedings of the 61st Annual Meeting of the Association for Computational Linguistics (Volume 1: Long Papers)*, pages 13641–13660, Toronto, Canada. Association for Computational Linguistics.

H. B. Mann and D. R. Whitney. 1947. On a Test of Whether one of Two Random Variables is Stochastically Larger than the Other. *The Annals of Mathematical Statistics*, 18(1):50 – 60.

Kirill Milintsevich, Kairit Sirts, and Gaël Dias. 2023. Towards automatic text-based estimation of depression through symptom prediction. *Brain Informatics*, 10(1):4.

Agnes Moors, Phoebe C Ellsworth, Klaus R Scherer, and Nico H Frijda. 2013. Appraisal theories of emotion: State of the art and future development. *Emotion review*, 5(2):119–124.

Flor Miriam Plaza-del Arco, Alba A. Cercas Curry, Amanda Cercas Curry, and Dirk Hovy. 2024. Emotion analysis in NLP: Trends, gaps and roadmap for future directions. In *Proceedings of the 2024 Joint International Conference on Computational Linguistics, Language Resources and Evaluation (LREC-COLING 2024)*, pages 5696–5710, Torino, Italia. ELRA and ICCL.

Emily Reif, Daphne Ippolito, Ann Yuan, Andy Coenen, Chris Callison-Burch, and Jason Wei. 2022. A recipe for arbitrary text style transfer with large language models. In *Proceedings of the 60th Annual Meeting of the Association for Computational Linguistics (Volume 2: Short Papers)*, pages 837–848, Dublin, Ireland. Association for Computational Linguistics.

Ashish Sharma, Inna W Lin, Adam S Miner, David C Atkins, and Tim Althoff. 2023a. Human–ai collaboration enables more empathic conversations in text-based peer-to-peer mental health support. *Nature Machine Intelligence*, 5(1):46–57.

Ashish Sharma, Kevin Rushton, Inna Lin, David Wadden, Khendra Lucas, Adam Miner, Theresa Nguyen, and Tim Althoff. 2023b. Cognitive reframing of negative thoughts through human-language model interaction. In *Proceedings of the 61st Annual Meeting of the Association for Computational Linguistics (Volume 1: Long Papers)*, pages 9977–10000, Toronto, Canada. Association for Computational Linguistics.

Sagarika Shreevastava and Peter Foltz. 2021. Detecting cognitive distortions from patient-therapist interactions. In *Proceedings of the Seventh Workshop on Computational Linguistics and Clinical Psychology: Improving Access*, pages 151–158, Online. Association for Computational Linguistics.

Justin S Tauscher, Kevin Lybarger, Xiruo Ding, Ayesha Chander, William J Hudenko, Trevor Cohen, and Dror Ben-Zeev. 2023. Automated detection of cognitive distortions in text exchanges between clinicians and people with serious mental illness. *Psychiatric services*, 74(4):407–410.

Enrica Troiano, Laura Ana Maria Oberlaender, Maximilian Wegge, and Roman Klinger. 2022. x-enVENT: A Corpus of Event Descriptions with Experiencer-specific Emotion and Appraisal Annotations. In *Proceedings of the Thirteenth Language Resources and Evaluation Conference*, pages 1365–1375.

Enrica Troiano, Laura Oberländer, and Roman Klinger. 2023. Dimensional modeling of emotions in text with appraisal theories: Corpus creation, annotation reliability, and prediction. *Computational Linguistics*, 49(1).

Bichen Wang, Pengfei Deng, Yanyan Zhao, and Bing Qin. 2023. C2D2 dataset: A resource for the cognitive distortion analysis and its impact on mental health. In *Findings of the Association for Computational Linguistics: EMNLP 2023*, pages 10149–10160, Singapore. Association for Computational Linguistics.

Yan Wang, Wei Song, Wei Tao, Antonio Liotta, Dawei Yang, Xinlei Li, Shuyong Gao, Yixuan Sun, Weifeng Ge, Wei Zhang, et al. 2022. A systematic review on affective computing: Emotion models, databases, and recent advances. *Information Fusion*, 83:19–52.

Maximilian Wegge, Enrica Troiano, Laura Ana Maria Oberlaender, and Roman Klinger. 2022. Experiencer-specific emotion and appraisal prediction. In *Proceedings of the Fifth Workshop on Natural Language Processing and Computational Social Science (NLP+CSS)*, pages 25–32.

Mengxi Xiao, Qianqian Xie, Ziyan Kuang, Zhicheng Liu, Kailai Yang, Min Peng, Weiguang Han, and Jimin Huang. 2024. Healme: Harnessing cognitive reframing in large language models for psychotherapy. *arXiv preprint arXiv:2403.05574*.

Gülin Yazici-Çelebi and Feridun Kaya. 2022. Interpersonal cognitive distortions and anxiety: The mediating role of emotional intelligence. *International Journal of Psychology and Educational Studies*, 9(3):741–753.

Hongli Zhan, Desmond Ong, and Junyi Jessy Li. 2023. Evaluating subjective cognitive appraisals of emotions from large language models. In *Findings of the Association for Computational Linguistics: EMNLP 2023*, pages 14418–14446.

Hongli Zhan, Allen Zheng, Yoon Kyung Lee, Jina Suh, Junyi Jessy Li, and Desmond C Ong. 2024. Large language models are capable of offering cognitive reappraisal, if guided. *arXiv preprint arXiv:2404.01288*.

Caleb Ziems, Minzhi Li, Anthony Zhang, and Diyi Yang. 2022. Inducing positive perspectives with text reframing. In *Proceedings of the 60th Annual Meeting of the Association for Computational Linguistics (Volume 1: Long Papers)*, pages 3682–3700, Dublin, Ireland. Association for Computational Linguistics.

Appendix

A Statistical analyses

Since the *Thinking Trap* dataset ensured a uniform distribution of data points across different distortion classes (Table 1), the appraisal values of texts from cognitive distortion categories outweigh those from the "no distortion" class, resulting in identical plots for the "Exclusive" (Figure 8(c)) and "All others" (Figure 8(a)) settings.

B Data Limitations and Corresponding Examples

The datasets used in this research also exhibit certain quality issues common to most datasets in the field. In the *Thinking Trap* dataset, some input thoughts are extremely short, making it difficult to assess them due to insufficient information. Some examples of such cases are provided below:

- I had a breakup, I am the cause of the breakup.
- I gained weight, I feel like I need to die to be happy.
- My diet is not working, I feel like a failure.

While these texts may provide some hints about potential cognitive distortions, both models and humans would struggle to accurately assess 21 different emotional appraisal dimensions.

C Label definitions

C.1 Emotion Appraisals

See Table 2 for the definitions of the 21 emotion appraisal dimensions used in this study.

C.2 Cognitive Distortions

See Table 3 for the definitions of the 14 cognitive distortion categories used in this study.

Dimension	Definition
Suddenness	The event was sudden or abrupt
Familiarity	The event was familiar
Event predictability	I could have predicted the occurrence of the event
Pleasantness	The event was pleasant
Unpleasantness	The event was unpleasant
Goal relevance	I expected the event to have important consequences for me
Situational responsibility	The event was caused by chance, special circumstances, or natural forces
Self responsibility	The event was caused by my own behavior
Others responsibility	The event was caused by somebody else's behavior
Anticipated consequence	I anticipated the consequences of the event
Goal support	I expected positive consequences for me
Urgency	The event required an immediate response
Self control	I was able to influence what was going on during the event
Others control	Someone other than me was influencing what was going on
Chance control	The situation was the result of outside influences of which nobody had control
Consequence acceptance	I anticipated that I would easily live with the unavoidable consequences of the event
Internal standards	The event clashed with my standards and ideals
External standards	The actions that produced the event violated laws or socially accepted norms
Attention	I had to pay attention to the situation
Not consider	I tried to shut the situation out of my mind
Effort	The situation required me a great deal of energy to deal with it

Table 2: List of appraisal dimensions considered in this research.

Dimension	Definition
Emotional reasoning	Treating your feelings like facts.
Overgeneralization	Jumping to conclusions based on one experience.
Disqualifying the positive	When something good happens, you ignore it or think it does not count.
Mind reading	Assuming that you know what someone else is thinking.
Labeling	Defining a person based on one action or characterstic.
Catastrophizing	Focusing on the worst/case scenario.
Personalizing	Taking things personally, or making them about you
All-or-nothing thinking	Thinking in extremes.
Fortune telling	Trying to predict the future. Focusing on one possibility and ignoring the other, more likely outcome
Negative feeling and emotion	Getting "stuck" on a distressing thought, emotion, or belief.
Blaming	Giving away your own power to other people.
Magnification	Exaggerating certain aspects of yourself, other people, or a situation while often simultaneously downplaying others.
Comparing and despairing	Comparing your worst to someone else's best.
Should statements	Setting unrealistic expectations of yourself.

Table 3: List of cognitive distortions considered in this research.

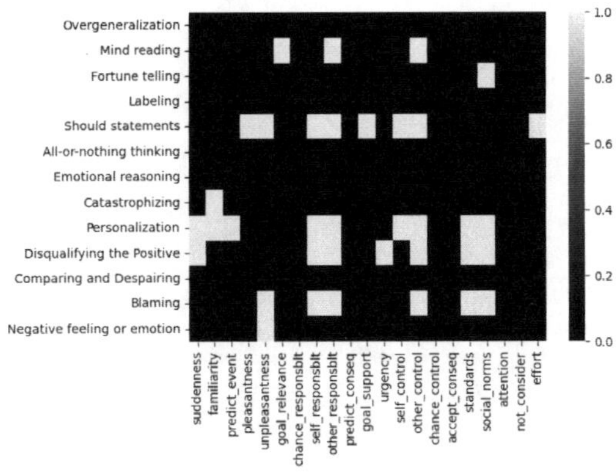

(a) All others

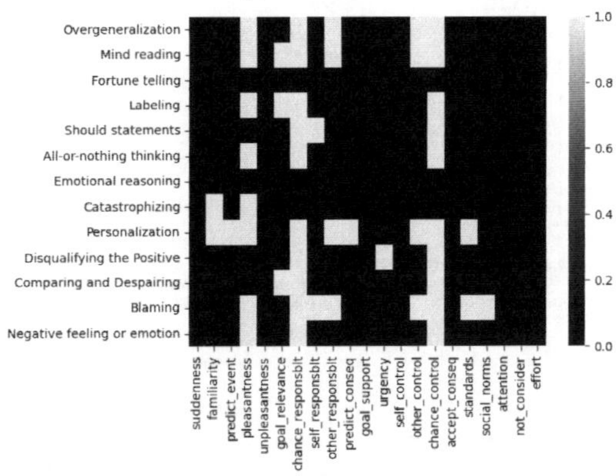

(b) No-distortion

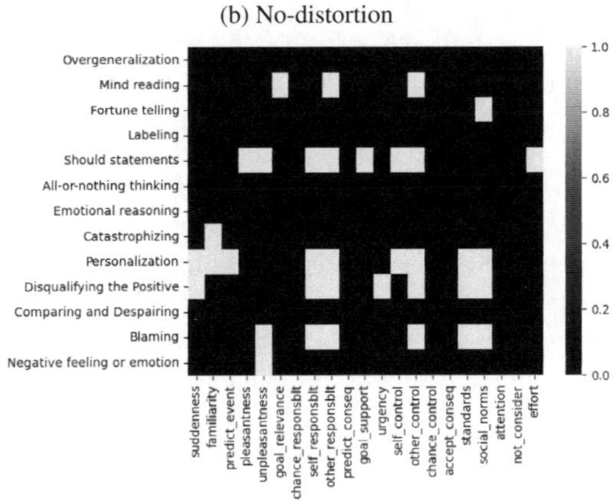

(c) Exclusive

Figure 8: Significance plot between distortions and appraisals for different definitions of negative distribution

Socratic Reasoning Improves Positive Text Rewriting

Anmol Goel[a]**, Nico Daheim**[a]**, Christian Montag**[b]**, Iryna Gurevych**[a]
[a]Ubiquitous Knowledge Processing Lab (UKP Lab)
Department of Computer Science and Hessian Center for AI (hessian.AI)
Technical University of Darmstadt
[b]Centre for Cognitive and Brain Sciences,
Institute of Collaborative Innovation,
University of Macau, Macau, China
www.ukp.tu-darmstadt.de

Abstract

Reframing a negative into a positive thought is at the crux of several cognitive approaches to mental health and psychotherapy that could be made more accessible by large language model-based solutions. Such reframing is typically non-trivial and requires multiple rationalization steps to uncover the underlying issue of a negative thought and transform it to be more positive. However, this rationalization process is currently neglected by both datasets and models which reframe thoughts in one step. In this work, we address this gap by augmenting open-source datasets for positive text rewriting with synthetically-generated Socratic rationales using a novel framework called SOCRATICREFRAME. SOCRATICREFRAME uses a sequence of question-answer pairs to rationalize the thought rewriting process. We show that such Socratic rationales significantly improve positive text rewriting for different open-source LLMs according to both automatic and human evaluations guided by criteria from psychotherapy research. We validate our framework and the synthetic rationalizations with expert judgements from domain experts and psychology students in an IRB-approved annotation study. Our findings highlight the potential of utilizing the synergy between LLM reasoning and established psychotherapy techniques to build assistive solutions for reframing negative thoughts.

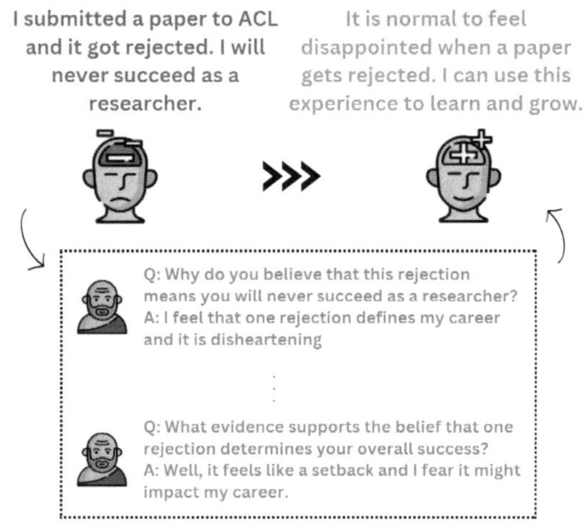

Figure 1: We use Socratic rationales consisting of question-answer pairs to improve positive text rewriting by verbalising the rewriting process. Here, we illustrate this: given a negative thought, we generate Socratic rationales grounded in Cognitive Behavioral Therapy to reframe the original thought into a positive thought.

1 Introduction

Negative thoughts can have a profound effect on human judgement and well-being. Oftentimes negative thoughts overshadow positive thoughts (Vaish et al., 2008), because they can be emotionally deep-rooted and triggering when brought to surface (Beck, 1979). *Cognitive Reframing* is a highly-validated Cognitive Behavioral Therapy (CBT) intervention technique that aims to address this by identifying and reframing negative thoughts into positive ones (Clark, 2013) and has proven useful in both clinical therapeutic and self-help settings (Williams, 2001). Therefore, reframing holds great potential as a strategy for clinicians to provide help in overcoming negative thoughts. Beyond this, teaching people to reframe their own thoughts as a self-guided mental health intervention could provide them with a coping strategy that does not rely on clinical help (Jorm et al., 2006). However, coming up with effective reframes can be challenging and training people, both for clinical and self-help settings, to effectively reframe negative thoughts is a time-consuming and laborious process. Altogether, this poses a risk of mental health improvements being slowed by a lack of availability of knowledge and clinicians.

Large Language Models (LLMs) hold great promise to overcome this lack of clinician support

Proceedings of the 10th Workshop on Computational Linguistics and Clinical Psychology (CLPsych 2025), pages 140–156
May 3, 2025

and self-help resources by providing assistive solutions and could also be used as a training resource. Notably, it has been shown that they can effectively guide text generation towards desired attributes like non-toxicity (Liu et al., 2021; Zheng et al., 2023), style (Reif et al., 2022; Tikhonov et al., 2019), or persona (Song et al., 2020), and could therefore provide a diverse set of reframes. Recent work has also proposed LLM-based style transfer (Krishna et al., 2020; Reif et al., 2022) and empathetic response generation (Cao et al., 2025) methods in downstream applications. This motivates this work, where we aim to improve the reframing performance of LLMs.

While LLMs hold great promise for mental health applications due to their step-by-step reasoning and rationalization capabilities (Kojima et al., 2022, inter alia), they require careful investigation in tandem with domain experts. Several factors need to be taken into consideration before deploying LLM-based applications for client or therapist-facing applications. LLM-based solutions should not only be consistent with the client's mental state, interpretable and avoid hallucinations, but importantly should also follow CBT guidelines.

Therefore, we aim to tackle this gap in literature by making the reasoning of LLMs explicit and include it as a supervision signal during training. We train LLMs for the positive text rewriting task by first generating a rationalization using the Socratic method[1], which uses questioning strategies for uncovering a person's beliefs and motivations, and has shown to be a useful tool for CBT-based cognitive reframing. Socratic reasoning typically uses probing questions to elicit alternative perspectives. Table 1 lists different types of Socratic questions that are common in literature and practice. We call our method SOCRATICREFRAME.

We benchmark SOCRATICREFRAME with various state-of-the-art LLMs for the task of positive text rewriting using both automatic and human evaluations with clinical experts. Our results show that explicitly using Socratic reasoning as a training signal improves the text rewriting performance of LLMs while staying interpretable and faithful to CBT guidelines. We conduct extensive analysis on the synthetically-generated Socratic rationales and find that the socratic rationales are both informative to the model and adhere to clinical standards according to a known Socratic-questioning evaluation scheme. Here, the rationales are rated as highly helpful and relevant for the reframing process.

We use the generated rationales to augment three existing cognitive reframing datasets from related works. We release all data and code publicly to enable subsequent research in the mental health domain, for example, for clinician training.

2 Background & Related Work

2.1 NLP and Mental Health

Research on NLP for mental health has primarily focused on utilizing linguistic features, as well as neural representations, to identify and analyze mental health conditions such as depression (Rinaldi et al., 2020; Yates et al., 2017), anxiety (Juhng et al., 2023; Wei et al., 2021), dyslexia (Björnsdóttir et al., 2023; Gala and Ziegler, 2016), autism (Cho et al., 2022; Goodkind et al., 2018), and schizophrenia (Mitchell et al., 2015; Sarioglu Kayi et al., 2017), among others. Typically, these studies rely on crowdsourced data or annotated social media posts to address the ethical and privacy concerns associated with medical data (Harrigian et al., 2020; Moßburger et al., 2020; Turcan and McKeown, 2019). Recently, efforts have been made to utilize datasets that are more representative of the interactions between a therapist and client in real world settings (Pérez-Rosas et al., 2017; Shapira et al., 2022; Howes et al., 2014; Lee et al., 2019; Cao et al., 2019; Tanana et al., 2015; Shreevastava and Foltz, 2021). Additionally, recent works have utilized synthetic data to enhance the performance of models in clinical (Kazi and Kahanda, 2019; Hiebel et al., 2023; Shim et al., 2021; Lindsay et al., 2022) and mental health (Wu et al., 2023b) settings. Chen et al. (2023b) propose Diagnosis-of-Thought prompting focusing only on diagnosing the type of cognitive distortion given a patient's speech. In contrast, we develop a novel training framework for improving LLM performance on cognitive reframing. Our work aims to contribute to the growing body of literature leveraging positive psychology (Sheng et al., 2023) and LLMs to improve performance on mental health tasks.

2.2 Socratic Questioning

The Socratic method has found wide use in pedagogy (Bautista, 2014) and psychotherapy (Braun et al., 2015), because it can improve understand-

[1]The terminology for the Socratic method is not consistent within the literature (Carey and Mullan, 2004). In the context of this work, we use the term Socratic reasoning to refer to question-answer sequences that follow the Socratic method.

Question Type	Description	Exemplars
Clarification	Questions to go deeper into a thought	"Why do you say that?"
Probing assumptions	Questions to make someone think about unquestioned beliefs	"What could we assume instead?"
Probing reasons and evidence	Questions digging into the reasoning behind a thought	"How did you know that...?"
Probing implications	Questions probing the consequences of a thought	"What is likely to happen if ...?"
Probing alternative viewpoints	Questions about other, equally valid viewpoints	"What is another way to look at it?"
Question about the question	Meta-questions about the question itself	"Why do you think I asked this?"

Table 1: The six types of Socratic questions with representative exemplars from (Paul and Elder, 2019). We synthetically generate Socratic questions spanning all types to improve cognitive reframing with (Section 3).

ing and enable alternative perspectives without being explicit or direct. Instead, it uses questioning strategies to leave room for exploration which is also helpful for other applications, such as tutoring (Macina et al., 2023b). Recent works have used the step-by-step nature of Socratic questioning to improve NLP methods. Ang et al. (2023) collect a large-scale dataset from Reddit annotated with a question type to train Socratic question generation models. Wu et al. (2023a) develop a benchmark of Socratic-inspired deductive reasoning patterns to train state-of-the-art textual entailment and question answering models. Pagnoni et al. (2023) propose Socratic pretraining to enable a question-driven approach for summarizing documents. The scaffolding of Socratic questioning also enables LLMs to solve complex problems by decomposing them into smaller sub-problems. Qi et al. (2023); Shridhar et al. (2022, 2023) use Socratic methods to improve LLM's performance on a variety of reasoning tasks, including math word problem solving and logical reasoning. Our work uses the Socratic method as the core principle to guide a language model while generating a reframed thought.

3 SocraticReframe

Cognitive reframing aims to reframe a negative thought into a positive thought. We approach this with LLMs that generate a positive thought conditioned on the negative thought and metadata. However, directly training an LLM to map positive to negative thought does not make the process behind the reframing explicit, which is important both for interpretability and model performance. Our method SocraticReframe overcomes this by training the model to verbalize a Socratic rationale before generating the reframing.

Formally, cognitive reframing transforms a negative thought $\mathbf{t} \in \mathcal{V}^*$ into a positive thought $\mathbf{r} \in \mathcal{V}^*$. Both are given as strings constructed from a (model) vocabulary $\mathcal{V}$. In addition to the negative thought, metadata $\boldsymbol{\psi} \in \mathcal{V}^*$ is often available which, for example, describes the person experiencing distress and the situation that caused it. One method for automatic reframing is using an autoregressive LLM with parameters $\boldsymbol{\theta}$ to model the distribution

$$p_{\boldsymbol{\theta}}(\mathbf{r} \mid \mathbf{t}, \boldsymbol{\psi}) = \prod_{n=1}^{|\mathbf{r}|} p_{\boldsymbol{\theta}}(r_n \mid \mathbf{t}, \boldsymbol{\psi}, \mathbf{r}_{<n}) \quad (1)$$

paired with a decoding strategy, such as, sampling or greedy decoding. The model can be both finetuned using paired data $\mathcal{D}_{\text{train}} = \{(\mathbf{t}_i, \mathbf{r}_i, \boldsymbol{\psi}_i)\}_{i=1}^N$ or used zero-shot with prompting. While intuitive, this does not allow the model to explicitly reason about and verbalize a rationale of the reframing. This is both less interpretable and might lead to overly-simplistic reframings.

Our proposed method SocraticReframe aims to overcome this by making the model rationalize the thought process behind a specific reframing by means of a sequence of Socratic question-answer pairs. This means that we introduce an additional variable $\mathbf{s} \in \mathcal{V}^*$ that is just a string of this question-answer sequence. The model is then tasked to first generate $\mathbf{s}$ and only then is allowed to generate $\mathbf{r}$. This is outlined in Figure 2. Hence, the model now becomes

$$p_{\boldsymbol{\theta}}(\mathbf{s} \circ \mathbf{r} \mid \mathbf{t}, \boldsymbol{\psi}) = \prod_{n=1}^{|\mathbf{s} \circ \mathbf{r}|} p_{\boldsymbol{\theta}}\big((s \circ r)_n \mid \mathbf{t}, \boldsymbol{\psi}, (\mathbf{s} \circ \mathbf{r})_{<n}\big), \quad (2)$$

where $\mathbf{s} \circ \mathbf{r}$ means that $\mathbf{s}$ is prepended to $\mathbf{r}$. For models that we finetune, this means that each training example is also augmented with a Socratic rationale $\mathbf{s}$ such that $\mathcal{D}_{\text{train}} = \{(\mathbf{t}_i, \mathbf{r}_i, \boldsymbol{\psi}_i, \mathbf{s}_i)\}_{i=1}^N$. During inference, $\mathbf{s}$ is not known and the model generates it before generating the positive thought. This forces the model to explicitly reason about why a specific positive thought is generated to reframe a corresponding negative thought which we expect to

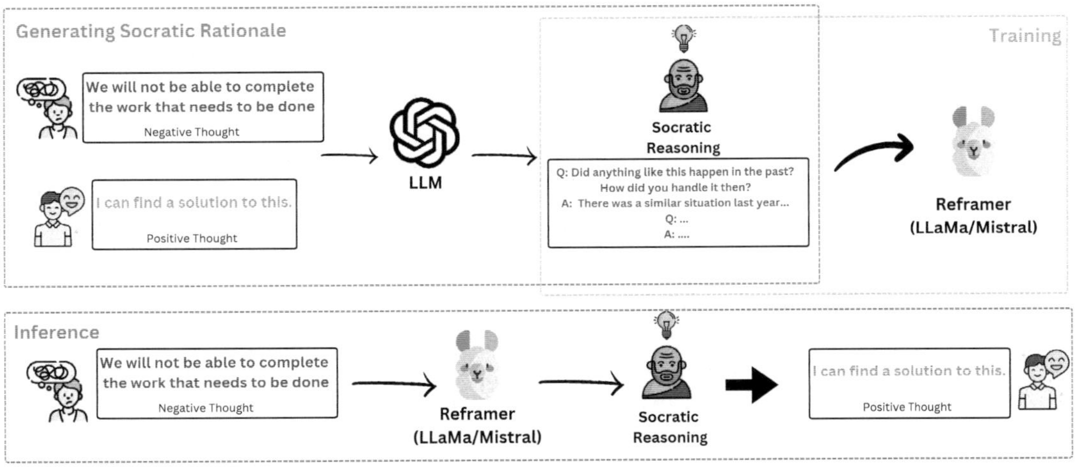

Figure 2: Detailed illustration of our framework SOCRATICREFRAME. First, we generate Socratic rationales using GPT-4 with a few-shot prompt. Then, we use the generated Socratic rationale to train models for cognitive reframing. During inference, the model generates the Socratic rationale before reframing the negative thought.

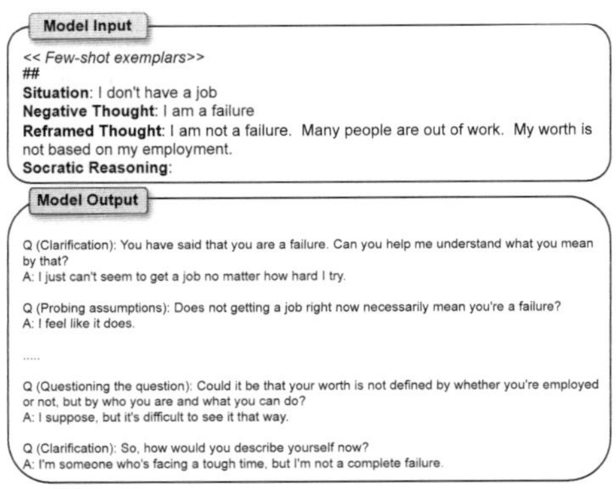

Figure 3: A sample Socratic rationale for an instance from COGREF (Sharma et al., 2023) generated with GPT-4 in a few-shot setting. We use clinical vignettes from psychology literature as few-shot exemplars.

improve both performance and interpretability. In the following section, we discuss how such Socratic rationales can reliably be synthetically generated for existing datasets by leveraging LLMs.

3.1 Generating Socratic Rationales

Our goal is to improve cognitive reframing with Socratic rationales. Such rationales can be time-consuming to write. Therefore, we use synthetically generated data which has recently been successfully used in text classification (Li et al., 2023), dialogue generation (Bao et al., 2023) and question-answering (Riabi et al., 2021), among others.

In particular, we few-shot prompt GPT-4 (Achiam et al., 2023) to generate Socratic question-answer pairs based on negative thought $\mathbf{t}$, positive thought $\mathbf{r}$, metadata ψ, and in-context exemplars of Socratic rationales from clinical literature (Padesky, 1993) to ensure high quality and groundedness. The generated Socratic rationale $\mathbf{s}$ then gives explicit reasoning steps for the particular reframing and can be used to augment the training data $\mathcal{D}_{\text{train}}$. This can be seen as a form of data augmentation and knowledge distillation into the model that is trained for cognitive reframing. As we will show in Section 6, the Socratic rationales are rated favourably by human annotators. An example rationale is shown in Figure 3. Details and examples are in Appendix A and Appendix D. In addition to generating rationales, we prompt GPT-4 to classify the specific type of each generated Socratic question according to the six types described in Table 1, namely "Clarification", "Probing Assumptions", "Probing reasons or evidence", "Probing implications", "Probing alternative viewpoints", and "Question". These can be used for further insights into the reframing process.

4 Empirical Analysis

We show the effectiveness of our method on multiple datasets for positive text rewriting that are outlined in Section 4.1. More details on the exact experimental set-up, such as the models used, are found in Section 4.2 and Appendix A and the used metrics are introduced in Section 4.3.

	POSREF	PATREF	COGREF
Train	6,679	5,249	400
Test	835	18,635	200
ψ	-	Persona	Situation

Table 2: Statistics of the used datasets.

4.1 Datasets

Table 2 describes the datasets used in this work which we further outline in the following.

We use three recently-released open-source datasets. First, Positive Psychology Frames (POSREF) (Ziems et al., 2022) which contains tweets where a hashtag has indicated stress. Each tweet is mapped to a corresponding reframed text which is grounded in a set of reframing strategies, for example, Growth Mindset, Optimism, Self-affirmation. POSREF does not contain any metadata. Then, we use Pattern Reframe (PATREF) (Maddela et al., 2023), a crowdsourced dataset of negative thoughts that are conditioned on personas. These personas describe the person experiencing the respective negative thought. Here, ψ contains the persona-specific information associated with each example. For instance, a persona from the dataset includes "My mother was a teacher. My favorite food is a salad. I enjoy nature. I teach a yoga class. I am single." Finally, we use Cognitive Reframe (COGREF) (Sharma et al., 2023) which is an expert-annotated dataset of situations, thoughts and reframes. Mental health practitioners were prompted with a situation, which can be used as metadata, to generate a negative and a reframed thought. For example, one situation from the dataset is "I participated in a hackathon and I lost".

4.2 Experimental Set-up

We use different LLMs that we either prompt or finetune. Namely, we prompt ChatGPT, LLaMa-2 7B (Touvron et al., 2023) and Mistral 7B (Jiang et al., 2023) and finetune LLaMa 7B and Mistral 7B for cognitive reframing. All models are finetuned using LoRA (Hu et al., 2021) due to computational constraints. For generating Socratic rationales, we use GPT-4. We use the transformers (Wolf et al., 2020) library for all our experiments.

We test the models on different setups. *Few Shot (FS):* Few shot exemplars of negative and positive thought are given to the model which are handcrafted from literature on cognitive reframing. *Chain of Thought (CoT):* The model is asked to think step-by-step and rationalize the reframed thought. [2] *Finetune (FT):* The model is finetuned using LoRA to reframe a negative thought without a Socratic rationale. *Socratic CoT (SoC_{CoT}):* The typical CoT prompt is modified by replacing "think step-by-step" with an instruction to generate Socratic rationale before generating the reframe. *Finetune with Socratic Rationales (SoC):* The model is finetuned using LoRA to reframe a negative thought by first rationalising it using Socratic questioning and then generating the positive reframing. The prompt template used for finetuning is shown in Figure 9.

4.3 Evaluating Cognitive Reframers

We aim to reframe a negative into a positive thought while preserving the original meaning and semantics. Hence, following Hu et al. (2022); Qu et al. (2023), we evaluate them along the criteria of Transfer Strength, which describes how well the thought was transferred to a positive one, and Content Preservation, which evaluates how well the original meaning of the thought was preserved. We describe how we measure both of them in the following.

Transfer strength defines how well negative sentiment is turned positive. To measure it, we first use a finetuned RoBERTa model to evaluate the sentiment scores of the original and reframed thoughts. Then, we use the pairwise difference of sentiment scores between the original and reframed thoughts and report the average, denoted by (ΔPos). We report the number of samples with an increase in positivity with respect to the original thought as accuracy. A decrease in the score from the original thought is considered a failed case (Acc). Sharma et al. (2023) show that people tend to prefer more empathetic reframes over overly positive reframes. Hence, we also report ΔEmp as the difference between empathy scores of original and reframed thoughts. The scores are computed using a pretrained RoBERTa empathy classifier. Our aim is to reframe a thought but also preserve the original meaning and not drastically change it. We rely on the following automatic metrics which have been widely used in the text reframing literature: BLEU (Papineni et al., 2002) and BLEURT (Sellam et al., 2020). Following all three datasets we consider

[2] We observed similar performance for zero-shot and few-shot CoT and therefore only report zero-shot CoT. We believe the similar performance in both setups stems from the nature of the reframing task where there is extra contextual information in the form of ψ and the LLMs are able to generalize well even without few-shot exemplars.

Dataset	Model	Content Preservation (↑)						Transfer Strength (↑)								
		BLEU			BLEURT			ΔPos			Acc. (%)			ΔEmp		
		LLaMa	Mistral	ChatGPT	LLaMa	Mistral	ChatGPT	LLaMa	Mistral	ChatGPT	LLaMa	Mistral	ChatGPT	LLaMa	Mistral	ChatGPT
POSREF	FS	12.6	13.1	15.1	0.53	0.55	0.57	0.54	0.49	0.54	89.15	90.10	92.00	0.58	0.66	0.70
	CoT	12.1	13.3	15.6	0.51	0.55	0.59	0.55	0.48	0.57	90.02	91.16	92.30	0.61	0.69	0.73
	FT	14.3	15.1	-	0.56	0.58	-	0.55	0.51	-	91.46	91.10	-	0.63	0.72	-
	SoC_{CoT}	13.9	13.6	14.5	0.49	0.51	0.55	0.51	0.49	0.51	89.82	90.42	92.12	0.55	0.59	0.71
	SoC	15.7	**15.9**	-	0.58	**0.61**	-	**0.58**	0.52	-	**92.45**	92.09	-	0.69	**0.76**	-
PATREF	FS	71.9	73.0	73.8	0.64	0.65	0.63	0.49	0.78	0.81	97.02	96.51	97.61	0.81	0.80	0.87
	CoT	72.0	73.5	73.9	0.64	0.65	0.65	0.48	0.77	0.83	97.23	96.01	97.60	0.83	0.81	0.89
	FT	72.3	74.1	-	0.66	0.67	-	0.51	0.80	-	97.67	96.79	-	0.89	0.85	-
	SoC_{CoT}	70.2	73.5	71.2	0.59	0.61	0.62	0.48	0.62	0.78	95.02	95.64	96.10	0.80	0.82	0.86
	SoC	**75.2**	75.1	-	**0.69**	0.68	-	0.52	**0.82**	-	**97.90**	97.00	-	**0.90**	0.88	-
COGREF	FS	27.1	27.9	30.0	0.59	0.61	0.63	0.67	0.68	0.69	90.12	90.94	91.90	0.79	0.81	0.88
	CoT	27.0	27.5	31.6	0.59	0.60	0.63	0.67	0.69	0.71	90.0	91.67	92.55	0.82	0.84	0.90
	FT	29.0	30.0	-	0.61	0.62	-	0.69	0.70	-	91.48	91.88	-	0.89	0.91	-
	SoC_{CoT}	26.4	27.1	30.6	0.55	0.59	0.60	0.65	0.68	0.70	90.44	91.54	92.0	0.80	0.85	0.89
	SoC	30.1	**31.7**	-	0.62	**0.63**	-	0.70	**0.72**	-	92.30	**92.56**	-	0.91	**0.94**	-

Table 3: Automatic evaluation results. FS=Few shot. CoT = Chain of Thought. FT=Finetune with LoRA. SoC=finetune with Socratic rationale. Items in **bold** represent the best performance. Note that we do not finetune ChatGPT. We observe that using Socratic rationales for fine-tuning models significantly improves the text rewriting performance across all datasets. Mean values over three runs are reported.

(Ziems et al., 2022; Maddela et al., 2023; Sharma et al., 2023), we report the BLEU scores for each pair of original and reframed thoughts. Since it has been shown that BLEU scores do not always correlate well with human judgements on semantic similarity, we also use a BERT-based BLEU variation, BLEURT-20 which is trained on synthetic samples to get accurate semantic similarity scores. We report the average over a dataset. Each score is a value between 0 and 1, ranging from no to complete semantic similarity.

4.4 Evaluating Socratic rationales

For Socratic rationales to be useful for a model, they should intuitively contain new salient information that is not contained in the negative thought. We use the recently-proposed information-theoretic metric REV (Chen et al., 2023a) to measure this. REV uses conditional $\mathcal{V}$-information to compute the usable information that can be extracted from a variable $(\mathbf{s})$ by a model to predict another variable $(\mathbf{r})$, conditioned on a third variable $(\mathbf{t}, \psi)$. In our setting, this measures the extra information provided by the Socratic rationales beyond what is contained in the original negative thought. More formally, we compute

$$\begin{aligned} \mathrm{REV}(\mathbf{t}, \mathbf{r}, \psi, \mathbf{s}) = & -\log p_{\theta'}(\mathbf{r} \mid \mathbf{t}, \psi) \\ & + \log p_{\theta}(\mathbf{r} \mid \mathbf{t}, \psi, \mathbf{s}), \end{aligned} \tag{3}$$

where θ' and θ are the parameters of two models trained to minimize cross-entropy, respectively. The REV metric can then be computed for an entire corpus $\mathcal{D}$ by averaging the pointwise (per-example) scores. While Chen et al. (2023a) use fine-tuned models on the specific datasets for computing the score, Lu et al. (2023) show that it can even be calculated directly from pretrained LLMs with the same effectiveness as using fine-tuned models (Lu et al., 2023).

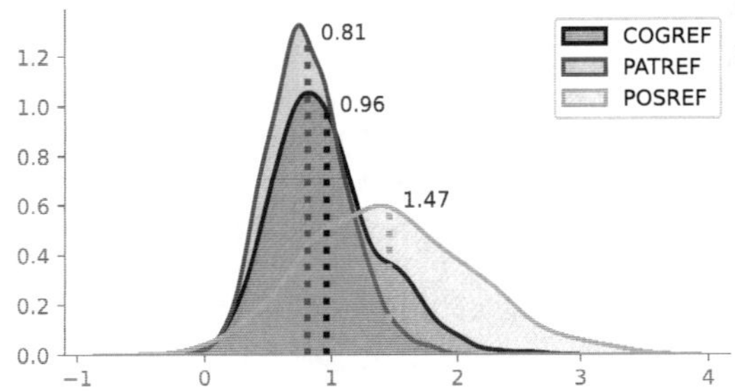

Figure 4: The Socratic rationales generated for all three datasets have positive REV scores, meaning that they indeed provide useful information for text rewriting. Mean of the example-wise REV values are plotted for all three datasets.

5 Results

In this section, we detail our results obtained with different strategies. First, our main results in Section 5.1 show that the Socratic rationales improve text rewriting. Then, we show that the rationales are indeed informative for the model in Section 4.4.

5.1 Main Results

Table 3 shows that using Socratic rationales consistently outperforms both vanilla finetuning and prompting strategies on all datasets. We show an example in Table 4 that highlights how the positive rewritings appear much more thoughtful, because the Socratic rationale forces the model to explicitly reason about the rewriting process. We empirically

Method	Reframed Thought
Mistral-FS	I will improve.
Mistral-CoT	I will improve and learn to be better.
Mistral-FT	I will learn from the feedback and grow.
Mistral-SoC	It is normal to feel disappointed. I can use this experience to learn and grow.

Table 4: Sample reframed thoughts generated by Mistral variants for the original thought: "I submitted a paper to ACL and it got rejected. I will never succeed as a researcher." The additional socratic rationale leads to a more detailed and well-founded reframing.

check whether the generated reframes are close in length to the ground truth reframes for all datasets. We observe that the average length of generated reframes with our best performing model on all datasets is only 5 tokens longer than the ground truth, on average. It is noteworthy that Socratic reasoning improves both the content preservation as well as transfer strength capabilities of smaller models beyond the performance of ChatGPT, despite the model being multiple times larger than LLaMa 7B and Mistral 7B. Across all datasets, we observe that Mistral performs better in content preservation while LLaMA performs better in the sentiment of the reframed thought. This could be due to the different pretraining datasets used for training the models. We also observe that our method consistently generates more empathetic reframes which is generally preferred by people (Sharma et al., 2023). We note that both positivity and empathy are lower for POSREF than other datasets, possibly because POSREF contains tweets which are known to be less positive in nature (Sokolova et al., 2017). We observe that using Socratic reasoning improves BLEU scores by almost 2 points on average when compared to finetuning without Socratic rationales. Distilling Socratic reasoning (from GPT-4) into smaller models helps to preserve the original meaning and improve the positive sentiment transfer. This is desirable for users and practitioners alike, because more positive sentiment will likely benefit users more and content preservation ensures that the reframing stays relevant for them.

5.2 Socratic Rationales are Informative

Similar to previous work for evaluating free-text rationales (Chen et al., 2023a), we use the pretrained GPT-Neo 2.7B (Gao et al., 2020) for computing REV. We report the average REV metric over each sample for all datasets. A REV value > 0 suggests Socratic rationales support reframing by providing additional information, while < 0 indicates otherwise. In Figure 4, all datasets have REV > 0, highlighting Socratic rationale's utility in enhancing reframing tasks. Table 5 shows the generated rationales for each of the dataset we consider. In particular, POSREF has the highest informativeness for the reframed thoughts. We attribute this to two main factors: First, POSREF lacks additional context like situations or personas, causing token dispersion where more information can lead to low values of REV. Second, POSREF samples from Reddit align better with GPT-Neo's training data than the crowdsourced data in the other datasets.

5.3 Socratic Rationales progressively lead to a better reframe

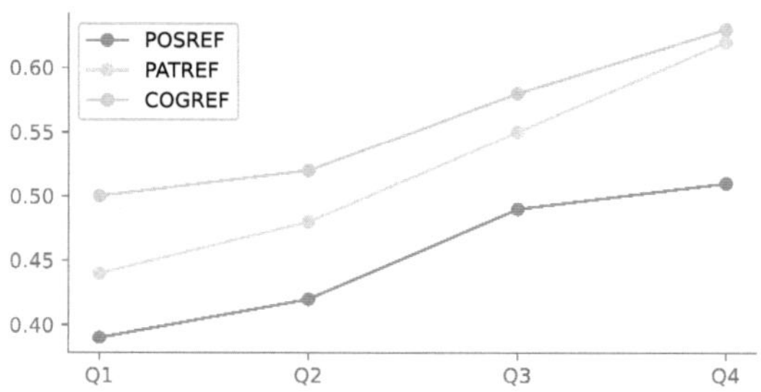

Figure 5: Sentiment scores for the generated Socratic question-answer pairs show that the synthetic rationales progressivley get better to lead to an improved reframe.

We partition Socratic rationales into four quarters sequentially, each representing 25% of the data, and calculate average sentiment scores for each quarter. Figure 5 reveals a consistent trend across all datasets: intermediate answers lead to increasingly positive questions, suggesting that they aid in iterative reframing.

6 Human Evaluation

6.1 Evaluating Reframed Thoughts

Prior cognitive reframing studies (Maddela et al., 2023; Sharma et al., 2023) use Likert scales for assessing aspects like fluency and readability. Yet, our focus lies on helping users overcome unwanted thoughts. Li et al. (2024) show that pairwise human preferences are well-defined. In a similar setup as Sharma et al. (2023), we compare 100 randomly selected SOCRATICREFRAME generated reframes against ChatGPT or a baseline model without Socratic rationales. After consenting to participate, two computer science graduate students were re-

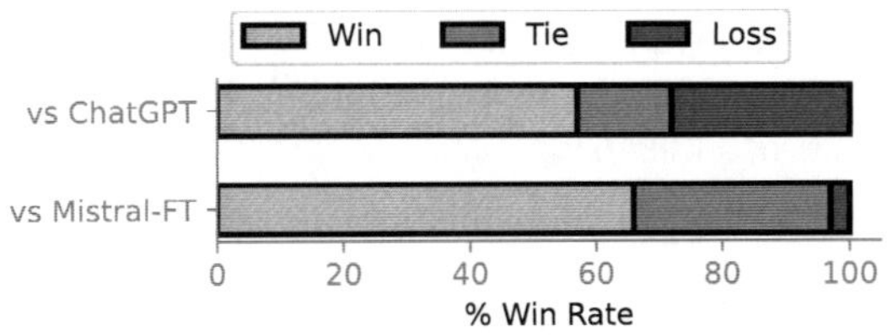

Figure 6: Human evaluation results for Mistral-SoC compared to Mistral-FT and ChatGPT. With a win rate of over 50% human annotators prefer reframes generated by our method to ChatGPT and simple fine-tuning.

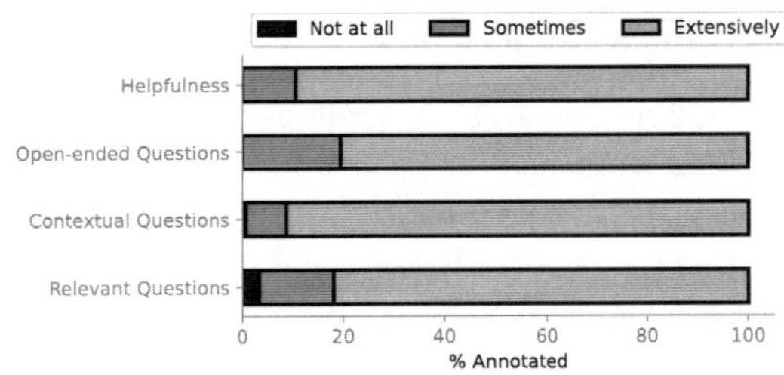

Figure 7: Human evaluation results for the Socratic rationales show that annotators judge the generated rationales as highly helpful and relevant for reframing negative thoughts.

cruited to conduct the comparisons. The win rate indicates a preference for Socratic reframes over the baseline by the raters. The participants were asked to select the reframed thought they find most relatable, helpful and memorable - a criteria for good reframes defined in Sharma et al. (2023).

As shown in Figure 6, Mistral-SoC outperforms Mistral-FT by a significant margin with a win rate of over 65% and tie rate of around 30%. We fit a Bradley-Terry Model (Bradley and Terry, 1952) on the preference data and obtain a strength of 1.3 for Mistral-SoC as compared to strengths -1.8 and 0.5 for Mistral-FT and ChatGPT, respectively. Intuitively, this means that Mistral-SoC will have a win rate of 95% over Mistral-FT and 67% over ChatGPT, confirming the effectiveness of our method.

6.2 Evaluating Socratic Rationales

To evaluate the synthetically generated rationales, we consider two key attributes: whether they represent the use of Socratic questioning in a clinically meaningful way and the general helpfulness of the questioning to overcome negative thoughts. Following Braun et al. (2015), we use the Socratic Questioning Scale (SQS) to evaluate the quality of generated rationales. SQS contains five Likert-scale questions evaluating the use of the Socratic method in a snippet. The final score is the sum of these ratings. Originally developed for therapy session transcripts, we include only three relevant questions for text evaluation: Open-endedness, Context, and Relevance.[3] The full questions are reported in Appendix C.1. Additionally, we consider *helpfulness* as a metric to assess whether or not the generated rationale is generally helpful in overcoming negative thoughts. Annotators rate the generated rationales on a 3-point Likert scale.

Expert Evaluation We recruit 14 mental health experts and clinical psychology graduate students for evaluating the quality of the Socratic rationales. We verify the quality of randomly sampled Socratic rationales with a clinical expert before conducting the study with psychology students. The annotation task was deemed exempt from an ethics review by an institutional review board. Amidei et al. (2019) suggest the use of Cronbach's α for checking the evaluation reliability in natural language generation tasks. We observe acceptably high values for helpfulness (0.67), diversity (0.69), reflectiveness (0.75) and relevance (0.65) of the questions in the rationale. We observe that raters validate the synthetically generated Socratic rationales as helpful to reframe the original thought. More than 50% of the raters also responded affirmatively on whether they are willing to use LLMs for cognitive reframing. More details are found in Appendix C.2.

Human Evaluations Three computer science graduate students annotated 100 randomly sampled Socratic rationales from the datasets. Figure 7 shows favorable ratings across all criteria. We obtained an Intra Class Correlation (ICC) of 0.806 for helpfulness and 0.601 for the SQS scale, indicating that while the SQS is challenging to evaluate, helpfulness is generally agreed upon by non-experts. These findings align with similar agreement scores in psychology (Braun et al., 2015). Additionally, a Pearson correlation coefficient of 0.72 ($p < 0.05$) between helpfulness and SQS ratings suggests that better Socratic questioning correlates with greater helpfulness in addressing negative thoughts.

[3]We note that in our evaluation criteria, "relevant" questions aim to tackle extrinsic hallucinations (generated text cannot be verified given the source) and "contextual" questions aim to tackle intrinsic hallucinations (generated text contradicts the source text), following the definitions from Ji et al. (2023).

7 Clinical Implications

Our findings suggest that large language models (LLMs) augmented with Socratic rationales can serve as valuable tools for mental health professionals in cognitive restructuring and psychotherapy. By breaking down the process of thought reframing into a structured sequence of rationalization steps, SOCRATICREFRAME aligns with established therapeutic approaches such as cognitive behavioral therapy (CBT) and Socratic questioning. This structured reasoning can assist clinicians by:

- **Enhancing Psychoeducation**: Mental health professionals can use LLM-assisted reframing to demonstrate thought restructuring techniques, making abstract cognitive principles more tangible for patients.
- **Supporting Self-Guided Interventions**: LLMs equipped with Socratic reasoning could supplement therapy by providing guided, interactive thought reframing outside of clinical sessions, helping patients develop cognitive flexibility in real time.
- **Reducing Cognitive Load for Clinicians**: – By automating initial rationalization steps, our approach may help therapists focus on more nuanced aspects of patient care, such as emotional processing and behavioral interventions, without sacrificing the rigor of cognitive restructuring.
- **Facilitating Training and Supervision**: Psychology students and early-career therapists could use LLM-generated Socratic rationales as a training aid, learning how to systematically guide patients through cognitive reframing with structured question-answer pairs.

Given these potential applications, our work highlights the importance of integrating AI-driven reasoning techniques with clinical expertise to create assistive tools that complement human-guided therapy rather than replace it. Future research should explore real-world deployment, patient engagement, and the ethical considerations of using LLM-based interventions in mental health contexts.

8 Conclusion & Future Work

In this work, we show that text rewriting models for cognitive reframing can be improved by using Socratic rationales to verbalize the reframing process. By releasing our code and data we hope to enable future research, for example, on incorporating the different question types that are annotated, or using the data in therapist training.

Limitations

While synthetic data generated by LLMs like GPT-4 offers a promising solution to data scarcity and privacy concerns, a drawback could be a potential lack of diversity and complexity. Our generated Socratic rationales may not fully capture all intricate nuances or patterns in authentic therapist dialogues while performing cognitive change with the Socratic method. A significant limitation when working with mental health and psychotherapy-based datasets is the use of heavily curated or crowd-sourced data. This can lead to an imbalance in the demographics and language of our datasets. Since the datasets we use are sourced from social media or crowdsourcing platforms, the quality of the negative thoughts might also be limited.

In addition to the reframed thoughts, the datasets considered in this work also contain the cognitive distortions associated with each negative thought. Cognitive distortions are thought patterns that lead to negative feelings. We believe information like cognitive distortions and reframing strategies can be utilized to further improve the quality of the Socratic rationales and subsequently enhance the text reframing performance.

Finally, our focus in this work was to improve text reframing which is an effective short-term in-the-moment strategy. However, we emphasize that assessing long-term outcomes is imperative and is a future research direction.

Ethical Statement

While this work focused on generating Socratic rationales to improve positive text rewriting, open-ended LLMs still carry the risk of generating harmful outputs. Therefore, we advise careful consideration before they are applied in practice. Unsupervised use of the Socratic questioning data as is, without consulting trained professionals, could be harmful and requires careful design and implementation before being applied in real-world settings.

Datasets Our work augments open-source positive text rewriting datasets, some of which include social media-derived data. Social media users are not representative of clinical populations, and self-reported distress lacks clinical verification. We mitigate this by incorporating expert evaluations to validate our approach. Public posts may contain sensitive disclosures. Our study adheres to IRB guidelines and we only use anonymized, public datasets, ensuring ethical data use. Social media data may reinforce biases or oversimplify mental health struggles. Our framework grounds reframing in psychotherapy principles, minimizing these risks. While social media data is valuable for scalable mental health interventions, expert oversight remains essential to ensure clinically sound applications.

IRB Approval We obtained approval from the Ulm University's Institutional Review Board. In the board's opinion, research using annotations and ratings on the anonymised data does not require consultation within the meaning of Article 23 of the Declaration of Helsinki, meaning that our study was exempt from further inquiry.

Acknowledgements

This research work has been funded by the German Federal Ministry of Education and Research and the Hessian Ministry of Higher Education, Research, Science and the Arts within their joint support of the National Research Center for Applied Cybersecurity ATHENE. This work was supported by the DYNAMIC center, which is funded by the LOEWE program of the Hessian Ministry of Science and Arts (Grant Number: LOEWE/1/16/519/03/09.001(0009)/98).

References

Josh Achiam, Steven Adler, Sandhini Agarwal, Lama Ahmad, Ilge Akkaya, Florencia Leoni Aleman, Diogo Almeida, Janko Altenschmidt, Sam Altman, Shyamal Anadkat, et al. 2023. Gpt-4 technical report. *arXiv preprint arXiv:2303.08774*.

Jacopo Amidei, Paul Piwek, and Alistair Willis. 2019. Agreement is overrated: A plea for correlation to assess human evaluation reliability. In *Proceedings of the 12th International Conference on Natural Language Generation*, pages 344–354, Tokyo, Japan. Association for Computational Linguistics.

Beng Heng Ang, Sujatha Das Gollapalli, and See-Kiong Ng. 2023. Socratic question generation: A novel dataset, models, and evaluation. In *Proceedings of the 17th Conference of the European Chapter of the Association for Computational Linguistics*, pages 147–165, Dubrovnik, Croatia. Association for Computational Linguistics.

Jianzhu Bao, Rui Wang, Yasheng Wang, Aixin Sun, Yitong Li, Fei Mi, and Ruifeng Xu. 2023. A synthetic data generation framework for grounded dialogues. In *Proceedings of the 61st Annual Meeting of the Association for Computational Linguistics (Volume 1: Long Papers)*, pages 10866–10882, Toronto, Canada. Association for Computational Linguistics.

Lowell Bautista. 2014. The socratic method as a pedagogical method in legal education. *University of Wollongong Faculty of Law, Humanities and the Arts - Papers*.

Aaron T Beck. 1979. *Cognitive therapy and the emotional disorders*. Penguin.

Marina Björnsdóttir, Nora Hollenstein, and Maria Barrett. 2023. Dyslexia prediction from natural reading of Danish texts. In *Proceedings of the 24th Nordic Conference on Computational Linguistics (NoDaLiDa)*, pages 60–70, Tórshavn, Faroe Islands. University of Tartu Library.

Ralph Allan Bradley and Milton E. Terry. 1952. Rank analysis of incomplete block designs: I. the method of paired comparisons. *Biometrika*, 39:324.

Justin D Braun, Daniel R Strunk, Katherine E Sasso, and Andrew A Cooper. 2015. Therapist use of socratic questioning predicts session-to-session symptom change in cognitive therapy for depression. *Behaviour research and therapy*, 70:32–37.

Huiying Cao, Yiqun Zhang, Shi Feng, Xiaocui Yang, Daling Wang, and Yifei Zhang. 2025. TOOL-ED: Enhancing empathetic response generation with the tool calling capability of LLM. In *Proceedings of the 31st International Conference on Computational Linguistics*, pages 5305–5320, Abu Dhabi, UAE. Association for Computational Linguistics.

Jie Cao, Michael Tanana, Zac Imel, Eric Poitras, David Atkins, and Vivek Srikumar. 2019. Observing dialogue in therapy: Categorizing and forecasting behavioral codes. In *Proceedings of the 57th Annual Meeting of the Association for Computational Linguistics*, pages 5599–5611, Florence, Italy. Association for Computational Linguistics.

Timothy A Carey and Richard J Mullan. 2004. What is socratic questioning? *Psychotherapy: theory, research, practice, training*, 41(3):217–226.

Hanjie Chen, Faeze Brahman, Xiang Ren, Yangfeng Ji, Yejin Choi, and Swabha Swayamdipta. 2023a. REV: Information-theoretic evaluation of free-text rationales. In *Proceedings of the 61st Annual Meeting of the Association for Computational Linguistics (Volume 1: Long Papers)*, pages 2007–2030, Toronto, Canada. Association for Computational Linguistics.

Zhiyu Chen, Yujie Lu, and William Wang. 2023b. Empowering psychotherapy with large language models: Cognitive distortion detection through diagnosis of thought prompting. In *Findings of the Association for Computational Linguistics: EMNLP 2023*, pages 4295–4304, Singapore. Association for Computational Linguistics.

Sunghye Cho, Riccardo Fusaroli, Maggie Rose Pelella, Kimberly Tena, Azia Knox, Aili Hauptmann, Maxine Covello, Alison Russell, Judith Miller, Alison Hulink, Jennifer Uzokwe, Kevin Walker, James Fiumara, Juhi Pandey, Christopher Chatham, Christopher Cieri, Robert Schultz, Mark Liberman, and Julia Parish-morris. 2022. Identifying stable speech-language markers of autism in children: Preliminary evidence from a longitudinal telephony-based study. In *Proceedings of the Eighth Workshop on Computational Linguistics and Clinical Psychology*, pages 40–46, Seattle, USA. Association for Computational Linguistics.

David A Clark. 2013. Cognitive restructuring. *The Wiley handbook of cognitive behavioral therapy*, pages 1–22.

Tim Dettmers, Artidoro Pagnoni, Ari Holtzman, and Luke Zettlemoyer. 2023. Qlora: efficient finetuning of quantized llms. In *Proceedings of the 37th International Conference on Neural Information Processing Systems*, NIPS '23, Red Hook, NY, USA.

Núria Gala and Johannes Ziegler. 2016. Reducing lexical complexity as a tool to increase text accessibility for children with dyslexia. In *Proceedings of the Workshop on Computational Linguistics for Linguistic Complexity (CL4LC)*, pages 59–66, Osaka, Japan. The COLING 2016 Organizing Committee.

Leo Gao, Stella Biderman, Sid Black, Laurence Golding, Travis Hoppe, Charles Foster, Jason Phang, Horace He, Anish Thite, Noa Nabeshima, et al. 2020. The pile: An 800gb dataset of diverse text for language modeling. *arXiv preprint arXiv:2101.00027*.

Adam Goodkind, Michelle Lee, Gary E. Martin, Molly Losh, and Klinton Bicknell. 2018. Detecting language impairments in autism: A computational analysis of semi-structured conversations with vector semantics. In *Proceedings of the Society for Computation in Linguistics (SCiL) 2018*, pages 12–22.

Keith Harrigian, Carlos Aguirre, and Mark Dredze. 2020. Do models of mental health based on social media data generalize? In *Findings of the Association for Computational Linguistics: EMNLP 2020*, pages 3774–3788, Online. Association for Computational Linguistics.

Nicolas Hiebel, Olivier Ferret, Karen Fort, and Aurélie Névéol. 2023. Can synthetic text help clinical named entity recognition? a study of electronic health records in French. In *Proceedings of the 17th Conference of the European Chapter of the Association for Computational Linguistics*, pages 2320–2338, Dubrovnik, Croatia. Association for Computational Linguistics.

Christine Howes, Matthew Purver, and Rose McCabe. 2014. Linguistic indicators of severity and progress in online text-based therapy for depression. In *Proceedings of the Workshop on Computational Linguistics and Clinical Psychology: From Linguistic Signal to Clinical Reality*, pages 7–16, Baltimore, Maryland, USA. Association for Computational Linguistics.

Edward J Hu, Phillip Wallis, Zeyuan Allen-Zhu, Yuanzhi Li, Shean Wang, Lu Wang, Weizhu Chen, et al. 2021. Lora: Low-rank adaptation of large language models. In *International Conference on Learning Representations*.

Zhiqiang Hu, Roy Ka-Wei Lee, Charu C. Aggarwal, and Aston Zhang. 2022. Text style transfer: A review and experimental evaluation. 24(1):14–45.

Ziwei Ji, Nayeon Lee, Rita Frieske, Tiezheng Yu, Dan Su, Yan Xu, Etsuko Ishii, Ye Jin Bang, Andrea Madotto, and Pascale Fung. 2023. Survey of hallucination in natural language generation. *ACM Computing Surveys*, 55(12):1–38.

Albert Q Jiang, Alexandre Sablayrolles, Arthur Mensch, Chris Bamford, Devendra Singh Chaplot, Diego de las Casas, Florian Bressand, Gianna Lengyel, Guillaume Lample, Lucile Saulnier, et al. 2023. Mistral 7b. *arXiv preprint arXiv:2310.06825*.

Anthony F Jorm, Nicholas B Allen, Colin P O'Donnell, Ruth A Parslow, Rosemary Purcell, and Amy J Morgan. 2006. Effectiveness of complementary and self-help treatments for depression in children and adolescents. *Medical journal of Australia*, 185(7):368–372.

Swanie Juhng, Matthew Matero, Vasudha Varadarajan, Johannes Eichstaedt, Adithya V Ganesan, and H. Andrew Schwartz. 2023. Discourse-level representations can improve prediction of degree of anxiety. In *Proceedings of the 61st Annual Meeting of the Association for Computational Linguistics (Volume 2: Short Papers)*, pages 1500–1511, Toronto, Canada. Association for Computational Linguistics.

Nazmul Kazi and Indika Kahanda. 2019. Automatically generating psychiatric case notes from digital transcripts of doctor-patient conversations. In *Proceedings of the 2nd Clinical Natural Language Processing Workshop*, pages 140–148, Minneapolis, Minnesota, USA. Association for Computational Linguistics.

Takeshi Kojima, Shixiang Shane Gu, Machel Reid, Yutaka Matsuo, and Yusuke Iwasawa. 2022. Large language models are zero-shot reasoners. *Advances in neural information processing systems*, 35:22199–22213.

Kalpesh Krishna, John Wieting, and Mohit Iyyer. 2020. Reformulating unsupervised style transfer as paraphrase generation. In *Proceedings of the 2020 Conference on Empirical Methods in Natural Language Processing (EMNLP)*, pages 737–762, Online. Association for Computational Linguistics.

Fei-Tzin Lee, Derrick Hull, Jacob Levine, Bonnie Ray, and Kathy McKeown. 2019. Identifying therapist conversational actions across diverse psychotherapeutic approaches. In *Proceedings of the Sixth Workshop on Computational Linguistics and Clinical Psychology*, pages 12–23, Minneapolis, Minnesota. Association for Computational Linguistics.

Junlong Li, Fan Zhou, Shichao Sun, Yikai Zhang, Hai Zhao, and Pengfei Liu. 2024. Dissecting human and llm preferences. *arXiv preprint arXiv:2402.11296*.

Zhuoyan Li, Hangxiao Zhu, Zhuoran Lu, and Ming Yin. 2023. Synthetic data generation with large language models for text classification: Potential and limitations. In *Proceedings of the 2023 Conference on Empirical Methods in Natural Language Processing*, pages 10443–10461, Singapore. Association for Computational Linguistics.

Hali Lindsay, Johannes Tröger, Mario Magued Mina, Philipp Müller, Nicklas Linz, Jan Alexandersson, and Inez Ramakers. 2022. Generating synthetic clinical speech data through simulated ASR deletion error. In *Proceedings of the RaPID Workshop - Resources and ProcessIng of linguistic, para-linguistic and extra-linguistic Data from people with various forms of cognitive/psychiatric/developmental impairments - within the 13th Language Resources and Evaluation Conference*, pages 9–16, Marseille, France. European Language Resources Association.

Alisa Liu, Maarten Sap, Ximing Lu, Swabha Swayamdipta, Chandra Bhagavatula, Noah A. Smith, and Yejin Choi. 2021. DExperts: Decoding-time controlled text generation with experts and anti-experts. In *Proceedings of the 59th Annual Meeting of the Association for Computational Linguistics and the 11th International Joint Conference on Natural Language Processing (Volume 1: Long Papers)*, pages 6691–6706, Online. Association for Computational Linguistics.

Sheng Lu, Shan Chen, Yingya Li, Danielle Bitterman, Guergana Savova, and Iryna Gurevych. 2023. Measuring pointwise $\mathcal{V}$-usable information in-context-ly. In *Findings of the Association for Computational Linguistics: EMNLP 2023*, pages 15739–15756, Singapore. Association for Computational Linguistics.

Jakub Macina, Nico Daheim, Sankalan Chowdhury, Tanmay Sinha, Manu Kapur, Iryna Gurevych, and Mrinmaya Sachan. 2023a. MathDial: A dialogue tutoring dataset with rich pedagogical properties grounded in math reasoning problems. In *Findings of the Association for Computational Linguistics: EMNLP 2023*, pages 5602–5621, Singapore. Association for Computational Linguistics.

Jakub Macina, Nico Daheim, Lingzhi Wang, Tanmay Sinha, Manu Kapur, Iryna Gurevych, and Mrinmaya Sachan. 2023b. Opportunities and challenges in neural dialog tutoring. In *Proceedings of the 17th Conference of the European Chapter of the Association for Computational Linguistics*, pages 2357–2372, Dubrovnik, Croatia. Association for Computational Linguistics.

Mounica Maddela, Megan Ung, Jing Xu, Andrea Madotto, Heather Foran, and Y-Lan Boureau. 2023. Training models to generate, recognize, and reframe unhelpful thoughts. In *Proceedings of the 61st Annual Meeting of the Association for Computational Linguistics (Volume 1: Long Papers)*, pages 13641–13660, Toronto, Canada. Association for Computational Linguistics.

Margaret Mitchell, Kristy Hollingshead, and Glen Coppersmith. 2015. Quantifying the language of schizophrenia in social media. In *Proceedings of the 2nd Workshop on Computational Linguistics and Clinical Psychology: From Linguistic Signal to Clinical Reality*, pages 11–20, Denver, Colorado. Association for Computational Linguistics.

Luis Moßburger, Felix Wende, Kay Brinkmann, and Thomas Schmidt. 2020. Exploring online depression forums via text mining: A comparison of Reddit and a curated online forum. In *Proceedings of the Fifth Social Media Mining for Health Applications Workshop & Shared Task*, pages 70–81, Barcelona, Spain (Online). Association for Computational Linguistics.

Christine A Padesky. 1993. Socratic questioning: Changing minds or guiding discovery. In *A keynote address delivered at the European Congress of Behavioural and Cognitive Therapies, London*, volume 24.

Artidoro Pagnoni, Alex Fabbri, Wojciech Kryscinski, and Chien-Sheng Wu. 2023. Socratic pretraining: Question-driven pretraining for controllable summarization. In *Proceedings of the 61st Annual Meeting of the Association for Computational Linguistics (Volume 1: Long Papers)*, pages 12737–12755, Toronto, Canada. Association for Computational Linguistics.

Kishore Papineni, Salim Roukos, Todd Ward, and Wei-Jing Zhu. 2002. Bleu: a method for automatic evaluation of machine translation. In *Proceedings of the 40th Annual Meeting of the Association for Computational Linguistics*, pages 311–318, Philadelphia, Pennsylvania, USA. Association for Computational Linguistics.

R. Paul and L. Elder. 2019. *The Thinker's Guide to Socratic Questioning*. Thinker's Guide Library. Foundation for Critical Thinking.

Verónica Pérez-Rosas, Rada Mihalcea, Kenneth Resnicow, Satinder Singh, and Lawrence An. 2017. Understanding and predicting empathic behavior in counseling therapy. In *Proceedings of the 55th Annual Meeting of the Association for Computational Linguistics (Volume 1: Long Papers)*, pages 1426–1435, Vancouver, Canada. Association for Computational Linguistics.

Jingyuan Qi, Zhiyang Xu, Ying Shen, Minqian Liu, Di Jin, Qifan Wang, and Lifu Huang. 2023. The art of SOCRATIC QUESTIONING: Recursive thinking with large language models. In *Proceedings of the 2023 Conference on Empirical Methods in Natural*

Language Processing, pages 4177–4199, Singapore. Association for Computational Linguistics.

Renyi Qu, Lyle Ungar, and João Sedoc. 2023. Conditioning on dialog acts improves empathy style transfer. In *Findings of the Association for Computational Linguistics: EMNLP 2023*, pages 13254–13271, Singapore. Association for Computational Linguistics.

Emily Reif, Daphne Ippolito, Ann Yuan, Andy Coenen, Chris Callison-Burch, and Jason Wei. 2022. A recipe for arbitrary text style transfer with large language models. In *Proceedings of the 60th Annual Meeting of the Association for Computational Linguistics (Volume 2: Short Papers)*, pages 837–848, Dublin, Ireland. Association for Computational Linguistics.

Arij Riabi, Thomas Scialom, Rachel Keraron, Benoît Sagot, Djamé Seddah, and Jacopo Staiano. 2021. Synthetic data augmentation for zero-shot cross-lingual question answering. In *Proceedings of the 2021 Conference on Empirical Methods in Natural Language Processing*, pages 7016–7030, Online and Punta Cana, Dominican Republic. Association for Computational Linguistics.

Alex Rinaldi, Jean Fox Tree, and Snigdha Chaturvedi. 2020. Predicting depression in screening interviews from latent categorization of interview prompts. In *Proceedings of the 58th Annual Meeting of the Association for Computational Linguistics*, pages 7–18, Online. Association for Computational Linguistics.

Efsun Sarioglu Kayi, Mona Diab, Luca Pauselli, Michael Compton, and Glen Coppersmith. 2017. Predictive linguistic features of schizophrenia. In *Proceedings of the 6th Joint Conference on Lexical and Computational Semantics (*SEM 2017)*, pages 241–250, Vancouver, Canada. Association for Computational Linguistics.

Thibault Sellam, Amy Pu, Hyung Won Chung, Sebastian Gehrmann, Qijun Tan, Markus Freitag, Dipanjan Das, and Ankur Parikh. 2020. Learning to evaluate translation beyond English: BLEURT submissions to the WMT metrics 2020 shared task. In *Proceedings of the Fifth Conference on Machine Translation*, pages 921–927, Online. Association for Computational Linguistics.

Natalie Shapira, Dana Atzil-Slonim, Rivka Tuval Mashiach, and Ori Shapira. 2022. Measuring linguistic synchrony in psychotherapy. In *Proceedings of the Eighth Workshop on Computational Linguistics and Clinical Psychology*, pages 158–176, Seattle, USA. Association for Computational Linguistics.

Ashish Sharma, Kevin Rushton, Inna Lin, David Wadden, Khendra Lucas, Adam Miner, Theresa Nguyen, and Tim Althoff. 2023. Cognitive reframing of negative thoughts through human-language model interaction. In *Proceedings of the 61st Annual Meeting of the Association for Computational Linguistics (Volume 1: Long Papers)*, pages 9977–10000, Toronto, Canada. Association for Computational Linguistics.

Xu Sheng, Fumiyo Fukumoto, Jiyi Li, Go Kentaro, and Yoshimi Suzuki. 2023. Learning disentangled meaning and style representations for positive text reframing. In *Proceedings of the 16th International Natural Language Generation Conference*, pages 424–430, Prague, Czechia. Association for Computational Linguistics.

Heereen Shim, Dietwig Lowet, Stijn Luca, and Bart Vanrumste. 2021. Synthetic data generation and multi-task learning for extracting temporal information from health-related narrative text. In *Proceedings of the Seventh Workshop on Noisy User-generated Text (W-NUT 2021)*, pages 260–273, Online. Association for Computational Linguistics.

Sagarika Shreevastava and Peter Foltz. 2021. Detecting cognitive distortions from patient-therapist interactions. In *Proceedings of the Seventh Workshop on Computational Linguistics and Clinical Psychology: Improving Access*, pages 151–158, Online. Association for Computational Linguistics.

Kumar Shridhar, Jakub Macina, Mennatallah El-Assady, Tanmay Sinha, Manu Kapur, and Mrinmaya Sachan. 2022. Automatic generation of socratic subquestions for teaching math word problems. In *Proceedings of the 2022 Conference on Empirical Methods in Natural Language Processing*, pages 4136–4149, Abu Dhabi, United Arab Emirates. Association for Computational Linguistics.

Kumar Shridhar, Alessandro Stolfo, and Mrinmaya Sachan. 2023. Distilling reasoning capabilities into smaller language models. In *Findings of the Association for Computational Linguistics: ACL 2023*, pages 7059–7073, Toronto, Canada. Association for Computational Linguistics.

Marina Sokolova, Vera Sazonova, Kanyi Huang, Rudraneel Chakraboty, and Stan Matwin. 2017. Studying positive speech on twitter. *arXiv preprint arXiv:1702.08866*.

Haoyu Song, Yan Wang, Wei-Nan Zhang, Xiaojiang Liu, and Ting Liu. 2020. Generate, delete and rewrite: A three-stage framework for improving persona consistency of dialogue generation. In *Proceedings of the 58th Annual Meeting of the Association for Computational Linguistics*, pages 5821–5831, Online. Association for Computational Linguistics.

Michael Tanana, Kevin Hallgren, Zac Imel, David Atkins, Padhraic Smyth, and Vivek Srikumar. 2015. Recursive neural networks for coding therapist and patient behavior in motivational interviewing. In *Proceedings of the 2nd Workshop on Computational Linguistics and Clinical Psychology: From Linguistic Signal to Clinical Reality*, pages 71–79, Denver, Colorado. Association for Computational Linguistics.

Alexey Tikhonov, Viacheslav Shibaev, Aleksander Nagaev, Aigul Nugmanova, and Ivan P. Yamshchikov. 2019. Style transfer for texts: Retrain, report errors, compare with rewrites. In *Proceedings of the*

2019 Conference on Empirical Methods in Natural Language Processing and the 9th International Joint Conference on Natural Language Processing (EMNLP-IJCNLP), pages 3936–3945, Hong Kong, China. Association for Computational Linguistics.

Hugo Touvron, Louis Martin, Kevin Stone, Peter Albert, Amjad Almahairi, Yasmine Babaei, Nikolay Bashlykov, Soumya Batra, Prajjwal Bhargava, Shruti Bhosale, et al. 2023. Llama 2: Open foundation and fine-tuned chat models. *arXiv preprint arXiv:2307.09288*.

Elsbeth Turcan and Kathy McKeown. 2019. Dreaddit: A Reddit dataset for stress analysis in social media. In *Proceedings of the Tenth International Workshop on Health Text Mining and Information Analysis (LOUHI 2019)*, pages 97–107, Hong Kong. Association for Computational Linguistics.

Amrisha Vaish, Tobias Grossmann, and Amanda Woodward. 2008. Not all emotions are created equal: the negativity bias in social-emotional development. *Psychological bulletin*, 134(3):383–403.

Jason Wei, Kelly Finn, Emma Templeton, Thalia Wheatley, and Soroush Vosoughi. 2021. Linguistic complexity loss in text-based therapy. In *Proceedings of the 2021 Conference of the North American Chapter of the Association for Computational Linguistics: Human Language Technologies*, pages 4450–4459, Online. Association for Computational Linguistics.

Chris Williams. 2001. Use of written cognitive–behavioural therapy self-help materials to treat depression. *Advances in Psychiatric Treatment*, 7(3):233–240.

Thomas Wolf, Lysandre Debut, Victor Sanh, Julien Chaumond, Clement Delangue, Anthony Moi, Pierric Cistac, Tim Rault, Remi Louf, Morgan Funtowicz, Joe Davison, Sam Shleifer, Patrick von Platen, Clara Ma, Yacine Jernite, Julien Plu, Canwen Xu, Teven Le Scao, Sylvain Gugger, Mariama Drame, Quentin Lhoest, and Alexander Rush. 2020. Transformers: State-of-the-art natural language processing. In *Proceedings of the 2020 Conference on Empirical Methods in Natural Language Processing: System Demonstrations*, pages 38–45, Online. Association for Computational Linguistics.

Yongkang Wu, Meng Han, Yutao Zhu, Lei Li, Xinyu Zhang, Ruofei Lai, Xiaoguang Li, Yuanhang Ren, Zhicheng Dou, and Zhao Cao. 2023a. Hence, socrates is mortal: A benchmark for natural language syllogistic reasoning. In *Findings of the Association for Computational Linguistics: ACL 2023*, pages 2347–2367, Toronto, Canada. Association for Computational Linguistics.

Zixiu Wu, Simone Balloccu, Ehud Reiter, Rim Helaoui, Diego Reforgiato Recupero, and Daniele Riboni. 2023b. Are experts needed? on human evaluation of counselling reflection generation. In *Proceedings of the 61st Annual Meeting of the Association for Computational Linguistics (Volume 1: Long Papers)*, pages 6906–6930, Toronto, Canada. Association for Computational Linguistics.

Andrew Yates, Arman Cohan, and Nazli Goharian. 2017. Depression and self-harm risk assessment in online forums. In *Proceedings of the 2017 Conference on Empirical Methods in Natural Language Processing*, pages 2968–2978, Copenhagen, Denmark. Association for Computational Linguistics.

Chujie Zheng, Pei Ke, Zheng Zhang, and Minlie Huang. 2023. Click: Controllable text generation with sequence likelihood contrastive learning. In *Findings of the Association for Computational Linguistics: ACL 2023*, pages 1022–1040, Toronto, Canada. Association for Computational Linguistics.

Caleb Ziems, Minzhi Li, Anthony Zhang, and Diyi Yang. 2022. Inducing positive perspectives with text reframing. In *Proceedings of the 60th Annual Meeting of the Association for Computational Linguistics (Volume 1: Long Papers)*, pages 3682–3700, Dublin, Ireland. Association for Computational Linguistics.

A Experimental details

We few-shot prompt GPT-4 with the default parameters from the OpenAI Python package. [4] We use three few-shot exemplars. Following similar works on synthetic data generation with LLMs (Macina et al., 2023a), we use temperature sampling with $T = 0.4$ and no top-k truncation for generating the reframes. For the fine-tuning experiments, following Dettmers et al. (2023), we apply LoRA to all linear layers; training for 5 epochs with ADAMW, a batch size of 8 and set the learning rate to 5e-4. All other hyperparameters take the default value in the configuration. We use NVIDIA A100 80GB for the training.

B Prompt Design

B.1 Generating Socratic Rationales

System Prompt for GPT-4

You are an expert in Cognitive Behavioral Therapy and Cognitive Restructuring, focusing on guided discovery. The term cognitive restructuring refers to the process of challenging, and changing, irrational thoughts. Socratic questioning is one technique to encourage this process. Therapists use Socratic questioning verbally by asking probing questions about their clients' irrational thoughts. As clients improve their awareness of irrational thoughts, they can begin to consciously question their own thoughts. The six types of Socratic Questions are: 1) Clarification, 2) Probing assumptions, 3) Probing reasons and evidence, 4) Questioning perspectives, 5) Probing implications and 6) Questioning the question. They can be used in any order and not all of them might be needed for a given client.

Figure 8: This system prompt is used to prime GPT-4 with a personality grounded in the Socratic method.

An example of a clinical vignette taken from Padesky (1993) used as a few-shot exemplar is:

```
Negative Thought: "I'm a complete failure in
    every way."
Positive Thought: "I am not a failure."
Socratic Questioning:
Q (Clarification): You look defeated when you
    say that. Do you feel defeated?
A: Yes. I'm no good.
Q (Probing assumptions): You say you are no good.
     Is it true that you haven't done anything
    at all good?
A: Nothing of importance.
Q (Probing reasons and evidence): How about for
    your children this week -- did you care for
    them at all?
A: Of course, I helped my wife put them to bed
    and took them to soccer practice.
Q (Questioning perspectives): Do you think that
    was important to them?
A: I suppose so.
Q (Probing implications): And did you do
    anything to make your wife happy this week?
A: She liked the fact that I came home from work
    on time.
Q (Probing implications): Would a "complete
    failure" be able to respond to his wife's
    request in such a successful way?
A: I guess not.
Q (Probing implications): So is it really
    accurate to say you are a complete failure
    in every way?
A: I suppose not.
Q (Clarification): So how do you feel now?
A: I guess a little better.
```

[4] https://github.com/openai/openai-python

The prompt template for GPT-4 is shown in Figure 3 and the system prompt is shown in Figure 8.

B.2 Finetuning

Prompt template used during finetuning

Input: Given a situation: «SITUATION» and the associated negative thought: «NEGATIVE THOUGHT», generate the Socratic rationale for guided discovery and reframing the negative thought to a positive thought.

Output:
«SOCRATIC RATIONALE»
«POSITIVE THOUGHT»

Figure 9: This template is used to finetune LLMs using the Socratic rationales.

The prompt template used for finetuning is shown in Figure 9.

C Human Evaluation Protocol

C.1 Evaluating Socratic Rationales

Socratic Questioning Scale (SQS) The following questions were rated by annotators on a 3-point Likert scale ranging from 1 (not at all) to 3 (extensively) :

- **How frequently were questions asked that help develop alternative perspectives?** - *This question aims to assess the frequency* *of inquiries that encourage the exploration of diverse viewpoints or opinions. Annotators should focus on identifying questions that prompt respondents to consider different angles, challenge assumptions, or think beyond the conventional narrative.*

- **Was the question answering focused on the emotions and situation of the person?** - *This question assesses whether the majority of answers provided focus on the emotional state and circumstances of the individual involved. Annotators should consider whether responses primarily address the feelings, experiences, or immediate context of the person in question.*
- **Were the questions open-ended and require thoughtful reflection?** - *This question is designed to evaluate whether the questions posed to respondents are open-ended and demand deep contemplation rather than eliciting simple, direct answers. Annotators should look for questions that encourage respondents to think critically and provide nuanced, reflective responses.*

Helpfulness Annotators rate the helpfulness of the question-answer pairs on a 3-point Likert scale ranging from 1 (not helpful at all) to 3 (very helpful). The following question was asked to the annotators:

How helpful was the questioning in general? *This question seeks to gauge the overall effectiveness and utility of the questions posed. Annotators should consider the extent to which the questions contributed to a meaningful discussion, elicited insightful responses, or facilitated a deeper exploration of the topic. Assessments should encompass the clarity, relevance, and engagement level of the questions.*

We plan to release the generated rationales to complement the original datasets under the Apache license.[5].

C.2 Validation with experts

We recruit psychology students for our study. The students were given credits in return of their annotation efforts. The average age of participants was 24.2 with 80% participants being females. The participants were shown three diverse samples from across the datasets and asked to rate the helpfulness and SQS, in the same setting as described in Appendix C.1. Additionally, participants were asked to answer the following question with either "Yes", "No" or "Not sure":

Would you use an LLM to help you restructure negative thoughts? *This question assesses the willingness and openness of psychology students towards incorporating Large Language Models (LLMs) into their therapeutic practices. Responses provide insights into whether LLMs are perceived as viable tools for cognitive restructuring, indicating their potential acceptance within psychological interventions.*

We observe that only 7% of the participants did not want to use LLMs for restructuring negative thoughts with 55% responding affirmatively and the rest being unsure. We plan to conduct a more large-scale annotation study with more experts and non-experts in the future.

C.3 Evaluating Reframes

For a given context and negative thought, annotators are shown two sample reframes and asked the following question:

> **Given the context and original negative thought, which reframed thought do you find more relatable, helpful and memorable (A vs B)?**

D Qualitative Examples

Table 5 shows qualitative examples of generated socratic rationales for each of the three datasets used in our study.

[5] `https://www.apache.org/licenses/LICENSE-2.0`

Dataset	Input	Generated Rationale
COGREF	Situation: Having problems coming up with words to write essay. Negative Thought: I will get a bad grade. Reframed Thought: I'm feeling anxious about this assignment. I can use skills to calm myself, try my best on the essay, ask for any appropriate help, and whatever grade I get I will know that I tried my hardest.	Q: You're worried about doing badly on this assignment. How have you done on past essays? A: I've gotten good grades, sometimes. But this one feels different. Q: Do you think getting a bad grade on this specific essay means you're generally bad at writing or at this subject overall? A: Well, I don't think so, but I'm unsure about this topic. Q: What evidence is leading you to believe that you will get a bad grade? A: I'm having trouble coming up with points to write about. Q: If a friend was in your position, what advice would you give them? A: I would probably say to not panic and just give it their best shot. Q: What would happen if you got a bad grade on this essay? How would that affect your overall academic achievement? A: One bad grade isn't going to ruin my overall performance I guess. Q: Can using some support like talking to your teacher or a study group help you in getting through this? A: That might be a good idea. Q: So given these factors, do you still think it's a definite that you will get a bad grade? A: No, I suppose it's not definite. Q: So, what will you do about your essay now? A: I think I'll start by focusing on writing a draft based on what I know and then ask for some help to see how I can improve it.
PATREF	Persona: my favorite band is rise against. my favorite team is the raptors. i usually go to bed late. i like sports. i am studying to be a dentist. Negative Thought: I am about to graduate and become a Dentist, but that means nothing. Anyone can do that. Reframed Thought: I am about to graduate and become a Dentist. Although it is possible for anyone to become a dentist, it is very difficult, and I am proud of myself.	Q: It sounds like you're doing something remarkable, yet you downplay its value. Do you genuinely believe becoming a dentist means nothing? A: It doesn't seem special when everyone else is doing similar things. Q: Do you think everyone could become a dentist? A: Maybe not everyone, but a lot of people could. Q: It required you to put in several years of hard study and practice, didn't it? A: Yes, I did work hard. Q: How many people do you think are willing or able to do what you did? A: I'm not sure, probably not that many. Q: So wouldn't that mean that your achievement is indeed quite significant and not something anyone could do? A: I guess so. Q: Isn't it possible then that your achievement of becoming a dentist is both valuable and hard-earned? A: Yes, it's starting to look that way. Q: How do you feel about your upcoming graduation now? A: I'm starting to recognize its importance and feel a bit proud.
POSREF	Negative Thought: This physics homework is the death of me every week. It's so hard. And then I have another physics exam tomorrow night. Reframed Thought: Physics is challenging, but is worth it. I have an exam tomorrow, which I hope to do well on.	Q: It sounds quite challenging, is Physics always this difficult for you or only certain aspects? A: Certain parts, like quantum mechanics. Q: So, is it accurate to state that the entire subject is difficult, or just the quantum mechanics part? A: I suppose just the quantum mechanics part. Q: Can you give me an example of a physics problem or concept you've been able to understand quite well? A: Yeah, I've been doing well in classical mechanics. Q: Do you think that possessing skills in one area of physics could potentially aid in understanding other areas, including quantum mechanics? A: Maybe. But quantum mechanics seems very different. Q: If classical mechanics and quantum mechanics are two aspects of the same discipline, isn't there a possibility that understanding one could help with the other? A: It's a possibility, I guess. Q: Do you think saying that the physics homework is the death of you is a fair and accurate statement? A: Probably not, it's just really frustrating sometimes. Q: What if you thought about the physics homework as a challenge rather than a death sentence? Could that change how you approach it? A: Maybe. I could try. Q: Now, how do you feel about the physics exam tomorrow? A: I'm nervous, but maybe if I look at it as a challenge, it might feel less overwhelming.

Table 5: Socratic rationales generated with GPT-4 for different types of inputs from the three datasets we consider.

Synthetic Empathy: Generating and Evaluating Artificial Psychotherapy Dialogues to Detect Empathy in Counseling Sessions

Daniel Cabrera Lozoya[1], Eloy Hernández Lúa[2], Juan Alberto Barajas Perches[2], Mike Conway[1], and Simon D'Alfonso[1]

[1]The University of Melbourne, Australia

[2]ITESM, Mexico

{dcabreralozo}@student.unimelb.edu.au

{jbperches, eloyhl}@exatec.tec.mx

{mike.conway, dalfonso}@unimelb.edu.au

Abstract

Natural language processing (NLP) holds potential for analyzing psychotherapy transcripts. Nonetheless, gathering the necessary data to train NLP models for clinical tasks is a challenging process due to patient confidentiality regulations that restrict data sharing. To overcome this obstacle, we propose leveraging large language models (LLMs) to create synthetic psychotherapy dialogues that can be used to train NLP models for downstream clinical tasks. To evaluate the quality of our synthetic data, we trained three multi-task RoBERTa-based bi-encoder models, originally developed by Sharma et al., to detect empathy in dialogues. These models, initially trained on Reddit data, were developed alongside EPITOME, a framework designed to characterize empathetic communication in conversations. We collected and annotated 579 therapeutic interactions between therapists and patients using the EPITOME framework. Additionally, we generated 10,464 synthetic therapeutic dialogues using various LLMs and prompting techniques, all of which were annotated following the EPITOME framework. We conducted two experiments: one where we augmented the original dataset with synthetic data and another where we replaced the Reddit dataset with synthetic data. Our first experiment showed that incorporating synthetic data can improve the F1 score of empathy detection by up to 10%. The second experiment revealed no substantial differences between organic and synthetic data, as their performance remained on par when substituted.

1 Introduction

Therapy transcripts offer rich insights into counseling sessions, capturing key details such as clients' concerns, emotional states, and therapeutic interventions (Lee et al., 2019; Imel et al., 2015). Natural language processing (NLP) models have shown great promise in analyzing these transcripts (Laricheva et al., 2024; Ewbank et al., 2020; Gaut et al., 2017). However, training such models demands substantial data, which is difficult to access due to the need to safeguard sensitive health information, and institutional barriers to obtaining clinical data (Lu et al., 2021; Aledavood et al., 2017).

Data Augmentation (DA) - a set of methods used for synthetic generation of training data - is a way to manage data scarcity when training machine learning models (Ansari and Saxena, 2024). The adoption and success of DA has mostly been in the computer vision field, whereas for NLP tasks it has exhibited a more limited impact when achieving performance gains (Maier Ferreira and Reali Costa, 2020). Traditionally, NLP-specific data augmentation approaches have relied on back-translation (Corbeil and Ghadivel, 2020) or performing simple operations to the original text, such as synonym replacements or random word insertion (Wei and Zou, 2019). However, performing simple transformations on existing text samples can lead to syntactic and semantic distortions of the text (Giridhara et al., 2019).

Generative language models have made a breakthrough in augmenting unstructured text data (Hagos et al., 2024). Models such as OpenAI's GPT series (OpenAI et al., 2024), have relied on sophisticated self-attention mechanisms to generate new data, rather than just performing local changes on the text. Related studies have started exploring the use of generative models for training few shot classifiers (Edwards et al., 2022), generating artificial text for enhancing intent classifiers (Sahu et al., 2022), and augmenting domain specific datasets to boost domain specific NLP tasks (Amin-Nejad et al., 2020). Although NLP models are an active area of research, the creation of synthetic datasets remains understudied in the mental health field.

In this research, we examined the viability of using LLMs for generating artificial counseling transcripts to enhance the performance of NLP models for clinical tasks. We trained the bi-encoder

Proceedings of the 10th Workshop on Computational Linguistics and Clinical Psychology (CLPsych 2025), pages 157–171
May 3, 2025 ©2025 Association for Computational Linguistics

model introduced in (Sharma et al., 2020), which is capable of recognizing empathetic dialogues and providing rationales that support its predictions. In therapy, empathy plays a crucial role and serves as a significant predictor of therapy outcomes (Elliott et al., 2018). It stands as one of the fundamental factors that contribute to establishing a strong working alliance between a psychotherapist and their client during a session, regardless of the specific therapeutic approach employed (Elliott et al., 2011). Research studies have shown that mental health professionals can enhance their empathetic responses through the provision of appropriate feedback (Benster and Swerdlow, 2020; Sharma et al., 2020). Hence, a model that can identify low empathetic dialogues could help therapists recognize areas where empathetic engagement could be improved. By leveraging the guidance and input provided by the model, therapists can refine their empathetic skills and give more supportive therapy sessions for their clients.

Our main contributions are as follows:

1. We present a methodology to generate and evaluate counselling transcripts for data augmentation purposes.
2. We demonstrate that our synthetic transcripts can effectively fine-tune state-of-the-art models, enabling them to surpass baseline models in text mining therapy transcripts.
3. We release the synthetic datasets used in this study to help future mental health research.

2 Related Work

The generation of realistic synthetic patient data has primarily concentrated on the production of electronic health records (EHRs). Before generative models were used for data augmentation purposes, many methods relied on rule-based methods (Ansari et al., 2021). Among the initial generative architectures used to augment EHR data, MedGan (Choi et al., 2017) introduced a generative adversarial network (GAN) designed to generate multi-label patient records. To improve the quality of the data generated by MedGan, medWgan and medBGan were developed (Baowaly et al., 2018). These models were based on the principles of Wasserstein GAN with gradient penalty (Gulrajani et al., 2017) and boundary-seeking GANs (Hjelm et al., 2018) respectively. It is worth noting that the previous models primarily concentrate on generating the structured data components of an EHR. Synthetic records frequently lack the inclusion of the unstructured text section, and when it is included, it is usually quite concise. For instance, in (Lee, 2018) their approach generates unstructured text that is limited to 18 tokens or less.

In addition to GANs, transformer-based models have also been used for medical data augmentation. Liu (2018) trained a transformer with memory-compressed attention to create EHRs, yielding promising results (1.76 in the perplexity per token and a 44.6 in the Rogue-1 metric). However, an evaluation to measure the data's quality for downstream task was not conducted. While previous research has primarily focused on augmenting data, limited attention has been given to evaluating its utility for training machine learning models. Wang et al. addressed this gap, in their study they employed synthetic data as supplementary training data for two biomedical NLP tasks: text classification and temporal relation extraction. Similarly, Lu et al. used a transformer-based model to train classifiers for patients readmission prediction.

While the majority of research has mostly concentrated on synthetic EHRs, there is also relevant work within the field of synthetic mental health data. One such example is found in (Ive et al., 2020), where they artificially generated discharge summaries from mental health providers. These summaries were utilized in a downstream NLP text classification task. Yet, there is still a scarcity of research focusing on the creation of synthetic data that mimics dialogues from therapy transcripts.

3 Method

3.1 Empathy Framework

To measure empathy in text-based conversations we used the EPITOME framework (Sharma et al., 2020), which establishes the following empathy dimensions:

1. **Emotional reactions** - entails the therapist expressing emotions such as warmth and compassion, in response to a patient's message.
2. **Interpretations** - involves the therapist conveying their comprehension and understanding of the emotions inferred from the patient's message.
3. **Explorations** - refers to the therapist's pursuit of a deeper understanding of the patient by

delving into unexpressed feelings and experiences that extend beyond the explicit content of their messages.

Each of these empathy dimensions can take a value of 0, indicating that the therapist is not expressing it at all; 1, indicating a weak degree of expression; or 2, indicating a strong expressing by the therapist.

3.2 Data sets

We gathered clinical therapy transcripts from the following data sources:

1. **MOST+ trial** - Moderated Online Social Therapy (MOST) is a youth-focused mental health web platform. In the MOST+ trial (Alvarez-Jimenez, et al., 2020), MOST was embedded within the online service of Australian youth mental health provider headspace, and provided an on-demand webchat service manned by headspace counsellors. A total of 200 therapy transcripts were gathered from this study. From this dataset we extracted 365 dialogue pairs between client and counsellor.

2. **Alexander Street Press** – The *Counseling and Psychotherapy Transcripts, Client Narratives, and Reference Works* (Alexander Street, 2009) contains 2,000 therapy session transcripts. From this dataset we gathered 214 dialogue pairs between patient and therapist.

3. **Mental health subreddits** We utilized the labeled Reddit dataset compiled by Sharma et al. (2020), which encompasses content from 55 subreddits dedicated to mental health. This dataset contains a total of 3,081 dialogue pairs between Reddit users that have been annotated using the EPITOME framework.

In close collaboration with therapists, we designed prompts for the LLMs to generate synthetic therapy transcripts. These prompts were crafted based on the EPITOME definitions of empathy, which were designed to characterize communication of empathy in text-based conversations. We developed a unique prompt for each of the three dimensions of empathy in EPITOME: Emotional Reactions, Interpretations, and Explorations. For a comprehensive list of all the prompts used, please refer to Appendix A. The prompts were used as inputs to the following models:

1. **Standalone LLM** The prompts were fed to a GPT-3 model (Brown et al., 2020) and a Falcon 7b model (Penedo et al., 2023), an LLM that was trained to follow complex instructions.

2. **LLM with verbal reinforcement learning** We used the Reflexion framework (Shinn et al., 2023) to reinforce GPT-3 and Faclon 7b through linguistic feedback. The linguistic feedback was designed in collaboration with a clinical psychologist. For a comprehensive list of all the linguistic feedback used, please refer to Appendix B.

For each model, we generated synthetic datasets and labeled them according to the EPITOME framework. In total, we produced 10,464 synthetic therapy dialogues. To evaluate the quality of our synthetic data, we compared the performance of an empathy classifier trained under two conditions: augmenting the Reddit dataset with synthetic data, and replacing portions of the Reddit dataset with synthetic data. The MOST+ trial and Alexander Street Press datasets served as the testing datasets.

3.3 Annotation Task and Process

3.3.1 Annotator training

Three authors of the paper annotated the datasets according to the EPITOME guidelines outlined in (Sharma et al., 2020). Each annotator completed a comprehensive training program consisting of nine one-hour coding sessions and received detailed manual feedback on 360 dialogue data points from a clinical psychologist.

3.3.2 Empathy Annotation

The annotators were presented with a dialogue pair extracted from a therapy transcript, involving a therapist and a patient. The annotators were tasked to identify the presence of the three empathy dimensions. For each dimension, they assigned labels of 0 (no communication), 1 (weak communication), or 2 (strong communication) to indicate the level of empathy conveyed in the therapist's response. The inter-annotator agreements for each dataset were as follows: 0.6719 for the synthetic transcripts from GPT-3, 0.6280 for the Alexander Street database, 0.6147 for the MOST+ transcripts, and 0.7822 for the Reddit dataset. These scores were calculated by averaging the pairwise Cohen's κ of all pairs of annotators, with each pair annotating more than 120 dialogue pairs per dataset.

	Data Source	None	Weak	Strong	Total
	Reddit	2,034	899	148	3,081
Emotional Reactions	Alexander	147	26	41	214
	MOST+	211	59	95	365
	Reddit	1,645	178	1,321	3,081
Interpretations	Alexander	94	76	44	214
	MOST+	180	116	69	365
	Reddit	2,600	104	377	3,081
Explorations	Alexander	131	60	23	214
	MOST+	156	141	68	365

Table 1: Empathy level distribution in datasets consisting of clinical therapy transcripts and dialogues from mental health support platforms

3.4 Model

For our empathy classifier we used the multi-task bi encoder developed by Sharma et al. (2020). This model was designed to evaluate the degree of empathy conveyed in a psychologist's response to a patient's message. This evaluation results in a numerical output, where a score of 2 signifies a strong communication of empathy, a score of 1 indicates a weak expression of empathy, and a score of 0 suggests the absence of empathy.

3.5 Experimental setup

To evaluate our synthetic data, we conducted two experiments: one where we augmented the Reddit dataset with synthetic data and another where we replaced portions of the Reddit dataset with synthetic data. In each experiment, we trained three bi-encoders, each designed to detect a type of empathy: emotional reaction, interpretation, or exploration.

The first experiment examined how adding synthetic data to the Reddit dataset affects model performance. We conducted 15 iterations, with the first iteration serving as a baseline containing no synthetic dialogues. In the following iterations, we incrementally added synthetic dialogues in batches of 30 data points, with the final iteration incorporating 420 synthetic dialogues. The dialogue pairs added to the Reddit dataset were evenly distributed across empathy levels.

The second experiment evaluated whether synthetic data could replace real data without compromising performance. In this experiment, we gradually substituted portions of the Reddit dataset with synthetic data while preserving the original empathy distribution. We conducted five iterations, each replacing 10% of the original data with synthetic data. The first iteration included 10% synthetic data, while the final iteration reached 50% replacement.

The testing dataset for all experiments consisted of 579 dialogue pairs from the Alexander Street Press and the MOST+ trial. For each experiment, we reported the accuracy and F1 score for the three components of empathy: exploration, interpretation, and emotional reaction.

To train the bi-encoders we used the default hyperparemeters proposed by Sharma et al. (2020). We trained the model for 4 epochs using a learning rate of 2×10^{-5}, and a batch size of 32. The computing infrastructure employed for training this model was an NVIDIA A100 GPU.

4 Results

In this section, we present the results of augmenting the Reddit dataset from (Sharma et al., 2020) with our synthetic data, as well as the results of partially substituting the Reddit dataset with synthetic data.

4.1 Reddit dataset augmentation

Figure 1 shows the accuracy and F1 score results of augmenting the Reddit dataset with synthetic data.

4.1.1 F1 scores

Training the bi-encoder models on the Reddit dataset resulted in F1 scores of 0.48, 0.32, and 0.58 for exploration, interpretation, and emotional reaction, respectively. Augmenting the Reddit dataset with 420 synthetic dialogues improved performance, resulting in F1 scores of 0.53, 0.48, and 0.59 for the same categories. This corresponds to an improvement of 0.05, 0.16, and 0.01, respectively. Notably, the highest F1 score for exploration, 0.57, was achieved with 360 synthetic data points, while for interpretation and emotional reaction, the model reached its peak F1 score of 0.60 with 390 synthetic data points.

4.1.2 Accuracy

Training the bi-encoder models on the Reddit dataset, resulted in accuracy scores of 0.64, 0.50, and 0.66 for exploration, interpretation, and emotional reaction, respectively. Augmenting the dataset with 420 synthetic dialogues improved performance, increasing accuracy to 0.66, 0.60, and 0.69 for the same categories. This corresponds to an improvement of 0.02, 0.10, and 0.03, respectively. Notably, the highest accuracy for exploration, 0.68, was achieved with 360 synthetic data points, while interpretation peaked at 0.61 with 360 additional synthetic data points, and emotional reaction reached its highest accuracy of 0.71 with 390 synthetic data points.

4.2 Reddit dataset substition

Figure 2 presents the accuracy and F1 score results of substituting the Reddit dataset with portions of synthetic data.

4.2.1 F1 score

The empathy dimension that showed the greatest improvement when replacing the Reddit dataset with synthetic data was interpretation. When 50% of the Reddit data was replaced with GPT-3-generated data using verbal reinforcement learning, the model achieved an F1-score of 0.43, compared to 0.32 when trained solely on the Reddit dataset.

For the emotional reaction metric, the quality of synthetic data generated by GPT-3 was comparable to that of the Reddit dataset. Their performance, rounded to two significant digits, remained consistent at 0.58 across all substitution percentages. Similarly, for the empathy exploration metric, performance remained similar across various substitution percentages, except in the 10% substitution test, where the Reddit dataset outperformed the synthetic data by 2%.

4.2.2 Accuracy

The empathy dimension that showed the greatest improvement when replacing the Reddit dataset with synthetic data was interpretation. When 50% of the Reddit data was replaced with GPT-3-generated data using verbal reinforcement learning, the model achieved an accuracy of 0.57, compared to 0.50 when trained solely on the Reddit dataset.

For the emotional reaction metric, the synthetic data generated by GPT-3 generally outperformed the Reddit dataset. The largest performance difference occurred with a 20% substitution, where GPT-3's reflexion-based data achieved a score of 0.72, surpassing the Reddit dataset's 0.67. For the empathy exploration metric, performance remained consistent across various substitution percentages, with a maximum difference of only 0.01.

5 Discussion

The data augmentation results reveal a notable trend: while adding synthetic data continues to improve performance, the rate of improvement decreases beyond a certain threshold. Specifically, for exploration, the impact of additional data slows after 90 data points, and for interpretation, after 150. This suggests that while synthetic data remains beneficial, its effectiveness diminishes over time, likely due to redundancy or a reduced introduction of novel information.

This finding has practical implications for dataset construction. Rather than indiscriminately increasing the volume of synthetic data, researchers should prioritize curating high-quality, diverse examples that fill specific gaps in the existing dataset. This targeted approach not only maximizes the impact of synthetic data but also reduces computational costs, training time, and, in the case of proprietary models like GPT-3, expenses associated with API usage. Notably, the synthetic data generated by the Falcon model also enhanced the model's performance when used to augment the training dataset. This is valuable since Falcon is licensed under Apache 2.0, unlike proprietary models that require paid access. Falcon LLM can be run locally, fine-tuned, and used without cost, offering an advantage for researchers seeking to generate synthetic data without financial constraints.

The substitution experiments demonstrate that synthetic data can replace portions of organic data without compromising performance. This suggests that synthetic data can serve as an alternative to organic data containing protected health information. This is beneficial when fine-tuning external models that require data to be sent to a third party, such as fine-tuning an OpenAI GPT model. By leveraging synthetic data, researchers can mitigate privacy concerns while maintaining, or even enhancing, model performance.

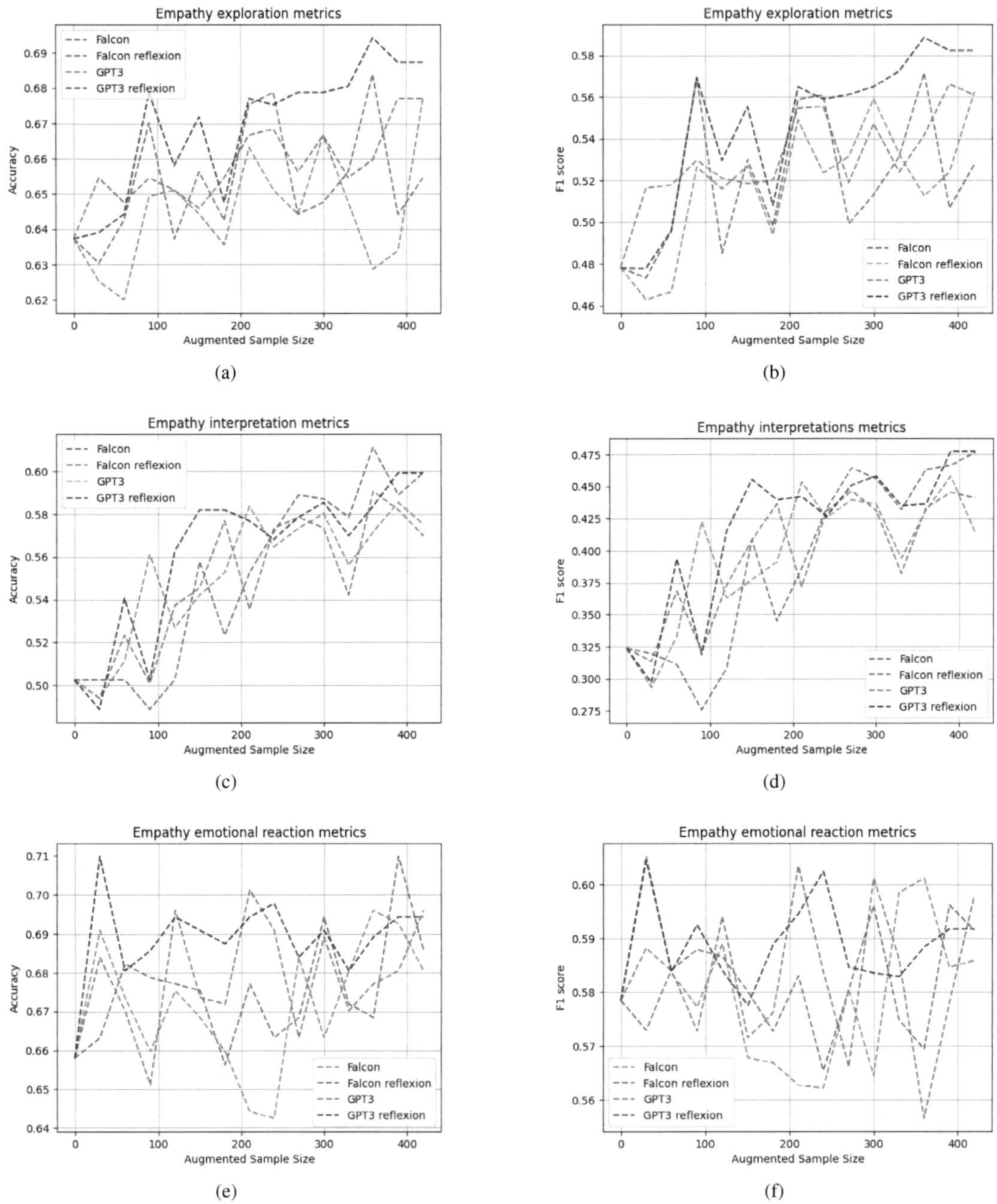

Figure 1: Accuracy and F1 scores for the three dimensions of empathy using synthetic data to augment the orignal Reddit data.

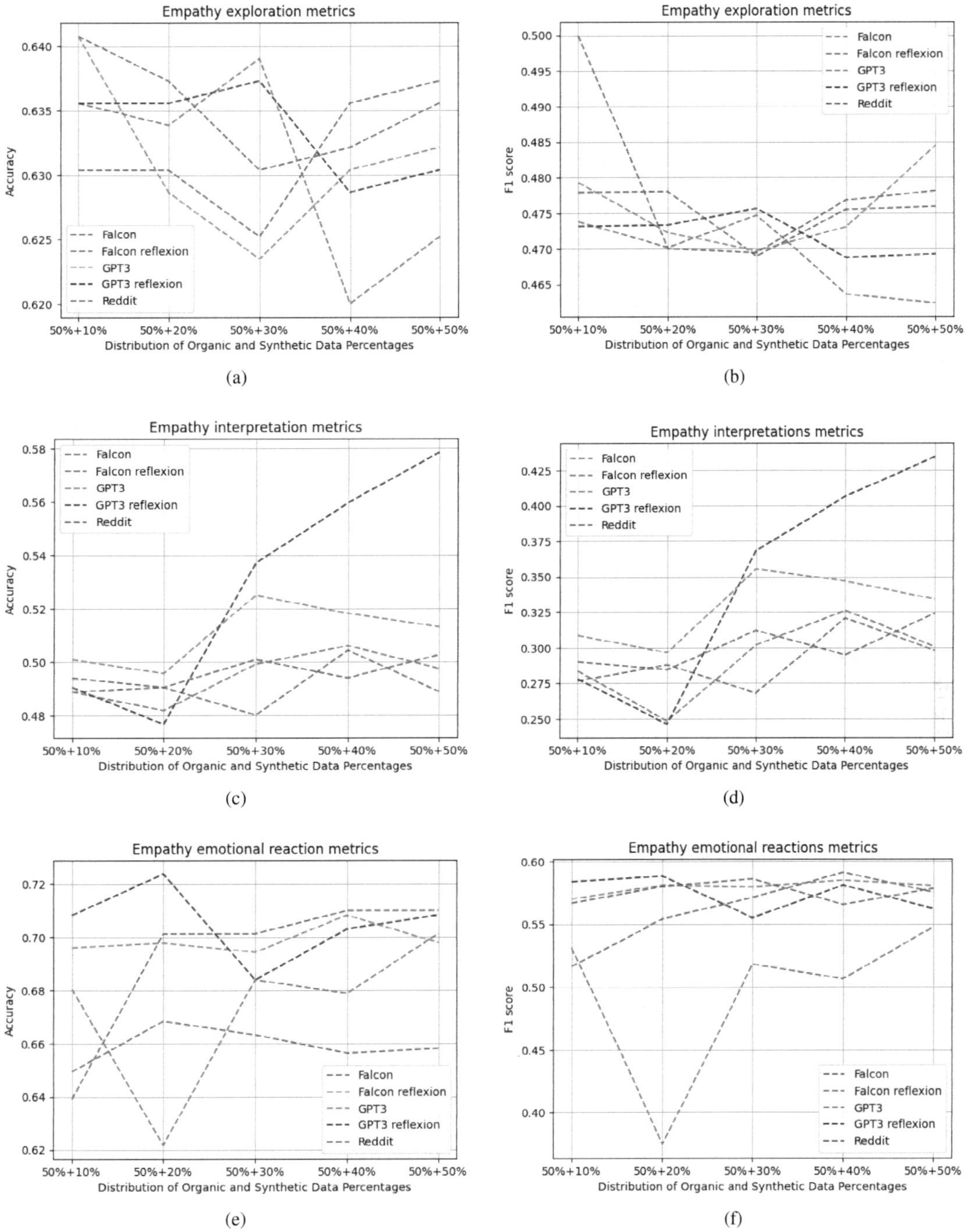

Figure 2: Accuracy and F1 scores for the three dimensions of empathy when substituting different percentages of the original Reddit data with synthetic data.

6 Conclusion

We generated synthetic datasets using LLMs and prompt engineering techniques, labeling them according to the EPITOME framework. To evaluate the impact of synthetic data, we trained bi-encoder models for empathy detection and measured their performance gains when augmenting the original dataset. Our results show that incorporating synthetic data improved the F1 score of empathy exploration detection by up to 10%. Notably, when replacing 50% of the original data with synthetic data, the interpretation dimension of empathy saw an 11% increase in F1 score. Meanwhile, the emotional reaction and exploration dimensions maintained consistent performance when substituting the original dataset entirely.

7 Ethical Considerations

While our results illustrate the advantages of using synthetic data to enhance NLP model performance, it is essential to acknowledge that LLMs can exhibit various biases in their outputs (Acerbi and Stubbersfield, 2023; Navigli et al., 2023). Therefore, a thorough examination is necessary to prevent the inadvertent propagation of such biases (Ayoub et al., 2024; Tao et al., 2024).

In the context of synthetic mental health data, assessing the presence of stereotypes in the generated texts is particularly critical (Lozoya et al., 2023). Research has shown that stereotypes and biases can negatively impact mental health treatment outcomes (Wirth and Bodenhausen, 2009; Chatmon, 2020). Future work should evaluate the extent to which synthetic dialogues reinforce or mitigate existing biases, particularly in the portrayal of different demographic groups and mental health conditions. This could involve conducting qualitative and quantitative analyses of the generated texts, comparing them to real-world clinical dialogues, and implementing bias-detection frameworks to identify and mitigate harmful stereotypes.

8 Limitations

Due to resource constraints, we limited the number of synthetic dialogues generated and labeled. Future research could explore the upper limits of performance improvement achievable with synthetic data, particularly for certain dimensions of empathy, such as interpretation, where the trend suggests that additional data may further enhance the model's performance.

Additionally, we only used 3 annotators to label the data, the annotators shared similar demographic features such as gender, age range, nationality, and educational background. This lack of diversity among annotators may have introduced biases into the dataset, as their perspectives and interpretations could be influenced by shared cultural and personal experiences. Future studies should consider employing a more diverse group of annotators to enhance the representativeness and generalizability of the labeled data.

Another limitation of our study, due to computational constraints, was that we only tested a 7B parameter model, rather than larger models that have demonstrated superior generative performance. Future work could explore the use of more advanced open-source LLMs, such as LLaMA 3 (Grattafiori et al., 2024) and Mistral (Jiang et al., 2023), to evaluate the quality of synthetic data. Additionally, testing newer techniques for prompt optimization could help improve the quality of the synthetic text we generate (Lozoya et al., 2024).

References

Alberto Acerbi and Joseph M. Stubbersfield. 2023. Large language models show human-like content biases in transmission chain experiments. *Proceedings of the National Academy of Sciences*, 120(44):e2313790120.

Talayeh Aledavood, Ana Maria Triana Hoyos, Tuomas Alakörkkö, Kimmo Kaski, Jari Saramäki, Erkki Isometsä, and Richard K Darst. 2017. Data collection for mental health studies through digital platforms: Requirements and design of a prototype. *JMIR Research Protocols*, 6(6):e110.

Alexander Street. 2009. Counseling and psychotherapy transcripts, client narratives, and reference works.

Ali Amin-Nejad, Julia Ive, and Sumithra Velupillai. 2020. Exploring transformer text generation for medical dataset augmentation. In *Proceedings of the Twelfth Language Resources and Evaluation Conference*, pages 4699–4708, Marseille, France. European Language Resources Association.

Gunjan Ansari, Muskan Garg, and Chandni Saxena. 2021. Data augmentation for mental health classification on social media. In *Proceedings of the 18th International Conference on Natural Language Processing (ICON)*, pages 152–161, National Institute of Technology Silchar, Silchar, India. NLP Association of India (NLPAI).

Gunjan Ansari and Chandni Saxena. 2024. Enhancing affective computing in NLP through data augmentation: Strategies for overcoming limited data availability. In *The Springer Series in Applied Machine Learning*, pages 201–216. Springer Nature Switzerland, Cham.

Noel F. Ayoub, Karthik Balakrishnan, Marc S. Ayoub, Thomas F. Barrett, Abel P. David, and Stacey T. Gray. 2024. Inherent bias in large language models: A random sampling analysis. *Mayo Clinic Proceedings: Digital Health*, 2(2):186–191.

Mrinal Kanti Baowaly, Chia-Ching Lin, Chao-Lin Liu, and Kuan-Ta Chen. 2018. Synthesizing electronic health records using improved generative adversarial networks. *Journal of the American Medical Informatics Association*, 26(3):228–241.

Lindsay L. Benster and Neal R. Swerdlow. 2020. Pathways to empathy in mental health care providers. *Advances in Health and Behavior*, 3(1):125–135.

Tom Brown, Benjamin Mann, Nick Ryder, Melanie Subbiah, Jared D Kaplan, Prafulla Dhariwal, Arvind Neelakantan, Pranav Shyam, Girish Sastry, Amanda Askell, Sandhini Agarwal, Ariel Herbert-Voss, Gretchen Krueger, Tom Henighan, Rewon Child, Aditya Ramesh, Daniel Ziegler, Jeffrey Wu, Clemens Winter, Chris Hesse, Mark Chen, Eric Sigler, Mateusz Litwin, Scott Gray, Benjamin Chess, Jack Clark, Christopher Berner, Sam McCandlish, Alec Radford, Ilya Sutskever, and Dario Amodei. 2020. Language models are few-shot learners. In *Advances in Neural Information Processing Systems*, volume 33, pages 1877–1901. Curran Associates, Inc.

Benita N. Chatmon. 2020. Males and mental health stigma. *American Journal of Men's Health*, 14(4).

Edward Choi, Siddharth Biswal, Bradley Malin, Jon Duke, Walter F. Stewart, and Jimeng Sun. 2017. Generating multi-label discrete patient records using generative adversarial networks. In *Proceedings of the 2nd Machine Learning for Healthcare Conference*, volume 68 of *Proceedings of Machine Learning Research*, pages 286–305. PMLR.

Jean-Philippe Corbeil and Hadi Abdi Ghadivel. 2020. Bet: A backtranslation approach for easy data augmentation in transformer-based paraphrase identification context. *arXiv preprint arXiv:2009.12452*.

Aleksandra Edwards, Asahi Ushio, Jose Camacho-collados, Helene Ribaupierre, and Alun Preece. 2022. Guiding generative language models for data augmentation in few-shot text classification. In *Proceedings of the Fourth Workshop on Data Science with Human-in-the-Loop (Language Advances)*, pages 51–63, Abu Dhabi, United Arab Emirates (Hybrid). Association for Computational Linguistics.

Robert Elliott, Arthur C. Bohart, Jeanne C. Watson, and Leslie S. Greenberg. 2011. Empathy. *Psychotherapy*, 48(1):43–49.

Robert Elliott, Arthur C. Bohart, Jeanne C. Watson, and David Murphy. 2018. Therapist empathy and client outcome: An updated meta-analysis. *Psychotherapy*, 55(4):399–410.

Michael P. Ewbank, Ronan Cummins, Valentin Tablan, Sarah Bateup, Ana Catarino, Alan J. Martin, and Andrew D. Blackwell. 2020. Quantifying the association between psychotherapy content and clinical outcomes using deep learning. *JAMA Psychiatry*, 77(1):35.

Garren Gaut, Mark Steyvers, Zac E. Imel, David C. Atkins, and Padhraic Smyth. 2017. Content coding of psychotherapy transcripts using labeled topic models. *IEEE Journal of Biomedical and Health Informatics*, 21(2):476–487.

Praveen Giridhara, Chinmaya Mishra, Reddy Venkataramana, Syed Bukhari, and Andreas Dengel. 2019. A study of various text augmentation techniques for relation classification in free text. In *Proceedings of the 8th International Conference on Pattern Recognition Applications and Methods*. SCITEPRESS - Science and Technology Publications.

Aaron Grattafiori, Abhimanyu Dubey, Abhinav Jauhri, Abhinav Pandey, Abhishek Kadian, Ahmad Al-Dahle, Aiesha Letman, Akhil Mathur, Alan Schelten, Alex Vaughan, Amy Yang, Angela Fan, Anirudh Goyal, Anthony Hartshorn, Aobo Yang, Archi Mitra, Archie Sravankumar, Artem Korenev, Arthur Hinsvark, Arun Rao, Aston Zhang, Aurelien Rodriguez, Austen Gregerson, Ava Spataru, Baptiste Roziere, Bethany Biron, Binh Tang, Bobbie Chern, Charlotte Caucheteux, Chaya Nayak, Chloe Bi, Chris Marra, Chris McConnell, Christian Keller, Christophe Touret, Chunyang Wu, Corinne Wong, Cristian Canton Ferrer, Cyrus Nikolaidis, Damien Allonsius, Daniel Song, Danielle Pintz, Danny Livshits, Danny Wyatt, David Esiobu, Dhruv Choudhary, Dhruv Mahajan, Diego Garcia-Olano, Diego Perino, Dieuwke Hupkes, Egor Lakomkin, Ehab AlBadawy, Elina Lobanova, Emily Dinan, Eric Michael Smith, Filip Radenovic, Francisco Guzmán, Frank Zhang, Gabriel Synnaeve, Gabrielle Lee, Georgia Lewis Anderson, Govind Thattai, Graeme Nail, Gregoire Mialon, Guan Pang, Guillem Cucurell, Hailey Nguyen, Hannah Korevaar, Hu Xu, Hugo Touvron, Iliyan Zarov, Imanol Arrieta Ibarra, Isabel Kloumann, Ishan Misra, Ivan Evtimov, Jack Zhang, Jade Copet, Jaewon Lee, Jan Geffert, Jana Vranes, Jason Park, Jay Mahadeokar, Jeet Shah, Jelmer van der Linde, Jennifer Billock, Jenny Hong, Jenya Lee, Jeremy Fu, Jianfeng Chi, Jianyu Huang, Jiawen Liu, Jie Wang, Jiecao Yu, Joanna Bitton, Joe Spisak, Jongsoo Park, Joseph Rocca, Joshua Johnstun, Joshua Saxe, Junteng Jia, Kalyan Vasuden Alwala, Karthik Prasad, Kartikeya Upasani, Kate Plawiak, Ke Li, Kenneth Heafield, Kevin Stone, Khalid El-Arini, Krithika Iyer, Kshitiz Malik, Kuenley Chiu, Kunal Bhalla, Kushal Lakhotia, Lauren Rantala-Yeary, Laurens van der Maaten, Lawrence Chen, Liang Tan, Liz Jenkins, Louis Martin, Lovish Madaan, Lubo Malo, Lukas

Blecher, Lukas Landzaat, Luke de Oliveira, Madeline Muzzi, Mahesh Pasupuleti, Mannat Singh, Manohar Paluri, Marcin Kardas, Maria Tsimpoukelli, Mathew Oldham, Mathieu Rita, Maya Pavlova, Melanie Kambadur, Mike Lewis, Min Si, Mitesh Kumar Singh, Mona Hassan, Naman Goyal, Narjes Torabi, Nikolay Bashlykov, Nikolay Bogoychev, Niladri Chatterji, Ning Zhang, Olivier Duchenne, Onur undefinedelebi, Patrick Alrassy, Pengchuan Zhang, Pengwei Li, Petar Vasic, Peter Weng, Prajjwal Bhargava, Pratik Dubal, Praveen Krishnan, Punit Singh Koura, Puxin Xu, Qing He, Qingxiao Dong, Ragavan Srinivasan, Raj Ganapathy, Ramon Calderer, Ricardo Silveira Cabral, Robert Stojnic, Roberta Raileanu, Rohan Maheswari, Rohit Girdhar, Rohit Patel, Romain Sauvestre, Ronnie Polidoro, Roshan Sumbaly, Ross Taylor, Ruan Silva, Rui Hou, Rui Wang, Saghar Hosseini, Sahana Chennabasappa, Sanjay Singh, Sean Bell, Seohyun Sonia Kim, Sergey Edunov, Shaoliang Nie, Sharan Narang, Sharath Raparthy, Sheng Shen, Shengye Wan, Shruti Bhosale, Shun Zhang, Simon Vandenhende, Soumya Batra, Spencer Whitman, Sten Sootla, Stephane Collot, Suchin Gururangan, Sydney Borodinsky, Tamar Herman, Tara Fowler, Tarek Sheasha, Thomas Georgiou, Thomas Scialom, Tobias Speckbacher, Todor Mihaylov, Tong Xiao, Ujjwal Karn, Vedanuj Goswami, Vibhor Gupta, Vignesh Ramanathan, Viktor Kerkez, Vincent Gonguet, Virginie Do, Vish Vogeti, Vítor Albiero, Vladan Petrovic, Weiwei Chu, Wenhan Xiong, Wenyin Fu, Whitney Meers, Xavier Martinet, Xiaodong Wang, Xiaofang Wang, Xiaoqing Ellen Tan, Xide Xia, Xinfeng Xie, Xuchao Jia, Xuewei Wang, Yaelle Goldschlag, Yashesh Gaur, Yasmine Babaei, Yi Wen, Yiwen Song, Yuchen Zhang, Yue Li, Yuning Mao, Zacharie Delpierre Coudert, Zheng Yan, Zhengxing Chen, Zoe Papakipos, Aaditya Singh, Aayushi Srivastava, Abha Jain, Adam Kelsey, Adam Shajnfeld, Adithya Gangidi, Adolfo Victoria, Ahuva Goldstand, Ajay Menon, Ajay Sharma, Alex Boesenberg, Alexei Baevski, Allie Feinstein, Amanda Kallet, Amit Sangani, Amos Teo, Anam Yunus, Andrei Lupu, Andres Alvarado, Andrew Caples, Andrew Gu, Andrew Ho, Andrew Poulton, Andrew Ryan, Ankit Ramchandani, Annie Dong, Annie Franco, Anuj Goyal, Aparajita Saraf, Arkabandhu Chowdhury, Ashley Gabriel, Ashwin Bharambe, Assaf Eisenman, Azadeh Yazdan, Beau James, Ben Maurer, Benjamin Leonhardi, Bernie Huang, Beth Loyd, Beto De Paola, Bhargavi Paranjape, Bing Liu, Bo Wu, Boyu Ni, Braden Hancock, Bram Wasti, Brandon Spence, Brani Stojkovic, Brian Gamido, Britt Montalvo, Carl Parker, Carly Burton, Catalina Mejia, Ce Liu, Changhan Wang, Changkyu Kim, Chao Zhou, Chester Hu, Ching-Hsiang Chu, Chris Cai, Chris Tindal, Christoph Feichtenhofer, Cynthia Gao, Damon Civin, Dana Beaty, Daniel Kreymer, Daniel Li, David Adkins, David Xu, Davide Testuggine, Delia David, Devi Parikh, Diana Liskovich, Didem Foss, Dingkang Wang, Duc Le, Dustin Holland, Edward Dowling, Eissa Jamil, Elaine Montgomery, Eleonora Presani, Emily Hahn, Emily Wood, Eric-Tuan Le, Erik Brinkman, Esteban Arcaute, Evan Dunbar, Evan Smothers, Fei Sun, Felix Kreuk, Feng Tian, Filippos Kokkinos, Firat Ozgenel, Francesco Caggioni, Frank Kanayet, Frank Seide, Gabriela Medina Florez, Gabriella Schwarz, Gada Badeer, Georgia Swee, Gil Halpern, Grant Herman, Grigory Sizov, Guangyi, Zhang, Guna Lakshminarayanan, Hakan Inan, Hamid Shojanazeri, Han Zou, Hannah Wang, Hanwen Zha, Haroun Habeeb, Harrison Rudolph, Helen Suk, Henry Aspegren, Hunter Goldman, Hongyuan Zhan, Ibrahim Damlaj, Igor Molybog, Igor Tufanov, Ilias Leontiadis, Irina-Elena Veliche, Itai Gat, Jake Weissman, James Geboski, James Kohli, Janice Lam, Japhet Asher, Jean-Baptiste Gaya, Jeff Marcus, Jeff Tang, Jennifer Chan, Jenny Zhen, Jeremy Reizenstein, Jeremy Teboul, Jessica Zhong, Jian Jin, Jingyi Yang, Joe Cummings, Jon Carvill, Jon Shepard, Jonathan McPhie, Jonathan Torres, Josh Ginsburg, Junjie Wang, Kai Wu, Kam Hou U, Karan Saxena, Kartikay Khandelwal, Katayoun Zand, Kathy Matosich, Kaushik Veeraraghavan, Kelly Michelena, Keqian Li, Kiran Jagadeesh, Kun Huang, Kunal Chawla, Kyle Huang, Lailin Chen, Lakshya Garg, Lavender A, Leandro Silva, Lee Bell, Lei Zhang, Liangpeng Guo, Licheng Yu, Liron Moshkovich, Luca Wehrstedt, Madian Khabsa, Manav Avalani, Manish Bhatt, Martynas Mankus, Matan Hasson, Matthew Lennie, Matthias Reso, Maxim Groshev, Maxim Naumov, Maya Lathi, Meghan Keneally, Miao Liu, Michael L. Seltzer, Michal Valko, Michelle Restrepo, Mihir Patel, Mik Vyatskov, Mikayel Samvelyan, Mike Clark, Mike Macey, Mike Wang, Miquel Jubert Hermoso, Mo Metanat, Mohammad Rastegari, Munish Bansal, Nandhini Santhanam, Natascha Parks, Natasha White, Navyata Bawa, Nayan Singhal, Nick Egebo, Nicolas Usunier, Nikhil Mehta, Nikolay Pavlovich Laptev, Ning Dong, Norman Cheng, Oleg Chernoguz, Olivia Hart, Omkar Salpekar, Ozlem Kalinli, Parkin Kent, Parth Parekh, Paul Saab, Pavan Balaji, Pedro Rittner, Philip Bontrager, Pierre Roux, Piotr Dollar, Polina Zvyagina, Prashant Ratanchandani, Pritish Yuvraj, Qian Liang, Rachad Alao, Rachel Rodriguez, Rafi Ayub, Raghotham Murthy, Raghu Nayani, Rahul Mitra, Rangaprabhu Parthasarathy, Raymond Li, Rebekkah Hogan, Robin Battey, Rocky Wang, Russ Howes, Ruty Rinott, Sachin Mehta, Sachin Siby, Sai Jayesh Bondu, Samyak Datta, Sara Chugh, Sara Hunt, Sargun Dhillon, Sasha Sidorov, Satadru Pan, Saurabh Mahajan, Saurabh Verma, Seiji Yamamoto, Sharadh Ramaswamy, Shaun Lindsay, Shaun Lindsay, Sheng Feng, Shenghao Lin, Shengxin Cindy Zha, Shishir Patil, Shiva Shankar, Shuqiang Zhang, Shuqiang Zhang, Sinong Wang, Sneha Agarwal, Soji Sajuyigbe, Soumith Chintala, Stephanie Max, Stephen Chen, Steve Kehoe, Steve Satterfield, Sudarshan Govindaprasad, Sumit Gupta, Summer Deng, Sungmin Cho, Sunny Virk, Suraj Subramanian, Sy Choudhury, Sydney Goldman, Tal Remez, Tamar Glaser, Tamara Best, Thilo Koehler, Thomas Robinson, Tianhe Li, Tianjun Zhang, Tim Matthews, Timothy Chou, Tzook Shaked, Varun Vontimitta, Victoria Ajayi, Victoria Montanez, Vijai Mohan, Vinay Satish Kumar, Vishal Mangla, Vlad Ionescu, Vlad Poenaru, Vlad Tiberiu Mihailescu,

Vladimir Ivanov, Wei Li, Wenchen Wang, Wenwen Jiang, Wes Bouaziz, Will Constable, Xiaocheng Tang, Xiaojian Wu, Xiaolan Wang, Xilun Wu, Xinbo Gao, Yaniv Kleinman, Yanjun Chen, Ye Hu, Ye Jia, Ye Qi, Yenda Li, Yilin Zhang, Ying Zhang, Yossi Adi, Youngjin Nam, Yu, Wang, Yu Zhao, Yuchen Hao, Yundi Qian, Yunlu Li, Yuzi He, Zach Rait, Zachary DeVito, Zef Rosnbrick, Zhaoduo Wen, Zhenyu Yang, Zhiwei Zhao, and Zhiyu Ma. 2024. The llama 3 herd of models. *arXiv preprint arXiv:2407.21783*.

Ishaan Gulrajani, Faruk Ahmed, Martin Arjovsky, Vincent Dumoulin, and Aaron Courville. 2017. Improved training of wasserstein gans. In *Proceedings of the 31st International Conference on Neural Information Processing Systems*, NIPS'17, page 5769–5779, Red Hook, NY, USA. Curran Associates Inc.

Desta Haileselassie Hagos, Rick Battle, and Danda B. Rawat. 2024. Recent advances in generative ai and large language models: Current status, challenges, and perspectives. *IEEE Transactions on Artificial Intelligence*, 5(12):5873–5893.

R Devon Hjelm, Athul Paul Jacob, Adam Trischler, Gerry Che, Kyunghyun Cho, and Yoshua Bengio. 2018. Boundary seeking GANs. In *International Conference on Learning Representations*.

Zac E. Imel, Mark Steyvers, and David C. Atkins. 2015. Computational psychotherapy research: Scaling up the evaluation of patient–provider interactions. *Psychotherapy*, 52(1):19–30.

Julia Ive, Natalia Viani, Joyce Kam, Lucia Yin, Somain Verma, Stephen Puntis, Rudolf N. Cardinal, Angus Roberts, Robert Stewart, and Sumithra Velupillai. 2020. Generation and evaluation of artificial mental health records for natural language processing. *npj Digital Medicine*, 3(1).

Albert Q. Jiang, Alexandre Sablayrolles, Arthur Mensch, Chris Bamford, Devendra Singh Chaplot, Diego de las Casas, Florian Bressand, Gianna Lengyel, Guillaume Lample, Lucile Saulnier, Lélio Renard Lavaud, Marie-Anne Lachaux, Pierre Stock, Teven Le Scao, Thibaut Lavril, Thomas Wang, Timothée Lacroix, and William El Sayed. 2023. Mistral 7b. *arXiv preprint arXiv:2310.06825*.

Maria Laricheva, Yan Liu, Edward Shi, and Amery Wu. 2024. Scoping review on natural language processing applications in counselling and psychotherapy. *British Journal of Psychology*.

Fei-Tzin Lee, Derrick Hull, Jacob Levine, Bonnie Ray, and Kathy McKeown. 2019. Identifying therapist conversational actions across diverse psychotherapeutic approaches. In *Proceedings of the Sixth Workshop on Computational Linguistics and Clinical Psychology*. Association for Computational Linguistics.

Scott H. Lee. 2018. Natural language generation for electronic health records. *npj Digital Medicine*, 1(1).

Peter J. Liu. 2018. Learning to write notes in electronic health records. *arXiv preprint arXiv:1808.02622*.

Daniel Lozoya, Alejandro Berazaluce, Juan Perches, Eloy Lúa, Mike Conway, and Simon D'Alfonso. 2024. Generating mental health transcripts with SAPE (Spanish adaptive prompt engineering). In *Proceedings of the 2024 Conference of the North American Chapter of the Association for Computational Linguistics: Human Language Technologies (Volume 1: Long Papers)*, pages 5096–5113, Mexico City, Mexico. Association for Computational Linguistics.

Daniel Cabrera Lozoya, Simon D'Alfonso, and Mike Conway. 2023. Identifying gender bias in generative models for mental health synthetic data. In *2023 IEEE 11th International Conference on Healthcare Informatics (ICHI)*, pages 619–626.

Qiuhao Lu, Dejing Dou, and Thien Huu Nguyen. 2021. Textual data augmentation for patient outcomes prediction. In *2021 IEEE International Conference on Bioinformatics and Biomedicine (BIBM)*. IEEE.

Taynan Maier Ferreira and Anna Helena Reali Costa. 2020. Deepbt and nlp data augmentation techniques: A new proposal and a comprehensive study. In *Intelligent Systems: 9th Brazilian Conference, BRACIS 2020, Rio Grande, Brazil, October 20–23, 2020, Proceedings, Part I*, page 435–449, Berlin, Heidelberg. Springer-Verlag.

Roberto Navigli, Simone Conia, and Björn Ross. 2023. Biases in large language models: Origins, inventory, and discussion. *J. Data and Information Quality*, 15(2).

OpenAI, Josh Achiam, Steven Adler, Sandhini Agarwal, Lama Ahmad, Ilge Akkaya, Florencia Leoni Aleman, Diogo Almeida, Janko Altenschmidt, Sam Altman, Shyamal Anadkat, Red Avila, Igor Babuschkin, Suchir Balaji, Valerie Balcom, Paul Baltescu, Haiming Bao, Mohammad Bavarian, Jeff Belgum, Irwan Bello, Jake Berdine, Gabriel Bernadett-Shapiro, Christopher Berner, Lenny Bogdonoff, Oleg Boiko, Madelaine Boyd, Anna-Luisa Brakman, Greg Brockman, Tim Brooks, Miles Brundage, Kevin Button, Trevor Cai, Rosie Campbell, Andrew Cann, Brittany Carey, Chelsea Carlson, Rory Carmichael, Brooke Chan, Che Chang, Fotis Chantzis, Derek Chen, Sully Chen, Ruby Chen, Jason Chen, Mark Chen, Ben Chess, Chester Cho, Casey Chu, Hyung Won Chung, Dave Cummings, Jeremiah Currier, Yunxing Dai, Cory Decareaux, Thomas Degry, Noah Deutsch, Damien Deville, Arka Dhar, David Dohan, Steve Dowling, Sheila Dunning, Adrien Ecoffet, Atty Eleti, Tyna Eloundou, David Farhi, Liam Fedus, Niko Felix, Simón Posada Fishman, Juston Forte, Isabella Fulford, Leo Gao, Elie Georges, Christian Gibson, Vik Goel, Tarun Gogineni, Gabriel Goh, Rapha Gontijo-Lopes, Jonathan Gordon, Morgan Grafstein, Scott Gray, Ryan Greene, Joshua Gross, Shixiang Shane Gu, Yufei Guo, Chris Hallacy, Jesse Han, Jeff Harris, Yuchen He, Mike Heaton, Johannes Heidecke, Chris

Hesse, Alan Hickey, Wade Hickey, Peter Hoeschele, Brandon Houghton, Kenny Hsu, Shengli Hu, Xin Hu, Joost Huizinga, Shantanu Jain, Shawn Jain, Joanne Jang, Angela Jiang, Roger Jiang, Haozhun Jin, Denny Jin, Shino Jomoto, Billie Jonn, Heewoo Jun, Tomer Kaftan, Łukasz Kaiser, Ali Kamali, Ingmar Kanitscheider, Nitish Shirish Keskar, Tabarak Khan, Logan Kilpatrick, Jong Wook Kim, Christina Kim, Yongjik Kim, Jan Hendrik Kirchner, Jamie Kiros, Matt Knight, Daniel Kokotajlo, Łukasz Kondraciuk, Andrew Kondrich, Aris Konstantinidis, Kyle Kosic, Gretchen Krueger, Vishal Kuo, Michael Lampe, Ikai Lan, Teddy Lee, Jan Leike, Jade Leung, Daniel Levy, Chak Ming Li, Rachel Lim, Molly Lin, Stephanie Lin, Mateusz Litwin, Theresa Lopez, Ryan Lowe, Patricia Lue, Anna Makanju, Kim Malfacini, Sam Manning, Todor Markov, Yaniv Markovski, Bianca Martin, Katie Mayer, Andrew Mayne, Bob McGrew, Scott Mayer McKinney, Christine McLeavey, Paul McMillan, Jake McNeil, David Medina, Aalok Mehta, Jacob Menick, Luke Metz, Andrey Mishchenko, Pamela Mishkin, Vinnie Monaco, Evan Morikawa, Daniel Mossing, Tong Mu, Mira Murati, Oleg Murk, David Mély, Ashvin Nair, Reiichiro Nakano, Rajeev Nayak, Arvind Neelakantan, Richard Ngo, Hyeonwoo Noh, Long Ouyang, Cullen O'Keefe, Jakub Pachocki, Alex Paino, Joe Palermo, Ashley Pantuliano, Giambattista Parascandolo, Joel Parish, Emy Parparita, Alex Passos, Mikhail Pavlov, Andrew Peng, Adam Perelman, Filipe de Avila Belbute Peres, Michael Petrov, Henrique Ponde de Oliveira Pinto, Michael, Pokorny, Michelle Pokrass, Vitchyr H. Pong, Tolly Powell, Alethea Power, Boris Power, Elizabeth Proehl, Raul Puri, Alec Radford, Jack Rae, Aditya Ramesh, Cameron Raymond, Francis Real, Kendra Rimbach, Carl Ross, Bob Rotsted, Henri Roussez, Nick Ryder, Mario Saltarelli, Ted Sanders, Shibani Santurkar, Girish Sastry, Heather Schmidt, David Schnurr, John Schulman, Daniel Selsam, Kyla Sheppard, Toki Sherbakov, Jessica Shieh, Sarah Shoker, Pranav Shyam, Szymon Sidor, Eric Sigler, Maddie Simens, Jordan Sitkin, Katarina Slama, Ian Sohl, Benjamin Sokolowsky, Yang Song, Natalie Staudacher, Felipe Petroski Such, Natalie Summers, Ilya Sutskever, Jie Tang, Nikolas Tezak, Madeleine B. Thompson, Phil Tillet, Amin Tootoonchian, Elizabeth Tseng, Preston Tuggle, Nick Turley, Jerry Tworek, Juan Felipe Cerón Uribe, Andrea Vallone, Arun Vijayvergiya, Chelsea Voss, Carroll Wainwright, Justin Jay Wang, Alvin Wang, Ben Wang, Jonathan Ward, Jason Wei, CJ Weinmann, Akila Welihinda, Peter Welinder, Jiayi Weng, Lilian Weng, Matt Wiethoff, Dave Willner, Clemens Winter, Samuel Wolrich, Hannah Wong, Lauren Workman, Sherwin Wu, Jeff Wu, Michael Wu, Kai Xiao, Tao Xu, Sarah Yoo, Kevin Yu, Qiming Yuan, Wojciech Zaremba, Rowan Zellers, Chong Zhang, Marvin Zhang, Shengjia Zhao, Tianhao Zheng, Juntang Zhuang, William Zhuk, and Barret Zoph. 2024. Gpt-4 technical report. *arXiv preprint arXiv:2303.08774*.

Guilherme Penedo, Quentin Malartic, Daniel Hesslow, Ruxandra Cojocaru, Alessandro Cappelli, Hamza Alobeidli, Baptiste Pannier, Ebtesam Almazrouei, and Julien Launay. 2023. The RefinedWeb dataset for Falcon LLM: outperforming curated corpora with web data, and web data only. *arXiv preprint arXiv:2306.01116*.

Gaurav Sahu, Pau Rodriguez, Issam Laradji, Parmida Atighehchian, David Vazquez, and Dzmitry Bahdanau. 2022. Data augmentation for intent classification with off-the-shelf large language models. In *Proceedings of the 4th Workshop on NLP for Conversational AI*, pages 47–57, Dublin, Ireland. Association for Computational Linguistics.

Ashish Sharma, Adam Miner, David Atkins, and Tim Althoff. 2020. A computational approach to understanding empathy expressed in text-based mental health support. In *Proceedings of the 2020 Conference on Empirical Methods in Natural Language Processing (EMNLP)*, pages 5263–5276, Online. Association for Computational Linguistics.

Noah Shinn, Federico Cassano, Ashwin Gopinath, Karthik R Narasimhan, and Shunyu Yao. 2023. Reflexion: language agents with verbal reinforcement learning. In *Thirty-seventh Conference on Neural Information Processing Systems*.

Yan Tao, Olga Viberg, Ryan S Baker, and René F Kizilcec. 2024. Cultural bias and cultural alignment of large language models. *PNAS Nexus*, 3(9).

Zixu Wang, Julia Ive, Sumithra Velupillai, and Lucia Specia. 2019. Is artificial data useful for biomedical natural language processing algorithms? In *Proceedings of the 18th BioNLP Workshop and Shared Task*, pages 240–249, Florence, Italy. Association for Computational Linguistics.

Jason Wei and Kai Zou. 2019. EDA: Easy data augmentation techniques for boosting performance on text classification tasks. In *Proceedings of the 2019 Conference on Empirical Methods in Natural Language Processing and the 9th International Joint Conference on Natural Language Processing (EMNLP-IJCNLP)*, pages 6382–6388, Hong Kong, China. Association for Computational Linguistics.

James H. Wirth and Galen V. Bodenhausen. 2009. The role of gender in mental-illness stigma: A national experiment. *Psychological Science*, 20(2):169–173.

A Synthetic Empathy Prompts

Table 2 presents the prompts used to generate synthetic dialogues, categorized by both empathy level and type.

B Reflexion prompts

As outlined by (Shinn et al., 2023), the reflection framework utilizes three LLMs working in tandem:

the actor, the evaluator, and the self-reflection component. In our experiments, the LLM-actor generates therapeutic dialogues using one of the prompts listed in Appendix A. The evaluator then assesses the generated dialogues based on the intended level of empathy, using the prompts from table 3. Following this evaluation, the self-reflection LLM provides feedback to the LLM-actor, enabling improvements in the therapeutic dialogue. The final, refined dialogue is then stored in the training dataset.

Category	Level	Description
Emotional Reactions	Strong	Write a dialogue between two individuals where one person (Person 1) seeks help while the other person (Person 2) provides emotional support. Person 2 should demonstrate strong communication skills by expressing empathy, warmth, compassion, and concern towards Person 1 after reading their message.
	Weak	Write a dialogue between two individuals in which one person (Person 1) seeks help, while the other (Person 2) responds with minimal empathy. However, Person 2 demonstrates weak communication skills by offering little compassion or emotional support, providing only indifferent or dismissive responses to Person 1's concerns.
	None	Write a dialogue between two individuals where one person (Person 1) seeks help while the other person (Person 2) provides no empathy at all. Person 2 only provides factual information or offensive and abusive responses showing no communication of empathy towards Person 1 after reading their message.
Interpretations	Strong	Write a dialogue between two individuals where one person (Person 1) seeks help while the other person (Person 2) provides emotional support. Person 2 should communicate an understanding of feelings and experiences inferred from Person 1's post, specifying the inferred feeling or experience or communicating understanding through descriptions of similar experiences.
	Weak	Write a dialogue between two individuals in which one person (Person 1) seeks help while the other (Person 2) provides emotional support. However, Person 2 demonstrates weak communication skills by offering only a minimal acknowledgment of Person 1's feelings and experiences, merely stating that they understand.
	None	Write a dialogue between two individuals where one person (Person 1) seeks help while the other person (Person 2) provides no empathy at all. Person 2 only provides factual information or offensive and abusive responses showing no communication of empathy towards Person 1 after reading their message.
Explorations	Strong	Write a dialogue between two individuals where one person (Person 1) seeks help while the other person (Person 2) provides emotional support. Person 2 should demonstrate strong communication skills by improving understanding of Person 1 by exploring the feelings and experiences not stated in the post, showing active interest in what the seeker is experiencing and feeling, and probing gently as an aspect of empathy.
	Weak	Write a dialogue between two individuals in which one person (Person 1) seeks help, while the other (Person 2) provides emotional support. However, Person 2 demonstrates weak communication skills by offering only a surface-level understanding of Person 1's feelings and experiences, merely restating or acknowledging what has already been expressed without deeper exploration.
	None	Write a dialogue between two individuals where one person (Person 1) seeks help while the other person (Person 2) provides no empathy at all. Person 2 only provides factual information or offensive and abusive responses showing no communication of empathy towards Person 1 after reading their message.

Table 2: Prompts for each type of empathy dimension

Category	Level	Description
Emotional Reactions	Strong	Evaluate whether Person 2 demonstrates strong communication skills by expressing empathy, warmth, compassion, and concern towards Person 1. Check if Person 2's responses include validating statements, supportive language, and expressions of care.
	Weak	Evaluate whether Person 2 provides a weak empathy while responding to Person 1. Check if Person 2 acknowledges the issue but provides little compassion or emotional support, with responses that are indifferent or dismissive.
	None	Evaluate whether Person 2 provides no empathy at all. Check if Person 2 responds with purely factual, indifferent, offensive, or abusive remarks, showing no concern for Person 1's emotions.
Interpretations	Strong	Evaluate whether Person 2 accurately infers and communicates an understanding of Person 1's feelings and experiences. Check if Person 2 explicitly states the inferred emotions or relates to similar experiences.
	Weak	Evaluate whether Person 2 provides only minimal acknowledgment of Person 1's feelings. Check if Person 2 states that they understand but does not elaborate on the emotions or experiences involved.
	None	Evaluate whether Person 2 provides no acknowledgment or interpretation of Person 1's feelings. Check if Person 2 responds with factual information, offensive, or abusive remarks without recognizing or addressing emotions.
Explorations	Strong	Evaluate whether Person 2 actively explores and probes Person 1's unstated feelings and experiences. Check if Person 2 asks questions, expresses curiosity, and deepens understanding by gently prompting further discussion.
	Weak	Evaluate whether Person 2 provides only surface-level responses without deep exploration of Person 1's emotions or experiences. Check if Person 2 merely acknowledges or restates what was already expressed without probing further.
	None	Evaluate whether Person 2 completely avoids exploring Person 1's emotions or experiences. Check if Person 2 provides only factual information, dismissive responses, or offensive and abusive remarks.

Table 3: Evaluation prompts for each type of empathy dimension

A Systematic Evaluation of LLM Strategies for Mental Health Text Analysis: Fine-tuning vs. Prompt Engineering vs. RAG

Arshia Kermani, Veronica Perez-Rosas, Vangelis Metsis
Department of Computer Science
Texas State University
San Marcos, TX 78666, USA
{arshia.kermani, vperezr, vmetsis}@txstate.edu

Abstract

This study presents a systematic comparison of three approaches for the analysis of mental health text using large language models (LLMs): prompt engineering, retrieval augmented generation (RAG), and fine-tuning. Using LLaMA 3, we evaluate these approaches on emotion classification and mental health condition detection tasks across two datasets. Fine-tuning achieves the highest accuracy (91% for emotion classification, 80% for mental health conditions) but requires substantial computational resources and large training sets, while prompt engineering and RAG offer more flexible deployment with moderate performance (40-68% accuracy). Our findings provide practical insights for implementing LLM-based solutions in mental health applications, highlighting the trade-offs between accuracy, computational requirements, and deployment flexibility.

1 Introduction

The increasing prevalence of mental health conditions, coupled with limited access to mental health professionals, has created an urgent need for scalable approaches to mental health assessment and support. Traditional diagnostic methods in this area often rely heavily on clinical interviews and self-reported questionnaires, which can be time-consuming, subject to human bias, and limited in their reach (Chung and Teo, 2023). Recent advances in large language models (LLMs) present promising opportunities to enhance mental health assessment through automated analysis of text-based data.

LLMs have demonstrated remarkable capabilities in understanding and generating human language, with recent models like GPT-4, LLaMA 2 (Touvron et al., 2023), and their derivatives achieving unprecedented performance across various natural language processing tasks (Brown et al., 2020). In the medical domain specifically, LLMs have shown potential in tasks ranging from clinical decision support to patient education and medical documentation (Thirunavukarasu et al., 2023). However, their application to mental health assessment presents unique challenges due to the nuanced nature of emotional expression and the critical importance of accuracy in clinical contexts.

Previous research has explored various approaches to leverage LLMs for mental health applications. Studies have investigated the use of zero-shot and few-shot prompt strategies for mental health text classification (Lamichhane, 2023), achieving moderate success in tasks such as detecting stress and depression. Other work has examined the potential of fine-tuned models for specific mental health tasks (Ezerceli and Dehkharghani, 2024), demonstrating improved performance through domain adaptation. However, there remains a significant gap in understanding the relative efficacy of different LLM deployment strategies for mental health assessment tasks.

Our study addresses this gap by conducting a systematic comparison of three distinct approaches for mental health text classification: prompt engineering (including both zero-shot and few-shot variants), retrieval augmented generation (RAG), and fine-tuning. We evaluated these approaches using two complementary datasets: the DAIR-AI Emotion dataset, comprising 20,000 tweets labeled with six basic emotions, and the Reddit SuicideWatch and Mental Health Collection (SWMH), which contains 54,412 posts related to various mental health conditions.

This work makes several key contributions to the field:

1. We provide the first comprehensive comparison of prompt engineering, RAG, and fine-tuning approaches for mental health text classification, offering insights into their relative strengths and limitations.

Proceedings of the 10th Workshop on Computational Linguistics and Clinical Psychology (CLPsych 2025), pages 172–180
May 3, 2025

2. We demonstrate the effectiveness of LLaMA 3-based models for mental health assessment tasks, achieving accuracy rates of up to 91% on emotion classification and 80% on mental health condition classification through fine-tuning.

3. We present practical insights into the implementation challenges and resource requirements of each approach, informing future applications in clinical settings.

Our findings have important implications for the development of automated mental health assessment tools, suggesting that while fine-tuning achieves the highest accuracy, both prompt engineering and RAG offer viable alternatives with different trade-offs in terms of computational resources and deployment flexibility. These results contribute to the broader goal of developing reliable, scalable tools to support mental health professionals and improve access to mental health assessment.

In the following sections, we present related work and background (Section 2), detail our methodology (Section 3), present our experimental results (Section 4), discuss their limitations (Section 7), and conclude in (Section 5).

2 Related Work

The intersection of large language models (LLMs) and mental health assessment represents a rapidly evolving field with significant potential for improving healthcare delivery. This section examines the current state of LLMs in healthcare applications and their specific developments in mental health contexts.

2.1 Large Language Models in Healthcare

Recent advances in LLMs have transformed their potential applications in healthcare (He et al., 2023). These models have demonstrated capabilities ranging from clinical decision support and medical documentation to patient education and healthcare communication (Thirunavukarasu et al., 2023). The emergence of domain-specific medical LLMs, such as Med-PaLM 2 and Clinical-Camel, has further enhanced their utility in healthcare settings by incorporating specialized medical knowledge and terminology (Singhal et al., 2025).

The use of LLMs in healthcare applications typically follows three main strategies: fine-tuning existing models, prompt engineering, and retrieval-augmented generation (RAG). Fine-tuning has shown particular promise in specialized medical tasks, with models achieving performance comparable to healthcare professionals in diagnostic scenarios (Singhal et al., 2025). Prompt engineering approaches have shown effectiveness in zero-shot and few-shot learning contexts, allowing flexible deployment without extensive retraining (Liu et al., 2023). RAG methods have emerged as a promising approach for grounding LLM responses with domain knowledge, thereby reducing hallucination and improving reliability (Lewis et al., 2020; Gao et al., 2023).

2.2 Mental Health Text Analysis

Mental health assessment presents unique challenges for automated analysis due to the subtle nature of emotional expression and the critical importance of accurate interpretation. Traditional approaches to the analysis of mental health text have relied on rule-based systems and classical machine learning techniques, often struggling to capture the nuanced context necessary for an accurate assessment (Kazdin, 2011).

Recent work has begun to explore the potential of LLMs for mental health applications. Studies have shown promising results in the detection of signs of depression, anxiety, and suicidal ideation from social media posts (Ma et al., 2024). The evolution of LLM capabilities has particular relevance for mental health research. Recent studies have shown that advanced LLM versions can provide human-level interpretations in qualitative coding tasks (Dunivin, 2024) and achieve accuracy comparable to mental health professionals in certain diagnostic contexts (Kim et al., 2024).

When applied to qualitative analysis, LLMs have demonstrated the ability to perform various analytical approaches, including thematic analysis, content analysis, and grounded theory, as validated by human experts (Xiao et al., 2023b; Rasheed et al., 2024). This suggests potential for enhancing, rather than replacing, traditional qualitative analysis methods in mental health research. However, these applications pose important challenges, including the need for accurate predictions given the critical nature of mental health assessment, concerns about privacy and data security, and the importance of maintaining therapeutic alliance in clinical settings (Byers et al., 2023).

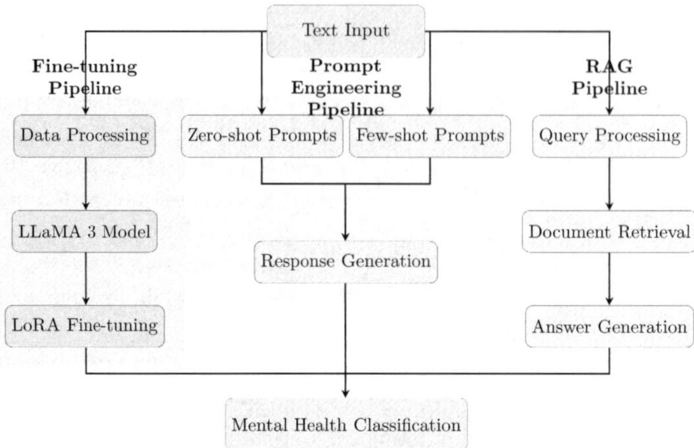

Figure 1: Overview of our experimental framework comparing three LLM deployment approaches (fine-tuning, prompt engineering, and RAG) for mental health text analysis. Each approach processes the same input data through distinct pipelines, enabling systematic comparison of their effectiveness.

2.3 Evaluation Frameworks

The evaluation of LLMs in mental health applications requires careful consideration of both model performance and clinical utility. Although traditional metrics such as accuracy and F1 scores provide important quantitative measures, they must be contextualized within the broader requirements of mental health assessment. Recent work has highlighted the importance of developing comprehensive evaluation frameworks that consider not only classification accuracy but also the ability of the model to provide interpretable and clinically relevant output (Xu et al., 2024).

The existing literature shows particular gaps in understanding the effectiveness of different LLM deployment strategies for mental health applications. Although studies have examined individual approaches, comprehensive comparisons of fine-tuning, prompt engineering, and RAG methods in mental health contexts remain limited. This gap is particularly significant given the practical considerations, such as the availability of resources and the computing requirements involved in deploying these different approaches in clinical settings.

Furthermore, the evaluation of LLMs in mental health applications must consider ethical implications and potential biases (Chancellor et al., 2019; Gallegos et al., 2024). This includes ensuring that models do not perpetuate existing biases in mental health diagnosis and that they maintain appropriate boundaries in therapeutic contexts. These considerations inform both the choice of evaluation metrics and the interpretation of results in mental health applications of LLMs.

3 Methodology

This study implements and evaluates three distinct approaches for LLMs in mental health text analysis: fine-tuning, prompt engineering, and retrieval-augmented generation (RAG). We utilize the LLaMA 3 model architecture, specifically the 8B parameter version, as our base model across all experiments to ensure fair comparison. Figure 1 presents the overall pipeline of our experimental framework. As shown in the Figure, our prompt engineering approach investigates both zero-shot and few-shot learning capabilities of the LLaMA 3 model for mental health text analysis.

3.1 Datasets

Our evaluation employs two complementary datasets that capture different aspects of mental health and emotional expression in text.

DAIR-AI Emotion Dataset The DAIR-AI (Saravia et al., 2018) dataset comprises `20,000` tweets labeled with one of six fundamental emotions: *joy*, *sadness*, *anger*, *fear*, *love*, or *surprise*. During our experiments, we maintain the original paper's data split: training set: 16,000 samples (80%); validation set: 2,000 samples (10%); test set: 2,000 samples (10%). Table 1 shows a sample of the DAIR-AI dataset.

Sample Text	Label
i can go from feeling so hopeless to so damned hopeful just from being around someone who cares and is awake	sadness
im grabbing a minute to post i feel greedy wrong	anger
i am ever feeling nostalgic about the fireplace i will know that it is still on the property	love
ive been taking or milligrams or times recommended amount and ive fallen asleep a lot faster but i also feel like so funny	surprise
i feel as confused about life as a teenager or as jaded as a year old man	fear
i have been with petronas for years i feel that petronas has performed well and made a huge profit	joy

Table 1: Samples and labels from the DAIR-AI Dataset.

Sample Text	Label
Wanting to skip my exam on Saturday because I'm so tired and mentally fried that a few days off might help.	Anxiety
Do other bipolar folks have problems with substance abuse? I've had overdoses and ended up in the ICU, and now I take my meds as prescribed.	Bipolar
Anonymous Entry: plz be nice. I've become a deteriorated husk of a person—hopefully this is my last moment of self-awareness.	Depression
I'm pretty sure my friend is suicidal; he keeps saying self-hating things like "I'm just a little emo prick." What do I do?	Suicide Watch

Table 2: Samples and labels from the SWMH Dataset.

Reddit SuicideWatch and Mental Health Collection (SWMH) The SWMH (Ji et al., 2021) dataset contains `54,412` Reddit posts that discuss various mental health conditions. Each post is labeled with one of the following categories: *depression*, *anxiety*, *bipolar disorder*, or *suicidal ideation*. The dataset is divided by the authors that published it as follows: training set: 34,824 samples (64%); validation set: 8,706 samples (16%); test set: 10,882 samples (20%). Table 2 presents a sample of the SWMH dataset.

3.1.1 Data Preprocessing

We conduct a preprocessing step on both datasets to ensure data quality and standardization. 1) Removal of URLs, user mentions, and special characters. 2) Standardization of text encoding to UTF-8. 3) Truncation of texts exceeding the model's maximum token limit (2048 tokens). 4) Verification of label consistency and removal of any samples with ambiguous or missing labels.

3.2 Experimental Setup

Our experiments are run on A100 GPU with 83.48 GB of RAM and 200 GB of disk space on Google Colab Pro+. We use the 8B parameter version of LLaMA 3 (Grattafiori et al., 2024), applying 4-bit quantization to optimize memory usage while preserving model performance. The base model configuration includes a `float16` precision for computational efficiency and a LLaMA tokenizer with right-padding and end-of-sequence tokens. During fine-tuning, we used the following hyperparameters: learning rate: 2e-4 with cosine schedule; weight decay: 0.001; batch size: 1 per device; gradient accumulation steps: 8; training epochs: 1; maximum steps: -1.

Classification evaluations are performed using F1 score, precision, and recall as our main metrics.

3.3 Fine-tuning

Our fine-tuning approach adapts the LLaMA 3 model to the specific requirements of mental health text analysis while maintaining computational efficiency. Figure 2 illustrates the fine-tuning architecture and process flow. To address the computational challenges of fine-tuning large models, we employed Low-Rank Adaptation (LoRA) (Hu et al., 2021). This approach significantly reduced the number of trainable parameters while maintaining model performance. Our LoRA configuration included a rank of 64 and an alpha scaling factor of 16.

3.4 Zero-Shot Prompting

We use zero-shot prompting to classify text without providing prior examples.

We used the following prompt template for both DAIR-AI and SWHM datasets, adjusting for the corresponding labels.

```
Analyze the emotional content in the following
text and classify it into exactly one of these
categories: joy, sadness, anger, fear, love,
or surprise. Provide only the category label as
output.

Text: [input_text]
```

3.5 Few-Shot Prompting

In our few-shot approach, we first select *random* examples and their corresponding labels from the training set and provide them as additional input

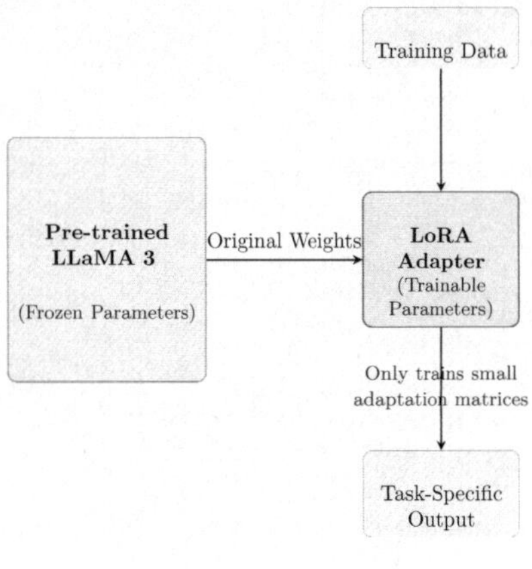

Figure 2: Architecture of our fine-tuning implementation, showing the integration of LoRA for parameter-efficient adaptation, the training process flow, and the evaluation pipeline.

to guide the model's reasoning. We included two examples for each label, ensuring balanced representation across classes. Example prompt:

```
Task: Classify the emotional content of text into
one of these categories: joy, sadness, anger, fear,
love, or surprise.

Example 1:
Text: "Finally got my dream job after months of
trying!"
Emotion: joy

Example 2:
Text: "I miss my old friends so much it hurts."
Emotion: sadness

Example 3:
Text: "How dare they treat people this way!"
Emotion: anger

[Additional examples...]

Now classify this text:
[input_text]
```

3.6 Retrieval-Augmented Generation (RAG)

The success of RAG models hinges on their capability to locate and retrieve pertinent examples and on the LLM's proficiency in effectively using the retrieved information. We believe that this is particularly helpful for the classification of mental health text, where additional context and examples can better inform the model. Since RAG can operate with much fewer training examples than is usually required to fine-tune an LLM model on a specific task, we consider RAG as a middle ground between few-shot prompting and fine-tuning.

We implement a RAG model that incorporates relevant contextual information derived from the training dataset during inference time. Figure 3 presents the overall architecture of our RAG implementation. It retrieves relevant examples from a knowledge base to be used during inference time to inform the model's decision, which are then added as part of the generation input.

3.6.1 Knowledge Base Construction

While implementing our model, we constructed a specialized knowledge base to support the retrieval process for each dataset:

Embedding Generation: We utilized the BAAI/bge-small-en-v1.5 (Xiao et al., 2023a) model to generate dense vector representations of training examples. The resulting embeddings are added to a vector database for storage and retrieval.

Vector Database: We use ChromaDB (Contributors, 2025) as our vector database. Our configuration includes cosine similarity as the distance metric, HNSW (Hierarchical Navigable Small World) as the indexing method, and category labels and source information as metadata storage.

3.6.2 Retrieval Process

During retrieval, we start by embedding the input query with the BAAI/bge-small-en-v1.5 model, then we selected the top-k nearest neighbors considering diverse examples across categories, finally, we form an unified context using the retrieved examples. The retriever returns the three most similar documents for each query, balancing context richness with computational efficiency.

Generation Component: The generation component combines the retrieved context with the input text to produce classification decisions. We use the following prompt template:

```
Review the following examples and context:

[Retrieved Context Documents]

Based on these examples, classify the emotional
content of the following text into one of these
categories: joy, sadness, anger, fear, love,
or surprise. Provide only the category label.

Text to classify: [input_text]
```

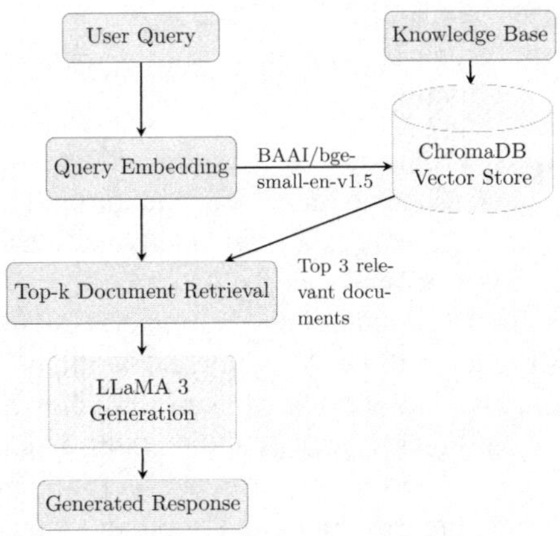

Figure 3: Architecture of our RAG model, illustrating the flow from input processing through retrieval and generation stages. The diagram shows how the system integrates embedded knowledge retrieval with LLM-based classification.

Method	DAIR-AI Emotion		SWMH	
	Accuracy	F1	Accuracy	F1
Fine-tuning	**91%**	**0.87**	**80%**	**0.81**
Zero-shot	49%	0.38	68%	0.67
Few-shot	39%	0.30	45%	0.57
RAG	47%	0.32	56%	0.45

Table 3: Performance comparison across methods and datasets. F1-scores (macro) provide a balanced measure of performance across all categories.

4 Results

Our experiments show significant performance variations across fine-tuning, prompt engineering, and retrieval augmented generation (RAG) approaches. Figure 4 presents the comparative performance in all methods.

Fine-tuning achieves the best performance, with accuracies of 91% and 80% on the DAIR-AI Emotion and SWMH datasets. Notably, zero-shot prompting emerged as the second-best performing approach, reaching 49% and 68% on each dataset, surpassing both the few-shot prompting and RAG. This suggests that carefully crafted prompts can effectively leverage the model's pre-trained knowledge for mental health text analysis, even without additional examples or context. However, we should clarify that in this study, we only used simple prompts as in the example shown in section 3.4 and kept them consistent throughout all experiments.

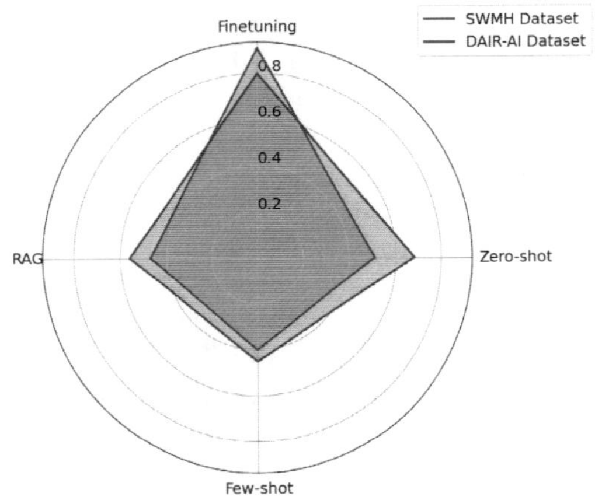

Figure 4: Performance comparison across different approaches for both datasets. The graph shows accuracy scores for fine-tuning, zero-shot prompting, few-shot prompting, and RAG methods.

Table 3 summarizes the classification performance in all methods and datasets. The performance disparity between the different approaches is particularly noteworthy, with the gap being more pronounced in emotion classification compared to the detection of mental health conditions.

Compared to prior work using relation networks (Ji et al., 2021), our approach demonstrates a 15.3% absolute improvement in classification accuracy and a significant boost in F1-score on the SWMH dataset. While the baseline relied on handcrafted sentiment and topic modeling features for classification, our fine-tuned LLaMA 3 model effectively captures the intricate language nuances in mental health discourse, yielding superior predictive performance. Furthermore, our zero-shot prompting approach (68% accuracy) surpassed the baseline's performance, suggesting that LLMs can generalize mental health-related text classification without requiring domain-specific feature engineering.

The greater advantage of the fine-tuning approach in the DAIR-AI Emotion dataset compared to the SWMH can be partially attributed to the dataset sizes (54.4K vs. 20K). A larger training set enables for a more effective fine-tuning, whereas this advantage diminishes and may even be reversed with smaller training sets.

4.1 Analysis of Best-Performing Methods

A more detailed evaluation of fine-tuning, zero-shot prompting, few-shot prompting, and RAG reveals distinct patterns in their effectiveness across different classification tasks. Table 4 presents a comprehensive comparison of these approaches in all classification categories.

The fine-tuned model demonstrated varying levels of performance across emotion categories. It achieved exceptional results for basic emotions such as *joy* and *sadness* (F1-scores of 0.94 and 0.95), followed by a strong performance for *anger* and *fear* (both 0.89). More complex emotional states proved more challenging, with *love* achieving an F1-score of 0.81 and *surprise* showing the lowest performance at 0.72. For mental health conditions, the model achieved the highest performance in detecting *anxiety* and *bipolar* disorder (F1-scores of 0.86 and 0.85, respectively) while maintaining robust performance for *depression* detection (F1: 0.79).

Zero-shot prompting showed notably strong performance in mental health condition detection, particularly for *depression* and *anxiety* (F1-scores of 0.70 and 0.74). However, its performance on emotion classification varied considerably. While achieving moderate results for *joy* and *sadness* (F1-scores of 0.56 and 0.58), it struggled significantly with more nuanced emotions like *love* and *surprise* (F1-scores of 0.25 and 0.26). The approach showed particularly low recall for *fear* detection despite high precision, indicating a conservative classification pattern for this category.

4.2 Analysis of Less Successful Methods

The evaluation of RAG and few-shot prompting revealed important insights about their practical limitations in mental health text analysis. Table 5 presents the key performance metrics for these approaches.

The RAG system achieved moderate performance levels (47% and 56% accuracy in DAIR-AI and SWMH, respectively), with effectiveness heavily dependent on retrieval quality. Performance was strongest when highly relevant context was successfully retrieved (64% accuracy) but dropped significantly with lower-quality retrievals (31% accuracy). Few-shot prompting showed unexpectedly lower performance compared to zero-shot approaches, suggesting that example-based prompting may introduce conflicting patterns that complicate the classification task in mental health contexts.

Our findings indicate that while RAG and few-shot prompting offer benefits in terms of interpretability and flexibility, their current implementations face significant challenges in achieving reliable performance for mental health text analysis task (Chung et al., 2023).

5 Conclusion

This study provided a systematic comparison of fine-tuning, prompt engineering, and retrieval augmented generation for mental health text classification. Fine-tuning showed superior performance, achieving 91% accuracy in emotion classification and 80% in the detection of mental health conditions, although at the cost of significant computational requirements. Zero-shot prompting emerged as a viable alternative, particularly for mental health condition detection (68% accuracy), suggesting that carefully designed prompts can effectively leverage pre-trained knowledge when fine-tuning is not feasible. However, both RAG and few-shot prompting showed limited effectiveness, with performance heavily dependent on retrieval quality and example selection.

These findings have important implications for developing automated mental health assessment tools. While fine-tuned models show promise for reliable screening applications, their varying performance across different emotional states and mental health conditions suggests current approaches may be better suited for initial assessment rather than definitive diagnosis.

Future research directions include the investigation of hybrid approaches that combine the strengths of multiple methods, the development of more efficient fine-tuning techniques, and the exploration of ways to improve the detection of nuanced psychological states. In addition, more work is needed to validate these approaches in clinical settings and across diverse populations.

6 Ethical Considerations

This study follows ethical guidelines on data usage, model reliability, and the responsible deployment of large language models (LLMs) for mental health evaluation. The datasets used in this research, DAIR-AI Emotion and SWMH, are publicly available, ensuring transparency and reproducibility. The SWMH dataset consists of publicly shared Reddit posts, while the DAIR-AI Emotion dataset contains labeled social media text. No personally identifiable information (PII) was processed, and no direct engagement with individuals was conducted.

Automated systems carry the risk of misclassifi-

Dataset	Category	Fine-tuning		Zero-shot		Few-shot		RAG	
		F1	Prec/Rec	F1	Prec/Rec	F1	Prec/Rec	F1	Prec/Rec
DAIR-AI	Joy	0.94	0.94/0.93	0.56	0.80/0.43	0.35	0.35/0.35	0.44	0.82/0.30
	Sadness	0.95	0.95/0.94	0.58	0.47/0.75	0.24	0.20/0.30	0.27	0.35/0.22
	Anger	0.89	0.88/0.91	0.46	0.39/0.57	0.36	0.30/0.45	0.29	0.41/0.23
	Fear	0.89	0.89/0.88	0.17	0.88/0.09	0.29	0.25/0.35	0.31	0.45/0.24
	Love	0.81	0.80/0.82	0.25	0.26/0.25	0.22	0.20/0.25	0.30	0.44/0.23
	Surprise	0.72	0.73/0.71	0.26	0.24/0.27	0.34	0.30/0.40	0.31	0.44/0.24
	Average	0.87	0.86/0.87	0.38	0.40/0.36	0.30	0.27/0.33	0.32	0.45/0.25
SWMH	Depression	0.79	0.78/0.80	0.70	0.59/0.84	0.52	0.74/0.40	0.40	0.45/0.36
	Anxiety	0.86	0.87/0.86	0.74	0.90/0.63	0.56	0.78/0.44	0.61	0.72/0.53
	Bipolar	0.85	0.87/0.83	0.62	0.88/0.48	0.63	0.80/0.53	0.40	0.45/0.36
	Suicide	0.75	0.75/0.75	0.61	0.68/0.56	0.55	0.64/0.48	0.39	0.44/0.35
	Average	0.81	0.82/0.81	0.67	0.70/0.64	0.57	0.75/0.47	0.45	0.52/0.40

Table 4: Detailed performance metrics for Fine-tuning, Zero-shot, Few-shot, and RAG approaches across all categories. Precision/Recall values are presented as Prec/Rec.

Method	DAIR-AI		SWMH	
	Acc.	Top Category	Acc.	Top Category
RAG	47%	Joy (0.44)	56%	Anxiety (0.61)
Few-shot	39%	Anger (0.36)	45%	Bipolar (0.63)

Table 5: Performance summary of RAG and few-shot approaches. The top Category shows the highest F1 score achieved for any single category.

cation, especially in sensitive areas such as depression and suicidal ideation. Any potential application of these models outside of research settings would require extensive validation, supervision by clinical professionals, and adherence to ethical and regulatory standards to avoid misinformation or unintended consequences.

7 Limitations

This study demonstrated the potential of large language models for psychological assessments but also showed a few limitations. Fine-tuning a model as extensive as LLaMA-3 8B required significant computational resources. This dependency on high-end resources limits the accessibility of our approach for researchers with constrained computational capacities. Furthermore, the models were trained and evaluated on the DAIR-AI Emotion and SWMH datasets, which, while diverse, may not fully capture the complexity and variability of real-world psychological text data. This could restrict the generalizability of the findings to other domains, languages, or text formats, e.g., short vs. long text. Additionally, our study does not address the practical integration of these tools into clinical workflows, which would require collaboration with domain experts and rigorous validation.

Addressing these limitations in future research could improve the accessibility, generalizability, and ethical applicability of LLM-based psychological assessment tools.

References

Tom Brown, Benjamin Mann, Nick Ryder, Melanie Subbiah, Jared D Kaplan, Prafulla Dhariwal, Arvind Neelakantan, Pranav Shyam, Girish Sastry, Amanda Askell, et al. 2020. Language models are few-shot learners. *Advances in neural information processing systems*, 33:1877–1901.

Morgan Byers, Mark Trahan, Erica Nason, Chinyere Eigege, Nicole Moore, Micki Washburn, and Vangelis Metsis. 2023. Detecting intensity of anxiety in language of student veterans with social anxiety using text analysis. *Journal of Technology in Human Services*, 41:1–21.

Stevie Chancellor, Michael L Birnbaum, Eric D Caine, Vincent MB Silenzio, and Munmun De Choudhury. 2019. A taxonomy of ethical tensions in inferring mental health states from social media. In *Proceedings of the conference on fairness, accountability, and transparency*, pages 79–88.

J. Chung and J. Teo. 2023. Single classifier vs. ensemble machine learning approaches for mental health prediction. *Brain Informatics*, 10.

Neo Christopher Chung, George Dyer, and Lennart Brocki. 2023. Challenges of large language models for mental health counseling. *Preprint*, arXiv:2311.13857.

Chroma Contributors. 2025. Chroma: The ai-native open-source embedding database. Accessed: 2025-02-12.

Zackary Okun Dunivin. 2024. Scalable qualitative coding with llms: Chain-of-thought reasoning matches human performance in some hermeneutic tasks. *arXiv preprint arXiv:2401.15170*.

Ozay Ezerceli and Rahim Dehkharghani. 2024. Mental disorder and suicidal ideation detection from social media using deep neural networks. *Journal of Computational Social Science*, pages 1–31.

Isabel O. Gallegos, Ryan A. Rossi, Joe Barrow, Md Mehrab Tanjim, Sungchul Kim, Franck Dernoncourt, Tong Yu, Ruiyi Zhang, and Nesreen K. Ahmed. 2024. Bias and fairness in large language models: A survey. *Computational Linguistics*, 50(3):1097–1179.

Yunfan Gao, Yun Xiong, Xinyu Gao, Kangxiang Jia, Jinliu Pan, Yuxi Bi, Yi Dai, Jiawei Sun, Qianyu Guo, Meng Wang, and Haofen Wang. 2023. Retrieval-augmented generation for large language models: A survey. *ArXiv*, abs/2312.10997.

Aaron Grattafiori, Abhimanyu Dubey, Abhinav Jauhri, et al. 2024. The llama 3 herd of models. *Preprint*, arXiv:2407.21783.

Kai He, Rui Mao, Qika Lin, Yucheng Ruan, Xiang Lan, Mengling Feng, and Erik Cambria. 2023. A survey of large language models for healthcare: from data, technology, and applications to accountability and ethics. *arXiv preprint arXiv:2310.05694*.

Edward J. Hu, Yelong Shen, Phillip Wallis, Zeyuan Allen-Zhu, Yuanzhi Li, Shean Wang, Lu Wang, and Weizhu Chen. 2021. Lora: Low-rank adaptation of large language models. *Preprint*, arXiv:2106.09685.

Shaoxiong Ji, Xue Li, Zi Huang, and Erik Cambria. 2021. Suicidal ideation and mental disorder detection with attentive relation networks. *Neural Computing and Applications*.

Alan E. Kazdin. 2011. Conceptualizing the challenge of reducing interpersonal violence and other public health problems: Behavioral and mental health disorders. *American Psychologist*, 66(7):621–639.

Jiyeong Kim, Kimberly G Leonte, Michael L Chen, John B Torous, Eleni Linos, Anthony Pinto, and Carolyn I Rodriguez. 2024. Large language models outperform mental and medical health care professionals in identifying obsessive-compulsive disorder. *NPJ Digital Medicine*, 7(1):193.

Bishal Lamichhane. 2023. Evaluation of chatgpt for nlp-based mental health applications. *Preprint*, arXiv:2303.15727.

Patrick Lewis, Ethan Perez, Aleksandra Piktus, Fabio Petroni, Vladimir Karpukhin, Naman Goyal, Heinrich Küttler, Mike Lewis, Wen-tau Yih, Tim Rocktäschel, et al. 2020. Retrieval-augmented generation for knowledge-intensive nlp tasks. *Advances in Neural Information Processing Systems*, 33:9459–9474.

Pengfei Liu, Weizhe Yuan, Jinlan Fu, Zhengbao Jiang, Hiroaki Hayashi, and Graham Neubig. 2023. Pre-train, prompt, and predict: A systematic survey of prompting methods in natural language processing. *ACM Computing Surveys*, 55(9):1–35.

Yingzhuo Ma, Yi Zeng, Tong Liu, Ruoshan Sun, Mingzhao Xiao, and Jun Wang. 2024. Integrating large language models in mental health practice: a qualitative descriptive study based on expert interviews. *Frontiers in Public Health*, 12:1475867.

Zeeshan Rasheed, Muhammad Waseem, Aakash Ahmad, Kai-Kristian Kemell, Wang Xiaofeng, Anh Nguyen Duc, and Pekka Abrahamsson. 2024. Can large language models serve as data analysts? a multi-agent assisted approach for qualitative data analysis. *arXiv preprint arXiv:2402.01386*.

Elvis Saravia, Hsien-Chi Toby Liu, Yen-Hao Huang, Junlin Wu, and Yi-Shin Chen. 2018. CARER: Contextualized affect representations for emotion recognition. In *Proceedings of the 2018 Conference on Empirical Methods in Natural Language Processing*, pages 3687–3697, Brussels, Belgium. Association for Computational Linguistics.

Karan Singhal, Tao Tu, Juraj Gottweis, Rory Sayres, Ellery Wulczyn, Mohamed Amin, Le Hou, Kevin Clark, Stephen R Pfohl, Heather Cole-Lewis, et al. 2025. Toward expert-level medical question answering with large language models. *Nature Medicine*, pages 1–8.

Arun James Thirunavukarasu, Darren Shu Jeng Ting, Kabilan Elangovan, Laura Gutierrez, Ting Fang Tan, and Daniel Shu Wei Ting. 2023. Large language models in medicine. *Nature medicine*, 29(8):1930–1940.

Hugo Touvron, Louis Martin, Kevin Stone, et al. 2023. Llama 2: Open foundation and fine-tuned chat models. *Preprint*, arXiv:2307.09288.

Shitao Xiao, Zheng Liu, Peitian Zhang, and Niklas Muennighoff. 2023a. C-pack: Packaged resources to advance general chinese embedding. *Preprint*, arXiv:2309.07597.

Ziang Xiao, Xingdi Yuan, Q Vera Liao, Rania Abdelghani, and Pierre-Yves Oudeyer. 2023b. Supporting qualitative analysis with large language models: Combining codebook with gpt-3 for deductive coding. In *Companion proceedings of the 28th international conference on intelligent user interfaces*, pages 75–78.

Xuhai Xu, Bingsheng Yao, Yuanzhe Dong, Saadia Gabriel, Hong Yu, James Hendler, Marzyeh Ghassemi, Anind K Dey, and Dakuo Wang. 2024. Mental-llm: Leveraging large language models for mental health prediction via online text data. *Proceedings of the ACM on Interactive, Mobile, Wearable and Ubiquitous Technologies*, 8(1):1–32.

Using LLMs to Aid Annotation and Collection of Clinically-Enriched Data in Bipolar Disorder and Schizophrenia

Ankit Aich[1, 5, 6], Avery Quynh[2], Pamela Osseyi[5], Amy Pinkham[3], Philip Harvey[4], Brenda Curtis[5], Colin Depp[2], Natalie Parde [1],
[1]Department of Computer Science, University of Illinois Chicago,
[2]University of California, San Diego,
[3]University of Texas Dallas, [4]University of Miami,
[5] National Institute on Drug Abuse, National Institutes of Health,
[6]School of Engineering and Applied Science, University of Pennsylvania

Abstract

Natural Language Processing (NLP) in mental health has largely focused on social media data or classification problems, often shifting focus from high caseloads or domain-specific needs of real-world practitioners. This study utilizes a dataset of 644 participants, including those with Bipolar Disorder, Schizophrenia, and Healthy Controls, who completed tasks from a standardized mental health instrument. Clinical annotators were used to label this dataset on five clinical variables. Expert annotations across five clinical variables demonstrated that contemporary language models, particularly smaller, fine-tuned models, can enhance data collection and annotation with greater accuracy and trust than larger commercial models. We show that these models can effectively capture nuanced clinical variables, offering a powerful tool for advancing mental health research. We also show that for clinically advanced tasks such as domain-specific annotation LLMs provide wrong labels as compared to a fine-tuned smaller model.

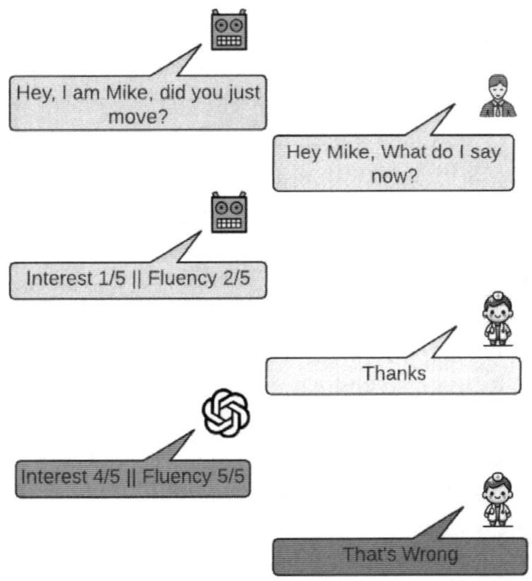

Figure 1: Our method creates a fine-tuned model. This model is able to directly interact with recruited participants to help them undertake established mental health instruments through turn-based tasks. It can annotate for clinical variables with low error. We see that commercial LLMs like GPT-4 / GPT-4o cannot annotate when it comes to clinical variables which are niche to a domain.

1 Introduction

The inherent complexity of mental health data presents significant challenges, even as the availability of AI systems designed to aid in its understanding and categorization continues to grow (Lee et al., 2021). AI-based systems have increasingly leveraged social media as a data source in the realm of mental healthcare, leading to the development of pre-trained models like MentalBERT (Ji et al., 2022) and initiatives to classify and detect various mental health phenomena, such as schizophrenia (Liu et al., 2022), disease progression (Birnbaum et al., 2019), depression (Kang et al., 2016), and stress (Winata et al., 2018).

In addition to the ethical issues surrounding the use of social media for clinical diagnoses, numerous other challenges persist. These include participant bias (Palacios-Ariza et al., 2023), issues with generalizability (Mitchell et al., 2015), and an overreliance on self-disclosure or non-clinical labels (Mitchell et al., 2015; Coppersmith et al., 2014).

Psychiatric disorders such as schizophrenia and bipolar disorder are often characterized by language deficiencies (Merrill et al., 2017). Individuals with these conditions may exhibit disorganized language comprehension and speech patterns (Kuperberg, 2010). Consequently, text or speech-based mental health instruments can be employed to assess individuals with medically validated diagnoses, thereby elucidating the effects of psychiatric disorders.

Proceedings of the 10th Workshop on Computational Linguistics and Clinical Psychology (CLPsych 2025), pages 181–192
May 3, 2025 ©2025 Association for Computational Linguistics

Previous efforts to apply AI in the context of schizophrenia and bipolar disorder have predominantly focused on automated diagnoses using smaller datasets (Sadeghi et al., 2021). These classification endeavors have encountered multiple challenges. For instance, social media data often results in non-clinical labels (Ernala et al., 2019), while the classification of clinical data is complicated by small datasets (Sadeghi et al., 2021), underutilization of records (Montazeri et al., 2022), and attempts to apply AI to multiple psychiatric disorders simultaneously (Chandran et al., 2019).

Moreover, the scarcity of robust data sources for mental health care and AI remains a significant barrier, as noted by Harrigian et al. (2021). The reliability of social media labels is further undermined over time due to evolving subjective annotation metrics (Harrigian and Dredze, 2022). To enhance the application of AI and language models in schizophrenia and bipolar disorder research, we propose a novel approach. This approach involves testing the efficacy of AI models in the context of data collection and annotation.

Our study starts with a dataset comprising 644 participants with established medical histories of schizo-affective disorder or schizophrenia (SZ), bipolar disorder (BD), or who are healthy controls (HC). These participants undergo a mental health instrument involving interviews conducted by expert clinicians (Patterson et al., 2001). We engaged two expert clinicians to annotate transcribed speech samples across five clinical variables. Importantly, we do not conduct automated diagnoses nor suggest that language models should be used for diagnostic purposes. Instead, we demonstrate how modern language models can assist in data collection and annotation.

The contributions of this paper are as follows:

- Extending a real-world dataset with expert clinical annotation, focusing on the language and speech deficiencies of individuals with bipolar disorder and schizophrenia.
- Creating a model that assists clinicians in maintaining dialogue with recruited participants for data collection purposes.
- Creating another model replicating clinical annotation of domain-specific variables with low error.
- Demonstrating that our models achieve low error rates and higher accuracy compared to commercial language models like GPT-4.

2 Data Collection and Labeling

We start by using the dataset introduced by Aich et al. (2022) in 2022. The data consists of transcribed texts from interviews with 644 participants. In the initial dataset, the authors recruited participants from three categories: participants with schizophrenia, participants with bipolar disorder, and healthy control groups. The diagnoses for subjects are all based on the DSM-V. The participants were in two simulated clinical tasks with expert clinicians to build the dataset. For task descriptions, please refer to appendix A.

We present a clinical annotation task to expand the dataset.

2.1 Clinical Annotation of Data

We collect clinical scores for our SSPA data. The SSPA instrument variables (Mausbach et al., 2008) are defined below. Annotators adhering to these definitions were found to have near-perfect agreement $\{\kappa \geq 0.85\}$ when labeling the presence of these variables (Patterson et al., 2001):

- **Interest/Disinterest:** Subjects with a relevant mental health condition show low engagement in SSPA tasks since brain functions are impaired.
- **Fluency:** Subjects with higher fluency use fewer filler words such as *umm*, *you know*, or *sooo*, and/or fewer long pauses during SSPA tasks.
- **Clarity:** Subjects with greater communication clarity exhibit stronger coherence in speech, both in how things were said and what was said. In lay terms, this variable describes how well subjects can get their point across.
- **Focus:** Subjects with greater focus can more solely concentrate on the task given to them without veering from their course. This variable also describes the subject's ability to focus on the interviewer and the current and overall task objectives.
- **Social Appropriateness:** Subjects with greater social appropriateness scores fare better socially with respect to the scene. They

react more appropriately to interview cues and are able to maintain increased composure during tasks.

These five SSPA scores are based on participants' interactions with the clinicians. Each of these scores is annotated for a subject in each scene. The scores are then averaged across the scene for the subject. A subject's total SSPA score is the average of their two scene scores. Scoring is performed manually by experts, achieving a high inter-class coefficient. As shown in prior work (Patterson et al., 2001), subjects' SSPA scores are significantly correlated with the presence or absence of schizophrenia/schizoaffective disorder ($p < 0.01$).[1]

For annotation and collection, there were two expert annotators. These were practicing clinicians and researchers in psychiatry. Each annotator reviews the entire transcript and labels all five scores. Gold standard labels are adjudicated by discussion among clinical experts. The SSPA is a well-established standardized test with the scoring metrics clearly defined. Cohen's Kappa κ for all clinical scores was $\kappa \geq 0.85$. For our work, we consider the final adjudicated gold standard labels.

3 Methods - Interview Sequence Generation

3.1 Context-Aware Interviewer

Our first specialized objective was to design a proof-of-concept context-aware interviewer to facilitate SSPA sessions. Currently the SSPA is administered by human clinicians with heavy case-loads. The US mental healthcare system is already heavily over burdened with a very low number of clinicians to a high number of patients (Coombs et al., 2021), potentially leading to mistakes and reduced efficiency. Having a trustworthy and viable agent can help alleviate some of this. To administer the SSPA in a language modeling setting understanding of context is important. Each response from an interviewer depends not only on the previous turn, but the entire dialogue history to that point, i.e. the entire context window of that string. As described previously, our SSPA data is represented as two sets of dialogues (lists of n utterances), one of which belongs to the patient P and the other of which belongs to the interviewer I: $P = \{P_0, P_1, ..., P_n\}$ and $I = \{I_0, I_1, ..., I_n\}$. Both are stored with associated timestamps indicating when utterances begin.

In a real world setting, it is expected that an interviewer has facilitated many interviews before, across people with bipolar disorder and schizophrenia as well as people with neither condition. It is also expected that in each complete dialogue turn $\{P_i, I_i\}$, the Interviewer response I_i is not only a response to the dialogue P_i but to the set of dialogues $\{P_0, I_0, P_1, I_1, ..., I_{i-1}, P_i\}$. The intuition is thus that the interviewer is responding not only to what was just uttered by the patient, but in a way that is suitable with the entire conversation so far, including all patient and interviewer utterances up to the most recent patient utterance.

3.2 Task Setup for Interview Experiment

In this section we describe our setup for the supervised fine-tuning (SFT) experiment. We model this task as a sequence to sequence problem. Our model is trained to generate an appropriate sequence of dialogue in response to dialogue sequences it has seen in such a way that it is aligned with that generated by a real-world interviewer. We train on 75% of all BD, HC, and SZ dialogues across both scenes. The input and outputs for the encoder-decoder forward pass are:

$$I \rightarrow Out = \begin{cases} P_0 \rightarrow I_0, & \text{if } n = 0 \\ P_0, I_0, ..., I_{i-1}, P_i \rightarrow I_i, & n = i \end{cases}$$

The equation above again emphasizes that to create input-output pairs we consider the dialogue history in addition to the most recent utterance. If we are at index 0 of a conversation, the interviewer's response is based directly on the the patient's utterance, but otherwise the interviewer response is based on the entire dialogue history between the patient and interviewer, until the *i-1*th interviewer utterance and the patient utterance P_i.

A schematic diagram for the SSPA language modeling process is shown in Figure 2. The model is fine-tuned until the loss drops from 1.64 initially to 0.1 after 15000 checkpoints and then results are calculated. To initialize training we provide the following source prefix:

```
You are an intelligent
interviewer see the examples
provided and learn to interview
a new patient
```

[1]Results are from a t-test taken comparing SSPA scores for schizophrenia and control group patients.

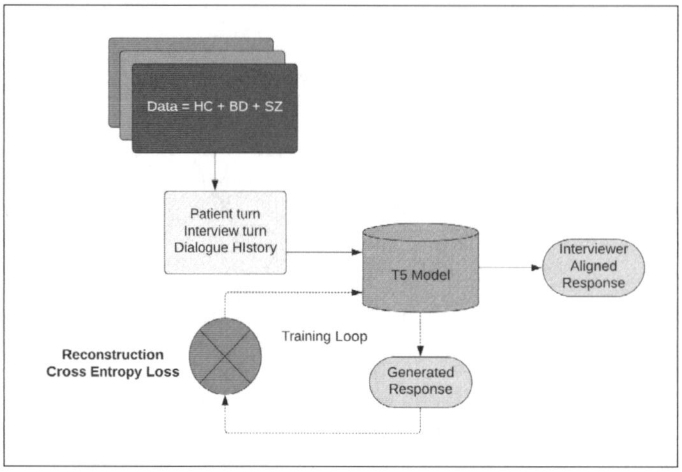

Figure 2: Interview model turns and dialogue history to calculate reconstruction loss and generate well aligned sequences towards the SSPA

We selected this prefix after experimenting with simpler versions (e.g., *"Interview a patient"* and *"Talk to a patient based on examples"*) and finding that the more complex final prefix was necessary to produce results well-aligned with reference interviews. This process does not involve in-context learning (ICL), prompt engineering, or tuning; it is a manually constructed prefix. Current literature suggests that better prefix descriptors, followed by improved training, yield superior results (Xue et al., 2022). Standard hyper-parameters were maintained at default values, and training was conducted on T4 GPUs.

The fine-tuned model was tested individually on all scenes and classes to simulate real-world interview conditions, where the interviewer focuses on a single scene and person, independent of prior interview training. To evaluate the quality of generated output, we computed syntactic similarity using ROUGE (Lin, 2004) scores, semantic similarity using cosine similarity, and alignment with human dialogue using BERTScore (Zhang et al., 2020).

3.3 Results: Generated Interview Quality

In Table 1 we present the results of the interview SFT experiment. To compute semantic similarity scores, we first encoded both the generated utterance and the corresponding gold expected utterance as word embeddings from the DeBERTA model owing to the model's increased ability to align with human speech (Zhang et al., 2020; He et al., 2021), and then calculated the cosine similarity between those embeddings. To compute syntactic similarity, we use ROUGE-1 to find overall unigram overlap and ROUGE-L to find the longest common subsequence overlap. We report precision, recall, and F1 scores for these two metrics using the ROUGE-SCORE package from the Python library.[2] We use the BERTScore (Zhang et al., 2020) metric directly, using a deberta model to vectorize the inputs to the metric generator (Zhang et al., 2020),[3] and report the precision, recall, and F1-score. According to the authors of the original paper, this model (deberta) offers the best understanding of the closeness of generated text to human intent. For all semantic metrics we use deberta as our choice of model since it has been consistently shown to outperform other encoder based popular choices like BERT or RoBERTa (He et al., 2021).

BERTScores, designed to capture intent and semantic similarity, are almost double the corresponding ROUGE scores for the same scenes. Recent studies have shown (Zhang et al., 2020; Hanna and Bojar, 2021) that BERTScore has two important properties. Firstly, it correlates with other summarization and similarity metrics (e.g., cosine similarity or BLEU score). Secondly, when a task becomes harder such as in our case, BERTScore accuracy peaks around 80% (Hanna and Bojar, 2021). Considering that our BERTScores for our task are close to 70% we can conclude that our model works at a high performance level. A better cosine simi-

[2] https://pypi.org/project/rouge-score/

[3] BERTScore needs users to specify which model to use to calculate metrics between two given strings. We use deberta for the same reasons cited earlier; i.e., studies have found it to generate text that more closely matches human speech.

Class × Scene	Semantic	Syntactic Similarity						Human Alignment		
	Cosine	ROUGE-1			ROUGE-L			BERTScore		
		P	R	F1	P	R	F1	P	R	F1
BD Scene_1	0.652	0.381	0.380	0.360	0.363	0.370	0.340	0.66	0.66	0.66
BD Scene_2	0.623	0.361	0.346	0.334	0.344	0.336	0.317	0.61	0.61	0.61
SZ Scene_1	0.634	0.331	0.316	0.301	0.328	0.314	0.300	0.63	0.64	0.63
SZ Scene_2	0.613	0.371	0.362	0.346	0.360	0.352	0.340	0.62	0.63	0.61
HC Scene_1	0.670	0.390	0.390	0.360	0.380	0.390	0.360	0.67	0.68	0.67
HC Scene_2	0.643	0.402	0.392	0.380	0.390	0.380	0.370	0.64	0.64	0.63

Table 1: Interview SFT Results. *P*=precision, and *R*=recall.

larity represents closeness in the embedding space of the vectors, whereas a good BERTScore tells us that the outputs are aligned with the reference sample.

However, even with a well-performing model, our ROUGE score is quite low. Some of this may be attributed to hallucinatory effects. For example, we observe that in one case while the interviewer in the original script says, e.g., "My name is INTERVIEWER," our model generates, e.g., "My name is NAME"—that is, a hallucinated name that was never previously mentioned in the dialogue. Thus, although this is structurally aligned with the reference, it differs in a key way that is best captured by ROUGE.

Another reason why our model exhibited lower syntactic than semantic performance may lie in disfluency. In our reference dialogues the interviewers often pause using filler words like *uh, uhh, okay,* or *mmhmm* to give the patients more time to speak. While our model thematically aligns decently well with these statements, its exact filler word matches are quite low. For example, we observe that the model also pauses but uses different filler words, or longer sequences of filler words, negatively affecting our ROUGE metric. However, throughout our observations, we can see that the model seems to understand the SSPA expectation of the interviewer role, even though we do not specify this in our SFT setup explicitly.

We qualitatively observe that the model appears capable at staying on-topic for the scene-specific task (e.g., generating content like *"Of course I will try to send someone over the fix the leak."*). It is interesting to observe that the model can discern the underlying task over long periods of training. Even without telling the model explicitly what the SSPA task involves, we can see that the model understands that a leaky pipe is at its core. This may suggest that LLMs are well-suited for tasks with better data and longer training (Min et al., 2022; Brown et al., 2020). While our alignment-based scores are not exceptionally high, this is still a strong starting benchmark for a nuanced task (Hanna and Bojar, 2021). The model captures close to 70 points of alignment with the intent of the actual interviewer. In the next phase we use this to generate annotator scores using another model to further progress the autonomous pipeline.

4 Methods - Annotation Generation

We also frame our SSPA score prediction task as a sequence to sequence task. Rather than simply predicting a sequence of scores, we also predict the label for which the score is being generated. In Section §2.1 we discussed what the five clinical variables are and how the scores are collected. In this score prediction task, the model learns to predict the score (SSPA clinical value) and the corresponding label. Therefore, the model predicts a sequence $Interest = XX, Fluency = YY$ rather than a simple distribution $4, 5, 3...$ This increases complexity, but helps us evaluate and walk towards a more explainable model. For this setting we use the interview dialogues from our source dataset and a `t5-base` LLM. We use the following prompt to generate scores:

```
You are an intelligent annotator
see the examples provided and
generate scores for each variable
```

The source prefix selection and other model parameters are kept the same as in the interview generation task described earlier in this paper. The model trains for 10000 checkpoints and the validation loss goes from 0.8 to 0.02 in our best performing model checkpoint. We calculate this reconstruction loss between the variable labels and scores that are annotated by our clinicians and the ones that are generated by our model. A standard cross-entropy loss function is used to find the loss. The model is trained on 75% of the data and validated on 5% of the data, with the remainder held out for testing.

4.1 Results: SSPA Score Prediction

The results of the SSPA score prediction model are presented in Table 2. The values represent the root mean squared error (RMSE) between the original annotated labels $Y = \{S_1, S_2, \ldots, S_n\}$ and the predicted labels $Y' = \{S'_1, S'_2, \ldots, S'_n\}$, where $Y \in \{\text{Interest}, \text{Fluency}, \text{Clarity}, \text{Focus}, \text{Social}\}$. The results indicate generally low error, with improved predictive performance in Scene 2 compared to Scene 1.

The model exhibits superior performance for the variables *Social* and *Focus*, which is anticipated as the SSPA predominantly evaluates social skills, and *Social* encapsulates social appropriateness. The variables *Focus*, *Clarity*, and *Fluency* are linguistically dependent, with the model performing best in this order, and the least effective for *Interest*. The higher RMSE for *Interest* can be attributed to its reliance on non-verbal cues such as body language, which are absent from our transcripts.

Overall, this standalone model demonstrates effective prediction capabilities. In the subsequent section, we illustrate the adaptation of our previous model from §3.1 into a chained pipeline, enabling SSPA interview transcripts to be scored with minimal RMSE differences compared to the standalone model.

5 Chained Model

So far in this paper we have created two standalone models: one in §3.1 that can learn from interviewers to appropriately interact with patients to facilitate the SSPA task, and the other in §4 that reads patient-interviewer transcripts and generates a sequence of SSPA scores for a patient. In this section we experiment with combining them. The primary motivation for this lies in anticipated real-world need, moving towards a seamless support tool for busy clinicians who may otherwise need to administer and score the SSPA manually. We create a chained model that (1) converses with the patient, and (2) predicts SSPA scores from the encounter.

We predict scores for dialogues that our model generated in §3.1. The input consists of the entire dialogue between the patient P and generated interviewer dialogues I_{gen}, forming the sequence $\{P_0, I_0, P_1, I_1, ...P_n, I_n\}$, where an interviewer dialogue $I_k \in \{I_{gen}\}$ acts as input and the model returns a sequence of five integer-valued scores, $\{S_1, S_2, .., S_5\}$, that quantify the SSPA variables defined in §2.1 (Interest, Fluency, Clarity, Focus, and Social).

5.1 Results: Chained Model

We present the results of the experiment in Table 3. The scores reported are the RMSE between the expected SSPA scores and the generated SSPA scores predicted for LLM-facilitated SSPA transcripts. Our first observation is the acute closeness to the stand-alone model scores (recall Table 2). This shows that even when LLM-based assistants are adapted in a chained end-to-end fashion, the results are similar to those observed using standalone models.

When we compare the difference between errors for Tables 2 and 3 we can see that the differences are quite low at both a variable level and a *class X scene* level. We can see in Tables 5 and 4 that on a per scene or per variable basis the differences are quite low with no significant difference [4]

6 Comparison with GPT Models

To compare the performance of our model against a large model like GPT, below we provide a baseline comparison between GPT-4, GPT-4o, and our method in replicating annotation tasks as detailed in §2.1. To get these labels we show GPT-4 the same de-identified data along with definitions of the clinical variables and ask it to label the data along these five categories. The results, presented in Table 6, illustrate the mean errors per class and scene, with statistical significance validated using the Wilcoxon signed-rank test. We find that GPT models show a high degree of error in comparison to our method when annotating clinical scores. This shows that a small fine-tuned model can outperform a large model like GPT with appropriate fine tuning.

Our experiments reveal two significant trends in our interview replication model: an intrinsic comprehension of tasks and the generation of unrelated

[4] A t-test between the RMSE scores per case (scene and class) and per variable shows the differences between score distributions for the standalone and chained models are not statistically significant ($p < 0.05$).

Class and Scene	RMSE					
	Interest	**Fluency**	**Clarity**	**Focus**	**Social**	**Avg. RMSE/Case**
BD Scene_1	1.36	1.10	1.04	0.97	1.06	1.10
BD Scene_2	1.09	1.11	1.14	1.15	1.12	1.12
SZ Scene_1	1.27	1.27	1.28	1.19	1.30	1.26
SZ Scene_2	1.22	1.10	1.13	1.10	1.07	1.12
HC Scene_1	1.28	1.36	1.35	1.33	1.33	1.33
HC Scene_2	0.84	0.78	0.68	0.84	0.68	0.76
Avg. RMSE/Var	1.17	1.12	1.10	1.09	1.09	N/A

Table 2: RMSE scores for standalone score prediction model, using original dataset. Avg-RMSE/Case represents the mean RMSE across a diagnostic group and scene. Avg-RMSE/Var represents the mean RMSE for that SSPA variable of the column.

Class and Scene	RMSE					
	Interest	**Fluency**	**Clarity**	**Focus**	**Social**	**Mean/Case**
BD Scene_1	1.28	1.12	1.07	0.97	1.06	1.10
BD Scene_2	1.39	1.11	1.14	1.18	1.10	1.18
SZ Scene_1	1.37	1.33	1.27	1.20	1.30	1.29
SZ Scene_2	1.33	1.13	1.12	1.15	1.10	1.16
HC Scene_1	1.33	1.37	1.27	1.30	1.28	1.31
HC Scene_2	0.83	0.78	0.75	0.92	0.75	0.80
Avg. RMSE/Var	1.25	1.14	1.10	1.12	1.09	N/A

Table 3: RMSE scores for the chained score prediction model. Interview sequences come from the generative model described in §3.1. Mean/Case represents the mean RMSE across a diagnostic group and scene. Avg-RMSE/Var represents the mean RMSE for that SSPA variable of the column.

information. Even without explicit task instructions, a well-constructed fine-tuning loop allows a smaller model to intuitively understand tasks, evidenced by the model's ability to identify tasks from indirect references. Despite the tendency of the model to hallucinate information such as names and dates, which typically impedes performance on tasks necessitating factual precision, our findings indicate that these hallucinations do not compromise task completion. For our annotation task, we maintained a sequence-to-sequence setup for predicting scores and variable labels, observing low error rates and consistent performance across both stand-alone and chained model setups.

7 Conclusion

This paper focused on an alternate purpose of LLMs in mental healthcare. Instead of classification or diagnostic problems we focus on a collaborative-LLM setup. We show that for real world clinical tasks, often involving complicated and nuanced variables, smaller and focused fine-tuning can help with data collection and annotation with relatively low error. We also show that such models can be chained together to create reliable and robust end-to-end data collection and annotation pipelines. We showed that modern LLMs such as GPT-4 or GPT-4o do not perform at the same level as a fine-tuned model on clinically nuanced tasks.

In mental health settings the expertise that clinicians bring cannot be replaced by LLM technology. Rather a collaborative approach where locally trained LLMs can learn from clinical labeling behavior without compromising data to external servers is a better way forward. Our findings indicate that language models can significantly assist clinicians in scaling data collection and labeling with high reliability, as evidenced by low error rates and high similarity scores. We anticipate that the clinical community will find our models ready for practical implementation, and our methods both translatable and adaptable to specific clinical tasks.

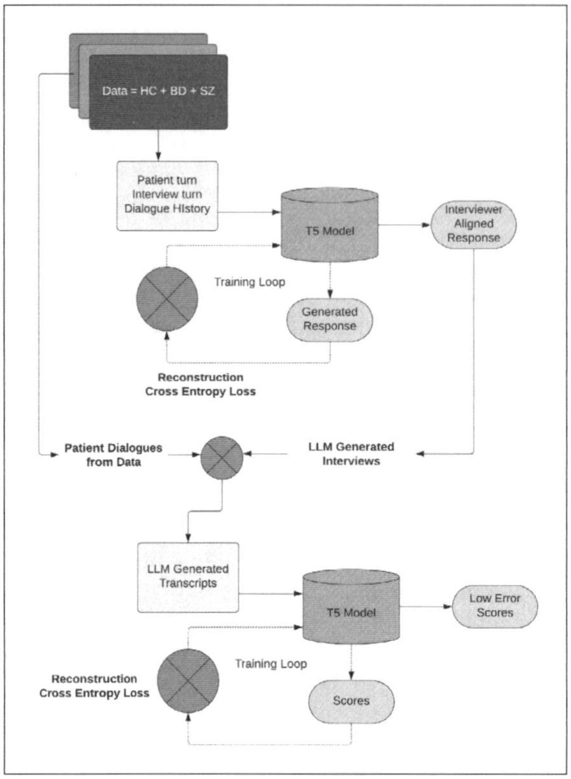

Figure 3: Chained Model Setup. Two standalone t5 models are chained by output and input. The Interview generator model works with patient dialogues to create LLM generated transcripts. This is fed into the score prediction model which outputs low error scores for the SSPA using a cross-entropy loss function.Picture resized for space limitations. Please zoom-in while reading review version.

BD Scene 1	BD Scene 2	SZ Scene 1	SZ Scene 2	HC Scene 1	HC Scene 2
0.0	0.06	0.03	0.04	0.2	0.04

Table 4: Difference of mean errors per case and scene.

Interest	Fluency	Clarity	Focus	Social
0.8	0.02	0.0	0.03	0.0

Table 5: Difference of mean errors per variable.

8 Limitations

In this study, we engaged 644 participants, which constitutes a relatively small sample size. While Patterson et al. (2001) originally identified eight variables in the SSPA, we selected only five for our analysis. The excluded variables were either unrelated to speech (e.g., personal grooming) or lacked expert raters due to their independence from the healthcare context (e.g., negotiation ability). We employed the T5 model for this task, primarily due to hardware constraints. Despite its smaller size, the T5 model demonstrated the capability to achieve relatively low error rates even with limited computational resources. This observation suggests that utilizing a larger model with more computational capacity could potentially reduce errors further. Furthermore, our study is limited to analyzing transcripts from audio recordings derived from the original dataset and does not incorporate multimodal aspects such as features of voice or audio.

Another limitation of our study concerns the use of a commercial language model, such as GPT-4/4o, exclusively for comparing annotations rather than conducting interviews. Although it could be argued that a commercial language model might

Scene/Class	GPT-4 Error	GPT-4o Error	Our Error	p 4	p - 4o
HC - Sc - 1	1.60	1.57	0.2	0.03	0.03
HC - Sc -2	1.70	1.66	0.04	0.03	0.03
SZ - Sc -1	1.44	1.50	0.03	0.03	0.03
SZ - Sc - 2	1.64	1.53	0.04	0.03	0.03
BD - Sc - 1	1.51	1.49	0.00	0.03	0.03
BD - Sc - 2	1.45	1.53	0.05	0.03	0.03

Table 6: Baseline Comparison with GPT4 and GPT4o. We can see that the p values comparing our error with GPT errors show a significant difference.

also be employed for interviews to compare performance outcomes, this approach raises significant ethical concerns. Firstly, most commercial language models do not possess adequate safeguards or specialized training to generate content that is safe for individuals with severe psychiatric conditions. Secondly, using such models could involve transmitting sensitive subject data and speech patterns to third-party systems, thereby raising serious ethical issues related to privacy and confidentiality. Consequently, commercial language models were restricted solely to annotation tasks using de-identified data.

The broader implications of these limitations merit careful consideration. Future work could explore how the currently excluded SSPA variables might be more formally defined and integrated into automated annotation pipelines using large language models (LLMs). While our study included 644 participants—a meaningful number in the context of psychiatric research—it remains relatively modest from the perspective of generalizability in AI applications for mental health. Nonetheless, this dataset represents one of the largest and medically validated corpora available for schizophrenia (SZ) and bipolar disorder (BD), laying essential groundwork for future model development. Due to hardware constraints, we employed the T5-base model, and it remains an open question whether scaling to larger variants (e.g., T5-XL or T5-XXL) would yield statistically significant performance improvements. We also limited the use of commercial models like GPT to annotation tasks only, in order to avoid exposing participants to unsupervised, third-party systems—particularly given the ethical concerns around deploying such models with vulnerable populations. Despite these constraints, our findings demonstrate that the speech patterns of trained psychiatrists can be reliably replicated with low error, opening the door to potential extensions such as modeling patient speech or augmenting data for low-resource clinical contexts. Finally, while we utilize established metrics such as BERTScore, ROUGE, cosine similarity, and RMSE, we acknowledge that these summarization benchmarks offer limited explainability with respect to individual-level communication quality. We encourage future work to incorporate more interpretable evaluation frameworks to deepen insight into both linguistic nuance and clinical relevance.

9 Ethical Concerns

This paper aims to demonstrate how modern language models can be deployed in clinical settings to collect and label data responsibly. We exclusively use labels that are well-established in clinical contexts. Importantly, this paper does not advocate for or implement the use of language models as diagnostic tools for mental health. We illustrate that markers of speech relevant to psychiatric healthcare can be predicted using language models. However, predicting variables like Interest or Focus should not be used or interpreted for unrelated tasks, such as advertising, targeted marketing, or any clinical purposes without appropriate expertise.

All data-related activities, including labeling, annotation, and sharing, were conducted with the approval of four independent academic Institutional Review Boards (IRBs). Participants in the original study provided informed consent. We adhere to all ethical codes established by the ACM and ACL. This paper involves numerous clinical experts in the labeling, adjudication, and language modeling processes, ensuring proper guidance and assistance. Using these models or concepts from this paper for non-clinical purposes or without expert guidance in clinical contexts is strictly prohibited.

References

Ankit Aich, Avery Quynh, Varsha Badal, Amy Pinkham, Philip Harvey, Colin Depp, and Natalie Parde. 2022. Towards intelligent clinically-informed language analyses of people with bipolar disorder and schizophrenia. In *Findings of the Association for Computational Linguistics: EMNLP 2022*, pages 2871–2887, Abu Dhabi, United Arab Emirates. Association for Computational Linguistics.

Michael Birnbaum, Sindhu Kiranmai Ernala, A. Rizvi, Elizabeth Arenare, Anna Van Meter, M. Choudhury, and J. Kane. 2019. Detecting relapse in youth with psychotic disorders utilizing patient-generated and patient-contributed digital data from facebook. *npj Schizophrenia*, 5.

Tom B. Brown, Benjamin Mann, Nick Ryder, Melanie Subbiah, Jared Kaplan, Prafulla Dhariwal, Arvind Neelakantan, Pranav Shyam, Girish Sastry, Amanda Askell, Sandhini Agarwal, Ariel Herbert-Voss, Gretchen Krueger, Tom Henighan, Rewon Child, Aditya Ramesh, Daniel M. Ziegler, Jeffrey Wu, Clemens Winter, Christopher Hesse, Mark Chen, Eric Sigler, Mateusz Litwin, Scott Gray, Benjamin Chess, Jack Clark, Christopher Berner, Sam McCandlish, Alec Radford, Ilya Sutskever, and Dario Amodei. 2020. Language models are few-shot learners. *Preprint*, arXiv:2005.14165.

David Chandran, Deborah Robbins, Chin-Kuo Chang, Hitesh Shetty, Jyoti Sanyal, Johnny Downs, Marcella Fok, Michael Ball, Richard Jackson, Robert Stewart, Hannah Cohen, Jentien Vermeulen, Frederike Schirmbeck, Lieuwe Haan, and Richard Hayes. 2019. Use of natural language processing to identify obsessive compulsive symptoms in patients with schizophrenia, schizoaffective disorder or bipolar disorder. *Scientific Reports*, 9:1–7.

Nicholas Coombs, Wyatt Meriwether, James Caringi, and Sophia Newcomer. 2021. Barriers to healthcare access among u.s. adults with mental health challenges: A population-based study. *SSM - Population Health*, 15:100847.

Glen Coppersmith, Mark Dredze, and Craig Harman. 2014. Quantifying mental health signals in Twitter. In *Proceedings of the Workshop on Computational Linguistics and Clinical Psychology: From Linguistic Signal to Clinical Reality*, pages 51–60, Baltimore, Maryland, USA. Association for Computational Linguistics.

Sindhu Kiranmai Ernala, Michael L. Birnbaum, Kristin A. Candan, Asra F. Rizvi, William A. Sterling, John M. Kane, and Munmun De Choudhury. 2019. Methodological gaps in predicting mental health states from social media: Triangulating diagnostic signals. In *Proceedings of the 2019 CHI Conference on Human Factors in Computing Systems*, CHI '19, page 1–16, New York, NY, USA. Association for Computing Machinery.

Michael Hanna and Ondřej Bojar. 2021. A fine-grained analysis of BERTScore. In *Proceedings of the Sixth Conference on Machine Translation*, pages 507–517, Online. Association for Computational Linguistics.

Keith Harrigian, Carlos Aguirre, and Mark Dredze. 2021. On the state of social media data for mental health research. In *Proceedings of the Seventh Workshop on Computational Linguistics and Clinical Psychology: Improving Access*, pages 15–24, Online. Association for Computational Linguistics.

Keith Harrigian and Mark Dredze. 2022. Then and now: Quantifying the longitudinal validity of self-disclosed depression diagnoses. In *Proceedings of the Eighth Workshop on Computational Linguistics and Clinical Psychology*, pages 59–75, Seattle, USA. Association for Computational Linguistics.

Pengcheng He, Jianfeng Gao, and Weizhu Chen. 2021. Debertav3: Improving deberta using electra-style pre-training with gradient-disentangled embedding sharing. *Preprint*, arXiv:2111.09543.

Shaoxiong Ji, Tianlin Zhang, Luna Ansari, Jie Fu, Prayag Tiwari, and Erik Cambria. 2022. Mental-BERT: Publicly Available Pretrained Language Models for Mental Healthcare. In *Proceedings of LREC*.

Keumhee Kang, Chanhee Yoon, and Eun Yi Kim. 2016. Identifying depressive users in twitter using multimodal analysis. In *2016 International Conference on Big Data and Smart Computing (BigComp)*, pages 231–238, Los Alamitos, CA, USA. IEEE Computer Society.

Gina Kuperberg. 2010. Language in schizophrenia part 1: An introduction. *Language and linguistics compass*, 4:576–589.

Ellen Lee, John Torous, Munmun Choudhury, Colin Depp, Sarah Graham, Ho-Cheol Kim, Martin Paulus, John Krystal, and Dilip Jeste. 2021. Artificial intelligence for mental health care: Clinical applications, barriers, facilitators, and artificial wisdom. *Biological Psychiatry: Cognitive Neuroscience and Neuroimaging*, 6.

Chin-Yew Lin. 2004. ROUGE: A package for automatic evaluation of summaries. In *Text Summarization Branches Out*, pages 74–81, Barcelona, Spain. Association for Computational Linguistics.

Tingting Liu, Salvatore Giorgi, Kenna Yadeta, H Andrew Schwartz, Lyle H Ungar, and Brenda Curtis. 2022. Linguistic predictors from facebook postings of substance use disorder treatment retention versus discontinuation. *The American journal of drug and alcohol abuse*, 48(5):573–585.

Brent Mausbach, Raeanne Moore, Christopher Bowie, Veronica Cardenas, and Thomas Patterson. 2008. A review of instruments for measuring functional recovery in those diagnosed with psychosis. *Schizophrenia bulletin*, 35:307–18.

Anne Merrill, Nicole Karcher, David Cicero, Theresa Becker, Anna Docherty, and John Kerns. 2017. Evidence that communication impairment in schizophrenia is associated with generalized poor task performance. *Psychiatry Research*, 249.

Sewon Min, Mike Lewis, Luke Zettlemoyer, and Hannaneh Hajishirzi. 2022. Metaicl: Learning to learn in context. *Preprint*, arXiv:2110.15943.

Margaret Mitchell, Kristy Hollingshead, and Glen Coppersmith. 2015. Quantifying the language of schizophrenia in social media. In *Proceedings of the 2nd Workshop on Computational Linguistics and Clinical Psychology: From Linguistic Signal to Clinical Reality*, pages 11–20, Denver, Colorado. Association for Computational Linguistics.

Mahdieh Montazeri, Mitra Montazeri, Kambiz Bahaadinbeigy, Mohadeseh Montazeri, and Ali Afraz. 2022. Application of machine learning methods in predicting schizophrenia and bipolar disorders: A systematic review. *Health Science Reports*, 6.

María Palacios-Ariza, Esteban Morales-Mendoza, Jossie Murcia, Rafael Arias, Germán Lara-Castellanos, Andrés Cely-Jiménez, Juan Rincón-Acuña, Marcos Araúzo-Bravo, and Jorge McDouall. 2023. Prediction of patient admission and readmission in adults from a colombian cohort with bipolar disorder using artificial intelligence. *Frontiers in psychiatry*, 14:1266548.

Thomas L Patterson, Sherry Moscona, Christine L McKibbin, Kevin Davidson, and Dilip V Jeste. 2001. Social skills performance assessment among older patients with schizophrenia. *Schizophrenia Research*, 48(2):351–360.

Delaram Sadeghi, Afshin Shoeibi, Navid Ghassemi, Parisa Moridian, Ali Khadem, Roohallah Alizadehsani, Mohammad Teshnehlab, Juan Gorriz, and Saeid Nahavandi. 2021. An overview on artificial intelligence techniques for diagnosis of schizophrenia based on magnetic resonance imaging modalities: Methods, challenges, and future works.

Genta Indra Winata, Onno Pepijn Kampman, and Pascale Fung. 2018. Attention-based lstm for psychological stress detection from spoken language using distant supervision. In *2018 IEEE International Conference on Acoustics, Speech and Signal Processing (ICASSP)*, pages 6204–6208.

Linting Xue, Aditya Barua, Noah Constant, Rami Al-Rfou, Sharan Narang, Mihir Kale, Adam Roberts, and Colin Raffel. 2022. ByT5: Towards a token-free future with pre-trained byte-to-byte models. *Transactions of the Association for Computational Linguistics*, 10:291–306.

Tianyi Zhang, Varsha Kishore, Felix Wu, Kilian Q. Weinberger, and Yoav Artzi. 2020. Bertscore: Evaluating text generation with bert. *Preprint*, arXiv:1904.09675.

A Task Description and Purpose

In this appendix we briefly describe the Social Skills Performance Assessment. We will talk about the purpose of the task, the task itself, and what we can gain from this task.

Task Motivation The Social Skills Performance Assessment, *abbrev. SSPA* is a mental-health instrument which serves as an indicator of social skill. The motivations, some of which we discussed in the introduction, is that people with psychiatric illnesses are more likely to show less cohesion and more disorganization in their speeches, as opposed to healthy control subjects. The SSPA task standardizes the way speech is measured for subjects with, and without psychiatric illnesses by having the subjects take on two tasks with expert clinicians.

For both tasks, the participants speak with a trained clinician. Their video and audio are recorded. Then transcribed. The labels mentioned in this paper were the clinicians rating the participants performance on the tasks to the two tasks.

Task Description There are two tasks to the SSPA. The first task is the neutral or friendly task, and the second task is the confrontational task.

The Friendly Task consists of the participant simulating a conversation as if they moved to a new neighborhood. They are asked to introduce themselves to the new neighbor. We observe people without psychiatric illnesses to briefly talk about 2-3 topics and stay consistent. People with illnesses tend to sway between 13-15 topics and are unable to concisely present thoughts.

The Confrontational Task consists of the participant complaining to their landlord after a leaky pipe has not been fixed for months. We observe that healthy controls are able to quickly articulate and talk only about the problem at hand. We observe that BD and SZ often talk about multiple different things and then talk about the problem given to them.

Task Outcomes Annotating clinical variables is a different task than classification. While these variables are not classifiers of psychiatric illnesses.

They are important features. These variables give clinicians and scientists much needed quantification in the field of life-long psychiatric illnesses. Therefore, it is imperative to bring modern technology to the equation and slowly make care and data collection accessible and efficient.

Prompt Details

This section describes the prompt that was used for GPT-4/4o to annotate the posts as described in Section §6.

> System Prompt - You are going to act as a clinical annotator. You will see a set of conversations between a doctor and a participant. You will also be told of a task. You need to return a python compatible list of five scores from a range of 1-5. Below I describe what these scores represent. Remember that for these scores 1 is lowest and 5 is highest.
>
> Interest - This score on 1-5 will describe how interested this person was in the conversation. Look at the participant's engagement in the conversation and rate this score.
>
> Fluency - This score on 1-5 will describe how fluent a person was. A person with more filler words will score lower.
>
> Clarity - This score on 1-5 will describe how clearly the subject was able to communicate their thoughts. A higher score shows better communication skills.
>
> Focus - This score on 1-5 will describe how concentrated the subject was on the task. A person who deviates off topic will score lower.
>
> Social Appropriateness - This score on 1-5 will describe how socially appropriate to the task this participant's score was. A higher score is more socially appropriate.
>
> Return the results as a list [] with five numbers for each of the scores above.
>
> User Prompt - In this task the participant has to introduce themselves as a new neighbor in the neighborhood.
>
> Doctor - Hey there, how are you? Participant - Hey I just moved. ...

Continues for entire conversation. The system prompt remains the same, while for each task the user-prompt changes.

Overview of the CLPsych 2025 Shared Task: Capturing Mental Health Dynamics from Social Media Timelines

Talia Tseriotou[1*], Jenny Chim[1*], Ayal Klein[3], Aya Shamir[3], Guy Dvir[3], Iqra Ali[1], Cian Kennedy[1], Guneet Singh Kohli[1], Anthony Hills[1], Ayah Zirikly[4], Dana Atzil-Slonim[3], Maria Liakata[1,2]
[1]Queen Mary University of London (UK), [2]The Alan Turing Institute (UK),
[3]Bar Ilan University (Israel), [4]Johns Hopkins University (US)
{t.tseriotou; c.chim; m.liakata}@qmul.ac.uk

Abstract

We provide an overview of the CLPsych 2025 Shared Task, which focuses on capturing mental health dynamics from social media timelines. Building on CLPsych 2022's longitudinal modeling approach, this work combines monitoring mental states with evidence and summary generation through four subtasks: (A.1) Evidence Extraction, highlighting text spans reflecting adaptive or maladaptive self-states; (A.2) Well-Being Score Prediction, assigning posts a 1 to 10 score based on social, occupational, and psychological functioning; (B) Post-level Summarization of the interplay between adaptive and maladaptive states within individual posts; and (C) Timeline-level Summarization capturing temporal dynamics of self-states over posts in a timeline. We describe key findings and future directions.

1 Introduction

Mental health concerns is a pressing global issue (WHO, 2022), necessitating solutions that both expand access to care and continuously monitor individuals over time, thereby reflecting the multi-faceted and dynamic nature of mental health.

Over the past decade, social media platforms have emerged as major venues where people openly discuss mental health, sharing experiences and emotional states that can span years (Coppersmith et al., 2014; Shing et al., 2018; Zirikly et al., 2019; Tsakalidis et al., 2022b). This abundance of user-generated data offers an unprecedented opportunity to monitor individuals longitudinally, providing early detection, prevention, and "just-in-time" interventions well before difficulties escalate.

While traditional NLP approaches to mental health centered on static classification tasks (e.g. depression detection in De Choudhury et al. (2013)), recent work has recognized the complexity of mental health trajectories as fluctuating dynamic states influenced by evolving contexts, interactions and psychological processes, emphasizing the need for longitudinal, context-rich models that capture how mood, behavior, and cognition fluctuate over time (Tsakalidis et al., 2022b; Tseriotou et al., 2023). Moreover, accounting for both maladaptive and adaptive states delivers a more nuanced picture of well-being while also uncovering factors that can lead to personalized interventions (Slonim, 2024).

CLPsych shared tasks have followed this trend, shifting from user-level classification (Coppersmith et al., 2014; Shing et al., 2018; Zirikly et al., 2019) to longitudinal tasks such as detecting "Moments of Change" (MoC) (Tsakalidis et al., 2022a) and evidence generation (Chim et al., 2024).

The CLPsych 2025 shared task combines longitudinal modeling in social media timelines with evidence generation, promoting humanly understandable rationales that support recognizing mental states as they dynamically change over time. Adopting the MIND transtheoretical framework (Slonim, 2024), we seek to identify both adaptive and maladaptive self-states in a users longitudinal data via the following tasks: (A.1) Evidence Extraction, highlighting text spans within posts that reflect adaptive or maladaptive states; (A.2) Well-Being Score Prediction, assigning a 1–10 rating indicative of individuals' social, occupational, and psychological functioning, informed by maladaptive and adaptive states; and (B-C) Summarization, capturing individuals' mental health progression at the post level (B) and across the entire timeline (C) on the basis of adaptive and maladaptive states.

Our dataset comprises Reddit-based user timelines from mental health related subreddits (MHS), with posts annotated by clinical experts following the MIND scheme, which captures how an individuals self-state evolves in response to personal challenges, life events, or social interactions. From a clinical perspective, this means not only detecting

*Denotes equal contribution.

Proceedings of the 10th Workshop on Computational Linguistics and Clinical Psychology (CLPsych 2025), pages 193–217
May 3, 2025

risk and symptoms but also identifying and tracking a person's strengths and coping abilities as they emerge and evolve. Summaries at various time resolutions further enhance explainability – critical for mental health professionals seeking clear, evidence-based insights. Computational challenges involve working with models that can process longitudinal data, incorporating and synthesizing appropriate evidence to generate rationales for the progression of an individual.

After providing a quick review of the landscape in NLP for mental health, focusing on temporality and explainability (§2), we describe the shared task (§3) and data annotation (§4.2). We discuss evaluation metrics (§5), methods by participating teams and results (§6.1), and conclude with an overview of key findings (§6.4), limitations, clinical implications and directions for future research (§7).

2 Related Work

2.1 NLP for Mental Health Applications

Explainability in mental health: Early work primarily focused on classification tasks, either at the document-level (Sawhney et al., 2022a) or user-level, with the latter addressing both static assessment of mental health conditions (Coppersmith et al., 2015; Shing et al., 2018; Zirikly et al., 2019; Sawhney et al., 2022b) and longitudinal monitoring of psychological states over time (Tsakalidis et al., 2022a,b; Hills et al., 2023).

Recent developments have shifted toward more fine-grained analysis and explainable approaches to mental health assessment. The 2023 eRisk Task focused on ranking sentences based on their relevance to depressive symptoms (Parapar et al., 2023), while Nguyen et al. (2022) developed BERT-based methods that incorporate PHQ-9 symptoms for improved interpretability in depression detection. Similarly, Nemesure et al. (2021) employed SHAP values (Lundberg and Lee, 2017) to explain predictions for anxiety and depression models, and Zirikly and Dredze (2022) leveraged PHQ-9 questions as auxiliary tasks to provide explanations for depression detection, evaluating performance on manually annotated text spans.

In the context of fostering interpretability, Garg (2024) annotated a dataset with highlighted text spans across various 'wellness' dimensions, while the CLPsych 2024 shared task explored Large Language Models (LLMs) to identify evidence supporting suicide risk assessments (Chim et al., 2024). This reflects the field's increasing emphasis on providing clinically meaningful explanations alongside predictions.

LLMs have been leveraged for mental health classification (Amin et al., 2023), data augmentation (Liyanage et al., 2023), and reasoning (Xu et al., 2023), demonstrating promise in detecting psychological indicators (Yang et al., 2023), extracting relevant evidence from text (Xu et al., 2024a), and generating clinically informed summaries (Song et al., 2024b). LLMs using instruction fine-tuning and Chain-of-Thought (CoT) prompting (Yang et al., 2023) have also been employed, though such approaches can pose risks of incorrect predictions and flawed reasoning, especially in complex conversations (Li et al., 2023).

Evidence extraction: Accurate span extraction is a crucial task in mental health assessment, enabling clinicians to identify and summarize the most relevant patient data for clinical evaluation. Prior work at the intersection of NLP and mental health have utilized LLMs to predict critical mental states and provide reasoning for predictions (Yang et al., 2024b; Xu et al., 2023, 2024b). Yet these approaches lack transparency and complex reasoning processes can lead to hallucination (Li et al., 2023).

LLMs and summarization: LLMs have been used to generate clinically meaningful summaries from social media posts (Song et al., 2024b, 2025; Sotudeh et al., 2022), summarize counseling sessions (Srivastava et al., 2022), generate structured medical reports from patient-doctor conversations(Adhikary et al., 2024; Michalopoulos et al., 2022), and summarize Mental State Examinations (MSE) (Mumtaz et al., 2024). However, uncertainties remain regarding the effectiveness of LLMs in generating contextually appropriate summaries, particularly in domains such as mental health (Klein et al., 2024; Asgari et al., 2024).

Temporal modeling: Most models have relied on recurrent neural networks without explicitly accounting for time intervals between posts (Tsakalidis et al., 2022a), or struggle to capture complex linguistic patterns over time (Bayram and Benhiba, 2022), despite the role of longitudinal linguistic features in mental health applications (Homan et al., 2022; Chim et al., 2025). Recent work has developed time-aware modeling approaches. Hills et al. (2023) introduced a Hawkes process-inspired approach capturing both temporal dynamics and linguistic context in user timelines, which was further

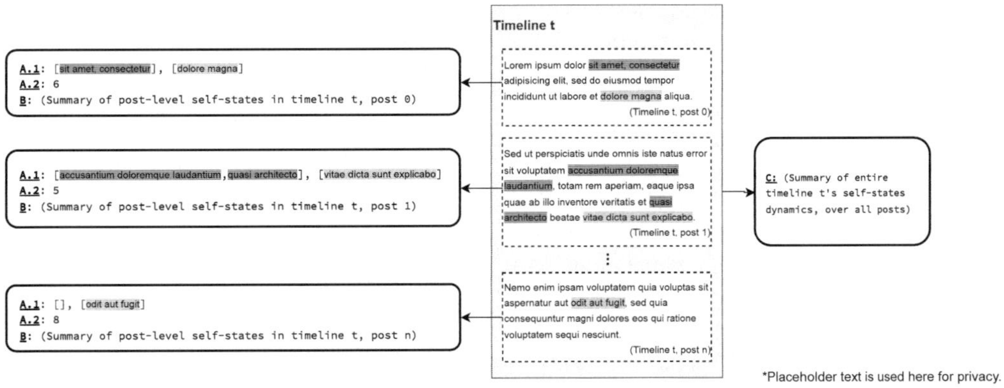

Figure 1: Participants are asked to tackle the tasks described in §3: identifying adaptive and maladaptive evidence (A.1), predicting well-being (A.2), and summarizing mental health dynamics at the post (B) and timeline level (C).

integrated into a hierarchical transformer architecture (Hills et al., 2024). Tseriotou et al. (2023) proposed sequential path signature networks to capture the temporal and linguistic progression in user posts, while Tseriotou et al. (2024b) introduced TempoFormer, which modifies the transformer architecture to account for relative temporal aspects between sequential data points, using time-sensitive rotary positional encodings. Such approaches have demonstrated superior performance in detecting subtle changes in mood and mental states by effectively modeling both linguistic and temporal context in social media posts.

2.2 Mental Health Self-State Dynamics

The MIND approach (Slonim, 2024) proposes a paradigm shift from categorical classification of trait-like psychopathology to modifiable intrapersonal dynamics. MIND provides a transtheoretical scheme that breaks down individuals experiences into core building blocks central to major therapeutic approaches, including cognitive-behavioral therapy (CBT), psychodynamic, interpersonal, relational, and experiential models. This highlights that human experience consists of multiple self-states that fluctuate and change over time (Beck et al., 2021; Bromberg, 2014; Stiles, 2001). Each self-state comprises identifiable elements characterized by specific combinations of Affect, Behaviour, Cognition, and Desire (ABCD) (Revelle, 2007) coactivated in a meaningful manner and directed either toward the self or others (Lazarus and Rafaeli, 2023). At a specific moment, one self-state may be dominant, while others, often complementary, may be subtler or remain in the background. Focusing on self-states in theory, research, and practice can reveal within-person adaptive and maladaptive states, as well as between-person differences in typical self-states.

Recent developments in NLP, specifically the emergence of LLMs, have demonstrated the capability to identify individuals' emotions (Mayer et al., 2024) and cognitions (Singh et al., 2024) from longitudinal mental health data. Nonetheless, no prior research has yet leveraged LLMs to systematically capture the complex interplay among ABCD elements especially as these manifest dynamically in adaptive and maladaptive self-states, a gap which this shared task directly seeks to address.

3 Task Definition and Instructions

We describe in detail the tasks introduced in §1 and provide an overview in Figure 1.

Task A consists of two sub-tasks: **Task A.1** involves identifying adaptive and maladaptive self-state evidence from an individual's post as a set of continuous spans. Each post can include either: (1) a single self-state (adaptive or maladaptive); (2) two complementary self-states (adaptive and maladaptive) or (3) evidence of neither. **Task A.2** requires rating the overall well-being of an individual on a scale from 1–10 based on GAF (Association et al., 2000). This score reflects the well-being of an individual based on three aspects: social, occupational and overall individual psychological functioning. While well-being scores are assigned at the post level, participants were encouraged to consider the sequence of previous posts and the extracted evidence from Task A.1 in this task.

Task B involves post-level summarization of self-states. Such summaries should capture the interplay between adaptive and maladaptive states manifesting in the post through identification of the central organizing aspects (ABCD) that drive the state and should provide the anchors for the summary. The expectation is to first identify the dominant self-state and then describe how core aspects influence the rest, emphasizing their evolution.

Task C involves summarizing self-states at the individuals' timeline-level. The focus should be on the temporal interplay between adaptive and maladaptive self-states, with emphasis on concepts such as flexibility, rigidity, improvement, and deterioration. When applicable, the temporal dynamics should capture changes in the dominant self-states and specifically how underlying changes in ABCD aspects contribute to potential transitions.

Ground truth data for all the above tasks were provided to the participants during training but not at test time. For Task A.1, additional information regarding each gold self-state evidence was made available in the training set only. Specifically, the types of evidence provided were under: Affect (A), Behavior towards the other (B-O), Behavior towards the self (B-S), Cognition of the other (C-O), Cognition of the self (C-S), and Desire/Expectation (D), along with further sub-categories stemming from each of the six categories. The full list of categories can be found in the Appendix A.2.

4 Data and Annotation

4.1 Data

We utilized the Reddit-New dataset originally introduced in the CLPsych 2022 shared task (Tsakalidis et al., 2022a). This dataset comprises user timelines extracted from various MHS from 2015 to 2021. Given the extensive nature of the MIND annotation scheme, annotating entire timelines proved to be prohibitively resource-intensive. To address this, we implemented a selective sampling strategy. Specifically, we reduced excessively long timelines by extracting subsets containing between 10 and 12 representative posts. Additionally, timelines of moderate length were preferentially sampled to balance feasibility with sufficient contextual richness. Beyond this length-based selection, timelines were chosen randomly, subject to two constraints:

- **User Uniqueness:** No user was represented by more than one timeline within the test set, and users appearing in the training set were explicitly excluded from the test set to ensure independence between training and evaluation data and to prevent potential data leakage that could inflate performance metrics.
- **Density Diversity:** Using the CLPsych 2022 annotations for mood switches and escalations (i.e. MoC), we define timeline 'density' as the proportion of posts labeled with MoC tags and use it for stratified sampling. This helps us to capture a diverse range of emotional fluctuation patterns and related mental health dynamics.

	Train	Test
# Timelines	30	10
# Posts	343	94
Avg. Tokens per Post	134.4	142.9
# Adaptive Evidences	399	145
# Maladaptive Evidences	526	171

Table 1: Dataset descriptive statistics.

The final dataset (see Table 1) contains timelines selected for length, content relevance, user uniqueness, and density distribution. This strategy maintains the longitudinal nature of the data while providing sufficient context for identifying adaptive and maladaptive self-states, as well as capturing the dynamics of psychological states over time. Furthermore, this enabled thorough annotations of detailed ABCD aspects in each post.

4.2 Annotation

Two Master's students in clinical psychology, both fluent in English, annotated the selected timelines using the MIND scheme (§2.2). Annotators received comprehensive training conducted by a clinical expert and ongoing supervision from a senior MA student with prior experience in annotation using the MIND scheme. Annotators underwent a preliminary training phase, during which they received iterative feedback and conducted reconciliation meetings to enhance consistency and interrater reliability.

Annotators followed a structured workflow. For each post, they first identified adaptive and maladaptive self-states. Within each identified self-state they annotated the present ABCD elements, selecting the most salient span as evidence for each element. Next, they assessed the individual's overall well-being based on GAF guidelines, considering both the specific post and the context of previous posts. They then composed a detailed summary for each post, specifying which self-state was dominant, the primary psychological dimension underpinning that self-state from the ABCD elements, and a description of the interplay between different elements constituting the self-state. This description considered temporality and causality to capture the evolving psychological dynamics within each post.

Beyond individual posts, annotators synthesized their insights at the timeline level, producing a comprehensive summary that mapped the interplay between adaptive and maladaptive self-states across the timeline. This included a description of how self-states dynamically changed (or remained stable) over time. Details about the annotation platform are specified in Appendix A.1.

Inter-annotator agreement was assessed using standard reliability metrics over 23 posts annotated by both annotators. For Task A.2 (Well-being Score), which involves numerical ratings, annotators demonstrated high agreement achieving a Pearson correlation coefficient (r) of 0.793 and an Intraclass Correlation Coefficient (ICC2) of 0.791, indicating high agreement. Additional evaluations of inter-annotator agreement using task-specific metrics detailed in Section 5 (BERTScore-based measures for Task A.1 and mean squared error for Task A.2), are provided in Appendix A.3.

5 Evaluation Metrics

5.1 Task A

Evidence Extraction. The main metric we consider is the recall of evidence spans. Recall is prioritized given that the costs of overlooking important evidence outweigh those of supplying excess evidence for the task of capturing mental health dynamics over time. Moreover, in our gold data, annotators selected the single most salient evidence span per self-state annotation (§4.2). As such, precision metrics could unfairly penalize valid predictions that simply differ from what the annotator considers as the most salient, whereas recall more accurately reflects performance.

Following Chim et al. (2024), for a given user, given predicted evidence spans H and gold evidence spans E, we average the maximum recall-oriented BERTScore (Zhang et al., 2020):

$$\text{Recall} = \frac{1}{|E|} \sum_{e \in E} \max_{h \in H} BERTScore(e, h)$$

We use `deberta-xlarge-mnli` to compute embeddings and apply rescaling as recommended by Zhang et al. (2020). In addition, we report a weighted version of recall, which is sensitive to predicted evidence lengths relative to gold evidence lengths. For a given user with gold evidence spans of cumulative token count n_{gold} and predicted spans with cumulative token count n_{pred}, if the predicted evidence spans are longer than the gold-standard ones, we apply weight w to the timeline-level recall:

$$w = \begin{cases} \frac{n_{\text{gold}}}{n_{\text{pred}}} & \text{if } n_{\text{pred}} > n_{\text{gold}} \\ 1 & \text{otherwise} \end{cases}$$

Well-being Score Prediction We evaluate well-being score predictions over all annotated posts using mean squared error (MSE), which appropriately penalizes larger errors and accommodates ordinal and continuous data:

$$\text{MSE} = \frac{1}{n} \sum_{i=1}^{n} (y_i - \hat{y}_i)^2$$

As supplementary metrics, we report MSE stratified by risk categories: serious impairment to functioning (1 to 4), impaired functioning (5 to 6), minimal impairment to functioning (7 to 10). Finally, based on these categories, we cast the task into a classification problem (serious, impaired, minimal) and report macro F1. This reflects a system's ability to identify well-being categories rather than exact scores, regardless of category prevalence.

5.2 Task B

Following prior work in general domain (Maynez et al., 2020) and mental health summarization (Song et al., 2024a), we leverage predictions from a natural language inference (NLI) model (Laurer et al., 2024) for summary evaluation.[1] We consider consistency to be the absence of contradiction. For each sentence in a submitted summary $s \in S$, we use the NLI model to compute its mean probability of contradicting each sentence in the corresponding gold-standard evidence summary $g \in G$, taking the gold sentence as premise and the submitted sentence as hypothesis:

$$\text{CS} = \tfrac{1}{|S| \cdot |G|} \textstyle\sum_{s \in S} \sum_{g \in G} (1 - \text{NLI}(\text{Contradict}|g, s))$$

To complement consistency, we also evaluate summaries by their contradiction to expert summaries. We expect there to be some natural contradictory information in most summaries, since

[1] https://huggingface.co/MoritzLaurer/DeBERTa-v3-large-mnli-fever-anli-ling-wanli

summarized evidence can include both risk factors and protective factors. We compute the contradiction score by averaging the *maximum* contradiction probability of a predicted sentence against gold evidence summary sentences:

$$\mathrm{CT} = \frac{1}{|S|} \sum_{s \in S} \max_{g \in G} \mathrm{NLI}(\mathrm{Contradict}|g, s)$$

To support post-hoc analysis, we assess whether generated summaries are supported by their corresponding supporting evidence spans. This is only informative if the assessed system actually uses predicted spans for post-level summarization.

$$\mathrm{EA} = \frac{1}{|H|} \sum_{h \in H} \max_{s \in S} \mathrm{NLI}(\mathrm{Entail}|h, s).$$

5.3 Task C

Following Task B, we evaluate timeline-level summaries primarily with mean consistency (CS), supplementing with contradiction (CT).

6 Teams & Results

6.1 Participating Teams

A total of 26 teams (69 members) completed the registration process (see Appendix B.1), with members of 3 teams having participated in a past CLPsych shared task. Out of these 26 teams, 14 (49 members) submitted output files for one or more tasks and 11 teams submitted a paper (Table 9). Teams who submitted solutions averaged 3.5 members while those who did not averaged 1.6, suggesting that having more members increased the chance of completion.

6.2 Baselines

A range of LLMs and smaller model baselines were provided along with the official team submissions' results. This allowed for a direct comparison of teams' solutions, given strong setups for each task. Baselines are presented below (with prompts in Appendix C). All LLM baselines used `Llama-3.1-8b-Instruct` (Grattafiori et al., 2024).[2]

Task A.1: For evidence extraction, two zero-shot prompting baselines and two smaller BART-based models were used, representing both single-post and window-based approaches with the latter taking into account the context of recent posts. BART-based models allowed showcasing the effect of fine-tuning for generation.

[2] https://huggingface.co/meta-llama/Llama-3.1-8B-Instruct

	All		**Adapt.**		**Maladapt.**	
Teams	**R**	**WR**	**R**	**WR**	**R**	**WR**
Aquarius	.51	.46	.50	.47	.52	.45
BLUE	.56	.39	.47	.40	.64	.38
BULUSI	.43	.37	.34	.34	.53	.40
CIOL	.25	.17	.23	.15	.26	.20
CSIRO-LT	.46	.43	.38	.38	.54	**.48**
EAIonFlux	.52	.47	.52	.48	.52	.46
ISM	.56	.45	.49	.46	.63	.44
MMKA	.60	.34	.52	.37	.68	.31
NoviceTrio	-.03	-.03	-.10	-.10	.05	.05
PsyMetric	.17	.17	.15	.15	.18	.18
ResBin	.47	.30	.26	.26	.68	.36
Seq2Psych	.28	.24	.25	.24	.31	.24
uOttawa	**.64**	**.50**	**.59**	**.54**	.68	.46
Zissou	.58	.32	.45	.31	**.71**	.34
Llama ZS Single-Post	.36	.34	.31	.29	.38	.41
Llama ZS 5-Post	.50	.26	.37	.25	.63	.27
BART Single-Post	.40	.38	.47	.46	.34	.30
BART 5-Post	.26	.26	.28	.28	.24	.24

Table 2: Results on Task A.1 (evidence extraction). We consider recall and weighted recall over all spans, adaptive spans only, and maladaptive spans only.

- Llama ZS Single-Post: Zero-shot prompting on each post by providing definitions of adaptive and maladaptive self-states and asking the LLM to generate an adaptive and a maladaptive evidence list.
- Llama ZS 5-Post: Same as above, but operating on each post along with the recent posting history from an individual's timeline (5 posts).
- BART Single-Post: BART (Lewis et al., 2020) fine-tuned separately for adaptive and maladaptive span generation on each post.
- BART 5-Post: BART fine-tuned as above but operating on a window of 5 posts separated by a [SEP] token (current post + 4 recent posts). Spans are based solely on the last post.

Task A.2: For well-being score prediction, two zero-shot prompting baselines and two smaller models were used. While one version of the models is single-post based, another version considered the context of recent posts, as is the case of the BiLSTM modeling the sequential aspect.

- Mode: Mode of training data scores (7).
- Llama ZS Single-Post: Zero-shot prompting on each post by providing definitions of each score for well-being prediction.
- Llama ZS 5-Post: Same as above, but operating on each post along with the recent posting history of the individual (5 posts).

- BERT Post-level: BERT model (Devlin et al., 2019) with a regression layer fine-tuned on the post-level for well-being score prediction, averaged over 5 seeds.
- BiLSTM 5-Post (BERT): BiLSTM operating on a window of 5 posts (current + 4 recent posts) for well-being score prediction, averaged over 5 seeds. For each post the [CLS] token of BERT representations is used.

Tasks B and C: Two zero-shot LLMs, with one version including an intermediate LLM generated summary, were used for the post and timeline summaries using prompts with clinical directions.

- Llama ZS Summary: Zero-shot prompting on single posts (Task B) and timelines (Task C). The model is instructed to identify the dominant and secondary self-states and highlight the central organizing ABCD aspects that drive the state along with their interplay, including guidance through definitions.
- Llama ZS w/ Intermediate Summary: A two-layer LLM approach following Song et al. (2024b), where first a post-level (Task B) or a timeline-level (Task C) concise summary is produced with zero-shot prompting, and then this summary is used as in *Llama ZS Summary* to generate the self-states summary.

6.3 Results

This section presents results and an overview of system submissions, focusing on the best run.[3]

System Characteristics The majority of submissions took a pipeline approach, using predictions from an earlier subtask to inform the next (e.g. use predicted evidences and scores to guide summarization). More than a third of teams used Retrieval Augmented Generation (RAG) through dense retrieving examples from the training set for in-context learning. Most used clinical information provided in the shared task description document in their prompts, and a few explored incorporating additional domain knowledge during feature extraction, prompt design, and data augmentation (Seq2Psych, CIOL, BLUE).

Model Characteristics A few teams employed Pretrained Language Models (PLMs), mostly for evidence extraction (MMKA, Seq2Psych). About a third used traditional approaches, such as KDE for sampling (ISM) and random forests for span classification (CIOL) and well-being score prediction (ResBin). All teams used LLMs in at least one task. LLMs have mostly been used to directly generate predictions, but also for feature extraction (Seq2Psych). Participants developed systems on private and self-hosted instances, without using Cloud APIs. All employed LLMs were generalist models, generally 9B or smaller in size (42%), and the majority can model long contexts of over 100k tokens (58%).

[3]For details of each submission and information about model families, sizes, and context lengths, see Appendix B.

Teams	MSE (↓) All	Min.	Imp.	Ser.	F1 (↑)
Aquarius	<u>2.01</u>	1.25	3.11	2.16	<u>0.37</u>
BLUE	2.26	2.06	3.69	**1.41**	**0.39**
BULUSI	**1.92**	**0.65**	<u>1.19</u>	3.04	0.35
CIOL	3.99	2.89	**0.49**	7.31	0.12
CSIRO-LT	2.04	<u>1.08</u>	3.68	1.82	0.34
EAIonFlux	2.08	2.11	3.71	<u>1.77</u>	0.32
ISM	2.76	2.74	5.00	1.93	0.32
MMKA	6.61	4.95	11.76	4.22	0.26
NoviceTrio	13.83	18.62	11.59	3.16	0.14
PsyMetric	3.23	3.28	6.63	2.52	0.30
ResBin	8.02	1.89	3.71	20.26	0.19
Seq2Psych	3.27	2.63	1.38	4.98	0.19
uOttawa	2.62	2.91	4.03	2.28	0.30
Zissou	3.14	3.09	4.32	2.91	0.34
Mode	7.30	0.47	1.31	19.20	0.13
Llama ZS Single-Post	4.22	3.20	3.66	4.67	0.26
Llama ZS 5-Post	4.46	7.06	3.20	1.67	0.27
BERT Post-level	2.90	2.81	2.32	3.39	0.14
BiLSTM 5-Post (BERT)	4.56	5.34	1.01	5.68	0.13

Table 3: Results on Task A.2 (well-being score prediction). In addition to overall MSE, performance on posts in different well-being score ranges are reflected by MSE computed over posts in the minimal impairment to functioning, impaired functioning, and serious impairment to functioning ranges, and macro F1.

Task A.1: Results for evidence identification are in Table 2. Instruction prompting with demonstrations proved effective, as shown in the system that achieved top recall and length-weighed recall (uOttawa). Most submissions followed this approach, although finetuned PLMs continue to be performant (MMKA). Systems that achieve high recall on adaptive tend to also perform well on maladaptive spans. Across the board, systems were better at identifying evidence for maladaptive self-states than adaptive ones, with the exception of EAIonFlux, which targets retrieval and achieves the same performance level on both self-state categories.

Task A.2: Results for well-being score prediction are in Table 3. The best-scoring system used an optimized weighted LLM ensemble (BULUSI). Sys-

	Task B			Task C	
	CS	**EA**	**CT (↓)**	**CS**	**CT (↓)**
Aquarius	.88	.69	.78	.92	.88
BLUE	**.91**	.59	**.53**	**.95**	.54
BULUSI	.87	.81	.81	.94	.71
CIOL	.61	**.81**	.97	.61	1.0
CSIRO-LT	-	-	-	-	-
EAIonFlux	.89	.76	.78	.91	.76
ISM	.86	.76	.78	.85	.83
MMKA	-	-	-	-	-
NoviceTrio	.69	.17	.89	.86	.60
PsyMetric	.70	.47	.56	.93	**.35**
ResBin	.76	.67	.84	.90	.82
Seq2Psych	-	-	-	-	-
uOttawa	.86	.70	.83	.94	.71
Zissou	.85	.74	.77	-	-
Llama ZS Summary	.88	-	.85	.88	.80
Llama ZS w/ Inter. Summary	.89	-	.84	.94	.58

Table 4: Results on Task B (post-level) and Task C (timeline-level). Summaries are assessed primarily on mean consistency to gold summaries (CS). We additionally report entailment by extracted evidences (EA) on post-level, and contradiction to gold summaries (CT).

tems that incorporate extracted evidence or jointly predict evidence and score tend to achieve better MSE (Aquarius, EAIonFlux), but tackling more than two subtasks in the same prompt remains challenging. Given this task's sequential nature, teams also explored using timeline-based features (CIOL) and person-contextualized modeling (Seq2Psych). Overall, systems that excel at posts in the impaired functioning range (5 to 6) also tend to excel at those indicating minimal impairment (7 to 10).

Tasks B & C: Table 4 shows results for post-level and timeline-level summarization. Almost all teams used LLMs, except team ResBin who fine-tuned long-context PLMs. Over half of the teams incorporated predictions from post-level tasks as additional well-being signals for summarization.

6.4 Performance Analysis and Discussion

Maladaptive vs Adaptive states: Figure 2 summarizes the post-level self-state summary performance across the best runs per team with respect to the number of labeled evidence spans in the test set. When the number of *adaptive* evidence spans in a post changes, the average team performance remains largely the same. By contrast, when the number of *maladaptive* evidence increases, performance increases. While this trend holds when the total number of spans increases, the mean consistency of the summaries clearly benefits from more maladaptive evidence spans. This uncovers model limitations; they more easily synthesize negative aspects compared to positive ones, potentially due to the latter being more subtle.

A closer look into the submissions in terms of *adaptive* self-state spans identification reveals that the top-performing teams either leverage large 70B LLMs with carefully selected demonstrations, through RAG or otherwise, or leverage the fine-tuning of much smaller models such as RoBERTa with data augmentation. By contrast for *maladaptive* self-states, while few-shot learning with 70B models and PLMs continued to work well, smaller LLM prompting in the range of 7-9B parameters achieved top performance. Furthermore, top systems perform better at capturing maladaptive (.71) compared to adaptive (.59) evidence. These results demonstrate that consistent with psychology literature (Baumeister et al., 2001), current LLMs, especially smaller ones, remain challenged by the task of identifying nuanced and subtle positive experiences, compared to negative experiences which are generally more salient and attention-grabbing.

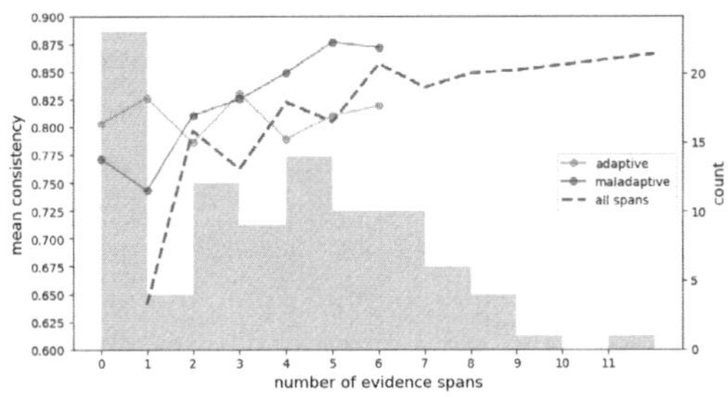

Figure 2: Post-level summarization performance in terms of average mean consistency across all teams with respect to the number of adaptive, maladaptive and total evidence spans (left) and histogram of posts per total evidence span (right).

Well-being Scores: Figure 3 provides different views to performance for well-being. As shown in the top boxplot, average MSE is the lowest (i.e. *best*) in the minimal impairment to functioning (scores 7-10) group. However, the middle line chart shows that the aggregate performance in terms of post-level self-state summarization for this group is *worse* compared to the group with serious impairment to functioning (1-4; left) and impaired functioning (5-6; middle). The bottom boxplot shows that posts in this group have the lowest median number of evidence spans. These results suggest that score prediction is differently impacted by the absence of self-state evidence compared to precise span extraction and summarization tasks;

posts with fewer evidence (and especially adaptive rather than maladaptive evidence) may be harder to summarize (Figure 2), but not necessarily harder to score on a well-being scale.

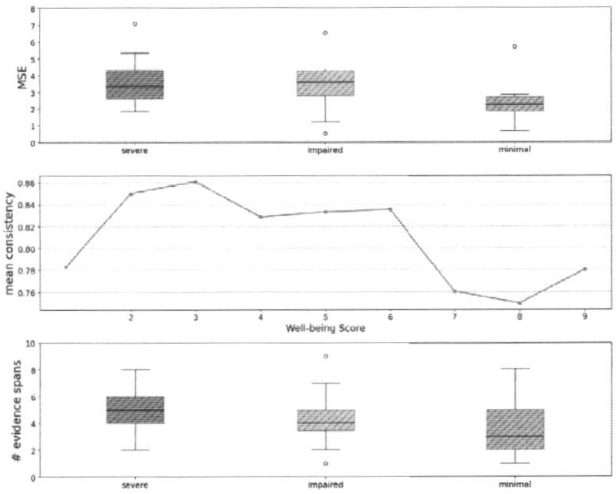

Figure 3: Distribution of average MSE per well-being score functioning bin per team (top), average mean consistency of post-level summary across well-being scores per team (middle), and number of labeled evidence spans per well-being bin (bottom).

Temporality and Well-being Score: In order to assess the temporal nature of well-being score's evolution in an individual's timeline, we compare this year's CLPsych annotations with CLPsych 2022's (Tsakalidis et al., 2022a) MoC annotations which are by definition longitudinal. For posts that do not present enough relevant information about an individual's mental state we assigned the latest recorded well-being score. We calculate the Spearman's correlation between the clinically annotated well-being score and the MoC seeing a statistically significant weak negative correlation (-.38). We further obtain a more longitudinal well-being score version by calculating the absolute well-being fluctuation between consecutive posts. The correlation of this variable with MoC is a stronger (significant) moderate correlation (.44). These results presented in Table 5, suggest that the currently provided well-being scores are of weak longitudinal nature, also manifesting through the lower performance of the BiLSTM baseline compared to the BERT Post-level baseline in Table 3.

Additionally, we append together statistics from the: current CLPsych 2025 Task A.2, CLPsych 2022 MoC and TalkLife MoC Dataset from Tseriotou et al. (2024a) in Table 6. Since the mean absolute well-being fluctuation is .91 and the standard deviation is .75, we define change in terms of

Variable	Spearman's Corr.	p-value
Well-being Score	-0.375	$4.8e^{-16}$
Well-being Fluctuation	0.440	$7.9e^{-12}$

Table 5: Spearman's correlation of well-being scores and fluctuation between consecutive posts with respect to corresponding *Moments of Change* labels.

fluctuations to be larger or equal to 2. As shown in the table, Reddit dataset changes are less frequent than TalkLife ones while their timelines span a longer period of time. Comparing this year's labels with other datasets', well-being score changes are considerably sparser than MoC (i.e. Switches and Escalations combined) and each change on average spans a longer time period, potentially limiting the degree of longitudinality of the well-being score prediction task. These findings may account for the lower performance of teams that attempted to tackle well-being score prediction in a temporal way (CIOL, Seq2Psych).

Dataset	Reddit (current) Well-being	Reddit MoC		TalkLife MoC	
	Change	Switch	Escalation	Switch	Escalation
Mean Point Time Diff.	4d 14hr 27min	2d 6hr 58min		6hr 51min 11sec	
Median Point Time Diff.	2d 19hr 11min	22hr 42min 55sec		59min 38sec	
Mean consecutive events	1.39	1.19	2.83	1.58	4.12
Median consecutive events	1	1	2	1	3
Mean events in timeline	2.30	1.60	3.85	1.77	4.03
Median events in timeline	2	1	2	1	1

Table 6: Well-being and MoC statistics of datasets on time and event length.

7 Conclusion

Expanding on previous shared tasks, we introduced a novel multi-task framework grounded in the transtheoretical MIND approach. Participants were asked to identify adaptive and maladaptive self-states (Task A.1), predict post-level well-being scores (Task A.2), and generate post- and timeline-level summaries that reflect psychological progression (Tasks B and C). Systems using LLMs were able to identify both adaptive and maladaptive states although an asymmetry was observed in favor of maladaptive states.

Future directions could address the more longitudinal nature of well-being by reformulating the task towards a more temporal one and exploring temporal models that focus on capturing sparser and more subtle changes over longer time periods as well as amplify the signal of adaptive behavior which is important in achieving and monitoring better therapeutic outcomes.

Limitations

As in the vast majority of prior work leveraging social media for individual-level mental health assessments, this year's shared task involves individuals who generated content in self-selected online communities. The present tasks were conducted using social media posts made on various mental health-related subreddits in the English language, by users who willingly self-disclosed their thoughts and feelings. Generalization of the approaches presented in this work to other contexts and in other languages remains an open area of research.

Annotation was performed over 40 relatively short timelines due to the annotation load for clinical experts. This potentially hinders the performance of smaller supervised models, still leaving open questions around their true potential. Additionally, although the well-being score was annotated on the post-level with full timeline content visibility, the longitudinal manifestation of individuals' well-being remains underexplored. Since the annotation process involved selection of the most salient available adaptive and maladaptive spans for each ABCD element, this task does not yet explore the more nuanced selection of additional evidence spans and their connection to one another.

Although the dynamic evolution of self-states was to some extent addressed in this work with respect to summarization, there is still need to explore such dynamic progression through the lenses of other tasks such as monitoring and dialogue tracking. Finally, multimodality, which provides important cues especially in the clinical setting in terms of the manifestation of self-states, remains for now a future direction.

Ethics

This year's tasks explored the prediction of well-being scores from online posts of users over time, as well as the extraction of adaptive and maladaptive evidence spans and further summarization of self-state information at the post and timeline levels. This multi-task framework is grounded in the MIND scheme (Slonim, 2024) that views human experience as consisting of self-states fluctuating over time. Each self-state constitutes of identifiable units characterized by specific combinations of Affect, Behavior, Cognition, and Desire (ABCD).

While the evidence extraction and summaries provide some guidance with respect to ABCD elements and maladaptive and adaptive states, this cannot be used for diagnostic purposes, especially without the involvement of human experts. Adaptive and maladaptive evidence extracted by such models should be reviewed by clinical experts or used in the loop to augment their capacity by efficiently presenting information to them.

Additionally, the task cannot make any claims about the potential evidence providing explanations for well-being scores. Rather, it forms a research direction towards making causal links between the two, paving the way towards language models that can better reason along their decision making process.

In terms of data, even though we are using publicly available content from Reddit, we prohibited its redistribution and the use of any third-party LLMs that would require sending (part of) the information to the provider's servers, to ensure protection of the sensitive content.

Acknowledgements

This work was supported by a UKRI/EPSRC Turing AI Fellowship to Maria Liakata (grant ref EP/V030302/1) and the Alan Turing Institute (grant ref EP/N510129/1). This work was supported by the Engineering and Physical Sciences Research Council [grant number EP/Y009800/1], through funding from Responsible Ai UK (KP0016) as a Keystone project lead by Maria Liakata. The shared task organizers would like to express their gratitude to the anonymous users of Reddit whose data feature in this year's shared task dataset; to the clinical experts from Bar-Ilan University who annotated the data for all tasks; to all team members for their participation; and to NAACL for its support for CLPsych.

References

Prottay Kumar Adhikary, Aseem Srivastava, Shivani Kumar, Salam Michael Singh, Puneet Manuja, Jini K Gopinath, Vijay Krishnan, Swati Kedia Gupta, Koushik Sinha Deb, and Tanmoy Chakraborty. 2024. Exploring the efficacy of large language models in summarizing mental health counseling sessions: benchmark study. *JMIR Mental Health*, 11:e57306.

Mostafa M Amin, Erik Cambria, and Björn W Schuller. 2023. Will affective computing emerge from foundation models and general ai? a first evaluation on chatgpt. *IEEE Intelligent Systems*, 38(2):15–23.

Anson Antony and Annika Marie Schoene. 2025. Retrieval-enhanced mental health assessment: Capturing self-state dynamics from social media using in-context learning. In *Proceedings of the Tenth Workshop on Computational Linguistics and Clinical Psychology*. Association for Computational Linguistics.

Elham Asgari, Nina Montana-Brown, Magda Dubois, Saleh Khalil, Jasmine Balloch, and Dominic Pimenta. 2024. A framework to assess clinical safety and hallucination rates of llms for medical text summarisation. *medRxiv*, pages 2024–09.

American Psychiatric Association et al. 2000. Diagnostic and statistical manual of mental disorders iv-tr washington. *DC: American Psychiatric Association*.

Abhin B and Renukasakshi V Patil. 2025. Transformer-based analysis of adaptive and maladaptive self-states in longitudinal social media data. In *Proceedings of the Tenth Workshop on Computational Linguistics and Clinical Psychology*. Association for Computational Linguistics.

Roy F Baumeister, Ellen Bratslavsky, Catrin Finkenauer, and Kathleen D Vohs. 2001. Bad is stronger than good. *Review of general psychology*, 5(4):323–370.

Ulya Bayram and Lamia Benhiba. 2022. Emotionally-informed models for detecting moments of change and suicide risk levels in longitudinal social media data. In *Proceedings of the Eighth Workshop on Computational Linguistics and Clinical Psychology*, pages 219–225, Seattle, USA. Association for Computational Linguistics.

Aaron T Beck, Molly R Finkel, and Judith S Beck. 2021. The theory of modes: Applications to schizophrenia and other psychological conditions. *Cognitive Therapy and Research*, 45:391–400.

Iz Beltagy, Matthew E. Peters, and Arman Cohan. 2020. Longformer: The long-document transformer. *arXiv:2004.05150*.

Xiao Bi, Deli Chen, Guanting Chen, Shanhuang Chen, Damai Dai, Chengqi Deng, Honghui Ding, Kai Dong, Qiushi Du, Zhe Fu, et al. 2024. Deepseek llm: Scaling open-source language models with longtermism. *arXiv preprint arXiv:2401.02954*.

Philip M Bromberg. 2014. *Standing in the spaces: Essays on clinical process trauma and dissociation*. Routledge.

Suchandra Chakraborty, Sudeshna Jana, Manjira Sinha, and Tirthankar Dasgupta. 2025. Self-state evidence extraction and well-being prediction from social media timelines. In *Proceedings of the Tenth Workshop on Computational Linguistics and Clinical Psychology*. Association for Computational Linguistics.

Callum Chan, Sunveer Khunkhun, Diana Inkpen, and Juan Antonio Lossio-Ventura. 2025. Prompt engineering for capturing dynamic mental health self states from social media posts. In *Proceedings of the Tenth Workshop on Computational Linguistics and Clinical Psychology*. Association for Computational Linguistics.

Jenny Chim, Julia Ive, and Maria Liakata. 2025. Evaluating synthetic data generation from user generated text. *Computational Linguistics*, 51(1):191–233.

Jenny Chim, Adam Tsakalidis, Dimitris Gkoumas, Dana Atzil-Slonim, Yaakov Ophir, Ayah Zirikly, Philip Resnik, and Maria Liakata. 2024. Overview of the CLPsych 2024 shared task: Leveraging large language models to identify evidence of suicidality risk in online posts. In *Proceedings of the 9th Workshop on Computational Linguistics and Clinical Psychology (CLPsych 2024)*, pages 177–190, St. Julians, Malta. Association for Computational Linguistics.

Glen Coppersmith, Mark Dredze, and Craig Harman. 2014. Quantifying mental health signals in twitter. In *Proceedings of the workshop on computational linguistics and clinical psychology: From linguistic signal to clinical reality*, pages 51–60.

Glen Coppersmith, Mark Dredze, Craig Harman, Kristy Hollingshead, and Margaret Mitchell. 2015. Clpsych 2015 shared task: Depression and ptsd on twitter. In *Proceedings of the 2nd workshop on computational linguistics and clinical psychology: from linguistic signal to clinical reality*, pages 31–39.

Munmun De Choudhury, Michael Gamon, Scott Counts, and Eric Horvitz. 2013. Predicting depression via social media. In *Proceedings of the international AAAI conference on web and social media*, volume 7, pages 128–137.

Louise Deleger, Qi Li, Todd Lingren, Megan Kaiser, Katalin Molnar, Laura Stoutenborough, Michal Kouril, Keith Marsolo, Imre Solti, et al. 2012. Building gold standard corpora for medical natural language processing tasks. In *AMIA Annual Symposium Proceedings*, volume 2012, page 144.

Jacob Devlin, Ming-Wei Chang, Kenton Lee, and Kristina Toutanova. 2019. Bert: Pre-training of deep bidirectional transformers for language understanding. In *Proceedings of the 2019 conference of the North American chapter of the association for computational linguistics: human language technologies, volume 1 (long and short papers)*, pages 4171–4186.

Muskan Garg. 2024. Wellxplain: Wellness concept extraction and classification in reddit posts for mental health analysis. *Knowledge-Based Systems*, 284:111228.

Aaron Grattafiori, Abhimanyu Dubey, Abhinav Jauhri, Abhinav Pandey, Abhishek Kadian, Ahmad Al-Dahle, Aiesha Letman, Akhil Mathur, Alan Schelten, Alex Vaughan, et al. 2024. The llama 3 herd of models. *arXiv preprint arXiv:2407.21783*.

Daya Guo, Dejian Yang, Haowei Zhang, Junxiao Song, Ruoyu Zhang, Runxin Xu, Qihao Zhu, Shirong Ma, Peiyi Wang, Xiao Bi, et al. 2025. Deepseek-r1: Incentivizing reasoning capability in llms via reinforcement learning. *arXiv preprint arXiv:2501.12948*.

Anthony Hills, Adam Tsakalidis, and Maria Liakata. 2023. Time-aware predictions of moments of change in longitudinal user posts on social media. In *International Workshop on Advanced Analytics and Learning on Temporal Data*, pages 293–305. Springer.

Anthony Hills, Talia Tseriotou, Xenia Miscouridou, Adam Tsakalidis, and Maria Liakata. 2024. Exciting mood changes: A time-aware hierarchical transformer for change detection modelling. In *Findings of the Association for Computational Linguistics: ACL 2024*, pages 12526–12537, Bangkok, Thailand. Association for Computational Linguistics.

Stephanie Homan, Marion Gabi, Nina Klee, Sandro Bachmann, Ann-Marie Moser, Martina Duri', Sofia Michel, Anna-Marie Bertram, Anke Maatz, Guido Seiler, Elisabeth Stark, and Birgit Kleim. 2022. Linguistic features of suicidal thoughts and behaviors: A systematic review. *Clinical Psychology Review*, 95:102161.

Md. Iqramul Hoque, Mahfuz Ahmed Anik, and Azmine Toushik Wasi. 2025. Ciol at clpsych 2025: Using large lanuage models for understanding and summarizing clinical texts. In *Proceedings of the Tenth Workshop on Computational Linguistics and Clinical Psychology*. Association for Computational Linguistics.

George Hripcsak and Adam S Rothschild. 2005. Agreement, the f-measure, and reliability in information retrieval. *Journal of the American medical informatics association*, 12(3):296–298.

Albert Q Jiang, Alexandre Sablayrolles, Antoine Roux, Arthur Mensch, Blanche Savary, Chris Bamford, Devendra Singh Chaplot, Diego de las Casas, Emma Bou Hanna, Florian Bressand, et al. 2024. Mixtral of experts. *arXiv preprint arXiv:2401.04088*.

Jihoon Kwon Sangmo Gu Yejin Kim Minkyung Cho Jy-yong Sohn Chanyeol Choi Junseong Kim, Seolhwa Lee. 2024. Linq-embed-mistral:elevating text retrieval with improved gpt data through task-specific control and quality refinement. Linq AI Research Blog.

Laerdon Kim. 2025. A baseline for self-state identification and classification in mental health data: Clpsych 2025 task. In *Proceedings of the Tenth Workshop on Computational Linguistics and Clinical Psychology*. Association for Computational Linguistics.

Ayal Klein, Jiayu Song, Jenny Chim, Liran Keren, Andreas Triantafyllopoulos, Björn W Schuller, Maria Liakata, and Dana Atzil-Slonim. 2024. Clinical insights from social media: Assessing summaries of large language models and humans.

Jan-Christoph Klie, Michael Bugert, Beto Boullosa, Richard Eckart de Castilho, and Iryna Gurevych. 2018. The inception platform: Machine-assisted and knowledge-oriented interactive annotation. In *Proceedings of the 27th International Conference on Computational Linguistics: System Demonstrations*, pages 5–9. Association for Computational Linguistics. Event Title: The 27th International Conference on Computational Linguistics (COLING 2018).

Moritz Laurer, Wouter Van Atteveldt, Andreu Casas, and Kasper Welbers. 2024. Less annotating, more classifying: Addressing the data scarcity issue of supervised machine learning with deep transfer learning and bert-nli. *Political Analysis*, 32(1):84–100.

Gal Lazarus and Eshkol Rafaeli. 2023. Modes: Cohesive personality states and their interrelationships as organizing concepts in psychopathology. *Journal of Psychopathology and Clinical Science*, 132(3):238.

Mike Lewis, Yinhan Liu, Naman Goyal, Marjan Ghazvininejad, Abdelrahman Mohamed, Omer Levy, Veselin Stoyanov, and Luke Zettlemoyer. 2020. Bart: Denoising sequence-to-sequence pre-training for natural language generation, translation, and comprehension. In *Proceedings of the 58th Annual Meeting of the Association for Computational Linguistics*, pages 7871–7880.

Yucheng Li, Bo Dong, Frank Guerin, and Chenghua Lin. 2023. Compressing context to enhance inference efficiency of large language models. In *Proceedings of the 2023 Conference on Empirical Methods in Natural Language Processing*, pages 6342–6353.

Yinhan Liu, Myle Ott, Naman Goyal, Jingfei Du, Mandar Joshi, Danqi Chen, Omer Levy, Mike Lewis, Luke Zettlemoyer, and Veselin Stoyanov. 2019. Roberta: A robustly optimized bert pretraining approach. *arXiv preprint arXiv:1907.11692*.

Chandreen Liyanage, Muskan Garg, Vijay Mago, and Sunghwan Sohn. 2023. Augmenting Reddit posts to determine wellness dimensions impacting mental health. In *The 22nd Workshop on Biomedical Natural Language Processing and BioNLP Shared Tasks*, pages 306–312, Toronto, Canada. Association for Computational Linguistics.

Scott M. Lundberg and Su-In Lee. 2017. A unified approach to interpreting model predictions. In *Proceedings of the 31st International Conference on Neural Information Processing Systems*, NIPS'17, page 47684777, Red Hook, NY, USA. Curran Associates Inc.

Yu A Malkov. 2018. Efficient and robust approximate nearest neighbor search using hierarchical navigable small world graphs. *IEEE transactions on pattern analysis and machine intelligence*, 42(4):824–836.

Tobias Mayer, Neha Warikoo, Amir Eliassaf, Dana Atzil-Slonim, and Iryna Gurevych. 2024. Predicting client emotions and therapist interventions in psychotherapy dialogues. In *Proceedings of the 18th Conference of the European Chapter of the Association for Computational Linguistics (Volume 1: Long Papers)*, pages 1463–1477.

Joshua Maynez, Shashi Narayan, Bernd Bohnet, and Ryan McDonald. 2020. On faithfulness and factuality in abstractive summarization. In *Proceedings of the 58th Annual Meeting of the Association for Computational Linguistics*, pages 1906–1919, Online. Association for Computational Linguistics.

George Michalopoulos, Kyle Williams, Gagandeep Singh, and Thomas Lin. 2022. MedicalSum: A guided clinical abstractive summarization model for generating medical reports from patient-doctor conversations. In *Findings of the Association for Computational Linguistics: EMNLP 2022*, pages 4741–4749, Abu Dhabi, United Arab Emirates. Association for Computational Linguistics.

Ummara Mumtaz, Awais Ahmed, and Summaya Mumtaz. 2024. Llms-healthcare : Current applications and challenges of large language models in various medical specialties.

Matthew D Nemesure, Michael V Heinz, Raphael Huang, and Nicholas C Jacobson. 2021. Predictive modeling of depression and anxiety using electronic health records and a novel machine learning approach with artificial intelligence. *Scientific reports*, 11(1):1980.

Thong Nguyen, Andrew Yates, Ayah Zirikly, Bart Desmet, and Arman Cohan. 2022. Improving the generalizability of depression detection by leveraging clinical questionnaires. In *Proceedings of the 60th Annual Meeting of the Association for Computational Linguistics (Volume 1: Long Papers)*, pages 8446–8459, Dublin, Ireland. Association for Computational Linguistics.

Javier Parapar, Patricia Martín-Rodilla, David E Losada, and Fabio Crestani. 2023. Overview of erisk 2023: Early risk prediction on the internet. In *International Conference of the Cross-Language Evaluation Forum for European Languages*, pages 294–315. Springer.

Jason Phang, Yao Zhao, and Peter Liu. 2023. Investigating efficiently extending transformers for long input summarization. In *Proceedings of the 2023 Conference on Empirical Methods in Natural Language Processing*, pages 3946–3961, Singapore. Association for Computational Linguistics.

Federico Ravenda, Fawzia-Zehra Kara-Isitt, Stephen Swift, Antonietta Mira, and Andrea Raballo. 2025. From evidence mining to meta-prediction: a gradient of methodologies for task-specific challenges in psychological assessment. In *Proceedings of the Tenth Workshop on Computational Linguistics and Clinical Psychology*. Association for Computational Linguistics.

William Revelle. 2007. Experimental approaches to the study of personality. *Handbook of research methods in personality psychology*, pages 37–61.

Anastasia Sandu, Teodor Mihailescu, Ana Sabina Uban, and Ana-Maria Bucur. 2025. Capturing the dynamics of mental well-being: Adaptive and maladaptive states in social media. In *Proceedings of the Tenth Workshop on Computational Linguistics and Clinical Psychology*. Association for Computational Linguistics.

Ramit Sawhney, Shivam Agarwal, Atula Tejaswi Neerkaje, Nikolaos Aletras, Preslav Nakov, and Lucie Flek. 2022a. Towards suicide ideation detection through online conversational context. In *Proceedings of the 45th international ACM SIGIR conference on research and development in information retrieval*, pages 1716–1727.

Ramit Sawhney, Atula Tejaswi Neerkaje, and Manas Gaur. 2022b. A risk-averse mechanism for suicidality assessment on social media. *Association for Computational Linguistics 2022 (ACL 2022)*.

Han-Chin Shing, Suraj Nair, Ayah Zirikly, Meir Friedenberg, Hal Daumé III, and Philip Resnik. 2018. Expert, crowdsourced, and machine assessment of suicide risk via online postings. In *Proceedings of the Fifth Workshop on Computational Linguistics and Clinical Psychology: From Keyboard to Clinic*, pages 25–36.

Gopendra Singh, Sai Vemulapalli, Mauajama Firdaus, and Asif Ekbal. 2024. Deciphering cognitive distortions in patient-doctor mental health conversations: A multimodal llm-based detection and reasoning framework. In *Proceedings of the 2024 Conference on Empirical Methods in Natural Language Processing*, pages 22546–22570.

Dana Atzil Slonim. 2024. Self-other dynamics (sod): A transtheoretical coding manual.

Jiayu Song, Mahmud Akhter, Dana Atzil Slonim, and Maria Liakata. 2025. Temporal reasoning for timeline summarisation in social media.

Jiayu Song, Jenny Chim, Adam Tsakalidis, Julia Ive, Dana Atzil-Slonim, and Maria Liakata. 2024a. Clinically meaningful timeline summarisation in social media for mental health monitoring.

Jiayu Song, Jenny Chim, Adam Tsakalidis, Julia Ive, Dana Atzil-Slonim, and Maria Liakata. 2024b. Combining hierachical VAEs with LLMs for clinically meaningful timeline summarisation in social media. In *Findings of the Association for Computational Linguistics: ACL 2024*, pages 14651–14672, Bangkok, Thailand. Association for Computational Linguistics.

Nikita Soni, Matthew Matero, Niranjan Balasubramanian, and H Andrew Schwartz. 2022. Human language modeling. In *Findings of the Association for Computational Linguistics: ACL 2022*, pages 622–636.

Nikita Soni, August Håkan Nilsson, Syeda Mahwish, Vasudha Varadarajan, H. Andrew Schwartz, and Ryan L. Boyd. 2025. Who we are, where we are: Mental health at the intersection of person, situation, and large language models. In *Proceedings of the Tenth Workshop on Computational Linguistics and Clinical Psychology*. Association for Computational Linguistics.

Sajad Sotudeh, Nazli Goharian, and Zachary Young. 2022. Mentsum: A resource for exploring summarization of mental health online posts.

Aseem Srivastava, Tharun Suresh, Sarah Peregrine, Lord, Md. Shad Akhtar, and Tanmoy Chakraborty. 2022. Counseling summarization using mental health knowledge guided utterance filtering.

William B Stiles. 2001. Assimilation of problematic experiences. *Psychotherapy: Theory, Research, Practice, Training*, 38(4):462.

Gemma Team, Morgane Riviere, Shreya Pathak, Pier Giuseppe Sessa, Cassidy Hardin, Surya Bhupatiraju, Léonard Hussenot, Thomas Mesnard, Bobak Shahriari, Alexandre Ramé, et al. 2024. Gemma 2: Improving open language models at a practical size. *arXiv preprint arXiv:2408.00118*.

Vu Tran and Tomoko Matsui. 2025. Team ism at clpsych 2025: Capturing mental health dynamics from social media timelines using a pretrained large language model with in-context learning. In *Proceedings of the Tenth Workshop on Computational Linguistics and Clinical Psychology*. Association for Computational Linguistics.

Adam Tsakalidis, Jenny Chim, Iman Munire Bilal, Ayah Zirikly, Dana Atzil-Slonim, Federico Nanni, Philip Resnik, Manas Gaur, Kaushik Roy, Becky Inkster, et al. 2022a. Overview of the clpsych 2022 shared task: Capturing moments of change in longitudinal user posts. In *Proceedings of the Eighth Workshop on Computational Linguistics and Clinical Psychology*, pages 184–198.

Adam Tsakalidis, Federico Nanni, Anthony Hills, Jenny Chim, Jiayu Song, and Maria Liakata. 2022b. Identifying moments of change from longitudinal user text. In *Proceedings of the 60th Annual Meeting of the Association for Computational Linguistics (Volume 1: Long Papers)*, pages 4647–4660.

Talia Tseriotou, Ryan Chan, Adam Tsakalidis, Iman Munire Bilal, Elena Kochkina, Terry Lyons, and Maria Liakata. 2024a. Sig-networks toolkit: Signature networks for longitudinal language modelling. In *Proceedings of the 18th Conference of the European Chapter of the Association for Computational Linguistics: System Demonstrations*, pages 223–237.

Talia Tseriotou, Adam Tsakalidis, Peter Foster, Terence Lyons, and Maria Liakata. 2023. Sequential path signature networks for personalised longitudinal language modeling. In *Findings of the Association for Computational Linguistics: ACL 2023*, pages 5016–5031.

Talia Tseriotou, Adam Tsakalidis, and Maria Liakata. 2024b. Tempoformer: A transformer for temporally-aware representations in change detection. In *Proceedings of the 2024 Conference on Empirical Methods in Natural Language Processing*, pages 19635–19653.

Zimu Wang, Hongbin Na, Rena Gao, Jiayuan Ma, Yining Hua, Ling Chen, and Wei Wang. 2025. From posts to timelines: Modeling mental health dynamics from social media timelines with hybrid llms. In *Proceedings of the Tenth Workshop on Computational Linguistics and Clinical Psychology*. Association for Computational Linguistics.

WHO. 2022. *World mental health report: Transforming mental health for all*. World Health Organization.

Xuhai Xu, Bingshen Yao, Yuanzhe Dong, Hong Yu, James Hendler, Anind K Dey, and Dakuo Wang. 2023. Leveraging large language models for mental health prediction via online text data. *arXiv preprint arXiv:2307.14385*.

Xuhai Xu, Bingsheng Yao, Yuanzhe Dong, Saadia Gabriel, Hong Yu, James Hendler, Marzyeh Ghassemi, Anind K. Dey, and Dakuo Wang. 2024a. Mental-llm: Leveraging large language models for mental health prediction via online text data. *Proc. ACM Interact. Mob. Wearable Ubiquitous Technol.*, 8(1).

Xuhai Xu, Bingsheng Yao, Yuanzhe Dong, Saadia Gabriel, Hong Yu, James Hendler, Marzyeh Ghassemi, Anind K Dey, and Dakuo Wang. 2024b. Mental-llm: Leveraging large language models for mental health prediction via online text data. *Proceedings of the ACM on Interactive, Mobile, Wearable and Ubiquitous Technologies*, 8(1):1–32.

An Yang, Baosong Yang, Beichen Zhang, Binyuan Hui, Bo Zheng, Bowen Yu, Chengyuan Li, Dayiheng Liu, Fei Huang, Haoran Wei, et al. 2024a. Qwen2. 5 technical report. *arXiv preprint arXiv:2412.15115*.

Kailai Yang, Shaoxiong Ji, Tianlin Zhang, Qianqian Xie, Ziyan Kuang, and Sophia Ananiadou. 2023. Towards interpretable mental health analysis with large language models. In *Proceedings of the 2023 Conference on Empirical Methods in Natural Language Processing*, pages 6056–6077, Singapore. Association for Computational Linguistics.

Kailai Yang, Tianlin Zhang, Ziyan Kuang, Qianqian Xie, Jimin Huang, and Sophia Ananiadou. 2024b. Mentallama: interpretable mental health analysis on social media with large language models. In *Proceedings of the ACM Web Conference 2024*, pages 4489–4500.

Tianyi Zhang, Varsha Kishore, Felix Wu, Kilian Q. Weinberger, and Yoav Artzi. 2020. Bertscore: Evaluating text generation with bert. In *International Conference on Learning Representations*.

Ayah Zirikly and Mark Dredze. 2022. Explaining models of mental health via clinically grounded auxiliary tasks. In *Proceedings of the Eighth Workshop on Computational Linguistics and Clinical Psychology*, pages 30–39.

Ayah Zirikly, Philip Resnik, Özlem Uzuner, and Kristy Hollingshead. 2019. CLPsych 2019 shared task: Predicting the degree of suicide risk in Reddit posts. In

Proceedings of the Sixth Workshop on Computational Linguistics and Clinical Psychology.

Appendix A Annotation

A.1 Annotation Interface

Figure 4 shows a screenshot of the annotation platform INCEpTION (Klie et al., 2018), which we adapted for our task. INCEpTION provides a user-friendly interface that enables annotators to efficiently assign labels and categories directly onto text segments. By customizing the annotation schema and label sets, we streamlined the annotation process, enhancing precision and consistency aligned with our research objectives.

A.2 MIND Framework

Table 10 shows categories and sub-categories within the framework.

A.3 Further IAA Measures

We complement the standard inter-annotator agreement measures reported in Section 4.2 with additional consistency metrics.

We first calculate relaxed pairwise F1 scores over spans identified in Task A.1 (State Evidence), following the previous CLPsych shared task (Chim et al., 2024) and established practices (Hripcsak and Rothschild, 2005; Deleger et al., 2012). In this relaxed metric, a minimal overlap of one token between spans is considered a match. Results are summarized in Table 7. These values indicate lower agreement compared to CLPsych 2024, likely due to the broader, more comprehensive nature of our task. Furthermore, we calculate the agreement between the annotators using the same metrics employed for system evaluation, recognizing that inter-annotator agreement serves as an essential reference point or upper bound for assessing system performance.

Span Type	F1
Adaptive spans	.51
Maladaptive spans	.58
Overall (micro-average)	**.56**

Table 7: Relaxed pairwise F1 agreement for Task A.1.

Our findings, detailed in Table 8, emphasize the complexity of the annotation tasks and underscore that achieving high agreement is challenging even for clinically informed annotators. Thus, achieving performance close to these inter-annotator agreement values can be considered as approaching the maximum attainable performance for these tasks.

Metric	Task	Value
BERTScore recall	A.1	0.469
BERTScore weighted recall	A.1	0.387
Mean squared error (MSE)	A.2	2.913
Macro F1	A.2	0.403

Table 8: Inter-annotator agreement over 23 posts using the system evaluation metrics. Note that recall measures were averaged across both comparison directions to reflect the symmetric nature of inter-annotator agreement.

Appendix B Participant Submissions

This section presents an overview of the registration process (§B.1), individual systems (§B.2) from each participating team and provides an overview of methods (§B.3).

B.1 Registration Process

The registration process consisted of three stages: a) completing an individual and a team registration through an online form, b) reading and signing a data sharing agreement, and c) receiving access instructions for training data stored in a password-protected compressed file. During stage a) the organizing team assisted participants looking for collaborators in the team-forming process. For b), the data sharing agreement asked the teams to determine the password-protected private storage of the data, while restricting explicit or implicit data distribution through third party LLM platforms.

Team	#Members	Task A	Task B	Task C	Paper submitted
Aquarius	5	2	2	2	✓
BLUE	3	3	3	3	✓
BULUSI	2	3	3	3	✓
CIOL	3	1	2	2	✓
CSIRO-LT	4	3	-	-	
EAIonFlux	2	2	2	2	✓
ISM	2	2	2	2	✓
MMKA	4	2	-	-	✓
NoviceTrio	3	1	1	1	
PsyMetric	2	1	1	1	
ResBin	2	1	1	1	✓
Seq2Psych	6	3	-	-	✓
uOttawa	4	3	3	3	✓
Zissou	7	1	1	-	✓
Total	**49**	**28**	**21**	**20**	**11**

Table 9: Team information and submissions for the CLPsych 2025 shared task.

Each team was allowed up to three submissions for the official team results. Additional submissions were allowed in order to facilitate ablation and further analysis by the teams. Upon receiving the submissions, results were returned within 24 hours based on our evaluation metrics (§5) on a test set of 10 timelines (§4). A summary of the team specifics including the number of submissions is provided in Table 9.

B.2 Individual Team Submissions

Aquarius Wang et al. (2025) integrated extracted evidence to guide well-being score prediction and summarization. For evidence span identification, they used fine-tuned Qwen2.5-7b (Yang et al., 2024a) to explore a sentence classification and a span generation approach. Then, they combined the content of each post with extracted evidences for well-being score prediction. For post-level and timeline-level summarization, the team employed Qwen2.5-32B, using as input post content, extracted evidences, predicted score(s), and a retrieved annotated example from the train set that had the highest embedding similarity.

BLUE Sandu et al. (2025) utilized a range of LLMs, prompting strategies, and machining learning approaches to the tasks. For evidence extraction they achieved the highest recall using Gemma 2 9B (Team et al., 2024) coupled with a default prompt, providing instruction for the task without including definitions of concepts or additional context, while the same model performs the best for well-being scoring using an expert prompt based on emotional, cognitive, and behavioral indicators. For post-level and timeline-level summarization, LLaMA 3.2 3B (Grattafiori et al., 2024) utilizing the default prompt performed best.

BULUSI For evidence extraction, Ravenda et al. (2025) formed candidate segments and then extracted the most relevant ones using retrieval based on the training evidence which were fed in three (22-72B) LLMs for consensus self-state classification with in-context learning. For well-being score prediction, the team explored three strategies for aggregating LLM predictions: an average ensemble, an Oracle-style meta-model, and an optimized weighted ensemble minimizing mean squared error while accounting for missing values. The optimized ensemble yielded the best result. Finally, for post and timeline-level summarization the team used post(s) content, predicted self-state segments, and retrieved top five relevant example posts to prompt the LLM.

CIOL Hoque et al. (2025) extracted evidence spans of adaptive and maladaptive self-states using Random Forest classifiers on thousands of TF-IDF features. For well-being score prediction they formulated a supervised approach through Gradient Boosting regression on sentiment and ratio-based features reflecting the relationship between adaptive and maladaptive evidence. For post-level summaries they DPO-finetuned Qwen2.5-7B-Instruct-1M (Yang et al., 2024a). This was followed by a few-shot prompting strategy guiding the model to identify the dominant self-state determining the ABCD elements based on evidence spans and well-being scores. This approach was extended on the timeline-level by fine tuning the post-level model above on timeline-level examples. Then they used this model to generate summaries based on a narrative arc analysis framework that treats each timeline as a psychological development trajectory. For an extension of the timeline-level summary their prompt directs the model to identify temporal self-state patterns and changes, highlighting key transitions between states.

EAIonFlux Antony and Schoene (2025) proposed systems based on vector similarity retrieval of relevant in-context demonstrations for LLM prompting. They used LLaMA 3.3 70B (Grattafiori et al., 2024) and experimented with different numbers of retrieved examples. They built one post-level and one timeline-level vector database (capturing temporal patterns) out of the training data embedding them through Linq-Embed-Mistral (Junseong Kim, 2024) to capture emotional content. Retrieval is based on cosine similarity with HNSW (Malkov, 2018) for fast nearest neighbor search. For each task, a task-specific module generates prompts and processes outputs tailored to the different objectives, and predictions from previous task(s) are integrated into the next ones.

ISM Tran and Matsui (2025) explored in-context learning with Llama-3-8B, using random sampling followed by Gaussian kernel density estimation to select training data instances as demonstration. The team jointly modeled post-level tasks in the same prompt, and focused on summarization only in the prompt for timeline-level generation.

MMKA Chakraborty et al. (2025) focused on Tasks A.1 and A.2. They fine-tuned a RoBERTa classification model (Liu et al., 2019) to extract adaptive and maladaptive self-states at the token level, augmenting the training data using `nlpaug` and implementing post-processing to obtain the most frequent label per sentence. For well-being score generation (with a justification generation),

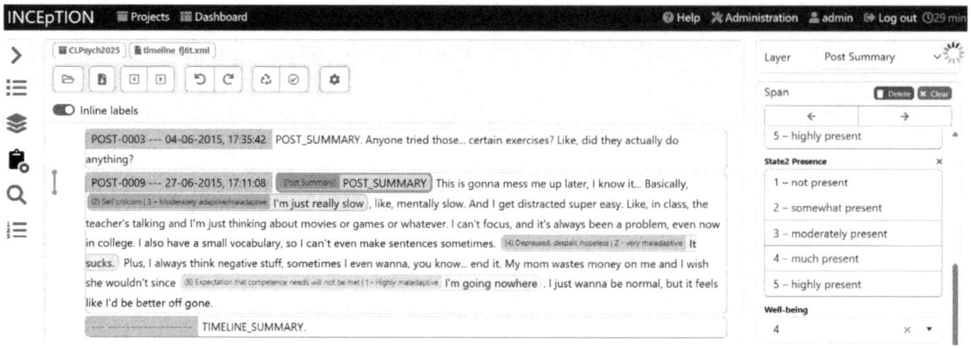

Figure 4: A screenshot from our annotation interface, leveraging the INCEpTION platform. Example timeline is reduced and paraphrased due to the sensitive nature of the data.

Category		Sub-Categories	
		Adaptive Example	Maladaptive Example
Affect	Type of emotion expressed by a person.	Calm/Laid back, Emotional Pain/Grieving, Content/Happy, Vigor/Energetic, Justifiable Anger/Assertive Anger, Proud	Anxious/Tense/Fearful, Depressed/Desperate/ Hopeless, Mania, Apathetic/Don't care/Blunted, Angry (Aggressive, Disgust, Contempt), Ashamed/Guilty
Behavior	**Behavior of the self with the Other (BO)** The person's main behavior(s) toward the other.	Relating behavior, Autonomous behavior	Fight or flight behavior, Overcontrolled/controlling behavior
	Behavior toward the Self (BS) The person's main behavior(s) toward the self.	Self-care behavior	Self-harm/Neglect/ Avoidance behavior
Cognition	**Cognition of the Other (CO)** The person's main perceptions of the other.	Perception of the other as related, Perception of the other as facilitating autonomy/ competence needs	Perception of the other as detached or over attached, Perception of the other as blocking autonomy needs
	Cognition of the Self (CS) The person's main self-perceptions.	Self-acceptance and self-compassion	Self-criticism
Desire	The person's main desire, need, intention, fear or expectation.	Relatedness, Autonomy and adaptive control, Competence, Self-esteem, Self-care	Expectation that relatedness need will not be met, Expectation that autonomy needs will not be met, Expectation that competence needs will not be met

Table 10: ABCD elements (Categories) with explanations, and their sub-categories.

they used RAG to select the top-k most similar posts to the current one using all-MiniLM-L6-v2 for embeddings. These were included for in-context learning with DeepSeek-7B (Bi et al., 2024). In post-hoc analysis, they found random forest regression yielding better results than the LLM approach.

ResBin B and Patil (2025) explored diverse approaches: they used Mixtral-8x7b (Jiang et al., 2024) for evidence extraction, obtained embeddings from PLMs such as MentalBERT to train random forest classifiers for well-being score prediction, fine-tuned Longformer (Beltagy et al., 2020) for post-level summarization with predicted evidences and post content as input, and fine-tuned Pegasus-X-Large (Phang et al., 2023) for timeline-level summarization with timeline content as input.

Seq2Psych Soni et al. (2025) focused on Task A, leveraging principled baseline features, such as Situational 8 DIAMONDS (S8D) and Person-Level Traits (PLT) including resilience quantification utilizing the Resilience through Language Modeling (ReLM) framework. For S8D, the team used the Deepseek-R1 (Guo et al., 2025) model with few-shot prompting to infer eight situational dimensions at the post-level. Using different feature combinations they fine-tuned the HaRT model (Soni et al., 2022) which processes temporal user language, to generate person-contextualized embeddings towards well-being score prediction and sentence-level binary adaptive/maladaptive classification.

uOttawa Chan et al. (2025) applied prompt engineering techniques with Llama-3.3-70B-Instruct to address all four subtasks. They compared variedly structured zero-shot, one-shot, and few-shot prompts, finding one-shot to be most performant for evidence extraction and post-level summarization, and few-shot to be best for well-being score prediction and timeline-level summarization.

Zissou Kim (2025) prompted a 4-bit quantized Gemma-2 9b (Team et al., 2024) with few-shot learning and presented an approach that explored the impact of preprocessing on span extraction. Each post was divided into sentences, identifying only the important sentences, and then classified as adaptive or maladaptive through prompting with self-state definitions. Providing the previous sentence context improved performance. For the other tasks, they generated post summaries and well-being scores based on the list of classifications and post text also with few-shot prompting.

B.3 Overview

We outline methods used in the *best run per team* in Table 11. For a complete picture of each team's approaches, including ablations, please refer to their respective paper. We consider whether a system:

- **LLM**: Uses a large language model.
- **PLM**: Uses a pretrained language model.
- **ML**: Uses traditional machine learning, focusing on algorithms (e.g. random forest) and excluding techniques (e.g. feature engineering).
- **RAG**: Uses retrieval augmented generation, focusing on automatic retrieval and excluding manually selected examples.
- **Pipeline**: Involves an approach where predictions from at least one task are used in predictions for another. Excludes joint modeling that happens in the same step.
- **Domain**: Involves explicit domain knowledge *beyond* what was provided in the shared task documents provided to all participants.
- **Temporal**: Involves explicit modeling of temporality/relationship between posts within a single timeline in Tasks A and B. Excludes cases of contextual modeling within an individual post (e.g. between sentences in one post).

Team	LLM	PLM	ML	RAG	Pipeline	Domain	Temporal
Aquarius	✓				✓		
BLUE	✓						
BULUSI	✓			✓	✓		
CIOL	✓		✓		✓	✓	✓
EAIonFlux	✓			✓	✓		
ISM	✓		✓	✓			
MMKA	✓	✓		✓			
ResBin	✓	✓	✓		✓		
Seq2Psych	✓	✓				✓	✓
uOttawa	✓						
Zissou	✓				✓		
Total	11	3	3	4	6	2	2

Table 11: Methods in each team's top submission.

LLMs Every team in this year's shared task used an LLM to tackle at least one subtask. Focusing on the *best run* per team, we categorize the types of models used, counting each model once per submission per team. We summarize the model type (Figure 5), context length (Figure 6), and size in terms of parameter count (Figure 7).

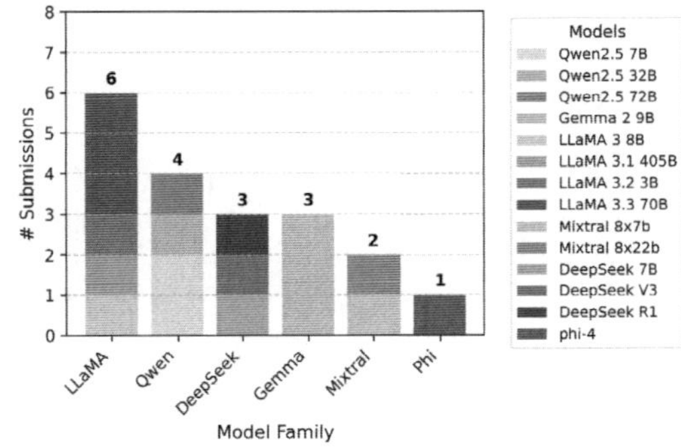

Figure 5: LLMs used in best runs of official submissions, grouped by model family and lineage.

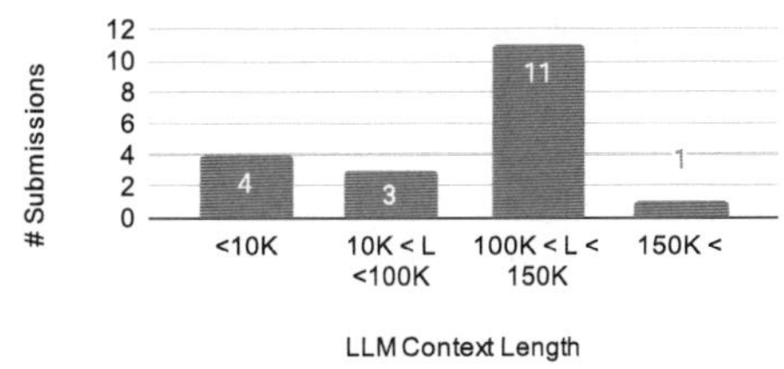

Figure 6: Maximum number of tokens that can be fed into the employed models.

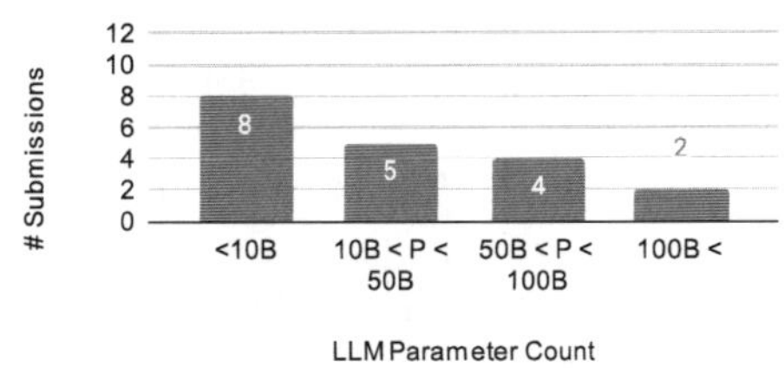

Figure 7: Size of employed models. We use the active parameter count for mixture-of-expert models.

Compared to the previous shared task which explored how well open LLMs can be leveraged to identify and synthesize textual evidences (Chim et al., 2024), we see noticeable increases in (1) model family diversity, (2) context length, and (3) use of retrieval augmented generation.

While LLaMA remains the most popular model family, many submissions leveraged alternatives such as Qwen and DeepSeek. The maximum theoretical context length that can be handled has increased from 32K to 1M tokens, and the majority of models used in the top submission this year have a context length in the 100K to 150K token range. These changes highlight the rapid technological progress in open LLMs and long context models, as well as opportunities to advance mental health modeling over longer horizons.

Appendix C Baselines

This section outlines implementation details of our baseline models (§6.2). In LLM-based methods, we employ `Llama 3.1-8B-Instruct`.

C.1 Task A.1

For the *Llama ZS Single-Post* and *Llama ZS 5-Post* baselines we used the prompts presented in Listings 1 and 2 respectively. The LLM generated outputs in JSON. We used top-p sampling with temperature (p=.9, t=.01), permitting decoding up to 550 new tokens.

After hyperparameter tuning the following parameters were used for *BART Single-Post* and *BART 5-Post*: learning rate = 2e−5, epochs = 10. BERTScore recall was used for best model selection. We used `BART-base`.[4]

C.2 Task A.2

For the *Llama ZS Single-Post* and *Llama ZS 5-Post* baselines we used the prompts presented in Listings 3 and 4 respectively. The LLM generated outputs in JSON. We use top-p sampling with temperature (p=.9, t=.01), permitting decoding up to 100 new tokens.

BERT Post-level was fine-tuned end-to-end with a regression head to predict well-being scores as a regression task. After hyperparameter tuning the following parameters were used for this model: learning rate = 2e−5, epochs = 15. We used `BERT-base-uncased` [5]. MSE was used for best model selection.

For the *BiLSTM 5-Post (BERT)* model, the following parameters were selected after hyperparameter tuning: num_layers= 1, dropout= .25, hidden_size= 100, learning rate = $1e{-4}$, epochs= 100. The `[CLS]` BERT representation token for each post was produced using off-the-shelf `BERT-base-uncased`. While the 5 latest posts (4 recent + current) were used for the BiLSTM sequence, only the score for the current post was predicted. In the absence of 4 posts in recent history padding was used. MSE was used for best model selection.

C.3 Task B & C

For Task B the *Llama ZS Summary* and *Llama ZS w/ Intermediate Summary* baselines we used the prompts presented in Listings 5 and 6 respectively. The corresponding prompts for Task C are presented in Listings 7 and 8 respectively. We use top-p sampling with temperature (p=.9, t=.1), permitting decoding up to 300 new tokens.

[4]https://huggingface.co/facebook/bart-base

[5]https://huggingface.co/google-bert/bert-base-uncased

```
Your goal is to identify and extract any sentence from the following social media
post which demonstrates an adaptive or maladaptive self-state of the user.

Definitions:
- **Adaptive Self-State** pertains to aspects of Affect, Behaviour, and Cognition
    towards the self or others, which is conducive to the fulfillment of basic
    desires/needs, such as relatedness, autonomy, and competence.
- **Maladaptive Self-State** pertains to aspects of Affect, Behaviour, and Cognition
     towards the self or others that hinder the fulfillment of basic desires/needs.

Instructions:
- Extract only the specific sentence spans from the post.
- Do not return these instructions or the entire post in your output.

Post Content: {post}
Output:

{
      "Adaptive": ["list of adaptive spans"],
      "Maladaptive": ["list of maladaptive spans"]
}
```

Listing 1: Prompt for Task A.1's Llama ZS Single-Post baseline.

```
Your goal is to identify and extract any sentence from the following social media
posts which demonstrates an adaptive or maladaptive self-state of the user.

Definitions:
- **Adaptive Self-State** pertains to aspects of Affect, Behaviour, and Cognition
    towards the self or others, which is conducive to the fulfillment of basic
    desires/needs, such as relatedness, autonomy, and competence.
- **Maladaptive Self-State** pertains to aspects of Affect, Behaviour, and Cognition
     towards the self or others that hinder the fulfillment of basic desires/needs.

Instructions:
- Extract only the specific sentence spans from the last post.
- Do not return these instructions or the entire post in your output.

Post Content: {posts}
Output:
{
      "Adaptive": ["list of adaptive spans"],
      "Maladaptive": ["list of maladaptive spans"]
}
```

Listing 2: Prompt for Task A.1's Llama ZS 5-Post baseline.

```
Your goal is to analyse and score the following social media post according to the
wellbeing scale below.

# Wellbeing Scale
- **10** No symptoms and superior functioning in a wide range of activities.
- **9** Absent or minimal symptoms (eg., mild anxiety before an exam), good
    functioning in all areas, interested and involved in a wide range of activities.
- **8** If symptoms are present, they are temporary and expected reactions to
    psychosocial stressors (eg., difficulty concentrating after family argument).
    Slight impairment in social, occupational or school functioning.
- **7** Mild symptoms (eg., depressed mood and mild insomnia) or some difficulty in
    social, occupational, or school functioning, but generally functioning well, has
     some meaningful interpersonal relationships.
- **6** Moderate symptoms (eg., panic attacks) or moderate difficulty in social,
    occupational or school functioning.
- **5** Serious symptoms (e.g., suicidal thoughts, severe compulsions) or serious
    impairment in social, occupational, or school functioning (eg., no friends,
    inability to keep a job).
- **4** Some impairment in reality testing or communication, or major impairment in
    multiple areas (withdrawal from social ties, inability to work, neglecting
    family, severe mood/thought impairment).
- **3** A person experiences delusions or hallucinations or serious impairment in
    communication or judgment or is unable to function in almost all areas (eg., no
    job, home, or friends).
- **2** In danger of hurting self or others (eg., suicide attempts; frequently
    violent; manic excitement) or may fail to maintain minimal personal hygiene or
    significant impairment in communication (e.g., incoherent or mute).
- **1** The person is in persistent danger of severely hurting self or others or
    persistent inability to maintain minimal personal hygiene or has attempted a
    serious suicidal act with a clear expectation of death.

Instructions:
- Only return the score for the entire post.
- Do not return these instructions or the entire post in your output.

Post Content: {post}
Output:
{
  "wellbeing scale": "score"
}
```

Listing 3: Prompt for Task A.2's Llama ZS Single-Post baseline.

```
Your goal is to analyse and score the following social media posts according to the
wellbeing scale below.

# Wellbeing Scale
- **10** No symptoms and superior functioning in a wide range of activities.
- **9** Absent or minimal symptoms (eg., mild anxiety before an exam), good
    functioning in all areas, interested and involved in a wide range of activities.
- **8** If symptoms are present, they are temporary and expected reactions to
    psychosocial stressors (eg., difficulty concentrating after family argument).
    Slight impairment in social, occupational or school functioning.
- **7** Mild symptoms (eg., depressed mood and mild insomnia) or some difficulty in
    social, occupational, or school functioning, but generally functioning well, has
     some meaningful interpersonal relationships.
- **6** Moderate symptoms (eg., panic attacks) or moderate difficulty in social,
    occupational or school functioning.
- **5** Serious symptoms (e.g., suicidal thoughts, severe compulsions) or serious
    impairment in social, occupational, or school functioning (eg., no friends,
    inability to keep a job).
- **4** Some impairment in reality testing or communication, or major impairment in
    multiple areas (withdrawal from social ties, inability to work, neglecting
    family, severe mood/thought impairment).
- **3** A person experiences delusions or hallucinations or serious impairment in
    communication or judgment or is unable to function in almost all areas (eg., no
    job, home, or friends).
- **2** In danger of hurting self or others (eg., suicide attempts; frequently
    violent; manic excitement) or may fail to maintain minimal personal hygiene or
    significant impairment in communication (e.g., incoherent or mute).
- **1** The person is in persistent danger of severely hurting self or others or
    persistent inability to maintain minimal personal hygiene or has attempted a
    serious suicidal act with a clear expectation of death.

Instructions:
- Only return the score for the entire last post.
- Do not return these instructions or the entire post in your output.

Post Content: {post}
Output:
{
  "wellbeing scale": "score"
}
```

Listing 4: Prompt for Task A.2's Llama 5-Post baseline.

```
Analyze the following social media post and identify the dominant self-state
(adaptive or maladaptive). Begin by determining which self-state is more dominant
and describe it first. For each self-state, highlight the central organizing aspect-
A (Affect), B (Behavior), C (Cognition), or D (Desire/Need)that drives the state.
Describe how this central aspect influences the other aspects, focusing on the
potential causal relationships between them. If the self-state is maladaptive,
explain how negative emotions, behaviors, or thoughts hinder psychological needs,
and if adaptive, explain how positive aspects support psychological needs. If both
adaptive and maladaptive states are present, describe each in turn. If only one
self-state is evident, focus solely on that. You must not make anything up. Keep the
description concise and only describe observations if they are fully supported by
the text.

Post Content: {post}
Summary:
```

Listing 5: Prompt for Task B's Llama ZS Summary baseline.

```
# Prompt 1 (General Summary)

Analyze the following social media post and generate the summary based on post
content. You must not make anything up. Keep the description concise and only
describe observations if they are fully supported by the text.

Post Content: {post}
Summary:


---------


# Prompt 2 (Self-State Analysis)

Analyze the following social media post summary and identify the dominant self-state
(adaptive or maladaptive). Begin by determining which self-state is more dominant
and describe it first. For each self-state, highlight the central organizing aspect-
A (Affect), B (Behavior), C (Cognition), or D (Desire/Need)that drives the state.
Describe how this central aspect influences the other aspects, focusing on the
potential causal relationships between them. If the self-state is maladaptive,
explain how negative emotions, behaviors, or thoughts hinder psychological needs,
and if adaptive, explain how positive aspects support psychological needs. If both
adaptive and maladaptive states are present, describe each in turn. If only one
self-state is evident, focus solely on that. You must not make anything up. Keep the
description concise and only describe observations if they are fully supported by
the text.

Post Summary: {post}
Final Summary:
```

Listing 6: Prompts for Task B's Llama ZS with Intermediate Summary baseline.

```
Generate a timeline-based summary analyzing the evolution of self-states across all
posts in chronological order. Emphasize the interplay between adaptive and
maladaptive self-states, focusing on temporal dynamics such as flexibility,
rigidity, improvement, and deterioration. Describe how the dominance of self-states
shifts over time, highlighting key emotional, cognitive, and behavioral changes that
contribute to these transitions. You must not make anything up. Keep the description
concise and only describe observations if they are fully supported by the text.

All Posts Content: {all_posts_concatenated}

Timeline Summary:
```

Listing 7: Prompt for Task C's Llama ZS Summary baseline.

```
# Prompt 1 (General Timeline Summary)

Generate a timeline-based summary analyzing all the posts in chronological order.
You must not make anything up. Keep the description concise and only describe
observations if they are fully supported by
the text.

All Posts Content: {all_posts_concatenated}

Timeline Summary:


---------


# Prompt 2 (Self-State Analysis over Timeline)

Generate a timeline-based summary analyzing the evolution of self-states across all
posts in chronological order. Emphasize the interplay between adaptive and
maladaptive self-states, focusing on temporal dynamics such as flexibility,
rigidity, improvement, and deterioration. Describe how the dominance of self-states
shifts over time, highlighting key emotional, cognitive, and behavioral changes that
contribute to these transitions. You must not make anything up. Keep the description
concise and only describe observations if they are fully supported by the text.

Post Summary: {timeline_summary}

Final Summary:
```

Listing 8: Prompts for Task C's Llama ZS with Intermediate Summary baseline.

A baseline for self-state identification and classification in mental health data: CLPsych 2025 Task

Laerdon Kim
Cornell University
Ithaca, NY
lyk25@cornell.edu

Abstract

We present a baseline for the CLPsych 2025 A.1 task: classifying self-states in mental health data taken from Reddit. We use few-shot learning with a 4-bit quantized Gemma 2 9B model (Gemma Team, 2024; Brown et al., 2020; Daniel Han and team, 2023) and a data preprocessing step which first identifies relevant sentences indicating self-state evidence, and then performs a binary classification to determine whether the sentence is evidence of an adaptive or maladaptive self-state. This system outperforms our other method which relies on an LLM to highlight spans of variable length independently. We attribute the performance of our model to the benefits of this sentence chunking step for two reasons: partitioning posts into sentences 1) broadly matches the granularity at which self-states were human-annotated and 2) simplifies the task for our language model to a binary classification problem. Our system placed third out of fourteen systems submitted for Task A.1, earning a test-time recall of 0.579.

1 Introduction

Evaluating the mental state of a patient takes careful analysis of textual data. Large language models (LLMs) have demonstrated strong ability to comprehend intention, perception, and cognition conferred by natural language. This extends to mental health tasks; for example, CLPsych 2024 demonstrates the ability of LLMs to accurately capture fragments of evidence justifying the classification of suicide risk based on online Reddit posting (Chim et al., 2024). We seek to provide information on how simple LLM systems respond to different forms of data preprocessing to scaffold a complex task like that of classifying self-states. We explain two primary strategies we employed to boost performance on this self-state evidence identification and classification task: a preprocessing step using LLMs to identify "important" spans which provide information about the user's psychological state, and a system using an LLM to identify specific spans which evidence an adaptive or maladaptive self-state.

2 Data

The training data provided by the CLPsych 2025 organizers consists of 30 JSON files each containing a Reddit user's timeline, totaling 343 posts overall. Each timeline entry consists of two levels of structure: a timeline level, which contains a summary (string). Within each timeline is a post level, which contains one or more posts, each with a unique post ID (string). Each post includes four fields: adaptive evidence, maladaptive evidence, summary, and well-being score. The evidence fields contains a list of strings which correspond to substrings within the post text.

Evaluation of submissions for Task A.1 was recall-oriented: system performance was calculated using an average of the maximum pairwise BERTScore (Zhang et al., 2020). For each predicted sentence, the highest BERTScore it achieves with any gold annotated sentence is taken, and these maximum scores are averaged. A secondary metric used to assess submissions was weighted recall, which recognized systems that had a cumulative number of annotated tokens more similar to the number of human-annotated tokens (Tseriotou et al., 2025).

3 Methods

For each of the methods described here, we use 4-bit quantized Gemma 2 9B, without fine-tuning. The prompts used to achieve these results are provided in Appendix A.

3.1 Baseline

To produce a baseline with our language model, we divide the post into sentences using spaCy and classify each sentence as adaptive or maladaptive.

Proceedings of the 10th Workshop on Computational Linguistics and Clinical Psychology (CLPsych 2025), pages 218–224
May 3, 2025

We provide definitions of adaptive and maladaptive self-states drawn from those provided in the task overview in our prompt.

3.2 Context

After initial runs of our baseline model, we sought to improve performance by providing the model with context (all previous sentences in the post) and using few-shot learning. We also add two examples of classification (one adaptive, one maladaptive) with a brief justification, and a detailed description of the MIND framework (Slonim, 2024).

3.3 Importance filtering

In order to increase the precision of our evidence extraction, we added a preprocessing step, using an LLM to first determine whether or not a sentence was "important" or not. We defined this as containing some reference to any one of six MIND self-state dimensions–affective, behavior-self, behavior-others, cognition-self, cognition-others, and desire (Slonim, 2024).

3.4 LLM span identification

After analyzing low-recall posts, we notice that many self-state spans are annotated at a sub-sentence level. We find that 70.2% of maladaptive and 68.7% of adaptive self-states were not sentence spans (defined as starting with a capital letter and ending with punctuation). 23.6% of adaptive spans and 19.2% of maladaptive spans are <7 words long. For example, one maladaptive span begins with a comma, explains that medical professionals are unable to help them, and does not end with punctuation. One adaptive span simply states that nobody can be perfect, a brief sentence less than seven words long.

In order to improve performance, we use the language model to identify self-states at a finer level, attempting to catch these sub-sentence spans. Our second method on Task A.1 separates the post into slightly larger contexts, and prompting our model to both identify and classify self-states independently. We use spaCy to again split sentences, and then merge them into 2-sentence groups (Honnibal et al., 2020). On each group, the model is then prompted to identify phrases at a sub-sentence level, and given the same information as the baseline in the prompt. The model returns a list of dictionaries containing substrings of the 2-sentence chunk and their predicted labels.

3.5 LLM span identification with adaptive recall boost

Low adaptive recall scores prompted us to experiment with explicitly steering the model to pay careful attention to subtle adaptive self-states embedded within sentences via prompting, noting that adaptive self-states may be hidden within seemingly maladaptive sentences, and encouraging the model to annotate as much of the chunk as possible. In addition, we modify the prompt to model this behavior in the examples, choosing larger substrings which collectively span over the entire chunk.

4 Results

4.1 Baseline results

Overall, our naive baseline method of classifying individual sentences of the post outperformed all methods except for Baseline + Context.

The addition of context provided a modest increase in adaptive self-state recall.

The addition of the importance filtering preprocessing step slightly degrades recall in exchange for improving weighted recall. On the sample training subset we use for evaluation in Table 1, the importance filtering reduces the number of spans considered from 370 to 232.

4.2 LLM span identification (LLM Span ID) results

Notably, our span identification system–which tasks our LLM with both identifying a self-state span and classifying it simultaneously–significantly increases maladaptive self-state recall by approximately 0.107 from the Baseline + Context + Importance, but also decreases adaptive self-state recall by 0.206.

Additionally, the LLM Span ID method offers weighted recall performance slightly below that achieved by importance filtering.

Our prompt steering in the LLM Span ID + Adaptive Boost row significantly improved adaptive recall by 0.107, at the expense of 0.033 points in maladaptive recall and losses in the weighted recall metric across both categories.

Table 1: Side-by-side comparison of methods, collected via a sample of five training set timelines. Overall vs. weighted metrics on separate rows. **A** indicates adaptive score; **M** indicates maladaptive score.

Method	Overall Recall Weighted Recall	Recall (A) Weighted Recall (A)	Recall (M) Weighted Recall (M)
Baseline	0.504 0.196	0.452 0.204	0.556 0.188
Baseline (Context)	**0.520** 0.201	**0.488** 0.204	0.553 0.197
Baseline (Context + Importance)	0.499 **0.244**	0.472 **0.266**	0.527 **0.221**
LLM Span ID	0.450 0.220	0.266 0.248	**0.634** 0.193
LLM Span ID (Adaptive Boost)	0.487 0.172	0.373 0.229	0.601 0.114

5 Discussion

5.1 Challenges in capturing adaptive expressions

The difficulty of capturing adaptive self-states is apparent in the results of Table 1. Adaptive recall consistently lags behind maladaptive recall. Observing human-annotated adaptive and maladaptive self-states reveals that adaptive self-states are generally much more subtle than their maladaptive counterparts. Annotations of adaptive self-states can contradict intuition; disappointment or anger can indeed signal an adaptive self-state if it reflects positive affective expression. Some adaptive self-states reference posters crying and breaking down into tears, or getting angry at others in their life like their partners. For an LM with limited context, it may be difficult to recognize such actions as adaptive signals.

For example, many adaptive self-states fall under categories of asking other posters for help, two-word interjections, or sentences describing the narrator's plans to do some common action, such as going to the store. A deep understanding of the poster's behavior overall is needed to assess whether or not the span signals adaptive thinking. In comparison, maladaptive self-states often reference self-harm, suicide, feelings of worthlessness–generally, these states contain some semantically similar terms.

In contrast, evidence reflecting maladaptive self-states is comparatively extreme, often explicitly referencing behaviors or perceptions ranging from self-harm to feelings of isolation.

While previous CLPsych tasks have demonstrated LLMs' strong performance in similar highlighting tasks with identifying evidence for suicide risk (Shing et al., 2018), the more subtle task of identifying patterns indicating psychological health is arguably more difficult (Zirikly et al., 2019; Tsakalidis et al., 2022).

5.2 Effects of context and importance filtering

Providing context and filtering irrelevant sentences improved recall and weighted recall, respectively.

We hypothesize the model can identify healthy changes in behavior more accurately with the user's context of more negative behavior. As discussed in the previous subsection, many maladaptive evidence spans are more apparent than adaptive spans–as a result, adaptive recall benefited more from the added information.

Our second addition, importance filtering, improves weighted recall, but lowers overall recall.

While many sentences in the provided training data bear no relevance to self-states, even very subtle references to users' sense of self-worth and asking for help from other members of the subreddit can qualify as adaptive self-states, for example. Subtler spans of evidence from either category are likely wrongfully discarded during this step.

Notably, the importance filtering step does not have the post's context–it inferences sentence-by-sentence only. Sentences which may seem irrelevant without context can become important with context, which may explain the degradation of recall.

Ultimately, we expect any filtering step to decrease recall to some extent.

5.3 Baseline vs. LLM Span ID

The superior performance of our original baseline system can be attributed to its more balanced recall across the two self-state categories. While the LLM Span ID strategy excels at identifying maladaptive self-states, it clearly fails to reliably identify adaptive evidence.

We propose that one potential reason for this divergence is the increased emotional intensity of maladaptive evidence spans compared to adaptive evidence spans. A manual qualitative analysis while randomly sampling pairs of adaptive and maladaptive spans corroborated our previous proposition that maladaptive evidence is generally more explicit. Span-based prompting may encourage the model to be more conservative, favoring strictly unambiguous phrases as evidence.

On the other hand, our baseline removes guesswork from this partitioning subtask: sentence-level pre-slicing simplifies the model's task to a binary classification problem at a fixed granularity. This aligns well with the sentence-level granularity of human annotation, and reduces the cognitive burden on the model compared to identifying and labeling variable-length sub-sentence spans. If a sentence is classified as adaptive, the more coarse sentence-level classification may provide a better match to evidence spans which are also at about sentence length.

6 Conclusion

We present a simple approach to the highlighting task presented in CLPsych 2025, centering LLMs in our system and using primarily prompting and data processing strategies to maximize our performance. By comparing two methods, baseline sentence classification and LLM span identification, we demonstrate how some performance variance can be elicited simply by structuring a task differently.

We hope our work provides some insight into the behavior of large language models when grappling with complex emotional dimensions.

Limitations

Our work has yet to explore a hybridized approach, potentially combining two distinct systems for adaptive and maladaptive classification. An adaptive span identification could be tuned to be more sensitive to subtler self-state dimension indications, whereas the maladaptive detection system could be designed more similarly to the high-performing vanilla LLM Span ID method. In addition, our preprocessing method of choosing 2-sentence long chunks for LLM span identification was not verified as an optimal choice–a 3-sentence sliding window may potentially be a better option, able to analyze each sentence in the context of sentences before and after it.

Ethics

This work was completed following the ACL code of ethics. Each team member completed a data usage agreement form and received the password-protected dataset securely. Data used was uploaded to the secure Cornell Information Science compute cluster, removed immediately following completion of the task. Models used for inference were entirely open-source. We have paraphrased examples from the dataset and removed our examples from the prompts in the appendix.

Acknowledgements

The author would like to thank the anonymous Reddit users whose posts are featured in this year's shared tasks, as well as the organizers of CLPsych at the Queen Mary University of London. In addition, the members of Team Zissou at Cornell University were invaluable to this submission: Dave Jung, Professor Cristian Danescu-Niculescu-Mizil, Professor Lillian Lee, Tushaar Gangavarapu, Sean Zhang, Son Tran, Vivian Nguyen, Luke Tao, Ethan Xia, and Yash Chatha.

References

Tom B. Brown, Benjamin Mann, Nick Ryder, Melanie Subbiah, Jared Kaplan, Prafulla Dhariwal, Arvind Neelakantan, Pranav Shyam, Girish Sastry, Amanda Askell, Sandhini Agarwal, Ariel Herbert-Voss, Gretchen Krueger, Tom Henighan, Rewon Child, Aditya Ramesh, Daniel M. Ziegler, Jeffrey Wu, Clemens Winter, Christopher Hesse, Mark Chen, Eric Sigler, Mateusz Litwin, Scott Gray, Benjamin Chess, Jack Clark, Christopher Berner, Sam McCandlish, Alec Radford, Ilya Sutskever, and Dario Amodei. 2020. Language models are few-shot learners.

Jenny Chim, Adam Tsakalidis, Dimitris Gkoumas, Dana Atzil-Slonim, Yaakov Ophir, Ayah Zirikly, Philip Resnik, and Maria Liakata. 2024. Overview of the CLPsych 2024 shared task: Leveraging large language models to identify evidence of suicidality risk in online posts. In *Proceedings of the 9th Workshop on Computational Linguistics and Clinical Psy-*

chology (CLPsych 2024), pages 177–190, St. Julians, Malta. Association for Computational Linguistics.

Michael Han Daniel Han and Unsloth team. 2023. Unsloth.

Gemma Team. 2024. Gemma 2: Improving open language models at a practical size.

Matthew Honnibal, Ines Montani, Sofie Van Landeghem, and Adriane Boyd. 2020. spaCy: Industrial-strength Natural Language Processing in Python.

Han-Chin Shing, Suraj Nair, Ayah Zirikly, Meir Friedenberg, Hal Daumé III, and Philip Resnik. 2018. Expert, crowdsourced, and machine assessment of suicide risk via online postings. In *Proceedings of the Fifth Workshop on Computational Linguistics and Clinical Psychology: From Keyboard to Clinic*, pages 25–36, New Orleans, LA. Association for Computational Linguistics.

Dana Atzil Slonim. 2024. Self-other dynamics (sod): A transtheoretical coding manual.

Adam Tsakalidis, Federico Nanni, Anthony Hills, Jenny Chim, Jiayu Song, and Maria Liakata. 2022. Identifying moments of change from longitudinal user text. In *Proceedings of the 60th Annual Meeting of the Association for Computational Linguistics (Volume 1: Long Papers)*, pages 4647–4660, Dublin, Ireland. Association for Computational Linguistics.

Talia Tseriotou, Jenny Chim, Ayal Klein, Aya Shamir, Guy Dvir, Iqra Ali, Cian Kennedy, Guneet Singh Kohli, Anthony Hills, Ayah Zirikly, Dana Atzil-Slonim, and Maria Liakata. 2025. Overview of the clpsych 2025 shared task: Capturing mental health dynamics from social media timelines. In *Proceedings of the 10th Workshop on Computational Linguistics and Clinical Psychology*. Association for Computational Linguistics.

Tianyi Zhang, Varsha Kishore, Felix Wu, Kilian Q. Weinberger, and Yoav Artzi. 2020. Bertscore: Evaluating text generation with bert.

Ayah Zirikly, Philip Resnik, Özlem Uzuner, and Kristy Hollingshead. 2019. CLPsych 2019 shared task: Predicting the degree of suicide risk in Reddit posts. In *Proceedings of the Sixth Workshop on Computational Linguistics and Clinical Psychology*, pages 24–33, Minneapolis, Minnesota. Association for Computational Linguistics.

A Appendix

A.1 Baseline

```
"""You are a professional psychologist.
    ↪ Given a social media post,
    ↪ classify whether or not a
    ↪ sentence demonstrates an adaptive
    ↪  or maladaptive self-state.
An adaptive self-state reflects aspects
    ↪ of the self that are flexible,
    ↪ non-ruminative, and promote well-
    ↪ being and optimal functioning.
A maladaptive self-state reflects
    ↪ internal states or perspectives
    ↪ that hinder an individual's
    ↪ ability to adapt to situations or
    ↪  cope with challenges effectively
    ↪ , potentially leading to
    ↪ emotional distress or behavioral
    ↪ problems.

Here is the sentence:
{post}"""
```

A.2 Baseline (Context)

```
"""You are a professional psychologist.
    ↪ Given a social media post,
    ↪ classify whether or not a
    ↪ sentence demonstrates an adaptive
    ↪  or maladaptive self-state.

An adaptive self-state reflects internal
    ↪  processes that are flexible,
    ↪ constructive, and promote
    ↪ emotional well-being, effective
    ↪ functioning, and psychological
    ↪ health.
A maladaptive self-state reflects
    ↪ internal processes that are rigid
    ↪ , ruminative, self-defeating, or
    ↪ harmful, and are often associated
    ↪  with emotional distress or
    ↪ impaired functioning.

To make your classification, use the
    ↪ ABCD framework for psychological
    ↪ self-states:

A. **Affect**     Type of emotional
    ↪ expression
   - Adaptive: calm, content, assertive,
       ↪ proud, justifiable pain/grief
   - Maladaptive: anxious, hopeless,
       ↪ apathetic, aggressive, ashamed
       ↪ , depressed

B. **Behavior**     Main behavioral
    ↪ tendencies
   - Toward Others (BO):
     - Adaptive: relational, autonomous
         ↪ behavior
     - Maladaptive: fight/flight
         ↪ response, controlling or
         ↪ overcontrolled behavior
   - Toward Self (BS):
     - Adaptive: self-care
     - Maladaptive: self-neglect,
         ↪ avoidance, self-harm

C. **Cognition**     Main thought
    ↪ patterns
   - Toward Others (CO):
     - Adaptive: perceiving others as
         ↪ supportive or related
     - Maladaptive: perceiving others as
         ↪  detached, overattached, or
```

```
        ↪ autonomy-blocking
    - Toward Self (CS):
      - Adaptive: self-compassion and
        ↪ acceptance
      - Maladaptive: self-criticism

D. **Desire**     Expressed needs, goals
    ↪ , intentions, or fears
    - Adaptive: desire for autonomy,
       ↪ relatedness, self-esteem, care
    - Maladaptive: fear that these needs
       ↪ w o n t  be met

Here are a couple of examples:
"--removed--"
This is maladaptive. It shows a
    ↪ bluntedness and apathic affective
    ↪  state.

"--removed--"
This is adaptive. The crying is not a
    ↪ sign of maladaptive self-state,
    ↪ rather it is a healthy sadness.

You will be shown:
1. The context of the post so far
2. The current sentence to classify

If the sentence clearly demonstrates one
    ↪  or more **maladaptive or
    ↪ adaptive self-state(s)** based on
    ↪  this framework, classify it
    ↪ accordingly.

Here is the post so far:
{context}

Here is the current sentence:
{sentence}

"""
```

A.3 Importance filtering

```
"""You are a professional psychologist.
    ↪ Given a social media post, decide
    ↪  whether or not the sentence is
    ↪ critically important. A sentence
    ↪ is critical if it evidences one
    ↪ of six things: it 1) expresses a
    ↪ distinct emotion (A), 2)
    ↪ expresses a person's interactions
    ↪  with another (B-O), 3) expresses
    ↪  a person's interactions with
    ↪ themselves (B-S), 4) expresses a
    ↪ person's perceptions of another (
    ↪ B-O) 5) expresses a person's
    ↪ perceptions of themselves, (C-O)
    ↪ or 6) expresses an explicit
    ↪ desire, need, intention, fear or
    ↪ expectation. (D) Not every
    ↪ sentence is important. If the
    ↪ sentence is critical, return True
    ↪ . If not, return False.

Now, it's your turn.
Here is how the post starts:
{post}"""
```

A.4 LLM Span ID

```
"""You are a professional psychologist.
    ↪ Your task is to analyze the
    ↪ following social media post and
    ↪ identify any phrases or subspans
    ↪ that reflect an **adaptive** or
    ↪ **maladaptive** self-state, even
    ↪ if they are mixed within the same
    ↪  sentence or paragraph.

An adaptive self-state reflects internal
    ↪  processes that are flexible,
    ↪ constructive, and promote
    ↪ emotional well-being, effective
    ↪ functioning, and psychological
    ↪ health. Pay close attention to
    ↪ subtle adaptive self-states
    ↪ within sentences.
A maladaptive self-state reflects
    ↪ internal processes that are rigid
    ↪ , ruminative, self-defeating, or
    ↪ harmful, and are often associated
    ↪  with emotional distress or
    ↪ impaired functioning.

To make your classification, use the
    ↪ ABCD framework for psychological
    ↪ self-states:

A. **Affect**     Type of emotional
    ↪ expression
    - Adaptive: calm, content, assertive,
       ↪ proud, justifiable pain/grief
    - Maladaptive: anxious, hopeless,
       ↪ apathetic, aggressive, ashamed
       ↪ , depressed

B. **Behavior**     Main behavioral
    ↪ tendencies
    - Toward Others (BO):
      - Adaptive: relational, autonomous
        ↪ behavior
      - Maladaptive: fight/flight
        ↪ response, controlling or
        ↪ overcontrolled behavior
    - Toward Self (BS):
      - Adaptive: self-care
      - Maladaptive: self-neglect,
        ↪ avoidance, self-harm

C. **Cognition**     Main thought
    ↪ patterns
    - Toward Others (CO):
      - Adaptive: perceiving others as
        ↪ supportive or related
      - Maladaptive: perceiving others as
        ↪  detached, overattached, or
        ↪ autonomy-blocking
    - Toward Self (CS):
      - Adaptive: self-compassion and
        ↪ acceptance
      - Maladaptive: self-criticism

D. **Desire**     Expressed needs, goals
    ↪ , intentions, or fears
    - Adaptive: desire for autonomy,
       ↪ relatedness, self-esteem, care
    - Maladaptive: fear that these needs
       ↪ w o n t  be met
```

```
Here is an example:
Sentences: --removed--
Predictions: [("adaptive", "--removed
    ↪ --"), ("maladaptive", "--removed
    ↪ --")]

These annotations are great. Firstly,
    ↪ this is because "--removed--" is
    ↪ rightly marked as adaptive    it
    ↪  demonstrates assertiveness, self
    ↪ -worth, and self-affirmation,
    ↪ aligning with adaptive affect and
    ↪  cognition of the self (CS). And
    ↪ "--removed--" is correctly
    ↪ labeled maladaptive    it
    ↪ reflects a sense of emotional
    ↪ abandonment and unmet relational
    ↪ needs, which maps onto
    ↪ maladaptive cognition of the
    ↪ other (CO), perceiving this ex as
    ↪  underattached.

Now, it's your turn.
Here is the context of the post so far:
{context}

Here is the current chunk of the post:
{chunk}
"""
```

A.5 LLM Span ID (Adaptive Boost)

```
"""You are a professional psychologist.
    ↪ Your task is to analyze the
    ↪ following social media post and
    ↪ identify any phrases or subspans
    ↪ that reflect an **adaptive** or
    ↪ **maladaptive** self-state, even
    ↪ if they are mixed within the same
    ↪  sentence or paragraph.

An adaptive self-state reflects internal
    ↪  processes that are flexible,
    ↪ constructive, and promote
    ↪ emotional well-being, effective
    ↪ functioning, and psychological
    ↪ health. Pay close attention to
    ↪ subtle adaptive self-states
    ↪ within sentences.
A maladaptive self-state reflects
    ↪ internal processes that are rigid
    ↪ , ruminative, self-defeating, or
    ↪ harmful, and are often associated
    ↪  with emotional distress or
    ↪ impaired functioning.

To make your classification, use the
    ↪ ABCD framework for psychological
    ↪ self-states:

A. **Affect**    Type of emotional
    ↪ expression
   - Adaptive: calm, content, assertive,
       ↪ proud, justifiable pain/grief
   - Maladaptive: anxious, hopeless,
       ↪ apathetic, aggressive, ashamed
       ↪ , depressed

B. **Behavior**    Main behavioral
    ↪ tendencies
   - Toward Others (BO):
     - Adaptive: relational, autonomous
         ↪ behavior
     - Maladaptive: fight/flight
         ↪ response, controlling or
         ↪ overcontrolled behavior
   - Toward Self (BS):
     - Adaptive: self-care
     - Maladaptive: self-neglect,
         ↪ avoidance, self-harm

C. **Cognition**    Main thought
    ↪ patterns
   - Toward Others (CO):
     - Adaptive: perceiving others as
         ↪ supportive or related
     - Maladaptive: perceiving others as
         ↪  detached, overattached, or
         ↪ autonomy-blocking
   - Toward Self (CS):
     - Adaptive: self-compassion and
         ↪ acceptance
     - Maladaptive: self-criticism

D. **Desire**    Expressed needs, goals
    ↪ , intentions, or fears
   - Adaptive: desire for autonomy,
       ↪ relatedness, self-esteem, care
   - Maladaptive: fear that these needs
       ↪  w o n t  be met

Here is an example:
Sentences: --removed--
Predictions: [("adaptive", "--removed
    ↪ --"), ("maladaptive", "--removed
    ↪ --")]

These annotations are great. Firstly,
    ↪ this is because "--removed--" is
    ↪ rightly marked as adaptive    it
    ↪  demonstrates assertiveness, self
    ↪ -worth, and self-affirmation,
    ↪ aligning with adaptive affect and
    ↪  cognition of the self (CS). And
    ↪ "--removed--" is correctly
    ↪ labeled maladaptive...

Now, it's your turn.
Your output should list any sentences
    ↪ that reflect either state.
    ↪ Sometimes, you will need to
    ↪ highlight a phrase inside a
    ↪  s e n t e n c e self-states can be
    ↪ subtle. You may return **multiple
    ↪ ** adaptive or maladaptive spans
    ↪ per chunk.
If a sentence seems neutral, mark it as
    ↪ adaptive. Try to annotate as much
    ↪  as  p o s s i b l e you should shoot
    ↪ for the highest recall possible.

Here is the context of the post so far:
{context}

Here is the current chunk of the post:
{chunk}
"""
```

Capturing the Dynamics of Mental Well-Being: Adaptive and Maladaptive States in Social Media

Anastasia Sandu[1], **Teodor Mihailescu**[1], **Ana Sabina Uban**[1,2], **Ana-Maria Bucur**[3,4]
[1]Faculty of Mathematics and Computer Science, [2]HLT Research Center,
[3]Interdisciplinary School of Doctoral Studies, University of Bucharest
[4]PRHLT Research Center, Universitat Politècnica de València
anastasiasandu777@gmail.com, teomihailescu@yahoo.com
auban@fmi.unibuc.ro, ana-maria.bucur@drd.unibuc.ro

Abstract

This paper describes the contributions of the BLUE team in the CLPsych 2025 Shared Task on Capturing Mental Health Dynamics from Social Media Timelines. We participate in all tasks with three submissions, for which we use two sets of approaches: an unsupervised approach using prompting of various large language models (LLM) with no fine-tuning for this task or domain, and a supervised approach based on several lightweight machine learning models trained to classify sentences for evidence extraction, based on an augmented training dataset sourced from public psychological questionnaires. We obtain the best results for summarization Tasks B and C in terms of consistency, and the best F1 score in Task A.2.

1 Introduction

The assessment of mental health through digital technologies is an increasingly important topic in both psychology and natural language processing. Digital mental health tools can support individuals in need and facilitate remote care, especially as the prevalence of mental disorders continues to rise while access to mental health services remains limited.[1] Most approaches for mental health assessment using online data are focused on performing binary classification for depression (Yates et al., 2017; Liu et al., 2023) or suicide risk (Coppersmith et al., 2018; Ramírez-Cifuentes et al., 2020; Lee et al., 2020), with few previous works focused on explainable mental health assessment (Wang et al., 2024; Bao et al., 2024; Uban et al., 2022). The CLPsych Workshop was the first to address the challenge of extracting evidence from social media data by proposing the task of highlighting evidence for mental disorders (Chim et al., 2024). This year's task builds on that foundation, shifting the focus to mental states, specifically adaptive and maladaptive states.

The Shared Tasks from CLPsych 2022 (Tsakalidis et al., 2022) and 2024 (Chim et al., 2024) focused on analyzing longitudinal user posts. The 2022 task focused on capturing moments of change from the social media timeline of a user, while the 2024 task aimed to extract evidence regarding the suicide risk of users. Similar to the shared task from this year, the 2024 edition included a summarization component, which required participants to provide textual summaries of the mental health dynamics throughout the entire timeline of the user. While the extraction of evidence from social media data is a relatively new task, it was previously modeled as a binary classification task for maladaptive states (Gollapalli et al., 2023, 2024).

In this paper, we present the contributions of the BLUE team to the CLPsych 2025 Shared Task: Capturing Mental Health Dynamics from Social Media Timelines (Tseriotou et al., 2025). Our approach relies on both classical machine learning algorithms and LLMs, merging established classification methods with recent advancements in the field. Our team achieved good results, scoring the highest in summarization Tasks B and C. Moreover, for highlighting evidence of adaptive and maladaptive states, as well as inferring the well-being score (Tasks A.1 and A.2), our team ranks fifth.

2 Data and Tasks

The data provided for this task consists of Reddit posts annotated by domain experts for self-states following the MIND Framework (Slonim, 2024). This dataset aligns with prior work in computational linguistics and clinical psychology, particularly studies on suicide risk assessment (Shing

[1]https://www.who.int/news/item/17-06-2022-who-highlights-urgent-need-to-transform-mental-health-and-mental-health-care

Proceedings of the 10th Workshop on Computational Linguistics and Clinical Psychology (CLPsych 2025), pages 225–234

et al., 2018; Zirikly et al., 2019) and longitudinal mental health analysis (Tsakalidis et al., 2022). The dataset allows for the evaluation of self-states and well-being, facilitating our contributions to mental health analysis.

The tasks for CLPsych 2025 are as follows:

Task A.1 Identification of adaptive and maladaptive self-states within each post.

Task A.2 Prediction of well-being score for each post, ranging from 1 (low well-being) to 10 (high well-being).

Task B Generation of a summary describing the interplay between adaptive and maladaptive self-states within a post.

Task C Generation of a timeline-level summary encapsulating the evolution of self-states across multiple posts from the same user.

3 Method

3.1 Machine learning approach

We used a machine learning method for Task A.1 to perform sentence classification and identify adaptive and maladaptive states in text using text embeddings and supervised classifiers. For extracting text embeddings, we use TF-IDF (Sparck Jones, 1972), BERT (Devlin et al., 2019) and SentenceTransformer (Reimers and Gurevych, 2019). XGBoost (Chen and Guestrin, 2016) was used for identifying maladaptive states, while Logistic Regression (Cox, 1958) was used for detecting adaptive ones. The hyperparameters used for the two classifiers are presented in Appendix A.4 and A.5. For feature extraction, we applied TF-IDF using the default configuration (Appendix A.6).

These models are trained on the labeled data provided for this task. To augment the dataset used for training, we include external psychological resources for better generalization. The external data sources used for maladaptive states include the items from the Young Schema Questionnaire (YSQ) (Young, 2003b), the Young Schema Questionnaire-Revised (YSQ-R) (Young, 2003a), and annotated texts from Liu et al. (2022). To enhance the dataset for adaptive states, we expanded it by including items from the Young Positive Schema Questionnaire (YPSQ) (Louis et al., 2018). Since the data for adaptive states was limited, we supplemented these samples by generating additional samples[2] using GPT-o1 (OpenAI, 2024b) and GPT-4o (OpenAI, 2024a), based on the YPSQ items.

Classification is performed at the sentence level using majority voting across multiple predictions, with Gaussian noise added to reduce overfitting.

3.2 LLMs approach

While the machine learning approach was applied only for Task A.1, LLMs were used for all tasks from CLPsych 2025. For the LLM-based approach, we implement a structured processing pipeline that uses multiple models. This pipeline consists of the following steps:

Post-Level Processing: Each post undergoes the following steps:

- **Evidence Extraction:** The model identifies adaptive and maladaptive self-state evidence in the post text.
- **Well-Being Prediction:** A well-being score is assigned to the post on a scale of 1 to 10, based on predefined psychological criteria.
- **Post Summary Generation:** The model summarizes the interplay between adaptive and maladaptive self-states within the post.

Timeline-Level Processing: After processing individual posts, the full timeline is analyzed:

- **Aggregation of Posts:** All posts from a single user are compiled into a coherent timeline.
- **Timeline Summary Generation:** The model generates a high-level summary describing the evolution of self-states over time.

This structured approach allows for a detailed analysis of self-states across individual posts and entire timelines, providing insights into psychological well-being and behavioral patterns.

In this approach, we rely on LLM prompting to solve the tasks. We use models in various families: Gemma 2 9B (Team et al., 2024), Mistral 7B (AI, 2024b), Llama 2 7B (Touvron et al., 2023), Llama 3.1 8B (Grattafiori et al., 2024), and Llama 3.2 3B (AI, 2024a). These models were selected based on their ability to process complex psychological

[2]We make the generated data available on github, together with the code used for the submissions: https://github.com/Teo1230/clpsych25-task

text while balancing computational efficiency and accuracy.

We experiment with three different types of prompts in our work. The first is the *default prompt*, which provides instructions for extracting evidence, predicting well-being scores, and summarizing information without including additional context or definitions of concepts. The second prompt, referred to as the *expert prompt*, guides the model to evaluate Reddit posts as if it were a psychology expert. This prompt includes definitions of adaptive and maladaptive states and asks the model to generate summaries based on key emotional, cognitive, and behavioral patterns. The third prompt, *structured summarization prompt*, focuses on generating summaries using the structure proposed by the CLPsych Shared Task. This emphasizes determining the dominant self-state and describing the interplay between adaptive and maladaptive self-states. The complete prompts can be found in Appendix A.

To ensure a balance between determinism and variability in generation, we experimented with different temperature settings across models. The temperature parameter controls the randomness of the model's responses: lower values make outputs more deterministic, while higher values introduce more diversity. For our experiments, we used a temperature of 0.7 for the Gemma 2 9B, Llama 3.1 8B, Llama 3.2 3B, and a temperature of 0.5 for Mistral 7B and Llama 2 7B. These settings were selected based on preliminary trials to balance consistency and adaptability in handling complex psychological discourse. All models used a top-k sampling of 40 and a top-p of 0.9. Our methodology relies on prompting rather than fine-tuning for several reasons:

Domain-Specific Adaptability Fine-tuning requires large domain-specific datasets and extensive computational resources, which may not generalize well to unseen cases. Prompting allows leveraging LLMs' broad pretraining without retraining.

Flexibility in Task Definitions By designing different prompts, we can easily modify task instructions without retraining models. This is crucial for a field like mental health, where criteria may evolve.

Clinical Interpretability Using well-defined prompts provides clearer interpretability compared to fine-tuned black-box models, making the approach more suitable for clinical applications where transparency is essential.

While fine-tuning could improve model specialization, it introduces several challenges, such as the need for large, annotated datasets specific to adaptive or maladaptive self-states, increased computational costs for training and inference, and the potential loss of generalizability across different domains. Given these limitations, our focus remains on optimizing prompting strategies while leveraging pre-trained LLMs.

3.3 Evaluation Metrics

Our system is evaluated using established metrics from prior CLPsych shared tasks (Zirikly et al., 2019; Tsakalidis et al., 2022):

Task A.1: Recall and Weighted Recall for adaptive and maladaptive self-state identification. The Recall metric is represented by the maximum recall-oriented BERTScore (Zhang et al., 2019).

Task A.2: Mean Squared Error (MSE) across different well-being categories and macro F1-score.

Task B: Consistency, maximum contradiction, and maximum entailment scores.

Task C: Mean consistency and maximum contradiction.

3.4 Submissions

In this section, we present our submissions for the CLPsych 2025 Shared Task (Table 1).

Submission 1 This submission consists of multiple LLMs within a common processing approach to analyze user timelines in Reddit posts, extracting psychological insights through structured tasks. **Gemma 2** is used for evidence extraction and well-being scoring (Tasks A.1 and A.2) using default prompts (Appendix A.1). Meanwhile, **LLaMA 3.2** generates summaries for the post (Task B) and timeline levels (Task C) using the default prompts.

Submission 2 This submission integrates both machine learning classifiers and LLMs. For evidence extraction (Task A.1), we employ supervised classifiers trained on labeled data. **XGBoost** is utilized to identify maladaptive states, while **Logistic Regression** is used for classifying adaptive states. TF-IDF is used for text representation. **Mistral** handles well-being scoring (Task A.2) with the default prompt. Summarization at both the post (Task B) and timeline levels (Task C) is performed using

Team BLUE	Task A.1 - Adaptive	Task A.1 - Maladaptive	Task A.2	Task B	Task C
Submission 1	Gemma 2_D	Gemma 2_D	Gemma 2_D	LLaMA 3.2_D	LLaMA 3.2_D
Submission 2	TF-IDF & LR	TF-IDF & XGB	Mistral_D	LLaMA 3.2_{SS}	LLaMA 3.2_{SS}
Submission 3	Gemma 2_E	Gemma 2_E	Gemma 2_E	LLaMA 3.1_D	LLaMA 3.1_D
Submission 4	TF-IDF & LR	BERT & XGBoost	$\text{Mistral}_{E,L}$	Mistral_E	Mistral_E
Submission 5	MiniLM & LR	MiniLM & XGBoost	LLaMA 3.1_D	LLaMA 2_D	LLaMA 2_D
Submission 6	Gemma $2_{D,L}$	Gemma $2_{D,L}$	LLaMA $3.2_{E,L}$	Gemma 2_D	Gemma $2_{D,L}$

Table 1: Approaches used for our team's submissions across all tasks. D denotes *default prompt*, E denotes *expert prompt*, SS - *structured summarization prompt*, and L denotes that LangChain was used for the prompt template.

Team BLUE	Task A.1: Recall	Task A.2: MSE	Task B: Mean Consistency	Task C: Mean Consistency
Submission 1	0.555	2.390	**0.910**	**0.946**
Submission 2	0.539	2.900	0.328	0.854
Submission 3	0.538	**2.260**	0.393	0.911
Submission 4	0.444	3.164	0.908	0.913
Submission 5	0.422	3.025	0.918	0.897
Submission 6	**0.569**	3.842	0.890	0.900
Ranking Team	5	5	1	1

Table 2: Final evaluation results for our team's submissions across all tasks.

LLaMA 3.2, leveraging the structured summarization prompt (Appendix A.3).

Submission 3 In this method, we again use multiple LLMs to analyze user timelines in Reddit posts. **Gemma 2** is used for evidence extraction and well-being scoring (Tasks A.1 and A.2) with the expert prompts (Appendix A.2). In addition, **LLaMA 3.1** generates summaries for the post (Task B) and timeline levels (Task C) using default prompts.

Submission 4 This submission combines machine learning and LLMs. For evidence extraction (Task A.1), we use **BERT** (bert-base-uncased) (Devlin et al., 2019) to generate sentence embeddings, which are then classified by **XGBoost** for detecting maladaptive states. Adaptive states are identified using **TF-IDF** features with **Logistic Regression**. For well-being scoring (Task A.2), we use **Mistral** with the expert prompt and LangChain (Chase, 2022). The same model and prompt are used for summarizing posts (Task B) and timelines (Task C).

Submission 5 For evidence extraction (Task A.1), we use the same method as in Submission 4, but instead of generating sentence embeddings with BERT, we experiment with **all-MiniLM-L6-v2**[3]. For well-being scoring (Task A.2), we use **LLaMA 3.1** with the default prompt, while **LLaMA 2** handles summarization (Tasks B and C), also employing the default prompt.

Submission 6 For evidence extraction (Task A.1), we use **Gemma 2** with the default prompt, implemented with LangChain. Well-being scoring (Task A.2) is handled by **LLaMA 3.2** with the expert prompt, also using LangChain. Summarization at both the post (Task B) and timeline levels (Task C) is performed with **Gemma 2**, using the default prompt for Task B and the default prompt with LangChain for Task C.

4 Results

Our system demonstrates strong performance across the tasks, particularly in summarization (Tasks B and C). The results of our submissions are presented in Table 2. The first three submissions in Table 2 are the official submissions for the CLPsych 2025 Shared Task, while the remaining three submissions are additional runs that were not submitted officially.

We performed best in summarization (Tasks B and C), achieving top consistency scores. For well-being scoring (Task A.2), Submission 3 achieved the smallest MSE of 2.260, placing us at rank 5. Submission 6 reached the highest recall on Task A.1 but had a higher MSE, indicating that while our methods are strong at summarizing timelines, they still struggle with the finer details of post-level scoring—especially when detecting adaptive signals.

We found it more challenging to extract adaptive

[3]https://huggingface.co/sentence-transformers/all-MiniLM-L6-v2

	Recall			Weighted Recall		
Team BLUE	**Overall**	**Adaptive**	**Maladaptive**	**Overall**	**Adaptive**	**Maladaptive**
Submission 1	0.555	**0.472**	0.639	0.392	0.400	0.384
Submission 2	0.539	0.298	**0.779**	0.239	0.285	0.192
Submission 3	0.538	0.414	0.662	0.389	0.351	**0.428**
Submission 4	0.444	0.298	0.589	0.326	0.286	0.365
Submission 5	0.422	0.303	0.540	0.334	0.291	0.376
Submission 6	**0.569**	0.457	0.681	**0.393**	**0.403**	0.382

Table 3: Evaluation results for Task A.1 for adaptive and maladaptive self-states.

evidence than maladaptive evidence, mostly because people tend to describe distress with clearer cues, while adaptive statements are often subtle and less standardized. Another factor is data imbalance: our original training dataset leaned heavily toward maladaptive examples, as prior work in mental health analysis has traditionally focused on distress or at-risk behaviors. To address this, we added data from the Young Positive Schema Questionnaire (YPSQ) (Louis et al., 2018) and generated more adaptive statements with GPT-o1 and GPT-4o. Although this helped balance the data and improve recall, identifying adaptive language is still challenging, which is reflected in the results presented in Table 3.

5 Conclusions and Future Work

The CLPsych 2025 Shared Task on Capturing Mental Health Dynamics from Social Media Timelines proposed a novel problem that has not been approached computationally in the past, related to adaptive and maladaptive states, in a variety of different tasks. Our team participated with three submissions using two sets of approaches: one based on prompting various LLMs for all tasks and a supervised approach based on classical machine learning models trained on the provided training data as well as external data, including expert-generated data (from relevant psychological questionnaires) and AI-generated. While we experiment and include in our submissions different kinds of LLM models and prompts, as well as different machine learning models for the second approach, our methods are relatively cheap and accessible, and our good results across tasks confirm that relatively simple approaches can be effective for identifying adaptive and maladaptive states in social media texts. In-context learning was minimal, with the only external knowledge provided to the models including a description of the scoring scheme for some of the prompts. The supervised approaches include classical machine learning algorithms (which performed better in this setting than pretrained transformers according to our preliminary experiments). All LLM models are general domain, with only Llama2 7B, Llama3.1 8B, Llama3.2 3B, Mistral 7B, Gemma 2 9B parameters, run using modest infrastructure. Using these relatively accessible approaches, we obtain competitive results compared to the other participants, with the best mean consistency score out of all teams in both summarization tasks (Tasks B and C), the 5th MSE score for well-being score (Task A.2) and the 5th Recall for Task A.1 related to evidence highlighting.

Future research should look at more diverse datasets to make sure our approaches work for different populations. Also, exploring specialized or fine-tuned LLMs that incorporate domain knowledge from psychology or clinical practice could further enhance both performance and interpretability in mental health tasks. Another direction is investigating more advanced or ensemble-based machine learning methods to improve the detection and classification of adaptive and maladaptive states.

Limitations

The data for this task primarily consists of Reddit posts, which may not accurately reflect the broader population. Social media often reveals biases related to factors such as gender and socioeconomic status, meaning our findings might not be applicable to all groups, particularly beyond American males (Gottfried, 2024). Furthermore, the collected posts may contain incomplete or misleading information, as users do not always provide factual or comprehensive details online. While the performance of our LLM-based solutions was limited by our infrastructure, our good results show that reasonable performance is achievable for this task, even with relatively small, generic LLMs.

Ethical Considerations

The data we used was obtained through a strict data agreement to ensure we adhered to ethical guidelines for handling sensitive information. We prioritize individual privacy and confidentiality by conducting all analyses locally, without using any external APIs that could compromise data security. We follow ethical research guidelines from Benton et al. (2017) for the sensitive data provided for this shared task. We recognize the potential impact of our findings on individuals facing mental health challenges. It is crucial to approach these analyses with sensitivity and to consider the broader societal implications of our work. Our goal is to make a positive contribution to mental health research while upholding ethical integrity.

Acknowledgements

This research was partially supported by the project "Romanian Hub for Artificial Intelligence - HRIA", Smart Growth, Digitization and Financial Instruments Program, 2021-2027, MySMIS no. 334906.

References

Meta AI. 2024a. Llama 3.2: Open foundation language models. https://github.com/meta-llama/llama-models.

Mistral AI. 2024b. Mistral models: High-performance open-weight language models.

Eliseo Bao, Anxo Pérez, and Javier Parapar. 2024. Explainable depression symptom detection in social media. *Health Information Science and Systems*, 12(1):47.

Adrian Benton, Glen Coppersmith, and Mark Dredze. 2017. Ethical research protocols for social media health research. In *Proceedings of the first ACL workshop on ethics in natural language processing*, pages 94–102.

Harrison Chase. 2022. LangChain. Software. Released on 2022-10-17.

Tianqi Chen and Carlos Guestrin. 2016. Xgboost: A scalable tree boosting system. In *Proceedings of the 22nd ACM SIGKDD International Conference on Knowledge Discovery and Data Mining (KDD)*, pages 785–794. ACM.

Jenny Chim, Adam Tsakalidis, Dimitris Gkoumas, Dana Atzil-Slonim, Yaakov Ophir, Ayah Zirikly, Philip Resnik, and Maria Liakata. 2024. Overview of the clpsych 2024 shared task: Leveraging large language models to identify evidence of suicidality risk in online posts. In *Proceedings of the 9th Workshop on Computational Linguistics and Clinical Psychology (CLPsych 2024)*, pages 177–190.

Glen Coppersmith, Ryan Leary, Patrick Crutchley, and Alex Fine. 2018. Natural language processing of social media as screening for suicide risk. *Biomedical informatics insights*, 10:1178222618792860.

D. R. Cox. 1958. The regression analysis of binary sequences. *Journal of the Royal Statistical Society: Series B (Methodological)*, 20(2):215–242.

Jacob Devlin, Ming-Wei Chang, Kenton Lee, and Kristina Toutanova. 2019. Bert: Pre-training of deep bidirectional transformers for language understanding. *Proceedings of NAACL-HLT*, pages 4171–4186.

Sujatha Gollapalli, Beng Ang, and See Kiong Ng. 2023. Identifying early maladaptive schemas from mental health question texts. In *Findings of the Association for Computational Linguistics: EMNLP 2023*, pages 11832–11843.

Sujatha Das Gollapalli, Beng Heng Ang, Mingzhe Du, and See-Kiong Ng. 2024. Counseling responses for mental health forum questions with early maladaptive schema prediction. In *ECAI 2024*, pages 2556–2563. IOS Press.

Jeffrey Gottfried. 2024. Americans' social media use. *Pew Research Center*, 31.

Aaron Grattafiori, Abhimanyu Dubey, Abhinav Jauhri, Abhinav Pandey, Abhishek Kadian, Ahmad Al-Dahle, Aiesha Letman, Akhil Mathur, Alan Schelten, Alex Vaughan, et al. 2024. The llama 3 herd of models. *arXiv preprint arXiv:2407.21783*.

Daeun Lee, Soyoung Park, Jiwon Kang, Daejin Choi, and Jinyoung Han. 2020. Cross-lingual suicidal-oriented word embedding toward suicide prevention. In *Findings of the Association for Computational Linguistics: EMNLP 2020*, pages 2208–2217.

X. Liu, Y. Zhang, and C. Tan. 2022. Identifying early maladaptive schemas from mental health question texts. *Proceedings of EMNLP*.

Yujian Liu, Laura Biester, and Rada Mihalcea. 2023. Improving mental health classifier generalization with pre-diagnosis data. In *Proceedings of the International AAAI Conference on Web and Social Media*, volume 17, pages 566–577.

J. P. Louis, A. M. Wood, G. Lockwood, M.-H. R. Ho, and E. Ferguson. 2018. Positive clinical psychology and schema therapy (st): The development of the young positive schema questionnaire (ypsq) to complement the young schema questionnaire 3 short form (ysq-s3). *Psychological Assessment*, 30(9):1199–1213.

OpenAI. 2024a. Gpt-4o system card.

OpenAI. 2024b. Openai o1 system card.

Diana Ramírez-Cifuentes, Ana Freire, Ricardo Baeza-Yates, Joaquim Puntí, Pilar Medina-Bravo, Diego Alejandro Velazquez, Josep Maria Gonfaus, and Jordi Gonzàlez. 2020. Detection of suicidal ideation on social media: multimodal, relational, and behavioral analysis. *Journal of medical internet research*, 22(7):e17758.

Nils Reimers and Iryna Gurevych. 2019. Sentence-bert: Sentence embeddings using siamese bert-networks. *Proceedings of EMNLP*, page 3982–3992.

Han-Chin Shing, Suraj Nair, Ayah Zirikly, Meir Friedenberg, Hal Daumé III, and Philip Resnik. 2018. Expert, crowdsourced, and machine assessment of suicide risk via online postings. In *Proceedings of the fifth workshop on computational linguistics and clinical psychology: from keyboard to clinic*, pages 25–36.

Dana Atzil Slonim. 2024. Self-other dynamics (sod): A transtheoretical coding manual.

Karen Sparck Jones. 1972. A statistical interpretation of term specificity and its application in retrieval. *Journal of Documentation*, 28(1):11–21.

Gemma Team, Morgane Riviere, Shreya Pathak, Pier Giuseppe Sessa, Cassidy Hardin, Surya Bhupatiraju, Léonard Hussenot, Thomas Mesnard, Bobak Shahriari, Alexandre Ramé, et al. 2024. Gemma 2: Improving open language models at a practical size. *arXiv preprint arXiv:2408.00118*.

Hugo Touvron, Louis Martin, Kevin Stone, Max Albert, Amjad Almahairi, Yasmine Babaei, Nikolay Bashlykov, Siddhartha Batra, Pallavi Bhargava, Shruti Bhosale, et al. 2023. Llama 2: Open foundation and fine-tuned chat models. *arXiv preprint arXiv:2307.09288*.

Adam Tsakalidis, Jenny Chim, Iman Munire Bilal, Ayah Zirikly, Dana Atzil-Slonim, Federico Nanni, Philip Resnik, Manas Gaur, Kaushik Roy, Becky Inkster, et al. 2022. Overview of the clpsych 2022 shared task: Capturing moments of change in longitudinal user posts. In *Proceedings of the Eighth Workshop on Computational Linguistics and Clinical Psychology*, pages 184–198.

Talia Tseriotou, Jenny Chim, Ayal Klein, Aya Shamir, Guy Dvir, Iqra Ali, Cian Kennedy, Guneet Singh Kohli, Anthony Hills, Ayah Zirikly, Dana Atzil-Slonim, and Maria Liakata. 2025. Overview of the clpsych 2025 shared task: Capturing mental health dynamics from social media timelines. In *Proceedings of the 10th Workshop on Computational Linguistics and Clinical Psychology*. Association for Computational Linguistics.

Ana-Sabina Uban, Berta Chulvi, and Paolo Rosso. 2022. Explainability of depression detection on social media: From deep learning models to psychological interpretations and multimodality. In *Early Detection of Mental Health Disorders by Social Media Monitoring: The First Five Years of the eRisk Project*, pages 289–320. Springer.

Yuxi Wang, Diana Inkpen, and Prasadith Kirinde Gamaarachchige. 2024. Explainable depression detection using large language models on social media data. In *Proceedings of the 9th Workshop on Computational Linguistics and Clinical Psychology (CLPsych 2024)*, pages 108–126.

Andrew Yates, Arman Cohan, and Nazli Goharian. 2017. Depression and self-harm risk assessment in online forums. In *Proceedings of the 2017 Conference on Empirical Methods in Natural Language Processing*, pages 2968–2978.

J. E. Young. 2003a. Young schema questionnaire - revised (ysq-r).

J. E. Young. 2003b. Young schema questionnaire (ysq).

Tianyi Zhang, Varsha Kishore, Felix Wu, Kilian Q Weinberger, and Yoav Artzi. 2019. Bertscore: Evaluating text generation with bert. In *International Conference on Learning Representations*.

Ayah Zirikly, Philip Resnik, Ozlem Uzuner, and Kristy Hollingshead. 2019. Clpsych 2019 shared task: Predicting the degree of suicide risk in reddit posts. In *Proceedings of the sixth workshop on computational linguistics and clinical psychology*, pages 24–33.

A Appendix

In this appendix, we provide the prompts used for analyzing self-states in Reddit posts. These prompts are designed to facilitate the extraction of adaptive and maladaptive self-state evidence, predict well-being scores, and generate summaries. The following prompts are used to guide the models in processing Reddit posts:

A.1 Default Prompts

- **Extract Evidence:**

    Given the following Reddit post, identify evidence of adaptive and maladaptive self-states. Extract text spans as JSON lists.
    Post: "{post_text}"
    Response format: {
    "adaptive_evidence": [<adaptive text spans>],
    "maladaptive_evidence": [<maladaptive text spans>]
    }

- **Predict Well-Being:**

    Given the following Reddit post, assign a well-being score from 1 (low) to 10 (high).

- **1**: The person is in persistent danger of severely hurting self or others...
- **2**: In danger of hurting self or others...
- **3**: A person experiences delusions or hallucinations...
- **4**: Some impairment in reality testing or communication...
- **5**: Serious symptoms (e.g., suicidal thoughts)...
- **6**: Moderate symptoms (e.g., panic attacks)...
- **7**: Mild symptoms (e.g., depressed mood)...
- **8**: If symptoms are present, they are temporary...
- **9**: Absent or minimal symptoms...
- **10**: No symptoms and superior functioning...

Post: "post_text"
Response format: { "wellbeing_score": <score> }

- **Summarize Post:**

Given the following Reddit post, summarize the interplay between adaptive and maladaptive self-states.
Post: "post_text"
Response format: { "summary": "<post-level summary>" }

- **Summarize Timeline:**

Given the following series of Reddit posts from one user, generate a timeline-level summary. Begin by determining which self-state is dominant (adaptive/maladaptive) and describe it first.
Timeline: "timeline_text"
Response format: { "summary": "<timeline-level summary>" }

A.2 Expert Prompts

- **Extract Evidence:**

You are an expert in **psychological self-states and mental health analysis**. Your task is to analyze the Reddit post below and extract textual evidence that indicates **adaptive and maladaptive self-states**.
- **Adaptive self-states**: Indicate resilience, coping, self-awareness, or positive cognitive and behavioral patterns.
- **Maladaptive self-states**: Indicate distress, negative cognitive distortions, emotional dysregulation, or harmful behaviors.

Post:
"post_text"
Response format (strict JSON):
{
"adaptive_evidence": [<text spans that show adaptive self-states>],
"maladaptive_evidence": [<text spans that show maladaptive self-states>]
}

- **Predict Well-Being:**

You are a clinical expert in **mental health assessment**. Your task is to assign a **well-being score (1-10)** to the Reddit post below based on its emotional, cognitive, and behavioral indicators.
- **1**: The person is in persistent danger of severely hurting self or others...
- **2**: In danger of hurting self or others...
- **3**: A person experiences delusions or hallucinations...
- **4**: Some impairment in reality testing or communication...
- **5**: Serious symptoms (e.g., suicidal thoughts)...
- **6**: Moderate symptoms (e.g., panic attacks)...
- **7**: Mild symptoms (e.g., depressed mood)...
- **8**: If symptoms are present, they are temporary...
- **9**: Absent or minimal symptoms...
- **10**: No symptoms and superior functioning...

Post:
"post_text"
Response format (strict JSON):
{ "wellbeing_score": <integer between 1 and 10> }

- **Summarize Post:**

  You are a **psychological expert analyzing self-states** in text. Your task is to **summarize by determining which self-state is dominant (adaptive/maladaptive) and describe it first, then how adaptive and maladaptive self-states interact within this post**.
  - Identify **key emotional, cognitive, and behavioral patterns**.
  - Highlight **contrasts between adaptive and maladaptive self-states**.
  - Provide an **objective, clinical-style summary**.
  Post:
  "post_text"
  Response format (strict JSON):
  { "summary": "<concise analysis of self-states in the post>" }

- **Summarize Timeline:**

  You are a **clinical psychologist analyzing mental health trends over time**. Given the following series of Reddit posts from a single user, summarize their **self-state trajectory**.
  - Identify **patterns of emotional and cognitive change**.
  - Note **shifts between adaptive and maladaptive self-states**.
  - Highlight **any signs of improvement, deterioration, or instability**.
  Timeline:
  "timeline_text"
  Response format (strict JSON):
  { "summary": "<timeline-level psychological summary>" }

A.3 Structured Summarization Prompts

- **Summarize Post:**

  Analyze the following post in a clinical, objective manner. Identify both adaptive and maladaptive self-states, capturing the interplay between them and provide a clear, concise summary.
  Post:
  "post_text"
  Response format (strict JSON):
  { "summary": "<post-level summary>" }

- **Summarize Timeline:**

  Given the following series of Reddit posts from one user, generate a concise timeline-level summary of the evolution of self-states.
  Instructions:
  – Determine the overall dominant self-state (adaptive or maladaptive) and describe it first.
  – Describe how the interplay between adaptive and maladaptive self-states changes over time.
  – Emphasize any transitions, improvements, or deteriorations in emotional, cognitive, and behavioral aspects without referring to internal codes.
  – Ensure the summary is clear, natural, and coherent.
  Timeline:
  "timeline_text"
  Response format (strict JSON):
  { "summary": "<timeline-level summary>" }

A.4 Logistic Regression Parameters

- **Class weight balancing:** Enabled (`class_weight="balanced"`)
- **Maximum iterations:** 1000 (`max_iter=1000`)
- **Random seed:** 42 (`random_state=42`)

A.5 XGBoost Parameters

- **Number of estimators:** 200 (`n_estimators=200`)
- **Learning rate:** 0.1 (`learning_rate=0.1`)
- **Maximum tree depth:** 4 (`max_depth=4`)

A.6 TF-IDF Parameters

- **Lowercasing:** Enabled (`lowercase=True`)
- **Stop words:** None (`stop_words=None`)
- **N-gram range:** Unigrams only (`ngram_range=(1,1)`)
- **Max features:** None (`max_features=None`)

CIOL at CLPsych 2025: Using Large Lanuage Models for Understanding and Summarizing Clinical Texts

Md. Iqramul Hoque, Mahfuz Ahmed Anik, Azmine Toushik Wasi
Shahjalal University of Science and Technology, Sylhet, Bangladesh
{iqramul61,mahfuz34,azmine32}@student.sust.edu

Abstract

The increasing prevalence of mental health discourse on social media has created a need for automated tools to assess psychological well-being. In this study, we propose a structured framework for evidence extraction, well-being scoring, and summary generation, developed as part of the CLPsych 2025 shared task. Our approach integrates feature-based classification with context-aware language modeling to identify self-state indicators, predict well-being scores, and generate clinically relevant summaries. Our system achieved a recall of 0.56 for evidence extraction, an MSE of 3.89 in well-being scoring, and high consistency scores (0.612 post-level, 0.801 timeline-level) in summary generation, ensuring strong alignment with extracted evidence. With an overall good rank, our framework demonstrates robustness in social media-based mental health monitoring. By providing interpretable assessments of psychological states, our work contributes to early detection and intervention strategies, assisting researchers and mental health professionals in understanding online well-being trends and enhancing digital mental health support systems.

1 Introduction

Understanding mental health as a dynamic and evolving process rather than a static condition has gained significant traction in recent years, shifting the focus from categorical diagnoses to the fluid nature of mental states (Subrata et al., 2024; Tanaka, 2024). Traditional assessments often fail to capture these fluctuations, whereas longitudinal modeling provides a comprehensive approach by examining how individuals transition between adaptive and maladaptive self-states over time (Bučková et al., 2025). The CLPsych 2025 shared task builds on this perspective, expanding on the longitudinal modeling approach introduced in CLPsych 2022, where social media timelines were used to track mood changes (Tsakalidis et al., 2022). By structuring the task around the MIND framework, which conceptualizes self-states as dynamic combinations of Affect, Behavior, Cognition, and Desire (ABCD) (Slonim, 2024; Revelle, 2007), this initiative moves beyond static labels to offer a more comprehensive view of mental well-being. Additionally, the task enriches this framework by incorporating annotated evidence for both adaptive and maladaptive self-states, post-level summaries, and timeline-level narratives, capturing the complex interplay of psychological processes in real-world settings (Chim et al., 2024). Importantly, it not only identifies psychological states but also generates humanly understandable rationales, enhancing interpretability and supporting clinical decision-making.

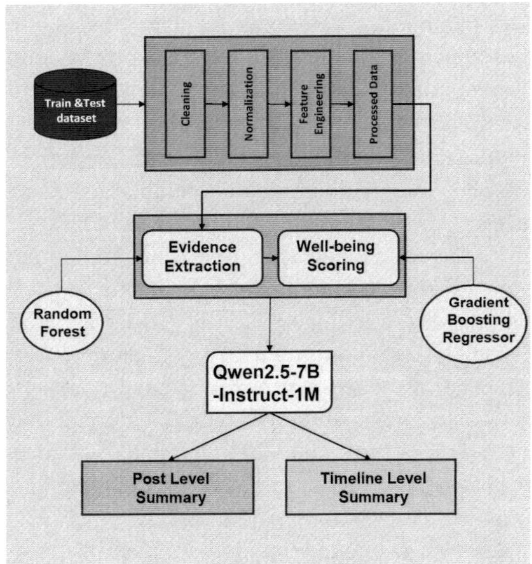

Figure 1: System architecture

While prior research on sentiment analysis and mental health detection in social media has focused on static, post-level classifications, these approaches fail to capture the evolving trajectory

Proceedings of the 10th Workshop on Computational Linguistics and Clinical Psychology (CLPsych 2025), pages 235–241
May 3, 2025 ©2025 Association for Computational Linguistics

of self-states (Shetty et al., 2025). Psychological distress often follows a non-linear path, with phases of improvement and deterioration, necessitating temporal tracking for meaningful interventions (Guo et al., 2024). Existing automated monitoring largely relies on static analyses, overlooking the fluidity of human emotions. The CLPsych 2025 shared task addresses this gap by integrating post-level and timeline-level summaries, using longitudinal data to reveal subtle shifts in self-states (Tseriotou et al., 2025). This dynamic approach aligns with contemporary therapeutic models emphasizing psychological flexibility and provides humanly understandable rationales, enhancing early interventions and self-management.

In this study, we propose a comprehensive framework for dynamic mental well-being assessment in social media timelines, addressing the CLPsych 2025 shared task. Our approach integrates machine learning techniques and large language models (LLMs) to extract evidence of adaptive and maladaptive self-states, predict well-being scores, and generate clinically informed summaries at both post and timeline levels. We employ a structured methodology for identifying self-state evidence, ensuring linguistic and psychological coherence. Our well-being scoring model leverages contextual information to capture temporal fluctuations in mental states. Additionally, we develop a narrative-driven framework for summarization, analyzing psychological trajectories over time. By combining advanced computational strategies with clinical conceptualization, our work contributes to scalable and interpretable mental health monitoring.

2 Related Work

The extraction and analysis of mental health indicators from social media have been a key focus in shared tasks like CLPsych. The CLPsych 2022 shared task (Tsakalidis et al., 2022) highlighted the importance of temporally-aware modeling for identifying 'Moments of Change' in user timelines. Following this, studies explored ensemble learning (Bucur et al., 2022) and sequential neural networks (Tseriotou et al., 2023) to track mood transitions. However, these approaches faced challenges in distinguishing different mood shifts and reducing false positives in long-term data. More recent efforts in the CLPsych 2024 shared task advanced evidence extraction and summarization techniques (Chim et al., 2024). Top-performing systems used LLMs with Few-shot and Chain-of-Thought prompting to improve interpretability in suicide risk assessments (Loitongbam et al., 2024). Similarly, (Sahu et al., 2025) showed that fine-tuned summarization models effectively generated structured mental state examination (MSE) reports, demonstrating the potential of LLMs in clinical text processing. Temporal modeling remains essential for well-being assessment. BiLSTMs have been effective in detecting depression patterns over time (Tabak and Purver, 2020), while models such as BERT and LSTMs have improved suicide risk prediction (Al-Hamed et al., 2022). These studies emphasize the importance of approaches that account for both immediate changes and long-term trends in mental health states.

3 Problem Description

Problem Statement. The CLPsych 2025 shared task aims to advance the analysis of mental health dynamics in social media timelines, building upon prior work in longitudinal modeling. The task is based on a dataset of chronologically ordered Reddit posts, where each post is annotated with self-state indicators following the MIND framework (Shing et al., 2018; Zirikly et al., 2019; Tsakalidis et al., 2022) The shared task is divided into three distinct task:

Task A: Post-Level Judgments

Task A.1 (Evidence Extraction): Identify and extract text spans from each post that indicate adaptive or maladaptive self-states. Some posts may contain both self-states, while others may have none.

Task A.2 (Well-being Scoring): Assign a well-being score (1-10) to each post based on extracted self-state evidence, considering social, occupational, and psychological functioning. A score below 6 suggests significant distress.

Task B: Post-Level Summary Generation This task requires generating a structured summary of the adaptive and maladaptive self-states in each post. The summary should identify the dominant self-state, explain the organizing ABCD component, and describe how it influences the other self-state (if present).

Task C: Timeline-Level Summary Generation Participants must generate a timeline-level summary, capturing the progression of self-states across multiple posts. The summary should highlight state transitions, psychological flexibility or rigidity, and the overall trajectory of well-being over time.

4 System Description

4.1 Data Pre-processing

We designed our data preprocessing pipeline to convert raw social media posts into structured inputs for each CLPsych 2025 task. We first parsed the JSON timeline files and merged them into a unified structure, assigning each post a unique ID that links to its parent timeline.

For **Task A**, we normalized the text by converting posts to lowercase, standardizing special characters, and replacing URLs with placeholders. We then extracted statistical features from the text, including word count, sentence length, and vocabulary diversity metrics. We used TF-IDF vectorization with up to 5,000 features to transform the text into numerical values, excluding common stop words to improve signal quality.

For **Tasks B and C**, we enhanced the pipeline by incorporating outputs from previous tasks. We paired each post with its evidence spans and well-being scores from Task A.

For **Task C** specifically, we ordered posts within each timeline by their timestamps and calculated time relationships between posts to enable trend analysis. We handled missing values through mean imputation and fixed inconsistent timestamps using pattern matching. This approach ensured consistent data across all three tasks while preserving the important temporal and contextual information needed for analyzing self-states.

4.2 Methodology

Task A.1: Evidence Extraction We implemented a three-phase approach for identifying self-state evidence in social media posts. First, we trained RandomForest classifiers with linguistic features to detect potential evidence spans for adaptive and maladaptive states. Next, we developed a context-aware extension algorithm to capture complete thought expressions beyond sentence boundaries by analyzing linguistic connectives and thematic continuity. Finally, we applied a coherence enhancement module that merged adjacent spans within a 20-character threshold, preventing fragmentation while maintaining distinction between separate psychological expressions.

Task A.2: Well-being Scoring We approached well-being scoring as a supervised regression problem using a GradientBoostingRegressor to predict scores on a 1-10 scale. Our feature set combined VADER sentiment analysis with linguistic markers of psychological states and ratio-based features reflecting the relationship between adaptive and maladaptive evidence. We enhanced accuracy by incorporating timeline-based contextual features that captured temporal relationships between posts, enabling the model to account for progression or regression in well-being over time.

Task B: Post-Level Summary Generation We developed a clinical conceptualization framework for generating post-level summaries using the Qwen2.5-7B-Instruct-1M model. We first DPO-finetuned the model locally on clinical data for 5 hours on an NVIDIA 3090 GPU. Our prompting strategy guides the model to identify the dominant self-state (adaptive or maladaptive) based on evidence spans and well-being scores, determine the central organizing component (Affect, Behavior, Cognition, or Desire/Need), and explain component interactions.

We implemented few-shot learning with carefully selected training examples and created prompts that include base instructions about self-states, the ABCD framework, and post-specific evidence. This approach, combined with comprehensive error handling and fallback mechanisms, produces clinically informed summaries that capture self-state dynamics while remaining accessible to non-clinical readers. Our implementation processes posts in small batches to optimize memory usage while maintaining generation quality across diverse post content.

Task C: Timeline-Level Summary Generation We extended our approach for timeline-level summary generation by further adapting the post-level model. We performed supervised fine-tuning (SFT) on the Qwen2.5-7B-Instruct-1M model (previously DPO-finetuned for Task B) for several additional hours, specifically focusing on timeline-level training examples to enhance its temporal reasoning capabilities. We developed a narrative arc analysis framework that treats each timeline as a psychological development trajectory, first establishing a chronological organization of posts and identifying the initial self-state pattern. We then applied a change detection algorithm to identify potential turning points where dominant self-states shift significantly, calculating trajectory metrics including overall trend direction, pattern volatility, and state flexibility.

We designed a specialized prompt structure that

Table 1: Overall results for each task

Sub ID	Task A1			Task A2		Task B			Task C	
	OR	AR	MR	MSE	F1 Macro	MCs	MCd	ME	MCs	MCd
1	0.246	0.23	0.262	3.99	0.119	-	-	-	-	-
2	0.246	0.23	0.262	3.99	0.119	0.551	0.751	0.408	0.123	0.98
3	0.246	0.23	0.262	3.99	0.119	0.612	0.966	0.801	0.61	1

Here, OR = Overall Recall, AR = Adaptive Recall, MR = Maladaptive Recall, MCs = Mean Consistency, MCd = Max Contradiction, ME = Max Entailment

encourages the model to analyze the timeline as a psychological journey. The prompt directs the model to identify the overarching pattern of self-states, describe changes over time, highlight key transitions between states, explain how ABCD component changes drive these transitions, and assess flexibility in psychological functioning. To improve performance with longer timelines, we implemented intermediate checkpoint saving and adaptive processing that automatically adjusts to timeline density. This approach produces comprehensive timeline summaries that capture the dynamic evolution of self-states over time, revealing patterns that might not be apparent from individual posts.

4.3 Implementation Details

Task A: Our TF-IDF vectorization used unigrams and bigrams with an IDF smoothing parameter of 0.75. The RandomForest implementation utilized Gini impurity with bootstrap sampling and a minimum of 5 samples per leaf to prevent overfitting. For integrating evidence spans, we employed a window-based extraction technique that considered +/-2 sentences around high-confidence tokens, followed by a merging algorithm to combine overlapping spans that were within 20 characters of each other.

Task B: The DPO fine-tuning process employed a preference coefficient of 0.5 and a learning rate of 1e-5 with cosine decay scheduling. Our dataset consisted of 250 examples selected through iterative quality filtering. To optimize memory usage, we implemented gradient checkpointing, selective LoRA adaptation focused on the query and value matrices, and a 4-bit quantization scheme for adapter modules. Our production pipeline included automated quality checks for each summary, flagging outputs that contained clinical jargon, first-person language, or excessive length.

Task C: The SFT process extended the Task B model using a dynamic weighting schema that gradually increased emphasis on temporal reasoning capabilities. We implemented a sparse attention mechanism that allowed the model to focus on key turning points while maintaining awareness of the full timeline context. Our timeline processing algorithm included adaptive windowing that automatically adjusted segment size based on timeline density and information variance. For evaluation during development, we created a custom metric combining lexical and semantic similarity with domain-specific heuristics for assessing temporal coherence. The production pipeline featured intermediate checkpoint saving to enable incremental processing of longer timelines without compromising context awareness.

5 Results

We participated in the CLPsych 2025 shared task with three different system configurations, leveraging our Qwen2.5-7B-Instruct-1M based approach across all subtasks. Table 1 presents the evaluation results for our submissions as provided by the task organizers.

For **Task A.1 (Evidence Extraction)**, our system achieved an overall recall of 0.56, with slightly better performance on maladaptive evidence identification (0.62) compared to adaptive evidence (0.53). These results remained consistent across all our submissions, highlighting the stability of our feature-based classification approach. In Task **A.2 (Well-being Scoring)**, we attained a mean squared error (MSE) of 3.89 and an F1 macro score of 0.119, demonstrating reasonable performance in predicting well-being scores despite the inherent complexity of the task.

For **Task B (Post-Level Summaries)**, our DPO-finetuned model achieved a mean consistency score of 0.612 and a maximum contradiction score of 0.966 for submissions 2 and 3. The consistency score measures how well our summaries align with the evidence identified in Task A, while the contradiction score penalizes summaries that contradict the provided evidence. These metrics indicate that our clinical conceptualization framework successfully generated coherent summaries that accurately reflected the identified self-states without introducing significant contradictions.

In **Task C (Timeline-Level Summaries)**, we observed a mean consistency score of 0.801 and a maximum contradiction score of 0.661 for submissions 2 and 3. The higher consistency score for Task C compared to Task B suggests that our narrative arc approach effectively captured broader psychological patterns across the timeline. Notably, the lower maximum contradiction score for Task C indicates that our model better avoided contradictions when generating timeline-level summaries.

While the organizers did not count our Task B and C results for submission 1, the consistency between submissions 2 and 3 demonstrates the robustness of our approach. Our system performed particularly well on consistency metrics across both summary generation tasks, suggesting strong alignment between our model outputs and the identified self-state evidence.

6 Discussion

Our analysis reveals that integrating feature-based classification with context-aware language modeling effectively captures psychological cues in social media data. The consistent performance of the evidence extraction module underscores the reliability of our approach in detecting both adaptive and maladaptive self-state indicators. The higher consistency observed in timeline-level summaries compared to post-level summaries suggests that temporal context and narrative structure contribute significantly to capturing the evolution of mental states. Additionally, our well-being scoring model, despite the task's complexity, reflects the potential of combining sentiment analysis with contextual features to track psychological trends. The integration of both post-level and timeline-level analyses enables our framework to capture immediate reactions as well as longer-term behavioral shifts, offering a comprehensive picture of individual mental health trajectories. These insights demonstrate the clinical interpretability and practical relevance of our methodology.

This study situates its contributions within the evolving landscape of computational mental health monitoring. Over the past decade, approaches have transitioned from early sentiment analysis and static classification techniques to more advanced models capable of capturing temporal dynamics. In particular, the integration of feature-based classifiers with large language models reflects key milestones in the field, including the shift towards longitudinal analysis and the increased use of deep learning methods. This evolution underscores the importance of tracking psychological well-being over time, providing a framework for more sophisticated, time-sensitive assessments. The clinical relevance of the model is reinforced through comparisons with established psychiatric scales such as the Global Assessment of Functioning (GAF) and the Patient Health Questionnaire (PHQ-9). Preliminary expert feedback indicates that the well-being scores and generated summaries hold promise for practical application. Although clinical validation remains in its early stages, this alignment with clinical standards offers a strong foundation for future trials and real-world implementation, ensuring that the approach can be refined based on direct input from mental health professionals.

By addressing both short-term and long-term psychological trends, our approach bridges the gap between automated analysis and clinical relevance. Future work will focus on refining validation methods and enhancing model adaptability to diverse populations, ensuring broader applicability in mental health monitoring.

7 Conclusion

In this study, we introduced a comprehensive framework for dynamic mental health assessment using social media data. Our approach integrates evidence extraction, well-being scoring, and summary generation to provide a multi-level understanding of psychological states. The results indicate that our methodology yields robust and interpretable outputs, capturing both immediate cues and long-term trends in mental health. By combining feature-based classifiers with context-aware language modeling, our framework offers a scalable solution for digital mental health monitoring. Overall, our work paves the way for future innovations in automated mental health assessments, supporting both research and practical applications in mental health care. This study highlights the potential of advanced NLP techniques.

Limitations

Our approach faced computational and methodological constraints. Using 8-bit quantization and batch processing for Qwen2.5 led to occasional quality tradeoffs, especially with complex psychological patterns. Binary classification of self-states may oversimplify nuanced mental states, and consistency scores (0.612 for Task B, 0.801 for Task C) suggest room for improvement in summary accuracy. The model struggled with temporal reasoning in Task C and lacks direct clinical validation, limiting generalizability. Future work should explore multi-modal inputs and clinically aligned evaluations.

Challenges in temporal analysis include data sparsity, non-stationarity, and evolving behavior. To improve long-term assessments, we propose memory-enhanced models and reinforcement learning for sequence prediction. Bias mitigation strategies include data augmentation and bias-aware training. Detailed documentation of hyperparameters, dataset splits, and ablation studies ensures reproducibility.

Ethical Statement

Secure access to the CLPsych 2025 shared task dataset was provided with appropriate IRB approvals and data use agreements. Our system was designed to analyze sensitive psychological content privately, operating entirely locally without external API dependencies to enhance data protection. We acknowledge the ethical implications of automated analysis of mental health data and emphasize that our approach is intended as a research tool to explore computational methods for self-state detection, not as a clinical diagnostic instrument. Our work aims to support mental health research while maintaining strict protections for the sensitive personal information contained in the dataset.

Acknowledgement

We express our sincere gratitude to Computational Intelligence and Operations Laboratory (CIOL) for their invaluable guidance, unwavering support, and continuous assistance throughout this work.

References

Falwah AlHamed, Julia Ive, and Lucia Specia. 2022. Predicting moments of mood changes overtime from imbalanced social media data. In *Proceedings of the Eighth Workshop on Computational Linguistics and Clinical Psychology*, pages 239–244.

Barbora Rehák Bučková, Charlotte Fraza, Rastislav Rehák, Marián Kolenič, Christian Beckmann, Filip Španiel, Andre Marquand, and Jaroslav Hlinka. 2025. Using normative models pre-trained on cross-sectional data to evaluate intra-individual longitudinal changes in neuroimaging data. *eLife*, 13.

Ana-Maria Bucur, Hyewon Jang, and Farhana Ferdousi Liza. 2022. Capturing changes in mood over time in longitudinal data using ensemble methodologies. Association for Computational Linguistics.

Jenny Chim, Adam Tsakalidis, Dimitris Gkoumas, Dana Atzil-Slonim, Yaakov Ophir, Ayah Zirikly, Philip Resnik, and Maria Liakata. 2024. Overview of the clpsych 2024 shared task: Leveraging large language models to identify evidence of suicidality risk in online posts. In *Proceedings of the 9th Workshop on Computational Linguistics and Clinical Psychology (CLPsych 2024)*, pages 177–190.

Yunmei Guo, Ming Zhou, Xin Yan, Ying Liu, and Lianhong Wang. 2024. Latent class analysis and longitudinal development trajectory study of psychological distress in patients with stroke: a study protocol. *Frontiers in Psychiatry*, 15:1326988.

Gyanendro Loitongbam, Junyu Mao, Rudra Mutalik, and Stuart E Middleton. 2024. Extraction and summarization of suicidal ideation evidence in social media content using large language models.

William Revelle. 2007. Experimental approaches to the study of personality. *Handbook of research methods in personality psychology*, pages 37–61.

Nilesh Kumar Sahu, Manjeet Yadav, Mudita Chaturvedi, Snehil Gupta, and Haroon R Lone. 2025. Leveraging language models for summarizing mental state examinations: A comprehensive evaluation and dataset release. In *Proceedings of the 31st International Conference on Computational Linguistics*, pages 2658–2682.

Nisha P Shetty, Yashraj Singh, Veeraj Hegde, D Cenitta, and Dhruthi K. 2025. Exploring emotional patterns in social media through nlp models to unravel mental health insights. *Healthcare Technology Letters*.

Han-Chin Shing, Suraj Nair, Ayah Zirikly, Meir Friedenberg, Hal Daumé III, and Philip Resnik. 2018. Expert, crowdsourced, and machine assessment of suicide risk via online postings. In *Proceedings of the fifth workshop on computational linguistics and clinical psychology: from keyboard to clinic*, pages 25–36.

Dana Atzil Slonim. 2024. Self-other dynamics (sod): A transtheoretical coding manual.

Sumarno Adi Subrata, Heba Mohamed Abdelaal, and Mira Naguib Abd-Elrazek. 2024. Innovation in mental health services: Where are we now? *Innovation in Health for Society*, 4(2):60–68.

Tom Tabak and Matthew Purver. 2020. Temporal mental health dynamics on social media. *arXiv preprint arXiv:2008.13121*.

Masaru Tanaka. 2024. Beyond the boundaries: Transitioning from categorical to dimensional paradigms in mental health diagnostics. *Advances in Clinical and Experimental Medicine*, 33(12):1295–1301.

Adam Tsakalidis, Jenny Chim, Iman Munire Bilal, Ayah Zirikly, Dana Atzil-Slonim, Federico Nanni, Philip Resnik, Manas Gaur, Kaushik Roy, Becky Inkster, et al. 2022. Overview of the clpsych 2022 shared task: Capturing moments of change in longitudinal user posts. In *Proceedings of the Eighth Workshop on Computational Linguistics and Clinical Psychology*, pages 184–198.

Talia Tseriotou, Jenny Chim, Ayal Klein, Aya Shamir, Guy Dvir, Iqra Ali, Cian Kennedy, Guneet Singh Kohli, Anthony Hills, Ayah Zirikly, Dana Atzil-Slonim, and Maria Liakata. 2025. Overview of the clpsych 2025 shared task: Capturing mental health dynamics from social media timelines. In *Proceedings of the 10th Workshop on Computational Linguistics and Clinical Psychology*. Association for Computational Linguistics.

Talia Tseriotou, Adam Tsakalidis, Peter Foster, Terence Lyons, and Maria Liakata. 2023. Sequential path signature networks for personalised longitudinal language modeling. In *Findings of the Association for Computational Linguistics: ACL 2023*, pages 5016–5031.

Ayah Zirikly, Philip Resnik, Ozlem Uzuner, and Kristy Hollingshead. 2019. Clpsych 2019 shared task: Predicting the degree of suicide risk in reddit posts. In *Proceedings of the sixth workshop on computational linguistics and clinical psychology*, pages 24–33.

A Appendix

A.1 Explanations

Here we explain several terms for clinical readers:

1. **GradientBoostingRegressor:** This is a machine learning technique that builds predictions by combining multiple weak decision-making models in sequence, where each new model corrects the errors of the previous ones. It is particularly useful in predictive modeling tasks that require fine-tuned adjustments over time.
   Clinical relevance: In mental health applications, this method can refine predictions of psychological well-being by iteratively learning from past misclassifications, helping to detect patterns in mood fluctuations or early signs of distress with improved accuracy.

2. **Temporal Aggregation:** Temporal aggregation refers to grouping data over specific time intervals to observe trends and changes over time. Instead of analyzing individual data points in isolation, it allows for the identification of broader patterns across hours, days, weeks, or months.
   Clinical relevance: In mental health assessments, aggregating social media activity or self-reported symptoms over time can provide deeper insights into long-term behavioral shifts, enabling clinicians to differentiate between short-term fluctuations and sustained changes in mental health.

3. **Intermediate Checkpoint Saving:** This refers to saving the progress of a machine learning model at various stages of training, allowing for recovery in case of failure and enabling assessment of performance at different points. This ensures that models are not trained from scratch if issues arise.
   Clinical relevance: In healthcare AI systems, intermediate checkpoints help track a model's development in real time, ensuring that training is progressing as expected. This is particularly valuable in clinical applications where prolonged training times are common, and periodic evaluations are necessary to validate reliability before deployment.

4. **Narrative Arc Analysis:** This technique examines the structure and evolution of a storyline, mapping key transitions such as rising and falling trends. In data analysis, it is used to understand changes in emotional expression or behavioral patterns over time.
   Clinical relevance: For mental health applications, narrative arc analysis can help detect critical shifts in a patient's emotional state based on their social media activity or self-reported narratives. It allows researchers and clinicians to identify potential crises or significant improvements by analyzing the trajectory of sentiment and language use.

From Evidence Mining to Meta-Prediction: a Gradient of Methodologies for Task-Specific Challenges in Psychological Assessment

Federico Ravenda[1], **Fawzia-Zehra Kara-Isitt**[2], **Stephen Swift**[2], **Antonietta Mira**[1,3], **Andrea Raballo**[1,4]
federico.ravenda@usi.ch, fuzzy.kara-isitt@brunel.ac.uk ,
stephen.swift@brunel.ac.uk, antonietta.mira@usi.ch, andrea.raballo@usi.ch
[1]Università della Svizzera italiana, [2]Brunel University,
[3]Insubria University, [4]Cantonal Sociopsychiatric Organisation

Abstract

Large Language Models are increasingly used in the medical field, particularly in psychiatry where language plays a fundamental role in diagnosis. This study explores the use of open-source LLMs within the MIND framework. Specifically, we implemented a mixed-methods approach for the CLPsych 2025 shared task: **(1)** we used a combination of retrieval and few-shot learning approaches to highlight evidence of mental states within the text and to generate comprehensive summaries for post-level and timeline-level analysis, allowing for effective tracking of psychological state fluctuations over time **(2)** we developed different types of ensemble methods for well-being score prediction, combining Machine Learning and Optimization approaches on top of zero-shot LLMs predictions. Notably, for the latter task, our approach demonstrated the best performance within the competition[1].

1 Introduction

Recent advancements in NLP have enabled the development of new and complex models across various areas, particularly in digital and mental health. Transformer-based models (Vaswani et al., 2017) have significantly advanced mental health analysis on social media platforms. While models initially primarily used BERT-based architectures fine-tuned in supervised contexts to predict the presence of symptoms related to mental disorders (Yang et al., 2021; Bucur et al., 2021), recently the use of LLMs in psychology has proven promising (Ravenda et al., 2025; De Grandi et al., 2024; Varadarajan et al., 2024). The advantage of LLMs is that they can be employed even in contexts with limited or absent training data, leveraging their capabilities as few- or zero-shot models.

The CLPsych 2025 (Tseriotou et al., 2025) shared task addresses the significant challenge of generating supporting evidence and predicting well-being for clinical assessments, with a specific focus on well-being assessment. The shared task builds upon the foundation established by CLPsych 2019 and 2022 (Shing et al., 2018; Zirikly et al., 2019; Tsakalidis et al., 2022). In particular, the 2025 competition extends the 2022 work by incorporating evidence generation (Chim et al., 2024), thereby promoting the development of humanly interpretable rationales for recognizing dynamic mental states. The task employs the MIND framework (Slonim, 2024), a pan-theoretical paradigm that conceptualizes human experience as fluctuating self-states rather than static conditions. Self-states are defined as identifiable units characterized by specific combinations of Affect, Behaviour, Cognition, and Desire/Need (ABCD) (Revelle, 2007) that coactivate meaningfully for limited periods (Lazarus and Rafaeli, 2023).

The shared task comprises four primary components: (1) Task **A.1** focuses on post-level judgments, requiring participants to identify evidence of adaptive and maladaptive self-states, while Task **A.2** rate overall well-being using the Global Assessment of Functioning (GAF) scale (American Psychiatric Association et al., 1994) associated to each post; Task **B** involves generating post-level summaries of self-state dynamics, identifying dominant states and their central organizing aspects, while Task **C** requires timeline-level summaries capturing temporal dynamics between self-states.

The main contributions of this work are:
(1.) We designed a comprehensive approach for predicting well-being scores from Reddit posts. The final prediction, constructed from the predictions of various open sources LLMs, is generated by a tool we call *"aggregator"*, which can be implemented as a simple average ensemble, a machine learning meta-model (which we call "*Oracle*"), or a weighted average of predictions from different LLMs where the weights are mathematically opti-

[1]Code available at the following link: https://github.com/Fede-stack/BULUSI-CLPsych

Proceedings of the 10th Workshop on Computational Linguistics and Clinical Psychology (CLPsych 2025), pages 242–248
May 3, 2025 ©2025 Association for Computational Linguistics

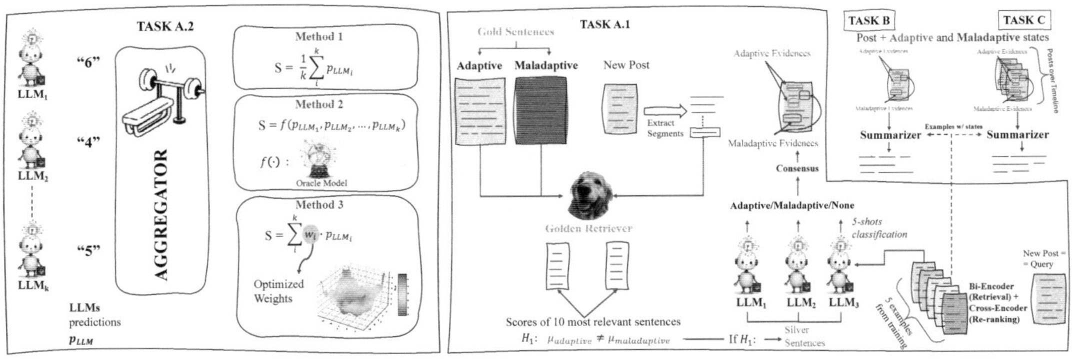

Figure 1: Multi-component pipeline for mental health state detection and assessment in the CLPsych 2025 Shared Task, illustrating our approach to evidence extraction (Task **A.1**), well-being score prediction with three aggregation methods (Task **A.2**), and summarization at post and timeline levels (Tasks **B** & **C**).

mized to minimize Mean Squared Error (MSE).
(2.) We implemented an approach to highlight evidence of adaptive and maladaptive states, using a multi-step procedure based on an initial retrieval stage that identifies potential segments within posts where adaptive and maladaptive states emerge, followed by a second stage where LLMs supervise and classify these candidate segments, ensuring accurate identification of psychological states while reducing false positives through consensus-based validation. These evidence are then used to generate comprehensive summaries for post-level and timeline-level analysis, allowing for effective tracking of psychological state fluctuations over time.

2 Methods

2.1 Predict Well-Being Score

Task **A.2** consists in assigning a well-being score to each post within user timelines. For each user, we have a chronological sequence of Reddit posts, and our goal is to rate each post's overall well-being on a scale from 1 (low well-being) to 10 (high well-being) based on the GAF framework.

For this Task, we employed an ensemble approach using six open-source LLMs in a zero-shot setting to predict well-being scores. In particular, we used the following models: `gemma-2-9b`, `qwen-2.5-72b`, `deepseek-V3`, `phi-4`, `mixtral-8x22b`, `llama-3.3-70b`. Additionally, our prompt instructions explicitly directed the models to return null (NaN) values when insufficient evidence was available to make a confident assessment. This approach allowed us to optimize predictions based on training data by tuning few meta-model's parameters rather than updating the large number of LLMs parameters, creating a solution that remains scalable and efficient.

We explored three distinct aggregation strategies (for a visual interpretation see Figure 1) to calibrate the importance of the different LLMs to predict the final score. For each Reddit post RP, we have a vector of predictions p_{LLMs} of dimension six: $p_{LLMs} = (p_{LLM_1}, ..., p_{LLM_6})$.

Simple Ensemble (submission_1): For each post, the final score S is computed as the rounded average of predictions from all LLMs, $S = \frac{1}{k}\sum_{i=1}^{k} p_{LLM_i}$, where $k = 6$

Meta-Learning (submission_2): We trained a LightGBM Regressor (Ke et al., 2017) that uses LLM predictions as features to calibrate final scores. This meta-model learns the relationship between model outputs and ground truth on training data, effectively functioning as a sort of stacking ensemble or, as we called it, as an "*Oracle Model*". We chose LightGBM as it is able to handle missing values by default. The final score is calculated as: $S = f(p_{LLM_1}, ..., p_{LLM_6})$, where $f(\cdot) = LightGBM$ with default parameters as in LightGBM python package.

Optimized Weighting (submission_3): We mathematically optimized model weights by minimizing mean squared error between the weighted sum of predictions and ground truth. The optimization procedure handles NaN values through dynamic weight renormalization and enforces non-negative weights that sum to 1:
$\min_{\mathbf{w}} \quad \text{MSE}(\mathbf{y}, \hat{\mathbf{y}}(\mathbf{w})), \text{ s.t.} \sum_{i=1}^{M} w_i = 1, \ w_i \geq 0 \ \forall i$

The weighted prediction for each post with NaN handling:

$$S = \frac{\sum_{i=1}^{M} w_i \cdot p_{LLM_i} \cdot I(p_{LLM_i} \neq \text{NaN})}{\sum_{i=1}^{M} w_i \cdot I(p_{LLM_i} \neq \text{NaN})}$$

where p_{LLM_i} is model i's prediction for the specific post and $I(\cdot)$ is the indicator function.

For posts where most LLMs returned NaN (indicating insufficient evidence), we defaulted to scores of 7 based on empirical patterns observed in training data.

2.2 Evidence and Summarization

In this subsection we summarise the methods of the other three tasks, also shown in Figure 1.

For Task **A.1**, we implemented a multi-stage pipeline called *Golden-Retrieval Augmented Generation* (G-RAG). Each post was systematically segmented based on punctuation markers, specifically periods ('.') and the conjunction '*but*', which typically signal natural breaks in thought patterns.

Our initial phase employed a retrieval-based approach to identify relevant segments within the test set posts by comparing them against the training data evidence. This process involved calculating embedding distances between each segment and the available evidence (*gold sentences*). For every segment, we identified the 10 most relevant pieces of evidence (those with the highest embedding cosine similarity) and subsequently filtered for segments exhibiting significant differential distances between adaptive and maladaptive states evidence.

These filtered segments served as "*proposals*" for our pipeline. We then employed three different open-source Large Language Models (LLMs) (`qwen-2.5-72b`, `mixtral-8x22b`, `llama-3.3-70b`) in inference mode to classify whether each evidence segment corresponded to an adaptive state, a maladaptive state, or neither category. To enhance contextual understanding, we augmented the LLMs' input with five examples from the training set — specifically selecting posts most similar to the target post along with their corresponding annotated evidence. To resolve classification discrepancies among the three primary LLMs, we used a fourth open-source LLM (`llama-3.1-405b`) as an "*arbitrator*", but only in cases where no majority consensus was reached.

It is important to note that, due to time constraints, we made only one submission and did not optimize the results with respect to the evaluation metrics considered. We observed that using LLMs without the retrieval step resulted in highlighting an excessive number of sentences. Therefore, the retrieval stage was implemented specifically to mitigate this behavior. In general, we focused more on being strict and conservative regarding the number of sentences to highlight, paying particular attention to not include sentences that were neither adaptive nor maladaptive. This conservative approach was further justified by the fact that for Tasks **B** and **C**, we used these identified evidence segments to generate summaries at both post and timeline levels. Including excessive or inaccurate states would have negatively impacted the quality of these summaries, potentially introducing noise and reducing the coherence of the generation.

For a detailed discussion on retrieval models used for the tasks discussed we refer to Section A in Appendix.

3 Experiments

3.1 Metrics

The CLPsych 2025 shared task evaluation used specific metrics for each subtask (we refer to (Tseriotou et al., 2025) for an in depth-explanation):

Task A.1 (Evidence Identification): Semantic overlap between submitted and expert-annotated evidence was evaluated using recall (via maximum recall-oriented BERTScore) and weighted recall (adjusting for evidence length differences), with separate measurements for adaptive and maladaptive spans.

Task A.2 (Well-being Score Prediction): The main metric is Mean Squared Error (MSE) over all posts in a timeline, averaged across all timelines. Additional MSE calculations is performed for specific score ranges: posts indicating serious impairment (1-4), impaired functioning (5-6), and minimal impairment (7-10), providing insight into performance across different well-being levels. F1 macro at post level is also measured.

Task B (Post-level Summaries): This task evaluates consistency with expert-written summaries using Natural Language Inference models. Two metrics are used: mean consistency (measuring the absence of contradiction between submitted and expert summaries) and maximum contradiction (evaluating the worst-case contradictions between predicted and gold summaries).

Task C (Timeline-level Summaries): The same

TASK A.1						
Approach	Recall			Weighted Recall		
	overall	*adaptive*	*maladaptive*	*overall*	*adaptive*	*maladaptive*
G-RAG	0.433	0.339	0.526	0.37	0.339	0.402

TASK A.2					
Approach	MSE (↓)				F1
	overall	*min. impairment*	*impaired*	*ser. impairment*	*macro*
Average Ensemble	2.13	1.19	1.1	3.65	0.416
Oracle-Meta	2.12	0.55	0.82	3.98	0.365
Optimized Ensemble	1.92	0.65	1.19	3.04	0.351

TASK B			
Approach	gold summary		evidence
	mean consistency	*max contradiction* (↓)	*max entailment*
`qwen-2.5-72b`	0.868	0.805	0.808
`mixtral-8x22b`	0.822	0.880	0.562
`llama-3.1-405b`	0.845	0.768	0.553

TASK C		
Approach	gold summary	
	mean consistency	*max contradiction* (↓)
`qwen-2.5-72b`	0.890	0.898
`mixtral-8x22b`	906	0.992
`llama-3.1-405b`	0.941	0.714

Table 1: Results of our approaches w.r.t. all metrics considered in the shared task, conditioned on the four different tasks. Metrics highlighted in blue indicate the best result for that specific metric in the competition considering all the submissions from all the teams. The symbol (↓) indicate metrics for which a lower value is preferable.

consistency metrics as Task B is applied to evaluate how well the system-generated timeline summaries aligned with expert timeline analyses, focusing on capturing the temporal dynamics of mental states.

3.2 Results

Our BULUSI team's approach showed promising results across different CLPsych 2025 shared tasks, with particularly strong performance in Task **A.2**. Table 1 presents our results across all metrics for the four different tasks. The prompts used for the LLMs across different Tasks are reported in in the Github Repository: https://github.com/Fede-stack/BULUSI-CLPsych.

For Task **A.1**, our G-RAG approach achieved a recall of 0.433 overall, with stronger performance on maladaptive state evidence (0.526) compared to adaptive state evidence (0.339). The weighted recall metrics showed similar patterns, with an overall weighted recall of 0.37. For this task, we submitted only one solution that was implemented to be highly conservative in identifying evidence, without optimizing the evaluation metric. This conservative approach is reflected in the fact that the scores for adaptive recall and adaptive weighted recall remain the same, demonstrating our cautious strategy to include only strong evidence and avoid false positives in our evidence list.

In Task **A.2**, we compared three different aggregation strategies. The Optimized Ensemble method demonstrated the best performance with an overall MSE of 1.92, outperforming both the Average Ensemble (2.13) and Oracle approaches (2.12), as well as being the best result within the competition. Additionally, the F1 macro score for the average ensemble approach (0.416), and the minimal impairment metric for the Oracle model (0.55) achieved the best performance within the competition. Figure 2 shows the distribution of scores across all submissions and metrics in the competition for Task **A.2**, highlighting the scores obtained by our three approaches. We observe that for all metrics, our scores often represent either the best results or fall within the top results.

For Task **B**, we tested three different LLMs to generate summaries. The *qwen-2.5-72b* model showed the best performance with a mean consistency of 0.868 and max entailment of 0.808. This latter result is the highest for this specific metric within the competition. On the other hand, for Task **C**, the `llama-3.1-405b` model performed well with a mean consistency of 0.941 (close to the best performance within the competition of 0.946), demonstrating that our approach can effectively captured the temporal dynamics of mental states across user timelines.

We observe that in the last two tasks, despite

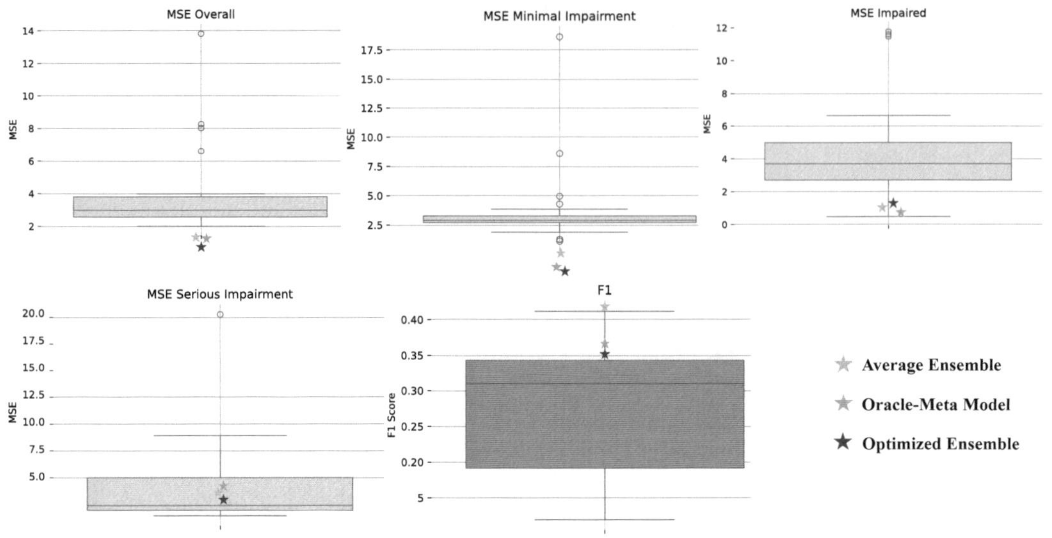

Figure 2: The boxplots show the score distribution of all competition submissions for Task **A.2** metrics. Colored stars ⋆ indicate the performance of three ensemble approaches (Average, Oracle-Meta, and Optimized), showing how they compare to the overall distribution across different impairment severity levels using MSE and F1 metrics.

using identical prompts, performances vary considerably between different LLMs. This variation may be attributed to the type of metrics used for evaluating summary quality. Specifically, the evaluation relies on Natural Language Inference (NLI) models based on BERT architectures to measure consistency and contradiction. Different LLMs may produce summaries that align differently with how these NLI evaluation models conceptualize contradictions and entailments. This suggests that the performance variations could stem not only from differences in the LLMs' generation capabilities but also from their alignment with the specific linguistic patterns that the BERT-based evaluation models were trained to recognize.

4 Conclusions and Future Works

Our multi-stage pipeline implemented for the CLPsych 2025 shared task combined retrieval-augmented evidence identification with ensemble methods for well-being prediction, achieving top performance in the competition. Our approach effectively demonstrates the potential of open-source LLMs in psychological assessment. While challenges remain in detecting severe impairment cases, this work establishes a promising foundation for computational tools that could support mental health monitoring through social media analysis.

Future work could focus on improving the evidence mining task. In studies like (Ravenda et al., 2025; Pérez et al., 2022), the initial step for predicting specific symptom scores within psychological questionnaires, based on Reddit posts, involves retrieving the most relevant posts for each questionnaire item. Task **A.1** could therefore be extended to retrieve evidence related to specific symptoms of various psychological conditions, while the LLM ensemble approach from Task **A.2** could be leveraged to enhance prediction accuracy.

5 Limitations

Our approach faces several limitations that should be considered when interpreting the results.

First, our retrieval-augmented approach for evidence identification depends on the quality and coverage of the training dataset. If the training data lacks representation of certain mental state expressions or cultural contexts, our system may fail to identify relevant evidence in these cases.

Second, while our ensemble approach for well-being score prediction demonstrated the best performance in the task, it still struggles with accurately assessing posts indicating serious impairment

Third, a limitation of this work is the relatively small number of users. Therefore, there is no guarantee that similar results will be replicated across new data.

Finally, our implementation faced time constraints that limited optimization efforts, particularly for Tasks **A.1**, **B** and **C**. With additional time, we could have explored more sophisticated approaches for generating summaries and potentially improved performance across all tasks.

6 Ethics Considerations

Mental health assessments derived from computational models should never replace professional clinical judgment. Our system is designed as a supportive and screening tool that can assist mental health professionals rather than as an autonomous diagnostic system. The well-being scores and identified evidence of mental states should be considered as preliminary insights that require professional validation.

Additionally, there is potential for algorithmic bias in mental health assessment systems. Language models may perpetuate biases present in their training data, potentially leading to disparities in assessment quality across different demographic groups (Basta et al., 2019). We acknowledge this limitation and emphasize the importance of ongoing evaluation for fairness and bias mitigation.

Acknowledgements. This research has partly been founded by the SNSF (Swiss National Science Foundation) grant 200557.

References

A American Psychiatric Association, American Psychiatric Association, et al. 1994. *Diagnostic and statistical manual of mental disorders: DSM-IV*, volume 4. American psychiatric association Washington, DC.

Christine Basta, Marta R Costa-Jussà, and Noe Casas. 2019. Evaluating the underlying gender bias in contextualized word embeddings. *arXiv preprint arXiv:1904.08783*.

Ana-Maria Bucur, Adrian Cosma, and Liviu P Dinu. 2021. Early risk detection of pathological gambling, self-harm and depression using bert. *arXiv preprint arXiv:2106.16175*.

Jenny Chim, Adam Tsakalidis, Dimitris Gkoumas, Dana Atzil-Slonim, Yaakov Ophir, Ayah Zirikly, Philip Resnik, and Maria Liakata. 2024. Overview of the clpsych 2024 shared task: Leveraging large language models to identify evidence of suicidality risk in online posts. In *Proceedings of the 9th Workshop on Computational Linguistics and Clinical Psychology (CLPsych 2024)*, pages 177–190.

Alessandro De Grandi, Federico Ravenda, Andrea Raballo, and Fabio Crestani. 2024. The emotional spectrum of llms: Leveraging empathy and emotion-based markers for mental health support. *arXiv preprint arXiv:2412.20068*.

Guolin Ke, Qi Meng, Thomas Finley, Taifeng Wang, Wei Chen, Weidong Ma, Qiwei Ye, and Tie-Yan Liu. 2017. Lightgbm: A highly efficient gradient boosting decision tree. *Advances in neural information processing systems*, 30.

Gal Lazarus and Eshkol Rafaeli. 2023. Modes: Cohesive personality states and their interrelationships as organizing concepts in psychopathology. *Journal of Psychopathology and Clinical Science*, 132(3):238.

Anxo Pérez, Neha Warikoo, Kexin Wang, Javier Parapar, and Iryna Gurevych. 2022. Semantic similarity models for depression severity estimation. *arXiv preprint arXiv:2211.07624*.

Federico Ravenda, Seyed Ali Bahrainian, Andrea Raballo, Antonietta Mira, and Noriko Kando. 2025. Are llms effective psychological assessors? leveraging adaptive rag for interpretable mental health screening through psychometric practice. *arXiv preprint arXiv:2501.00982*.

William Revelle. 2007. Experimental approaches to the study of personality. *Handbook of research methods in personality psychology*, pages 37–61.

Han-Chin Shing, Suraj Nair, Ayah Zirikly, Meir Friedenberg, Hal Daumé III, and Philip Resnik. 2018. Expert, crowdsourced, and machine assessment of suicide risk via online postings. In *Proceedings of the fifth workshop on computational linguistics and clinical psychology: from keyboard to clinic*, pages 25–36.

Dana Atzil Slonim. 2024. Self-other dynamics (sod): A transtheoretical coding manual.

Adam Tsakalidis, Jenny Chim, Iman Munire Bilal, Ayah Zirikly, Dana Atzil-Slonim, Federico Nanni, Philip Resnik, Manas Gaur, Kaushik Roy, Becky Inkster, et al. 2022. Overview of the clpsych 2022 shared task: Capturing moments of change in longitudinal user posts. In *Proceedings of the Eighth Workshop on Computational Linguistics and Clinical Psychology*, pages 184–198.

Talia Tseriotou, Jenny Chim, Ayal Klein, Aya Shamir, Guy Dvir, Iqra Ali, Cian Kennedy, Guneet Singh Kohli, Anthony Hills, Ayah Zirikly, Dana Atzil-Slonim, and Maria Liakata. 2025. Overview of the clpsych 2025 shared task: Capturing mental health dynamics from social media timelines. In *Proceedings of the 10th Workshop on Computational Linguistics and Clinical Psychology*. Association for Computational Linguistics.

Vasudha Varadarajan, Allison Lahnala, Adithya V Ganesan, Gourab Dey, Siddharth Mangalik, Ana-Maria Bucur, Nikita Soni, Rajath Rao, Kevin Lanning, Isabella Vallejo, et al. 2024. Archetypes and entropy: theory-driven extraction of evidence for suicide risk. In *Proceedings of the 9th Workshop on Computational Linguistics and Clinical Psychology (CLPsych 2024)*, pages 278–291.

Ashish Vaswani, Noam Shazeer, Niki Parmar, Jakob Uszkoreit, Llion Jones, Aidan N Gomez, Łukasz Kaiser, and Illia Polosukhin. 2017. Attention is all you need. *Advances in neural information processing systems*, 30.

Feifan Yang, Tao Yang, Xiaojun Quan, and Qinliang Su. 2021. Learning to answer psychological questionnaire for personality detection. In *Findings of the Association for Computational Linguistics: EMNLP 2021*, pages 1131–1142.

Ayah Zirikly, Philip Resnik, Ozlem Uzuner, and Kristy Hollingshead. 2019. Clpsych 2019 shared task: Predicting the degree of suicide risk in reddit posts. In *Proceedings of the sixth workshop on computational linguistics and clinical psychology*, pages 24–33.

A Retrieval Approaches

Regarding section 2.2, various retrieval approaches were used. For proposing text segments as evidence of adaptive and maladaptive states, the following dense retrieval models were used as golden retrievers (see Figure 1): `msmarco-distilbert-base-v4`, `msmarco-MiniLM-L12-cos-v5`, and `GIST-large-Embedding-v0`. Each of these returned a list of evidence, obtained as described in Section 2.2, and the union of all unique evidence items was then taken as "*proposals*".

For Task **B** and **C** in Section 2.2, when processing a new test post, we identified the 5 most similar posts from the training set to provide examples that would assist the LLM in generation. To select these 5 most relevant posts, we first retrieved the 50 most relevant posts using a zero-shot retrieval approach, `contriever`, and then obtained the 5 most similar posts from this initial set by using a re-ranking model, specifically `ms-marco-MiniLM-L-6-v2`. The retrieved posts were added to the LLMs prompts as examples to follow.

From Posts to Timelines: Modeling Mental Health Dynamics from Social Media Timelines with Hybrid LLMs

Zimu Wang[1,2,*], **Hongbin Na**[3,*], **Rena Gao**[4], **Jiayuan Ma**[5], **Yining Hua**[6], **Ling Chen**[3], **Wei Wang**[1]
[1]School of Advanced Technology, Xi'an Jiaotong-Liverpool University
[2]University of Liverpool [3]University of Technology Sydney
[4]The University of Melbourne [5]The University of Sydney [6]Harvard University
Zimu.Wang19@student.xjtlu.edu.cn, Hongbin.Na@student.uts.edu.au

Abstract

Social media data is recognized for its usefulness in the early detection of mental disorders; however, there is a lack of research focused on modeling individuals' longitudinal mental health dynamics. Moreover, fine-tuning large language models (LLMs) on large-scale, annotated datasets presents challenges due to privacy concerns and the difficulties on data collection and annotation. In this paper, we propose a novel approach for modeling mental health dynamics using hybrid LLMs, where we first apply both classification-based and generation-based models to identify adaptive and maladaptive evidence from individual posts. This evidence is then used to predict well-being scores and generate post-level and timeline-level summaries. Experimental results on the CLPsych 2025 shared task demonstrate the effectiveness of our method, with the generative-based model showing a marked advantage in evidence identification.

1 Introduction

Mental disorders have emerged as a critical global challenge, being recognized as one of the leading contributors to illness and disability (Hua et al., 2024; Na et al., 2025). The World Health Organization[1] (WHO) reports that over 25% individuals will experience mental or neurological disorders in their lifetime. This phenomenon has been further exacerbated by COVID-19, leading to significant increases in anxiety and depression (Penninx et al., 2022), underscoring the urgent need for enhanced monitoring systems to facilitate early intervention.

Despite this phenomenon, mental health services remain undertreated and under-resourced, particularly in low- and middle-income countries. Social media platforms, such as X[2] and Reddit[3], offer significant potential for the early detection of mental disorders, as users regularly express their thoughts, emotions, and behaviors on these platforms. By leveraging machine learning algorithms, especially those utilizing large language models (LLMs), to analyze this data, it becomes possible to identify patterns indicative of disorders like depression or anxiety, facilitating earlier interventions (Shing et al., 2018; Tsakalidis et al., 2022a,b; Chim et al., 2024; Wang et al., 2024a,b; Qian et al., 2024). However, these methods are often limited to individual posts, with the longitudinal modeling of individuals' mental health dynamics largely overlooked in prior research. Moreover, due to privacy concerns and the challenges associated with collecting and annotating mental health data, fine-tuning LLMs on large-scale, curated annotated datasets remains challenging. As a result, prompt engineering are emerged as a promising and valuable line for mental health-related research (Peng et al., 2023; Na et al., 2024; Ma et al., 2025).

In this paper, we introduce a novel approach to modeling mental health dynamics from social media using hybrid LLMs, where the tasks explored include *Adaptive/Maladaptive Evidence Identification*, *Overall Well-being Rating*, and *Post-level and Timeline-level Summaries*. Specifically, in accordance to the prompts organized in Figure 3, we first leverage both classification-based and generation-based models with LLMs to identify adaptive and maladaptive evidence from individual posts (Figure 1). This evidence is then integrated to predict users' well-being scores and generate post-level and timeline-level summaries (Figure 2). The evidence identification and well-being rating tasks are performed using fine-tuned LLMs based on Qwen2.5-7B (Yang et al., 2025), while the summaries are generated using Qwen2.5-32B through in-context learning (ICL, Brown et al., 2020). In-context examples are selected from the training set based on the highest post similarity with the

*Equal contribution.
[1]https://www.who.int/
[2]https://x.com/
[3]https://www.reddit.com/

Proceedings of the 10th Workshop on Computational Linguistics and Clinical Psychology (CLPsych 2025), pages 249–255
May 3, 2025

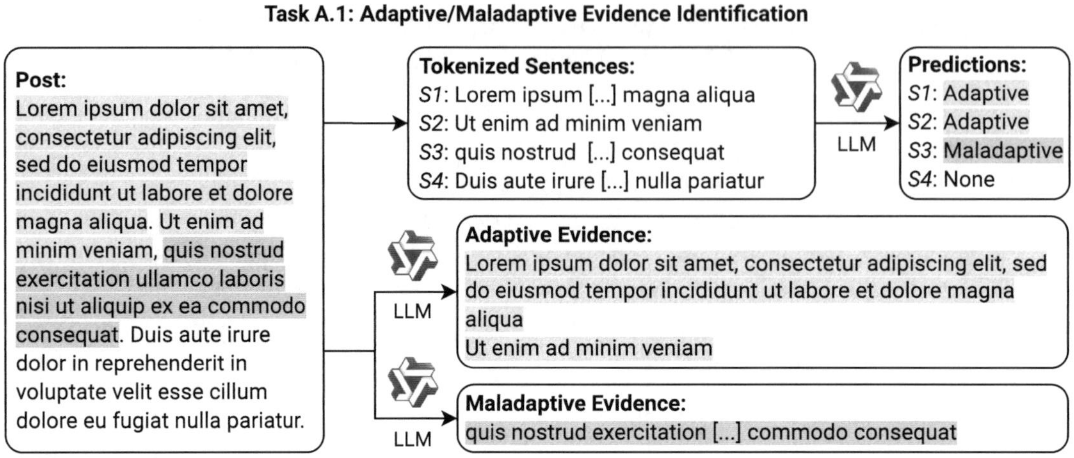

Figure 1: Framework for **Task A.1** using classification-based and generation-based models, with the post replaced by *lorem ipsum* for illustrative purposes.

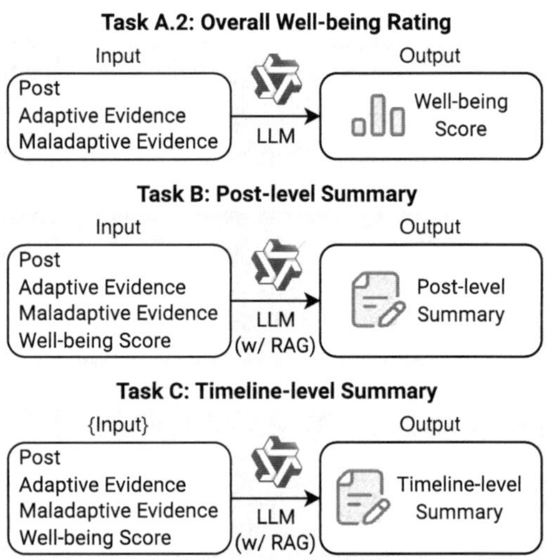

Figure 2: Frameworks for **Tasks A.2, B, and C**.

BGE-Large embedding model (Xiao et al., 2024). Experimental results on the CLPsych 2025 shared task (Tseriotou et al., 2025) highlight the effectiveness of our method, and the generative-based model demonstrates a significant advantage in evidence identification. Both the generative-based model and the classification-based model achieve similar performance when their extracted evidence is integrated into subsequent tasks.

2 Methodology

2.1 Evidence Identification

We begin by identifying the adaptive and maladaptive evidence in each post, where each post may contain a single self-state, two complementary self-states, or neither, represented by continuous spans extracted from the post. To accomplish this, we employ classification-based and generation-based models for this process:

Classification-based Model. Given a post p consisting of multiple sentences, we first tokenize the text into individual sentences $\{t_1, \ldots, t_M\}$, where M represents the total number of sentences in the post. We then apply a fine-tuned LLM, denoted as $\mathcal{M}_C$, to classify each sentence into an estimated label $\hat{e} \in \mathcal{E}$ conditioned on an instruction I_C:

$$\hat{e} = \arg\max_{e} P(e|t_i, I_C, \mathcal{M}_C), \quad (1)$$

where $\mathcal{E} = \{\text{ADAPTIVE}, \text{MALADAPTIVE}, \text{NONE}\}$. t_i ($i \in [1, M]$) represents an input tokenized sentence. Intuitively, this method is highly dependent on the results of sentence tokenization, and may not be well-suited for cases involving complementary self-states, where both adaptive and maladaptive evidence might coexist.

Generation-based Models. In this approach, we leverage two LLMs, $\mathcal{M}_A$ and $\mathcal{M}_M$, each independently trained to identify adaptive and maladaptive evidence, respectively, enabling the direct generation of target evidence within each post p:

$$E_A = \mathcal{M}_A(p, I_A), \quad E_M = \mathcal{M}_M(p, I_M), \quad (2)$$

where I_A and I_M represent instructions for extracting adaptive and maladaptive evidence, while E_A and E_M denote the corresponding lists of sentences containing each type of evidence.

2.2 Overall Well-being Rating

The well-being score, derived from the Global Assessment of Functioning (GAF, American Psychiatric Association et al., 1994), measures an individual's overall functioning across three key domains: social functioning, occupational functioning, and psychological well-being. In this work, we utilize a fine-tuned LLM, denoted as $\mathcal{M}_W$, to predict the well-being score s based on a given post p and its corresponding adaptive and maladaptive evidence, E_A and E_M, respectively, as follows:

$$s = \mathcal{M}_W(p, E_A, E_M, I_W), \tag{3}$$

where I_W represents the instruction for predicting the well-being score.

2.3 Post-level and Timeline-level and Summaries

Next, we generate a post-level summary that captures the interaction between adaptive and maladaptive states identified in each post. Previous research has revealed that prompting larger LMs yields superior summarization performance than fine-tuning smaller models (Thulke et al., 2024); therefore, we leverage ICL (Brown et al., 2020) that conditions LLMs with few-shot demonstrations in producing more effective summaries. To identify the most suitable in-context example, given a candidate post p and a set of annotated posts $\mathcal{D} = \{p'_1, \ldots, p'_N\}$ belonging to multiple users, we first utilize an embedding model $Emb(\cdot)$ to generate the embeddings for the posts, and then locate the post p' that exhibits the highest semantic similarity to p:

$$\boldsymbol{v}_p = Emb(p), \tag{4}$$

$$\boldsymbol{v}_{p'_i} = Emb(p'_i) \quad \forall p'_i \in \mathcal{D}, \tag{5}$$

$$p' = \arg\max_{p'_i \in \mathcal{D}} \frac{\boldsymbol{v}_p \cdot \boldsymbol{v}_{p'_i}}{||\boldsymbol{v}_p|| \cdot ||\boldsymbol{v}_{p'_i}||}, \tag{6}$$

where $\boldsymbol{v}_{(\cdot)}$ denotes the embedding of a given post. Afterwards, we generate the summary m based on the post p and its corresponding evidence E_A and E_M, incorporating the retrieved $\{p', E'_A, E'_M, m'\}$ with an LLM $\mathcal{M}$ and an instruction I_{PS}:

$$m = \mathcal{M}(p, E_A, E_M, p', E'_A, E'_M, m', I_{PS}). \tag{7}$$

For the timeline-level summary, the generation process follows a similar approach to the post-level summary. However, instead of individual posts, all posts associated with each user are concatenated to identify the most relevant in-context example. Additionally, evidence from all posts is incorporated during the generation process.

3 Experiments

3.1 Dataset

The CLPsych 2025 shared task (Tseriotou et al., 2025) integrates longitudinal modeling of social media timelines with evidence generation, offering annotated evidence for both adaptive and maladaptive self-states, as well as a score representing the overall well-being reflected in each post. It also provides post-level summaries that capture the interaction between adaptive and maladaptive self-states within individual posts, and timeline-level summaries that offer clinical insights, along with a dynamic narrative of mental state fluctuations and trajectories over time. This task is organized around the MIND framework (Slonim, 2024), a pan-theoretical model that conceptualizes human experience as a series of self-states that evolve and fluctuate over time.

3.2 Baselines and Evaluation Metrics

Task A.1. The baselines for the task *Adaptive/ Maladaptive Evidence Identification* include a zero-shot Llama 3.1-8B (Grattafiori et al., 2024) and a fine-tuned BART-Large (Lewis et al., 2020) model. The input for both models consists of either a single post or a window of five consecutive posts. Experimental results were evaluated using recall-oriented BERTScore (Zhang et al., 2020) and weighted recall metrics computed over adaptive and maladaptive spans.

Task A.2. The baselines for *Overall Well-being Rating* include zero-shot Llama 3.1-8B and a fine-tuned BERT model (Devlin et al., 2019), where the input for both models consists of either a single post or a window of five consecutive posts. Metrics for this task include Mean Squared Error (MSE), computed for each post within a timeline and then averaged across all timelines. The MSE for posts that indicate serious impairment (1 to 4), impaired (5 to 6), or minimal impairment (7 to 10) to functioning were also calculated. Macro F1-scores were also evaluated based on the aforementioned classes and their corresponding ranges.

Tasks B and C. The baselines for the *Post-level and Timeline-level Summary* tasks include a zero-shot Llama 3.1-8B model, with an intermediate post-level summary also utilized to generate a self-state summary. The evaluation metrics encompass mean consistency, maximum contradiction, and maximum entailment.

Model	Overall		Adaptive		Maladaptive	
	R	W	R	W	R	W
Llama 3.1	35.8	33.7	30.6	29.3	38.2	41.1
w/ Win.	49.6	26.2	36.5	25.2	**62.7**	27.2
BART	40.4	38.2	47.3	46.4	33.6	29.9
w/ Win.	26.0	25.8	28.2	27.9	23.8	23.7
Ours (C.)	34.1	31.4	24.9	24.9	43.3	37.8
Ours (G.)	**50.7**	**45.6**	**49.9**	**46.5**	51.6	**44.6**

Table 1: Experimental results of our proposed method against baselines on Task A.1 (*Adaptive/Maladaptive Evidence Identification*). "R" and "W" denote recall and weighted recall; "C." and "G." denote classification-based and generation-based models; *w/ Win.* represents the incorporation of post windows.

Model	MSE↓	M-S	M-I	M-M	F1
Llama 3.1	4.22	4.67	3.66	3.20	25.5
w/ Windows	4.46	1.67	3.20	7.07	27.4
BERT	2.90	3.38	2.32	2.81	13.9
w/ Windows	4.56	5.68	**1.01**	5.34	13.5
Ours (w/ Class.)	**2.01**	1.25	3.11	**2.16**	**36.6**
Ours (w. Gen.)	2.17	**1.23**	3.60	2.31	34.3

Table 2: Experimental results of our proposed method against baselines on Task A.2 (*Overall Well-being Rating*). "M-S", "'M-I", and "M-M" denote MSE across serious impairment, impaired, and minimal impairment.

3.3 Experiment Setup

We utilized two distinct LLMs for different tasks in our research. For Tasks A.1 and A.2, we fine-tuned Qwen2.5-7B (Yang et al., 2025) on the relevant datasets using LoRA (Hu et al., 2022). During the fine-tuning process, we configured the number of epochs to 10, the batch size to 2, and set the gradient accumulation steps to 8. For Tasks B and C, we used Qwen2.5-32B as the base model, and we leveraged BGE-Large (`EN-v1.5`, Xiao et al., 2024) as the embedding model to select the in-context example with the highest similarity to the target post. All experiments were conducted on 2 NVIDIA L20 graphics cards.

3.4 Experimental Results

Tables 1, 2, 3, and 4 present the performance of our approach compared to the baselines for Tasks A.1, A.2, B, and C, respectively. From the results, we observed that our method outperformed the majority of the baselines except on Task C. Among the two evidence identification models we proposed, the generative-based model proved to be significantly more effective due to its more accurate identification of evidence locations. However, we also found that both approaches performed comparably when their extracted evidence was incorporated into subsequent tasks. For Task C, as indicated in Table 4, none of the methods outperformed the baselines. This underscores a significant limitation of current LLMs in long-term, timeline-level summarization under standard few-shot prompting, pointing to a promising avenue in future research, such as incorporating post windows into the summarization process, as evidenced by the baseline results.

Model	Mean Con.	Max Con.↓	Max Ent.
Llama 3.1	88.0	84.8	–
w/ Windows	**89.1**	83.6	–
Ours (w/ Class.)	82.9	80.8	**75.0**
Ours (w/ Gen.)	88.0	**78.1**	69.2

Table 3: Experimental results of our proposed method against baselines on Task B (*Post-level Summary).* "Mean Con.", "Max Con.", and "Max Ent." denote mean consistency, maximum contradiction, and maximum entailment.

Model	Mean Con.	Max Con.↓
Llama 3.1	87.8	79.9
w/ Windows	**94.0**	**58.0**
Ours (w/ Class.)	91.4	78.5
Ours (w/ Gen.)	91.5	87.6

Table 4: Experimental results of our proposed method against baselines on Task C (*Timeline-level Summary).*

4 Conclusions and Future Work

We introduced a novel approach to modeling mental health dynamics from social media using hybrid LLMs, where the tasks explored include *Adaptive/-Maladaptive Evidence Identification*, *Overall Well-being Rating*, and *Post-level and Timeline-level Summary*. Specifically, we first leveraged both classification-based and generation-based models with LLMs to identify adaptive and maladaptive evidence from individual posts. This evidence was then integrated to predict users' well-being scores and generate post-level and timeline-level summaries. Experimental results on the CLPsych 2025 shared task highlighted the effectiveness of our method, and the generative-based model demonstrated a significant advantage in evidence identification. In the future, we will dedicate on proposing more advanced models for generating timeline-level summaries, such as incorporating post windows into the summarization process.

Limitations

Our study has two primary limitations: (1) Due to time constraints, we evaluated our approach using two state-of-the-art LLMs, Qwen2.5-7B and Qwen2.5-32B, while more established models such as Llama3.1-8B/70B were not included in our experiments; (2) Our method, regardless of whether the evidence was obtained through classification-based or generation-based models, did not outperform the baseline models when generating timeline-level summaries. Future work could address this limitation, potentially by incorporating post windows into the summarization process, as evidenced by the baseline results.

Ethical Considerations

We discuss the ethical considerations and broader impact of this work here: (1) **Intellectual Property:** Our approach is applied to the CLPsych 2025 shared task, adhering to the data access form and ensuring compliance with data protection protocolsensuring responsible data handling practices. All illustrative examples, including those in figures and prompts, have been replaced by *lorem ipsum* to respect data confidentiality. (2) **Intended Use.** This approach is designed for research purposes focused on understanding mental health patterns over time through social media timelines. It includes identifying adaptive and maladaptive evidence, predicting overall well-being scores, and summarizing posts and timelines. (3) **Misuse Risks**. This method is not intended for processing sensitive, personal, or non-consensually obtained data. Furthermore, the output generated is inherently dependent on the input text and should not be used to support financial, political, or clinical decision-making without appropriate human oversight and ethical approval.

References

A American Psychiatric Association, American Psychiatric Association, et al. 1994. *Diagnostic and statistical manual of mental disorders: DSM-IV*, volume 4. American psychiatric association Washington, DC.

Tom Brown, Benjamin Mann, Nick Ryder, Melanie Subbiah, Jared D Kaplan, Prafulla Dhariwal, Arvind Neelakantan, Pranav Shyam, Girish Sastry, Amanda Askell, Sandhini Agarwal, Ariel Herbert-Voss, Gretchen Krueger, Tom Henighan, Rewon Child, Aditya Ramesh, Daniel Ziegler, Jeffrey Wu, Clemens Winter, Chris Hesse, Mark Chen, Eric Sigler, Mateusz Litwin, Scott Gray, Benjamin Chess, Jack Clark, Christopher Berner, Sam McCandlish, Alec Radford, Ilya Sutskever, and Dario Amodei. 2020. Language models are few-shot learners. In *Advances in Neural Information Processing Systems*, volume 33, pages 1877–1901.

Jenny Chim, Adam Tsakalidis, Dimitris Gkoumas, Dana Atzil-Slonim, Yaakov Ophir, Ayah Zirikly, Philip Resnik, and Maria Liakata. 2024. Overview of the CLPsych 2024 shared task: Leveraging large language models to identify evidence of suicidality risk in online posts. In *Proceedings of CLPsych*, pages 177–190.

Jacob Devlin, Ming-Wei Chang, Kenton Lee, and Kristina Toutanova. 2019. BERT: Pre-training of deep bidirectional transformers for language understanding. In *Proceedings of NAACL-HLT*, pages 4171–4186.

Aaron Grattafiori, Abhimanyu Dubey, Abhinav Jauhri, Abhinav Pandey, Abhishek Kadian, Ahmad Al-Dahle, Aiesha Letman, Akhil Mathur, Alan Schelten, Alex Vaughan, et al. 2024. The llama 3 herd of models. *Preprint*, arXiv:2407.21783.

Edward J Hu, yelong shen, Phillip Wallis, Zeyuan Allen-Zhu, Yuanzhi Li, Shean Wang, Lu Wang, and Weizhu Chen. 2022. LoRA: Low-rank adaptation of large language models. In *Proceedings of ICLR*.

Yining Hua, Fenglin Liu, Kailai Yang, Zehan Li, Hongbin Na, Yi han Sheu, Peilin Zhou, Lauren V. Moran, Sophia Ananiadou, Andrew Beam, and John Torous. 2024. Large language models in mental health care: a scoping review. *Preprint*, arXiv:2401.02984.

Mike Lewis, Yinhan Liu, Naman Goyal, Marjan Ghazvininejad, Abdelrahman Mohamed, Omer Levy, Veselin Stoyanov, and Luke Zettlemoyer. 2020. BART: Denoising sequence-to-sequence pre-training for natural language generation, translation, and comprehension. In *Proceedings of ACL*, pages 7871–7880.

Jiayuan Ma, Hongbin Na, Zimu Wang, Yining Hua, Yue Liu, Wei Wang, and Ling Chen. 2025. Detecting conversational mental manipulation with intent-aware prompting. In *Proceedings of COLING*, pages 9176–9183.

Hongbin Na, Yining Hua, Zimu Wang, Tao Shen, Beibei Yu, Lilin Wang, Wei Wang, John Torous, and Ling Chen. 2025. A survey of large language models in psychotherapy: Current landscape and future directions. *Preprint*, arXiv:2502.11095.

Hongbin Na, Tao Shen, Shumao Yu, and Ling Chen. 2024. Multi-session client-centered treatment outcome evaluation in psychotherapy. *Preprint*, arXiv:2410.05824.

Hao Peng, Xiaozhi Wang, Jianhui Chen, Weikai Li, Yunjia Qi, Zimu Wang, Zhili Wu, Kaisheng Zeng, Bin Xu, Lei Hou, and Juanzi Li. 2023. When does in-context learning fall short and why? a study on specification-heavy tasks. *Preprint*, arXiv:2311.08993.

Brenda WJH Penninx, Michael E Benros, Robyn S Klein, and Christiaan H Vinkers. 2022. How covid-19 shaped mental health: from infection to pandemic effects. *Nature medicine*, 28(10):2027–2037.

Lu Qian, Yuqi Wang, Zimu Wang, Haiyang Zhang, Wei Wang, Ting Yu, and Anh Nguyen. 2024. Domain-specific guided summarization for mental health posts. *Preprint*, arXiv:2411.01485.

Han-Chin Shing, Suraj Nair, Ayah Zirikly, Meir Friedenberg, Hal Daumé III, and Philip Resnik. 2018. Expert, crowdsourced, and machine assessment of suicide risk via online postings. In *Proceedings of CLPsych*, pages 25–36.

Dana A Slonim. 2024. Self-other dynamics (sod): A transtheoretical coding manual.

David Thulke, Yingbo Gao, Rricha Jalota, Christian Dugast, and Hermann Ney. 2024. Prompting and fine-tuning of small llms for length-controllable telephone call summarization. In *Proceedings of FLLM*, pages 305–312.

Adam Tsakalidis, Jenny Chim, Iman Munire Bilal, Ayah Zirikly, Dana Atzil-Slonim, Federico Nanni, Philip Resnik, Manas Gaur, Kaushik Roy, Becky Inkster, Jeff Leintz, and Maria Liakata. 2022a. Overview of the CLPsych 2022 shared task: Capturing moments of change in longitudinal user posts. In *Proceedings of CLPsych*, pages 184–198.

Adam Tsakalidis, Federico Nanni, Anthony Hills, Jenny Chim, Jiayu Song, and Maria Liakata. 2022b. Identifying moments of change from longitudinal user text. In *Proceedings of ACL*, pages 4647–4660.

Talia Tseriotou, Jenny Chim, Ayal Klein, Aya Shamir, Guy Dvir, Ali Iqra, Cian Kennedy, Guneet Singh Kohli, Anthony Hills, Ayah Zirikly, Dana Atzil-Slonim, and Maria Liakata. 2025. Overview of the clpsych 2025 shared task: Capturing mental health dynamics from social media timelines. In *Proceedings of CLPsych*.

Yuqi Wang, Zimu Wang, Nijia Han, Wei Wang, Qi Chen, Haiyang Zhang, Yushan Pan, and Anh Nguyen. 2024a. Knowledge distillation from monolingual to multilingual models for intelligent and interpretable multilingual emotion detection. In *Proceedings of WASSA*, pages 470–475.

Zimu Wang, Wei Wang, Qi Chen, Qiufeng Wang, and Anh Nguyen. 2024b. Generating valid and natural adversarial examples with large language models. In *Proceedings of CSCWD*, pages 1716–1721.

Shitao Xiao, Zheng Liu, Peitian Zhang, Niklas Muennighoff, Defu Lian, and Jian-Yun Nie. 2024. C-pack: Packed resources for general chinese embeddings. In *Proceedings of SIGIR*, page 641–649.

An Yang, Baosong Yang, Beichen Zhang, Binyuan Hui, Bo Zheng, Bowen Yu, Chengyuan Li, Dayiheng Liu, Fei Huang, Haoran Wei, et al. 2025. Qwen2.5 technical report. *Preprint*, arXiv:2412.15115.

Tianyi Zhang, Varsha Kishore, Felix Wu, Kilian Q. Weinberger, and Yoav Artzi. 2020. Bertscore: Evaluating text generation with bert. In *Proceedings of ICLR*.

A Prompts

Classification-based Evidence Identification

```
Determine whether the following sentence demonstrates an adaptive state (1), a maladaptive state (2), or
 neither (0).
{Post Sentence}
```

Generation-based Evidence Identification

```
Identify all sentences or phrases from the following text that demonstrate adaptive states.
{Post Text}
```

Overall Well-being Rating

```
Based on the following text and the provided evidence of adaptive and maladaptive states, assess the
author's mental well-being. Provide a score from 1 to 10, where 1 indicates extremely poor and 10
indicates excellent.
Text: {Post Text}
Adaptive Evidence: {Adaptive Evidence}
Maladaptive Evidence: {Maladaptive Evidence}
```

Post-level Summary

```
Summarize the following post, considering the evidence of adaptive and maladaptive states and the well-
being score.

Post:
{Post Text for In-context Example}
Adaptive Evidence:
{Adaptive Evidence for In-context Example}
Maladaptive Evidence:
{Maladaptive Evidence for In-context Example}
Well-being Score:
{Well-being Score for In-context Example}
Summary:
{Summary for In-context Example}

Post:
{Post Text}
Adaptive Evidence:
{Adaptive Evidence}
Maladaptive Evidence:
{Maladaptive Evidence}
Well-being Score:
{Well-being Score}
Summary:
```

Figure 3: Prompts designed for each tasks. The prompt for timeline-level summary closely follows the structure of the one for post-level summary but integrates the information of multiple posts, with the incorporation of information from multiple posts to generate a summary of the entire timeline.

Prompt Engineering for Capturing Dynamic Mental Health Self-States from Social Media Posts

Callum Chan[1], **Sunveer Khunkhun**[1],
Diana Inkpen[1], **Juan Antonio Lossio-Ventura**[2]
[1]School of Electrical Engineering and Computer Science, University of Ottawa, Canada
[2]Machine Learning Core, National Institute of Mental Health,
National Institutes of Health, USA
{cchan073, skhun073, dinkpen}@uottawa.ca, juan.lossio@nih.gov

Abstract

With the advent of modern Computational Linguistic techniques and the growing societal mental health crisis, we contribute to the field of Clinical Psychology by participating in the CLPsych 2025 shared task. This paper describes the methods and results obtained by the uOttawa team's submission (which included a researcher from the National Institutes of Health in the USA, in addition to three researchers from the University of Ottawa, Canada). The task consists of four subtasks focused on modeling longitudinal changes in social media users' mental states and generating accurate summaries of these dynamic self-states. Through prompt engineering of a modern large language model (Llama-3.3-70B-Instruct), the uOttawa team placed first, sixth, fifth, and second, respectively, for each subtask, amongst the other submissions. This work demonstrates the capacity of modern large language models to recognize nuances in the analysis of mental states and to generate summaries through carefully crafted prompting.

1 Introduction

Large Language Models (LLMs) have been explored in the mental health domain for tasks such as analyzing emotional states, sentiment, and depression (Xu et al., 2024a; Yang et al., 2024; Xin et al., 2024; Malgaroli et al., 2025). Their ability to process language at scale presents potential benefits for automating various processes in Clinical Psychology. Prompt engineering helps optimize LLM performance across domains (Liu et al., 2023), including biomedical and clinical applications (Hu et al., 2024; Sivarajkumar et al., 2024), where zero-shot and few-shot techniques have been investigated. These recent studies and our observations suggest that tailoring prompts to specific tasks can help improve performance in mental and clinical Natural Language Processing (NLP) tasks (Xu et al., 2024b), such as evaluating changes in individuals' mental well-being over time (Owen et al., 2023). Analyzing shifts in mental states over time provides valuable insights into overall mental health (Tsakalidis et al., 2022). This shared task builds on previous efforts by incorporating the generation of easily interpretable summaries, encouraging the Computational Linguistics community to further explore the dynamics of self-states. In this work, we use Llama-3.3-70B-Instruct (Grattafiori et al., 2024) to demonstrate how modern LLMs can identify textual indicators of adaptive and maladaptive self states, and generate summaries of social media posters' mental state as it changes over time. An adaptive self-state is characterized by aspects of an individual's mental state which facilitates the realization of basic needs and desires. Contrarily, a maladaptive self-state is characterized by aspects which impede the realization of basic needs and desires (Slonim, 2024).

2 Shared Task

The CLPsych 2025 shared task consists of four subtasks (Tseriotou et al., 2025). The first one (Task A.1) is to identify and extract textual evidence (spans) of adaptive and maladaptive self states of individual social media posts. The second (Task A.2) is to generate a well-being score from 1 to 10 of each post. The third (Task B) is to generate a summary for each post that describes the evidence contributing to the dominant self state, the non-dominant self state and the interplay between the two. The final task (Task C) is to generate a summary with similar requirements to Task B, but over a timeline of posts from a given user.

3 Dataset

The dataset was developed by the shared tasks organizers over the previous several years (Shing et al., 2018; Tsakalidis et al., 2022; Zirikly et al., 2019; Tseriotou et al., 2025). Thirty timelines of posts

Proceedings of the 10th Workshop on Computational Linguistics and Clinical Psychology (CLPsych 2025), pages 256–267
May 3, 2025

were provided to the shared task participants this year to use as training data, totaling 343 posts in all. This training dataset is fully annotated with gold evidence of present self states, well-being scores, post-level self-state summaries, and timeline-level self-state summaries. Finally, ten additional timelines, totaling 94 posts in all, was provided to the participants as test data. These timelines were used in the evaluation of our proposed methods.

4 Methods

To address the four subtasks of the CLPsych 2025 shared task, we employed prompt engineering techniques on a local instance of the Llama-3.3-70B-Instruct model. For Task A.2, that expects numeric predictions, we also implemented linear and multinomial logistic regression classifiers for various types of embeddings.

4.1 Prompt Engineering Strategies

We employed three distinct prompting strategies in our official submissions: zero-shot prompts, structured prompts with one example, and structured prompts with multiple examples. For each strategy, separate prompt templates were tailored for each of the four tasks. Thus, each strategy consisted of four prompt templates (totaling 12 templates for our official submissions). The templates included the following features: task requirements, definition of terms such as self state, adaptive and maladaptive, and guidelines to format responses. The prompts for Task A.2 included explanations of how to characterize well-being. The prompts for Task C included the responses from Task B, in order to generate the post timeline summary. Table 1 shows the key features of each strategy. Appendix A.3 presents examples of the structured contextual one-shot prompt used for tasks A.1, A.2, B, and C.

4.2 Regression Models on Embeddings

Our additional approach for Task A.2 consisted of two stages, which we describe below.

Embeddings: Embeddings were generated using various transformer-based LLMs. In the first stage, each post was passed to these models to generate embeddings, which are numerical vectors of hidden dimension d. Each LLM tokenizes the post and converts each token into vectors based on the context. The final vector for each post is obtained by averaging the token vectors. We used the base and large variants of BERT (Devlin et al., 2019), RoBERTa (Zhuang et al., 2021), as well as MentalBERT (Ji et al., 2022) and MentalRoBERTa (Ji et al., 2022), which are BERT and RoBERTa models pretrained on additional mental health-related data. We also incorporated SBERT (Sentence BERT) (Reimers and Gurevych, 2019) and LLaMA-3.3-70B-Instruct. For comparison, we included traditional techniques like TF-IDF and Bag-of-Words representations. Each embedding had a different context length and dimension depending on the model used. For more details, see Table 3 in Appendix A.1.

Regression: We trained linear (LR) and multinomial logistic regression (MLR) models on the embedding vectors of all posts. The models were trained and evaluated using 5-fold cross-validation, with hyperparameters optimized through grid search. For MLR, data was stratified by well-being score, and the loss function was adjusted for class imbalance by weighting errors. For LR, the output was rounded to the nearest integer.

4.3 Evaluation Metrics

Performance of each submission was evaluated using task-specific metrics specified and applied by the shared task organizers (Tseriotou et al., 2025). For Task A.1, recall and weighted recall was computed using BERTScore (Zhang et al., 2020). Incorrectly predicted empty span lists received a score of 0. Additionally, separate recall scores were provided for adaptive only spans and maladaptive only spans. Task A.2 was evaluated using the Mean Squared Error (MSE) and Macro-F1 across all posts. Additional MSE scores were also provided per well-being severity class (serious: 1–4, impaired: 5–6, minimal: 7–10), with incorrect null predictions being penalized by the maximum error. Tasks B and C assessed summary quality using a Natural Language Inference (NLI) model to measure mean consistency (absence of contradiction) and maximum contradiction between submitted and gold summaries; incorrectly predicted null summaries defaulted to 0. Task C applied these metrics at the timeline level.

5 Results

The results of our submissions to the CLPsych 2025 shared task demonstrate the effectiveness of prompt engineering in leveraging LLMs for mental health analysis. The performance of our methods across

	Contextual Zero-Shot Prompting (uOttawa 1)	**Structured Contextual One-Shot Prompting** (uOttawa 2)	**Structured Contextual Few-Shot Prompting** (uOttawa 3)
Prompt Structure	Prompts were loosely structured, with less explicit delineation between sections such as objectives, definitions, and output guidelines. Domain-specific terms (e.g., adaptive and maladaptive self states) were included into the prompts to provide context.	Prompts were finely structured, including clear delineations for task objectives, definitions of key terms, output guidelines, and one annotated example extracted from the training data.	Prompts were structured similarly to the second approach, with delineations for task objectives, definitions, and output guidelines. However, this submission included multiple examples: seven examples for Tasks A.1, A.2, and B, and three examples for Task C.
Approach	This strategy relied on the LLM's ability to infer task requirements and generate responses without explicit examples. The prompts were designed to guide the model in identifying textual evidence of self states (Task A.1), generating well-being scores (Task A.2), and creating summaries for individual posts (Task B), and timelines (Task C).	This strategy was designed to improve the LLM's understanding of the task by providing a single example for each subtask. The example served as a reference for the model to better align its responses with the desired format and content for more accurate outputs.	By providing multiple examples, this strategy aimed to further refine the LLM's ability to generate accurate and contextually appropriate responses. The examples were carefully selected to cover a range of self states, post lengths and timeline lengths in an effort to limit its reliance on existing knowledge and to prepare the LLM for a diverse set of possible test data.

Table 1: Prompting Strategies: Key Features.

the four subtasks is presented in Table 2.[1]

We also included scores for a baseline method using zero-shot learning with simple prompts and a smaller model (Llama-3.1-8B-Instruct), to assess the effectiveness of using the larger model (70B). Additionally, we included the results we obtained for Task A.2 using prompt engineering and linear regression.

Task A.1: Identification of Self States The uOttawa team achieved strong results in identifying self states, with uOttawa_2 (one-shot prompting) performing the best among our submissions and among all the submitted runs by the shared task participants. It achieved an overall recall of 0.637 (adaptive: 0.594, maladaptive: 0.681) and a weighted recall of 0.498 (adaptive: 0.542, maladaptive: 0.455). The results for the uOttawa_2 and uOttawa_3 submissions were better than those of uOttawa_1, highlighting the importance of structured prompts for this task. Surprisingly, the few-shot learning did not outperform the one-shot learning.

Task A.2: Well-Being Score For this task, uOttawa_3 (few-shot prompting) achieved the lowest overall MSE of 2.62, with strong performance across impairment levels (minimal: 2.91, impaired: 4.03, serious: 2.28). This demonstrates the effectiveness of providing multiple examples for accurate well-being prediction. The lowest MSE achieved by participating teams was 1.920. We ran additional experiments for this task (S4, S5, S6, and more), with better results as shown in Table 2. With S4 and S5, we obtained the lowest MSE, even compared to the official results obtained by other teams. Also, a linear regression classifier using Llama-3.3-70B embeddings achieved a competitive MSE score of 2.015. See Appendix A.2 for more details on the additional experiments.

Task B: Post-Level Summaries In post-level summary generation, uOttawa_2 (one-shot prompting) achieved the highest mean consistency (0.860),

[1]Lower scores are better for the metrics MSE and Max Contradiction.

Task	Metric	Baseline	Official Submissions S1	S2	S3	Additional Submissions S4	S5	S6
A.1	Recall ↑	0.405	0.469	**0.637**	<u>0.550</u>	-	-	-
	Weighted Recall ↑	0.214	0.386	**0.498**	<u>0.455</u>	-	-	-
A.2	MSE ↓	4.682	2.830	3.430	2.620	**1.673**	<u>1.693</u>	2.015
	Macro F1 ↑	0.286	0.355	**0.378**	0.302	<u>0.361</u>	<u>0.361</u>	0.348
B	Mean Consistency ↑	0.780	0.773	**0.860**	<u>0.859</u>	-	-	-
	Max Contradiction ↓	0.815	**0.756**	0.832	<u>0.804</u>	-	-	-
C	Mean Consistency ↑	0.897	<u>0.926</u>	0.918	**0.943**	-	-	-
	Max Contradiction ↓	<u>0.747</u>	0.794	0.751	**0.714**	-	-	-

Table 2: Performance of the uOttawa team across the four subtasks. The best scores are **bolded**, and the runners-up are <u>underlined</u>. The baseline was based on zero-shot prompts with Llama-3.1-8B-Instruct (small model). S1–S6 represent the file submissions we made for evaluation. S1, S2, and S3 correspond to uOttawa_1, uOttawa_2, and uOttawa_3, respectively. S4 and S5 were based on prompt engineering slightly different from the initial (S1, S2, and S3), with S4 containing 4 examples per class of well-being and S5 without examples. S6 was based on linear regression with Llama-3.3-70B-Instruct embeddings.

indicating coherent and consistent summaries. The few-shot strategy (uOttawa_3) also performed well, with a mean consistency of 0.859. The highest mean consistency achieved by a team was 0.910.

Task C: Timeline-Level Summaries For timeline-level summaries, uOttawa_3 (few-shot prompting) achieved the highest mean consistency (0.943) and the lowest max contradiction score (0.714), demonstrating its ability to capture longitudinal dynamics effectively. The highest mean consistency achieved by a team was 0.946, only a little higher than ours.

Overall Performance The uOttawa team placed first in Task A.1, sixth in Task A.2, fifth in Task B, and second in Task C in the shared task. These results showcase the potential of prompt engineering to enhance LLM performance in nuanced mental health analysis tasks. The one-shot and few-shot strategies proved effective in guiding the model.

6 Discussion

The results of team uOttawa's methods demonstrate competitive performance when applied to the CLPsych 2025 shared task.

For task A.1, the one-shot strategy outperformed zero-shot and few-shot approaches, achieving the highest recall (0.637) and weighted recall (0.498). The one-shot strategy's superior performance likely stems from how span identification benefits from precise, unambiguous guidance. A single well-chosen example appears sufficient to demonstrate what constitutes adaptive/maladaptive evidence, while avoiding the potential confusion that multiple examples might introduce. Few-shot prompts risk including borderline cases that could confuse the model's judgment, whereas zero-shot lacks any concrete reference points altogether. This suggests that for evidence extraction tasks, one carefully selected example may serve as an ideal template, providing just enough context to ensure consistent span detection without overcomplicating the prompt structure.

For task A.2, the few-shot strategy achieved the lowest MSE (2.62), demonstrating the value of multiple examples for fine-grained predictions. However, the relatively high MSE scores across all submissions suggest that well-being scoring remains challenging for strategies that do not rely on fine-tuning on training data. Therefore, we conducted additional experiments based on prompt engineering and regression models on embeddings (S4, S5, and S6 in Table 2), which improved MSE (1.673, 1.693, and 2.015 respectively, see appendix A.2), outperforming our initial results as well as those of the other participants. The relative success of few-shot prompting in numerical scoring could be explained by its ability to demonstrate the contextual nature of well-being assessments through multiple examples. By showing how similar phrases (e.g., "I'm exhausted") might receive different scores depending on surrounding context, the few-shot approach may help the model develop a more nuanced scoring rubric. However, the even stronger performance of regression models suggests an important limitation of prompting strategies for numerical tasks - they may ultimately be less effective than approaches that can directly learn statistical patterns from embeddings. This could indicate that numerical scoring depends more on quantitative feature recognition than qualitative example-based learning.

For task B, the minimal difference between one-shot (0.860 MSE) and few-shot (0.859) approaches suggests diminishing returns for additional examples in summary generation. While zero-shot (0.773) trailed both approaches, its relatively strong performance indicates that the base model already possesses substantial summarization capability. The pattern reveals a hierarchy of effectiveness: one-shot learning provides optimal guidance for most cases (balancing structure and simplicity), few-shot learning offers slight contradiction reduction (0.804 vs. 0.832) for complex posts, and zero-shot learning serves as a competent but less reliable baseline. This implies that while a single example sufficiently anchors the task, the choice between approaches could prioritize consistency over edge-case handling or the other way around.

For task C, the few-shot strategy excelled, achieving the highest mean consistency (0.943) and lowest max contradiction score (0.714). Few-shot prompting's strong performance in timeline analysis may reflect the fundamentally different cognitive demands of longitudinal reasoning compared to single-instance tasks. The multiple examples likely help the model recognize various temporal patterns and transitional relationships that would be difficult to convey with just one example. This could explain why the additional context proves so valuable: it may allow the model to build a more comprehensive understanding of how mental states evolve over time, including recognizing triggers (such as job loss) and their typical emotional consequences. The one-shot approach's limitation here suggests that temporal reasoning may require exposure to multiple case examples to be effective.

Overall, our results demonstrate the effectiveness of prompt engineering in guiding LLMs for mental health analysis. The one-shot strategy excels in tasks requiring precise identification and summarization, while the few-shot strategy is better for nuanced tasks. The zero-shot strategy, while competitive, consistently underperformed, highlighting the importance of examples and structured guidance.

7 Clinical Applications

Our work presents powerful state-of-the-art methods to the greater clinical community. Not only do these approaches achieve impressive results, they are also very accessible and can be easily implemented by those with little background in machine learning or artificial intelligence. One such example is the self state monitoring of consenting high-risk social media users. Using users' post history and new posts, social media administrators could use these strategies to automatically flag high-risk users showing signs of degrading well-being and an increasing dominant maladaptive self state. Summaries generated by our methods can then be used to guide more personalized intervention strategies instead of generic responses (for example, offering specific tailored advice to manage stress instead of merely suggesting to contact a mental health hotline).

8 Conclusion and Future Work

We have showcased how using structured prompts with one or a few examples can lead to very good results when detecting and summarizing mental states from social posts.

Future work includes the continuation of development for the few shot learning approach by exploring different numbers of examples and the careful selection of the most relevant examples embedded within the prompts. We should also experiment with different types and various sizes of LLMs. These could also be further pre-trained or fine-tuned on data from the mental health domain. Finally, for Task A.2 that outputs a numeric score, we plan to model a sequence of decisions and that uses features extracted from previous posts at each step.

Limitations

Our experiments are limited to the type of social media data available for the shared task, focusing exclusively on English-language posts.

Additionally, we tested only one type of LLM, Llama, using a small version for the baseline and a larger version for the main method. Additionally, we experimented with several regression models, each using different text embeddings for Task A.2. We scored the posts of each user in sequence, but did not condition the prediction for the current post on the previous prediction. Such a strategy would be useful to detect extended periods of a user exhibiting a dominant adaptive or maladaptive self state.

Furthermore, due to the limited time frame during which our submissions could be scored, we could not perform detailed ablation studies to analyze the specific aspects of our prompts which contributed to performance, nor explore alternative methods such as hyperparameter tuning of temperature or top_p.

Finally, biases introduced during the process of prompt engineering may skew the responses of the LLM and our results. One source of bias stems from our previous experiences and projects using the Llama model and influenced the way in which we structured our prompts and tuned them. Another source comes from the way we presented the definitions of adaptive and maladaptive self state in our prompts. This contextual information could distort the LLM's understanding of these terms and the task it is presented with.

Ethics

The data was collected from public sources and anonymized. However, it is still sensitive, since mental health status labels were assigned. For accessing the data, we signed the data sharing agreement for the shared task and complied with all the clauses therein. This ensures proper use of the data, solely for research purposes, as well as secure storage of the data. As requested by the shared task organizers, we did not use ChatGPT or other closed-source models that could use this data for further training or model refinement.

To preserve the privacy of the social media posters used in this shared task, this work should not be replicated using the data referenced throughout this paper. This work's contributions are merely the ideas it presents. While modern computation linguistic tools provide powerful means of mental health monitoring and assessment, we encourage the greater Artificial Intelligence community to take a measured approach when dealing with sensitive user data to ensure its privacy.

References

Jacob Devlin, Ming-Wei Chang, Kenton Lee, and Kristina Toutanova. 2019. BERT: Pre-training of deep bidirectional transformers for language understanding. In *Proceedings of the 2019 Conference of the North American Chapter of the Association for Computational Linguistics: Human Language Technologies, Volume 1 (Long and Short Papers)*, pages 4171–4186, Minneapolis, Minnesota. Association for Computational Linguistics.

Aaron Grattafiori, Abhimanyu Dubey, Abhinav Jauhri, Abhinav Pandey, Abhishek Kadian, Ahmad Al-Dahle, Aiesha Letman, Akhil Mathur, Alan Schelten, Alex Vaughan, Amy Yang, Angela Fan, Anirudh Goyal, Anthony Hartshorn, Aobo Yang, Archi Mitra, Archie Sravankumar, Artem Korenev, Arthur Hinsvark, and 542 others. 2024. The llama 3 herd of models. *Preprint*, arXiv:2407.21783.

Yan Hu, Qingyu Chen, Jingcheng Du, Xueqing Peng, Vipina Kuttichi Keloth, Xu Zuo, Yujia Zhou, Zehan Li, Xiaoqian Jiang, Zhiyong Lu, and 1 others. 2024. Improving large language models for clinical named entity recognition via prompt engineering. *Journal of the American Medical Informatics Association*, 31(9):1812–1820.

Shaoxiong Ji, Tianlin Zhang, Luna Ansari, Jie Fu, Prayag Tiwari, and Erik Cambria. 2022. MentalBERT: Publicly available pretrained language models for mental healthcare. In *Proceedings of the Thirteenth Language Resources and Evaluation Conference*, pages 7184–7190, Marseille, France. European Language Resources Association.

Pengfei Liu, Weizhe Yuan, Jinlan Fu, Zhengbao Jiang, Hiroaki Hayashi, and Graham Neubig. 2023. Pretrain, prompt, and predict: A systematic survey of prompting methods in natural language processing. *ACM Comput. Surv.*, 55(9).

Matteo Malgaroli, Katharina Schultebraucks, Keris Jan Myrick, Alexandre Andrade Loch, Laura Ospina-Pinillos, Tanzeem Choudhury, Roman Kotov, Munmun De Choudhury, and John Torous. 2025. Large language models for the mental health community: framework for translating code to care. *The Lancet Digital Health*.

David Owen, Dimosthenis Antypas, Athanasios Hassoulas, Antonio F Pardiñas, Luis Espinosa-Anke, Jose Camacho Collados, and 1 others. 2023. Enabling early health care intervention by detecting depression in users of web-based forums using language models: longitudinal analysis and evaluation. *JMIR AI*, 2(1):e41205.

Nils Reimers and Iryna Gurevych. 2019. Sentence-BERT: Sentence embeddings using Siamese BERT-networks. In *Proceedings of the 2019 Conference on Empirical Methods in Natural Language Processing and the 9th International Joint Conference on Natural Language Processing (EMNLP-IJCNLP)*, pages 3982–3992, Hong Kong, China. Association for Computational Linguistics.

Han-Chin Shing, Suraj Nair, Ayah Zirikly, Meir Friedenberg, Hal Daumé III, and Philip Resnik. 2018. Expert, crowdsourced, and machine assessment of suicide risk via online postings. In *Proceedings of the Fifth*

Workshop on Computational Linguistics and Clinical Psychology: From Keyboard to Clinic, pages 25–36, New Orleans, LA. Association for Computational Linguistics.

Sonish Sivarajkumar, Mark Kelley, Alyssa Samolyk-Mazzanti, Shyam Visweswaran, and Yanshan Wang. 2024. An empirical evaluation of prompting strategies for large language models in zero-shot clinical natural language processing: Algorithm development and validation study. *JMIR Medical Informatics*, 12.

Dana Atzil Slonim. 2024. Self-Other Dynamics (SOD): A Transtheoretical Coding Manual.

Adam Tsakalidis, Jenny Chim, Iman Munire Bilal, Ayah Zirikly, Dana Atzil-Slonim, Federico Nanni, Philip Resnik, Manas Gaur, Kaushik Roy, Becky Inkster, Jeff Leintz, and Maria Liakata. 2022. Overview of the CLPsych 2022 shared task: Capturing moments of change in longitudinal user posts. In *Proceedings of the Eighth Workshop on Computational Linguistics and Clinical Psychology*, pages 184–198, Seattle, USA. Association for Computational Linguistics.

Talia Tseriotou, Jenny Chim, Ayal Klein, Aya Shamir, Guy Dvir, Iqra Ali, Cian Kennedy, Guneet Singh Kohli, Anthony Hills, Ayah Zirikly, Dana Atzil-Slonim, and Maria Liakata. 2025. Overview of the clpsych 2025 shared task: Capturing mental health dynamics from social media timelines. In *Proceedings of the 10th Workshop on Computational Linguistics and Clinical Psychology*. Association for Computational Linguistics.

Alison W Xin, Dylan M Nielson, Karolin Rose Krause, Guilherme Fiorini, Nick Midgley, Francisco Pereira, and Juan Antonio Lossio-Ventura. 2024. Using large language models to detect outcomes in qualitative studies of adolescent depression. *Journal of the American Medical Informatics Association*, page ocae298.

Xuhai Xu, Bingsheng Yao, Yuanzhe Dong, Saadia Gabriel, Hong Yu, James Hendler, Marzyeh Ghassemi, Anind K. Dey, and Dakuo Wang. 2024a. Mental-llm: Leveraging large language models for mental health prediction via online text data. *Proc. ACM Interact. Mob. Wearable Ubiquitous Technol.*, 8(1).

Xuhai Xu, Bingsheng Yao, Yuanzhe Dong, Saadia Gabriel, Hong Yu, James Hendler, Marzyeh Ghassemi, Anind K. Dey, and Dakuo Wang. 2024b. Mental-llm: Leveraging large language models for mental health prediction via online text data. *Proc. ACM Interact. Mob. Wearable Ubiquitous Technol.*, 8(1).

Kailai Yang, Tianlin Zhang, Ziyan Kuang, Qianqian Xie, Jimin Huang, and Sophia Ananiadou. 2024. Mentallama: Interpretable mental health analysis on social media with large language models. In *Proceedings of the ACM Web Conference 2024*, WWW '24, page 4489–4500, New York, NY, USA. Association for Computing Machinery.

Tianyi Zhang, Varsha Kishore, Felix Wu, Kilian Q. Weinberger, and Yoav Artzi. 2020. Bertscore: Evaluating text generation with bert. *Preprint*, arXiv:1904.09675.

Liu Zhuang, Lin Wayne, Shi Ya, and Zhao Jun. 2021. A robustly optimized BERT pre-training approach with post-training. In *Proceedings of the 20th Chinese National Conference on Computational Linguistics*, pages 1218–1227, Huhhot, China. Chinese Information Processing Society of China.

Ayah Zirikly, Philip Resnik, Özlem Uzuner, and Kristy Hollingshead. 2019. CLPsych 2019 shared task: Predicting the degree of suicide risk in Reddit posts. In *Proceedings of the Sixth Workshop on Computational Linguistics and Clinical Psychology*, pages 24–33, Minneapolis, Minnesota. Association for Computational Linguistics.

A Appendix

A.1 Embeddings for Task A.2

Table 3 provides an overview of the context length and hidden dimension sizes for the LLMs used to generate embeddings for the additional experiments for Task A2, as well as the configurations for traditional techniques such as BoW and TF-IDF. The listed models, including BERT, SBERT, RoBERTa, and Llama, employ varying context lengths and dimensionalities, which are important factors for their performance in subsequent regression analysis.

Model	Context Length	Dimension
BERT-base	512	768
BERT-large	512	1024
RoBERTa	512	768
MentalBERT	512	768
MentalRoBERTa	512	768
SBERT	384	768
Llama-3.3-70B-Instruct	128K	8192
Bag-of-Words	N/A	3000
TF-IDF	N/A	3000

Table 3: Context length and hidden dimension sizes for the LLMs used to generate embeddings, along with traditional techniques like TF-IDF and Bag-of-Words.

A.2 Additional Results for Task A.2

We present results for additional experiments for Task A.2 in Tables 4 and 2. They were scored by the organizers after the shared task submission date. In our first additional round, we generated text embeddings with several methods (Bag-of-Words, TF-IDF, BERT, SBERT, and Llama 3.3 70B). Then, we trained multinomial logistic regression (MLR) classifiers to produce a class (1, 2, ... 10) and linear regression (LR) classifiers to output a numeric value that was rounded up to the closest integer between 1 and 10. The scores for this additional submission are presented in Table 4. We observe improved results compared to our official submissions (for S6).

Our second effort to improve our scores for Task A.2 was to revisit our strategy of prompt engineering. The prompt templates from our official submissions were further refined, included additional information and guidance for the LLM to respond with only one token. This additional submission included results from a 4-Shot and 0-Shot variant of this prompt template. These scores are presented in Table 2 as S4 and S5. This approach proved very effective and ranked the best amongst all teams' official submissions for Task A.2.

Model	Type	MSE $\downarrow$	MacroF1 $\uparrow$
BOW	MLR	3.844	0.267
	LR	4.216	0.167
TF-IDF	MLR	4.426	0.248
	LR	3.812	0.226
BERT	MLR	4.379	0.270
SBERT	MLR	4.649	0.250
	LR	<u>3.229</u>	<u>0.302</u>
Llama-3.3-70B-Instr	MLR	4.111	0.236
	LR	**2.015**	**0.348**

Table 4: Results for the regression methods for Task A.2. The "Model" column names the language model from which embeddings were extracted to train the regression model. The best scores are **bolded**, runners-up are <u>underlined</u>.

A.3 Examples of Structured Prompt

We present examples of structured prompt that we used for Task A.1, A.2, B, and C in the uOttawa_2 submission (one-shot learning). These are shown in tables 5, 6, 7 and 8 respectively. The example post, its evidence of adaptive and maladaptive self states, well-being score, and post summaries have been redacted to preserve the privacy of the training data.

Task:
Your task is to identify evidence of adaptive and maladaptive self-states from a post (input text). Each post can include either: (1) a single self-state (adaptive or maladaptive); (2) two complementary self-states (adaptive and maladaptive) or (3) evidence of neither an adaptive or maladaptive state. For each self-state (adaptive or maladaptive), the evidence is a set of continuous spans of text from the post.

—

Definitions:
Self-states constitute identifiable units characterized by specific combinations of Affect, Behavior, Cognition, and Desire/Need (ABCD dimensions) that tend to be coactivated in a meaningful manner for limited periods of time.
- An adaptive self-state pertains to aspects of Affect, Behaviour, and Cognition towards the self or others, which is conducive to the fulfillment of basic desires/needs (D), such as relatedness, autonomy and competence.
- A maladaptive self-state pertains to aspects of Affect, Behaviour, and Cognition towards the self or others, that hinder the fulfillment of basic desires/needs (D).

ABCD dimensions:
1. Affect (A): The type of emotion expressed by the person.
- Adaptive Examples: Calm/Laid back, Emotional Pain/Grieving, Content/Happy, Vigor/Energetic, Justifiable, Anger/Assertive Anger, Proud.
- Maladaptive Examples: Anxious/Tense/Fearful, Depressed/Despair/Hopeless, Mania, Apathetic/Don't care/Blunted, Angry (Aggressive, Disgust, Contempt), Ashamed/Guilty.

2. Behavior of the self with the Other (BO) : The person's main behavior(s) toward the other
- Adaptive Examples: Relating behavior, Autonomous behavior
- Maladaptive Examples: Fight or fight behavior, Overcontrolled/controlling behavior

3. Behavior toward the Self (BS): The person's main behavior(s) toward the self
- Adaptive Examples: Self-care behavior
- Maladaptive Examples: Self-harm, Neglect, Avoidance behavior

4. Cognition of the Other (CO): The person's main perceptions of the other
- Adaptive Examples: Perception of the other as related, Perception of the other as facilitating autonomy needs
- Maladaptive Examples: Perception of the other as detached or over attached, Perception of the other as blocking autonomy needs

5. Cognition of the Self (CS): How the person perceives themselves.
- Adaptive Examples: Self-acceptance and self-compassion
- Maladaptive Examples: Self-criticism

6. Desire (D): The person's main desire, need, intention, fear or expectation
- Adaptive Examples: Relatedness, Autonomy and adaptive control, Competence, Self-esteem, Self-care
- Maladaptive Examples: Expectation that relatedness need will not be met, Expectation that autonomy needs will not be met, Expectation that competence needs will not be met

—

Guidelines for Output:
- Responses Section: Provide answers for both self-states under the headings '### Section Adaptive' and '### Section Maladaptive'.
- Each section should list adaptive and maladaptive self-states, respectively, with supporting text spans.
- Begin each extracted text span with a dash ('-').
- If no adaptive or maladaptive self-state is found, create both sections but leave them empty; do not include any dashes ('-').
- Finally, do not include any additional information; only the text spans are needed.

—

Example:

Input text:
Redacted example post.

Output text:

Section Adaptive
- Redacted adaptive evidence #1 from example post.
- Redacted adaptive evidence #2 from example post.

Section Maladaptive
- Redacted maladaptive evidence #1 from example post.
- Redacted maladaptive evidence #2 from example post.

—

Analyze the following input text based on the given criteria.

Input Text:
{INCLUDE_TEXT (*POST*) }

Output Text:

Table 5: One-Shot Structured Prompt Template for Task A.1.

Task:
Your task is to rate the overall well-being present in the post on a scale from 1 (low well-being) to 10 (high well-being). The score is based on GAF (American Psychiatric Association, 2000), and reflects how well an individual has been doing based on three key domains: Social functioning (school,friendships), occupational functioning (work) and an individual's overall psychological functioning. The clinical cutoff score is 6, meaning that individuals scoring below 6 may be experiencing significant distress

—

Definitions:
Here is an example of the wellbeing scale:
10: No symptoms and superior functioning in a wide range of activities
9: Absent or minimal symptoms (eg., mild anxiety before an exam), good functioning in all areas, interested and involved in a wide range of activities.
8: If symptoms are present, they are temporary and expected reactions to psychosocial stressors (eg., difficulty concentrating after family argument). Slight impairment in social, occupational or school functioning.
7: Mild symptoms (eg., depressed mood and mild insomnia) or some difficulty in social, occupational, or school functioning, but generally functioning well, has some meaningful interpersonal relationships.
6: Moderate symptoms (eg., panic attacks) or moderate difficulty in social, occupational or school functioning.
5: Serious symptoms (e.g., suicidal thoughts, severe compulsions) or serious impairment in social, occupational, or school functioning (eg., no friends, inability to keep a job).
4: Some impairment in reality testing or communication, or major impairment in multiple areas (withdrawal from social ties, inability to work, neglecting family, severe mood/thought impairment).
3: A person experiences delusions or hallucinations or serious impairment in communication or judgment or is unable to function in almost all areas (eg., no job, home, or friends).
2: In danger of hurting self or others (eg., suicide attempts; frequently violent; manic excitement) or may fail to maintain minimal personal hygiene or significant impairment in communication (e.g., incoherent or mute)
1: The person is in persistent danger of severely hurting self or others or persistent inability to maintain minimal personal hygiene or has attempted a serious suicidal act with a clear expectation of death.
0: Unable to assess well-being.

—

Guidelines for Output:
- Provide each answer as "Well-being score: <number>", where '<number>' is the well-being score you assign.
- If a well-being score cannot be provided, answer as "Well-being score: 0".
- Do not include any additional information or explanations - only the score is needed.

—

Example:

Input text:
Redacted example post.

Output text:

Well-being score: Redacted

—

Analyze the following input text based on the given criteria.

Input Text:
{INCLUDE_TEXT (*POST*) }

Output text:

Table 6: One-Shot Structured Prompt Template for Task A.2.

Task:
Your task is to summarize self-states for the social media post below. Specifically, generate a summary of the interplay between adaptive and maladaptive states identified in the post. Begin by determining which self-state is dominant (adaptive/maladaptive) and describe it first. For each self-state, identify the central organizing aspect (A, B, C, or D) that drives the state and structure the summary around it. Describe how this central aspect influences the rest, emphasizing potential causal relationships between them. Then, proceed to the second self-state and follow the same approach. If the post contains only one self-state (either adaptive or maladaptive), summarize only that state. Note that the summary does not need to explicitly highlight A, B, C, or D, but should aim to naturally integrate these elements into the description.

—

Definitions:
Self-states constitute identifiable units characterized by specific combinations of Affect, Behavior, Cognition, and Desire/Need (ABCD dimensions) that tend to be coactivated in a meaningful manner for limited periods of time.

- An adaptive self-state pertains to aspects of Affect, Behaviour, and Cognition towards the self or others, which is conducive to the fulfillment of basic desires/needs (D), such as relatedness, autonomy and competence.
- A maladaptive self-state pertains to aspects of Affect, Behaviour, and Cognition towards the self or others, that hinder the fulfillment of basic desires/needs (D).

ABCD dimensions:
1. Affect (A): The type of emotion expressed by the person.
- Adaptive Examples: Calm/Laid back, Emotional Pain/Grieving, Content/Happy, Vigor/Energetic, Justifiable, Anger/Assertive Anger, Proud.
- Maladaptive Examples: Anxious/Tense/Fearful, Depressed/Despair/Hopeless, Mania, Apathetic/Don't care/Blunted, Angry (Aggressive, Disgust, Contempt), Ashamed/Guilty.

2. Behavior of the self with the Other (BO) : The person's main behavior(s) toward the other
- Adaptive Examples: Relating behavior, Autonomous behavior
- Maladaptive Examples: Fight or fight behavior, Overcontrolled/controlling behavior

3. Behavior toward the Self (BS): The person's main behavior(s) toward the self
- Adaptive Examples: Self-care behavior
- Maladaptive Examples: Self-harm, Neglect, Avoidance behavior

4. Cognition of the Other (CO): The person's main perceptions of the other
- Adaptive Examples: Perception of the other as related, Perception of the other as facilitating autonomy needs
- Maladaptive Examples: Perception of the other as detached or over attached, Perception of the other as blocking autonomy needs

5. Cognition of the Self (CS): How the person perceives themselves.
- Adaptive Examples: Self-acceptance and self-compassion
- Maladaptive Examples: Self-criticism

6. Desire (D): The person's main desire, need, intention, fear or expectation
- Adaptive Examples: Relatedness, Autonomy and adaptive control, Competence, Self-esteem, Self-care
- Maladaptive Examples: Expectation that relatedness need will not be met, Expectation that autonomy needs will not be met, Expectation that competence needs will not be met

—

Guidelines for Output:
- Response Section: Provide an answer under the headings '### Summary:'.
- Format the answer as a single paragraph, making it clear and consise.
- The summary should be no more than 6 sentences. - Ensure the summary captures the main points without unnecessary details.

—

Example:

Input text:
Redacted example post.

Output text:

Summary:
Redacted summary.

—

Analyze the following input text based on the given criteria.

Input Text:
{INCLUDE_TEXT (*POST*) }

Output Text:

Table 7: One-Shot Structured Prompt Template for Task B.

Task:
Your task is to summarize self-states for each timeline, given the summaries for each post on the timeline. Specifically, generate a summary focusing on the Interplay between adaptive and maladaptive self-states along the timeline. Emphasize temporal dynamics focusing on concepts such as flexibility, rigidity, improvement, and deterioration. If applicable, describe the extent to which the dominance of the self-states changes over time and how changes in aspects (Affect, Behavior, Cognition, and Desire) contribute to these transitions.

—

Definitions:
Self-states constitute identifiable units characterized by specific combinations of Affect, Behavior, Cognition, and Desire/Need (ABCD dimensions) that tend to be coactivated in a meaningful manner for limited periods of time.

- An adaptive self-state pertains to aspects of Affect, Behaviour, and Cognition towards the self or others, which is conducive to the fulfillment of basic desires/needs (D), such as relatedness, autonomy and competence.
- A maladaptive self-state pertains to aspects of Affect, Behaviour, and Cognition towards the self or others, that hinder the fulfillment of basic desires/needs (D).

ABCD dimensions:
1. Affect (A): The type of emotion expressed by the person.
- Adaptive Examples: Calm/Laid back, Emotional Pain/Grieving, Content/Happy, Vigor/Energetic, Justifiable, Anger/Assertive Anger, Proud.
- Maladaptive Examples: Anxious/Tense/Fearful, Depressed/Despair/Hopeless, Mania, Apathetic/Don't care/Blunted, Angry (Aggressive, Disgust, Contempt), Ashamed/Guilty.

2. Behavior of the self with the Other (BO) : The person's main behavior(s) toward the other
- Adaptive Examples: Relating behavior, Autonomous behavior
- Maladaptive Examples: Fight or fight behavior, Overcontrolled/controlling behavior

3. Behavior toward the Self (BS): The person's main behavior(s) toward the self
- Adaptive Examples: Self-care behavior
- Maladaptive Examples: Self-harm, Neglect, Avoidance behavior

4. Cognition of the Other (CO): The person's main perceptions of the other
- Adaptive Examples: Perception of the other as related, Perception of the other as facilitating autonomy needs
- Maladaptive Examples: Perception of the other as detached or over attached, Perception of the other as blocking autonomy needs

5. Cognition of the Self (CS): How the person perceives themselves.
- Adaptive Examples: Self-acceptance and self-compassion
- Maladaptive Examples: Self-criticism

6. Desire (D): The person's main desire, need, intention, fear or expectation
- Adaptive Examples: Relatedness, Autonomy and adaptive control, Competence, Self-esteem, Self-care
- Maladaptive Examples: Expectation that relatedness need will not be met, Expectation that autonomy needs will not be met, Expectation that competence needs will not be met

—

Guidelines for Output:
- Response Section: Provide an answer under the headings '### Timeline Summary:'.
- Format the answer as a single paragraph, making it clear and consise.
- The summary should be no more than 6 sentences. - Ensure the timeline summary captures the main points without unnecessary details.

—

Example:

Input text:
A chronologically ordered sequence of summarized posts from timeline

Output text:

Timeline Summary:
A timeline summary.

—

Analyze the following input text based on the given criteria.

Input Text:
{INCLUDE_TEXT (*Summaries of posts on specific timeline from Task B*) }

Output Text:

Table 8: One-Shot Structured Prompt Template for Task C.

Retrieval-Enhanced Mental Health Assessment: Capturing Self-State Dynamics from Social Media Using In-Context Learning

Anson Antony and **Annika M. Schoene**
The Institute for Experiential AI, Boston, MA
{a.antony, a.schoene}@northeastern.edu

Abstract

This paper presents our approach to the CLPsych 2025 (Tseriotou et al., 2025) shared task, where our proposed system implements a comprehensive solution using In-Context Learning (ICL) with vector similarity to retrieve relevant examples that guide Large Language Models (LLMs) without specific fine-tuning. We leverage ICL to analyze self-states and mental health indicators across three tasks. We developed a pipeline architecture using Ollama, where we are running Llama 3.3 70B locally and specialized vector databases for post- and timeline-level examples. We experimented with different numbers of retrieved examples (k=5 and k=10) to optimize performance. Our results demonstrate the effectiveness of ICL for clinical assessment tasks, particularly when dealing with limited training data in sensitive domains. The system shows strong performance across all tasks, with particular strength in capturing self-state dynamics.

1 Introduction

Mental health disorders affect approximately 970 million people worldwide, with depression and anxiety among the leading causes of disability globally.(World Health Organization, 2022) Social media is one of the many spaces where individuals often share aspects of their psychological well-being, seek support, and sometimes express distress. Given the widespread use of these platforms, they have been studied as potential sources of insight into mental health trends at scale.

CLPsych 2025 focuses on capturing mental health dynamics from social media timelines, viewing human experience as consisting of self-states that fluctuate over time. In this paper, we propose ICL to detect self-states and make the following contributions:

- A cascading framework that models mental health assessment across three progressive levels: evidence identification, post dynamics, and timeline patterns.
- A dual-granularity retrieval system (post-level and timeline-level) showing how optimal retrieval parameters (k=5, k=10) vary by assessment task complexity.

This approach allows us to leverage domain expertise without specific fine-tuning, which is particularly valuable when dealing with limited training data in sensitive domains like mental health.

2 Related Work

Mental health assessments on social media have gained significant attention in recent years. Previous CLPsych shared tasks have explored various aspects of mental health analysis, including longitudinal modeling of mood changes (Tsakalidis et al., 2022) and evidence generation for suicidality risk (Zirikly et al., 2019; Shing et al., 2018). *ICL* has emerged as a powerful technique for leveraging large language models without task-specific fine-tuning (Brown et al., 2020). By providing relevant examples within the prompt, ICL enables models to learn from demonstrations rather than parameter updates. Recent work by Uluslu et al. (2024) has shown the effectiveness of integrating emotional information retrieval with ICL for detecting suicidality risk, achieving top performance in the CLPsych 2024 shared task. *Retrieval-Augmented Generation* approaches and vector databases enhance LLM performance on specialized tasks by retrieving information beyond parametric knowledge (Lewis et al., 2020). This is particularly valuable in clinical domains, where accuracy and evidence-based reasoning are crucial. Similar cascading architectures have been effective in legal judgment prediction (Chalkidis et al., 2022) and medical diagnosis (Wang et al., 2023), refining insights through sequential processing. Framework-guided retrieval

Proceedings of the 10th Workshop on Computational Linguistics and Clinical Psychology (CLPsych 2025), pages 268–278
May 3, 2025

has also improved educational applications (Liu et al., 2023), where pedagogical principles inform example selection.

3 Task Description

This iteration of CLPsych 2025 analyzes social media timelines to capture mental health dynamics. Each social media timeline consists of chronologically ordered posts by the same individual, with each post potentially containing evidence of adaptive or maladaptive self-states.

Task A: Post-level Judgments Task A consists of two subtasks, where the detailed prompts can be found in Appendix A.1:

*Task A.1:*Identifying evidence of adaptive and maladaptive self-states in posts, which requires extracting spans of text that provide evidence for different types of self-states and is evaluated using recall-oriented BERTScore metrics.

Task A.2: Rating overall well-being on a scale from 1 (low well-being) to 10 (high well-being), which is evaluated using Mean Squared Error (MSE) and F1 Macro score.

Task B: Post-level Summaries Task B involves generating a summary of the interplay between adaptive and maladaptive self-states identified in each post. This requires determining which self-state is dominant and identifying the central organizing aspect (A, B, C, or D) that drives the state. Summaries are evaluated using Mean Consistency and Max Contradiction metrics based on Natural Language Inference models (see prompt in Appendix A.2).

Task C: Timeline-level Summaries Task C requires generating a summary focusing on the interplay between adaptive and maladaptive self-states along the entire timeline (see full prompt in Appendix A.3). This involves emphasizing temporal dynamics, such as flexibility, rigidity, improvement, and deterioration. Evaluation uses the same consistency-based metrics as Task B.

4 System Description

Our system implements a comprehensive approach to all three tasks using ICL with local LLM inference via Ollama. We utilized Llama 3.3 70B as our primary language model, running it locally through Ollama to maintain data privacy and control over the inference process. The system consists of three main components: vector database creation, task-specific processing, and result integration.

System Architecture Figure 1 presents the overall architecture of our system. The architecture consists of five main layers, where the system consists of multiple layers, each serving a distinct function in processing social media data. The *Data Layer* provides the training dataset, comprising social media posts annotated by experts. These posts are then transformed into vector representations through the *Embedding Layer*, which employs Linq-Embed-Mistral (Junseong Kim, 2024) to capture emotional content. The *Vector and ICL Processing Layer* integrates specialized vector databases for posts and timelines, facilitating interactions with Llama 3.3 70B. At the *Tasks Layer*, task-specific modules (A, B, C) generate prompts and process outputs tailored to different analytical objectives. Finally, the *Output Layer* structures and organizes the final outputs for each task, ensuring clarity and usability. The architecture enables experimentation with different k values for ICL, as shown by the parametrized connection between the vector databases and task modules.

Vector Database Foundation We built two specialized vector databases for efficient data organization and retrieval. The post-level database stores individual posts with annotations, evidence spans, well-being scores, and summaries, enabling detailed analysis. The timeline-level database captures broader temporal patterns by storing timeline representations, providing a comprehensive view of psychological trends.

Both databases utilize the Linq-Embed-Mistral embedding model, chosen for its strong performance on the Massive Text Embedding Benchmark (MTEB) (Muennighoff et al., 2023). This model effectively captures semantic relationships between posts with similar psychological states. We measured similarity using cosine similarity between normalized embeddings and indexed vectors with HNSW (Malkov and Yashunin, 2018) for fast approximate nearest neighbor search. Instead of a distance threshold, we retrieved a fixed top-k (k=5 or k=10) nearest neighbors per query via ChromaDB, optimizing retrieval speed and quality for real-time ICL operations.

ICL Framework: Our approach follows a structured process across all tasks. First, the input, whether a post or a timeline, is embedded using

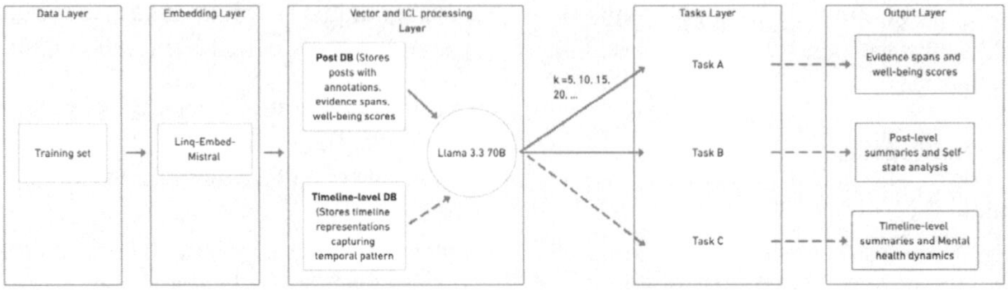

Figure 1: Overview of System architecture.

Linq-Embed-Mistral. The system then queries the vector database to identify the most similar examples, which are subsequently formatted into a demonstration section. A detailed prompt is then constructed, incorporating task definitions, the ABCD framework, relevant example demonstrations, and the target input. This prompt is sent to Ollama, which runs Llama 3.3 70B for inference. Finally, the model's response is processed to extract the required outputs, ensuring task-specific insights are effectively derived.

The complete prompt templates used for each task are provided in Appendix A.

We experimented with different values of k (the number of examples retrieved for in-context learning), specifically k=5 and k=10, resulting in two separate submissions to the shared task. This allowed us to evaluate the impact of example quantity on model performance.

Task A Implementation: Our system retrieves similar posts from the vector database to serve as examples, guiding the LLM in identifying evidence spans and assessing well-being. The prompt is designed to include detailed definitions of self-states and the ABCD framework (see Appendix A.1). The TaskAWithICL class processes each post by first locating k similar posts in the vector database (k=5 or k=10, depending on the configuration). These examples are then formatted into the prompt before querying Llama 3.3 70B with a structured input. Finally, the system extracts the evidence spans and well-being scores from the LLM's response, ensuring accurate assessment of self-state indicators.

Task B Implementation: Our system builds on Task A's outputs and retrieves posts with high-quality summaries to serve as examples. While Task A retrieval is based solely on post content similarity, Task B employs a more selective approach. It queries the same vector database but applies additional filtering to prioritize examples that have both evidence annotations and existing summaries, ensuring higher quality demonstrations for the summarization task. As detailed in Appendix A.2, the prompt instructs the LLM to determine the dominant self-state, identify the central organizing aspect, and generate a cohesive paragraph summary.

Task C Implementation For Task C, our system generates a structured timeline representation to capture temporal mental health dynamics. We implement the following components:

- **Timeline metrics**: We calculate duration (days between first and last post) and posting frequency (posts per week) using date parsing functions that handle multiple formats
- **Well-being statistics**: We compute average scores, range (min/max), and trend analysis (improving, declining, fluctuating, or stable) using post-level well-being assessments from Task A
- **Self-state pattern analysis**: We identify predominant psychological patterns by counting adaptive, maladaptive, and mixed states across posts, classifying timelines as "Predominantly Adaptive," "Predominantly Maladaptive," "Mixed," or "Balanced"
- **Chronological mapping**: We create a sequence of posts with associated self-states, ordered by date when available, to track psychological evolution

Our system retrieves similar timeline patterns from our timeline-level vector database using cosine similarity between embeddings generated by Linq-Embed-Mistral. We specifically prioritize retrieving examples with high-quality summaries to serve as effective demonstrations. The system then constructs a prompt that instructs the LLM to focus on temporal dynamics such as flexibility, rigidity, improvement, and deterioration (see Appendix A.3). This structured approach helps the model generate coherent summaries that capture the evolution of self-states over time.

Example Output: Task B and C To illustrate the clinical relevance of our system, we provide example outputs in Appendix B. The Task B example demonstrates how our system identifies and summarizes the interplay between adaptive and maladaptive self-states within a single post, organizing the analysis around the dominant affect-driven maladaptive state while acknowledging a secondary adaptive state.

The Task C example effectively captures the temporal dynamics of self-states across a three-month period, highlighting the transition from predominantly maladaptive to increasingly adaptive states. It identifies key transition points (therapy engagement) and describes the specific ABCD elements that change over time (affect, cognition, and behavior). This structured analysis demonstrates the system's ability to synthesize complex psychological patterns across multiple posts, providing clinically relevant insights about a user's mental health trajectory.

Integrated Workflow: Our system implements an integrated workflow in which each task builds upon the previous one. Task A identifies evidence and well-being with ICL guidance. Task B then generates post summaries using Task A's outputs and ICL examples. Finally, Task C synthesizes a timeline analysis by integrating all previous outputs and applying timeline-level ICL. This cascading approach enables the system to conduct increasingly complex psychological assessments without requiring task-specific fine-tuning. Additionally, the prompts for each stage (see Appendix A) progressively increase in complexity and scope, ensuring a structured and scalable assessment process.

Error Mitigation Mechanisms To address the risk of cascading errors in our pipeline, we implemented several safeguards:

- **Quality filtering**: Our implementation ensures high-quality evidence identification through prompt instructions that require "exact text spans from the post, without modifications" and structured JSON output validation
- **Consistency checking**: Our system compares post summaries with identified evidence, ensuring Task B outputs align with Task A findings before generating timeline-level summaries
- **Similarity-based retrieval**: Our vector database retrieves the most semantically relevant examples using the specialized Linq-Embed-Mistral model, enhancing the quality and relevance of in-context examples
- **Format verification**: We implement regex pattern matching to validate structured timeline representations before processing, ensuring consistent input formatting across tasks
- **Exception handling**: Robust try-except blocks throughout the implementation prevent crashes when encountering unexpected data formats, providing graceful degradation

These mechanisms help reduce error amplification through the pipeline, though a human-in-the-loop validation would further enhance reliability in clinical applications. Future implementations could incorporate clinician feedback at key decision points.

Computational Considerations Running Llama 3.3 70B locally via Ollama requires substantial computational resources. Our implementation includes several practical considerations to balance performance and accessibility:

- **Model flexibility**: Our system architecture allows specifying different Ollama models through command-line arguments (as seen in our '–model' parameter), enabling users to select models based on their available hardware
- **Controlled batch processing**: We implement timed delays between processing files and posts (using 'time.sleep()') to prevent system overload, with configurable pause durations
- **Progressive task structure**: Our cascading pipeline allows running different components independently (Tasks A, B, and C), enabling incremental processing on systems with limited resources
- **Efficient vector retrieval**: We leverage ChromaDB's HNSW indexing for similarity search operations, making example retrieval faster and more resource-efficient

These design choices allow for deployment across various hardware configurations, though users should consider the performance trade-offs when using smaller models for complex timeline analysis tasks.

5 Results

Table 1 presents our submitted team results (EAIon-Flux) compared to other participating systems. We submitted two different configurations—one with k=5 and another with k=10 for the number of examples retrieved during in-context learning—to evaluate the impact of example quantity on performance. Interestingly, our results showed that the configuration with k=10 generally outperformed k=5 for Tasks A.1 and B, suggesting that more examples provide better guidance for these complex tasks. However, for Tasks A.2 and C, the difference was less pronounced, indicating that well-being assessment and timeline-level summarization may be less sensitive to the number of examples.

Our system performed particularly well on Tasks B and C, demonstrating the effectiveness of our ICL approach with Llama 3.3 70B in generating coherent and clinically meaningful summaries. The relatively small difference in mean consistency metrics compared to other systems suggests that our approach effectively captures mental health dynamics at both the post and timeline levels.

6 Discussion

Our results demonstrate ICL's effectiveness for clinical assessment tasks with limited training data. The proposed system leveraged vector similarity retrieval using Linq-Embed-Mistral embeddings to identify semantically similar examples that reflected psychological patterns, which was crucial for nuanced mental health assessment. Model capabilities were enhanced through Llama 3.3 70B, which demonstrated strong reasoning abilities for complex psychological concepts, enabling the generation of clinically meaningful outputs. Example optimization experiments with k=5 and k=10 showed that incorporating more examples improved performance in Tasks A.1 and B, auiding the model's comprehension of intricate self-state patterns. To enhance clinical knowledge integration, prompts were structured using the ABCD framework (detailed in Appendix A.1), guiding the model toward more accurate psychological assessments. The system followed a cascading architecture, mirroring clinical workflows by allowing tasks to build upon previous insights without requiring task-specific fine-tuning. Lastly, privacy-preserving inference was ensured through local deployment via Ollama, maintaining data privacy while upholding performance quality.

7 Conclusion

In this paper, we presented our approach to the CLPsych 2025 shared task, which focuses on capturing mental health dynamics from social media timelines. Our system implements In-Context Learning with vector similarity to retrieve relevant examples that guide Llama 3.3 70B without specific fine-tuning. The results demonstrate the effectiveness of our approach for clinical assessment tasks, particularly when dealing with limited training data in sensitive domains. Our system shows strong performance across all tasks, with particular strength in capturing self-state dynamics at both the post and timeline levels.

Future work could explore several promising directions: developing specialized psychological embeddings to improve on our current Linq-Embed-Mistral implementation; implementing diversity-aware example selection strategies beyond simple vector similarity; integrating explainability features that highlight influential text spans; incorporating human-in-the-loop validation for error prevention; conducting comprehensive fairness evaluations across demographic groups; and extending to multimodal analysis for more holistic assessment. These enhancements would improve both technical performance and clinical utility, moving toward more equitable, transparent tools for mental healthcare support.

Ethical Considerations and Limitations

This work raises a number of important ethical considerations. All data used in this study was provided as part of the CLPsych 2025 shared task and has been properly de-identified to protect user privacy. No additional data collection was performed. While our approach prioritizes privacy and security by running models locally through Ollama rather than sending sensitive data to external API services, we acknowledge that automated mental health assessment tools should only be used as supportive aids and not as replacements for professional clinical judgment. Additionally, we want to emphasize that any practical deployment would require ex-

Task	Metric	EAIonFlux_1 (k=5)	Delta	EAIonFlux_2 (k=10)	Delta	Best System (Score)
Task A.1	Recall$^{\uparrow}$	0.498	0.139	0.517	0.120	uOttawa (0.637)
	Weighted Recall$^{\uparrow}$	0.480	0.018	0.471	0.027	uOttawa (0.498)
Task A.2	MSE$^{\downarrow}$	2.08	0.16	2.87	0.95	BULUSI (1.920)
	F1 Macro$^{\uparrow}$	0.321	0.072	0.320	0.073	BLUE (0.393)
Task B	Mean Consistency$^{\uparrow}$	0.884	0.026	0.888	0.022	BLUE (0.910)
	Max Contradiction$^{\downarrow}$	0.780	0.247	0.782	0.249	BLUE (0.533)
Task C	Mean Consistency$^{\uparrow}$	0.906	0.040	0.913	0.033	BLUE (0.946)
	Max Contradiction$^{\downarrow}$	0.774	0.420	0.760	0.406	PsyMetric (0.354)

$^{\uparrow}$ Higher values are better. $^{\downarrow}$ Lower values are better.
Delta shows the absolute difference between our system and the best system for each metric.

Table 1: Results of EAIonFlux submissions compared to the best-performing systems in the CLPsych 2025.

tensive clinical validation, careful consideration of bias, and appropriate safeguards to prevent misuse and comply with regulatory standards. We also recognize that computational models of mental health states may reflect biases present in training data. While our in-context learning approach aims to mitigate some biases by explicitly incorporating clinical frameworks, more work is needed to ensure fair and equitable performance across diverse populations.

While our system demonstrated strong performance across all tasks, several limitations should be noted:

Dependency on Example Quality The effectiveness of our ICL approach depends heavily on the quality and representativeness of the examples in the vector database. Our implementation prioritizes examples with human-verified summaries when available, as seen in the timeline similarity retrieval method in Task C, but future versions should incorporate more sophisticated filtering to eliminate potentially misleading examples.

Computational Requirements The use of vector databases and running Llama 3.3 70B locally requires substantial computational resources, which could limit accessibility. Our code includes configurable parameters for model selection and batch processing delays, but the core implementation still requires high-end hardware for optimal performance, potentially creating barriers to adoption in resource-constrained environments.

Limited Clinical Validation While our system was evaluated against expert annotations, broader clinical validation would be necessary before any real-world deployment. The shared task evaluation metrics may not fully capture all aspects of clinical utility, and real-world application would require additional validation studies with mental health professionals.

Potential for Hallucination LLMs can sometimes generate plausible-sounding but incorrect information, which is particularly concerning in clinical contexts. Although our prompts explicitly instruct the model to "Include only EXACT text spans from the post, without any modifications," we observed that the model sometimes struggled with adhering to this constraint. To address these issues, our implementation includes:

- Structured JSON response formats that constrain the model's outputs
- JSON response cleaning methods that validate and sanitize model outputs
- Explicit instructions in prompts to reference only content present in the input text
- Post-processing that validates evidence spans against original post content

Despite these measures, hallucination remains a challenge requiring ongoing research and potential integration of human oversight in critical applications.

Cultural and Demographic Biases The system may inherit biases present in the training data of the underlying LLMs, which could affect its performance across different demographic groups. Mental health expressions vary across cultures, and our current approach does not explicitly account for these differences. For example, the ABCD framework may not adequately capture culturally-specific expressions of psychological distress that fall outside Western clinical paradigms. Our vector database implementation does not include specific mechanisms to ensure diverse representation across cultural contexts.

Cascading Error Propagation and Lack of Human Oversight Our cascading architecture, while efficient, creates the potential for error propagation through the pipeline and subsequently severe ethical risks. Our code analysis revealed that errors in evidence identification from Task A directly affect the input to Tasks B and C, as seen in the data flow between the task implementations. While our implementation includes exception handling and validation steps, it lacks explicit mechanisms for detecting or correcting propagated errors. Future work should explore incorporating human-in-the-loop validation checkpoints between stages to prevent error cascades and to provide corrective feedback that could further improve the system's accuracy and reliability.

A Task-Specific Prompts

This appendix contains the core prompt templates used for each task in our system. These prompts were dynamically combined with retrieved examples during in-context learning. The system message instructing the model to act as "an expert in clinical psychology analyzing social media posts" was consistent across all tasks.

A.1 Task A Prompt: Post-level Evidence Identification and Well-being Assessment

The following is the complete Task A prompt implementation with in-context learning examples as used in our code:

> **System Message:** You are an expert in clinical psychology analyzing social media posts.
>
> **User Message:**
> You are analyzing social media posts for the CLPsych 2025 shared task. Your task is to:
>
> 1. Identify evidence of adaptive and maladaptive self-states in the post. 2. Rate the overall well-being presented in the post on a scale from 1 (low) to 10 (high).
>
> ## Definitions of Self-States
> Self-states constitute identifiable units characterized by specific combinations of Affect, Behavior, Cognition, and Desire/Need (ABCD) that tend to be coactivated in a meaningful manner for limited periods of time.
>
> - An adaptive self-state pertains to aspects of Affect, Behavior, and Cognition towards the self or others, which is conducive to the fulfillment of basic desires/needs (D), such as relatedness, autonomy and competence.
>
> - A maladaptive self-state pertains to aspects of Affect, Behavior, and Cognition towards the self or others, that hinder the fulfillment of basic desires/needs (D).
>
> ## ABCD Elements with Examples
>
> ### Affect: Type of emotion expressed by a person
> - Adaptive Examples: Calm/Laid back, Emotional Pain/Grieving, Content/Happy, Vigor/Energetic, Justifiable Anger/Assertive Anger, Proud. - Maladaptive Examples: Anxious/Tense/Fearful, Depressed/Desperate/Hopeless, Mania, Apathetic/Don't care/Blunted, Angry (Aggressive, Disgust, Contempt), Ashamed/Guilty.
>
> ### Behavior
>
> #### Behavior of the self with the Other (B-O): The person's main behavior(s) toward the other
> - Adaptive Examples: Relating behavior, Autonomous behavior - Maladaptive Examples: Fight or flight behavior, Overcontrolled/controlling behavior
>
> #### Behavior toward the Self (B-S): The person's main behavior(s) toward the self
> - Adaptive Examples: Self-care behavior - Maladaptive Examples: Self-harm/Neglect/Avoidance behavior
>
> ### Cognition
>
> #### Cognition of the Other (C-O): The person's main perceptions of the other
> - Adaptive Examples: Perception of the other as related, Perception of the other as facilitating autonomy needs - Maladaptive Examples: Perception of the other as detached or over attached, Perception of the other as blocking autonomy needs
>
> #### Cognition of the Self (C-S): The person's main self-perceptions
> - Adaptive Examples: Self-acceptance and self-compassion - Maladaptive Examples: Self-criticism
>
> ### Desire: The person's main desire, need, intention, fear or expectation
> - Adaptive Examples: Relatedness, Autonomy and adaptive control, Competence, Self-esteem, Self-care - Maladaptive Examples: Expectation that relatedness need will not be met, Expectation that autonomy needs will not be met, Expectation that competence needs will not be met
>
> ## Well-being Scale (1-10)
>
> - 10: No symptoms and superior functioning in a wide range of activities - 9: Absent or minimal symptoms (e.g., mild anxiety before an exam), good functioning in all areas, interested and involved in a wide range of activities. - 8: If symptoms are present, they are temporary and expected reactions to psychosocial stressors (e.g., difficulty concentrating after family argument). Slight impairment in social, occupational or school functioning. - 7: Mild symptoms (e.g., depressed mood and mild insomnia) or some difficulty in social, occupational, or school functioning, but generally functioning well, has some meaningful interpersonal relationships. - 6: Moderate symptoms (e.g., panic attacks) or moderate difficulty

in social, occupational or school functioning. - 5: Serious symptoms (e.g., suicidal thoughts, severe compulsions) or serious impairment in social, occupational, or school functioning (e.g., no friends, inability to keep a job). - 4: Some impairment in reality testing or communication, or major impairment in multiple areas (withdrawal from social ties, inability to work, neglecting family, severe mood/thought impairment). - 3: A person experiences delusions or hallucinations or serious impairment in communication or judgment or is unable to function in almost all areas (e.g., no job, home, or friends). - 2: In danger of hurting self or others (e.g., suicide attempts; frequently violent; manic excitement) or may fail to maintain minimal personal hygiene or significant impairment in communication (e.g., incoherent or mute) - 1: The person is in persistent danger of severely hurting self or others or persistent inability to maintain minimal personal hygiene or has attempted a serious suicidal act with a clear expectation of death.

The clinical cutoff score is 6, meaning that individuals scoring below 6 may be experiencing significant distress.

Similar Examples for Reference
Here are some examples of similar posts with their annotations:

Example 1:

Post: "{example_post_1}"

Annotation: { "adaptive_evidence": { "A": { "highlighted_evidence": "{adaptive_evidence_span}", "Category": "{adaptive_category}" }, ... }, "maladaptive_evidence": {...}, "well_being_score": {score} }

Example 2:

Post: "{example_post_2}"

Annotation: { ... }

[ADDITIONAL EXAMPLES UP TO k=5 OR k=10]

Please analyze the target post following a similar approach to these examples, but make your own assessment based on the specific content.

Post to Analyze

Here is the post to analyze: "{post_content}"

Response Format

Respond in JSON format with the following structure: { "adaptive_evidence": { // Include only the categories where evidence is found "A": { "highlighted_evidence": "exact text span", "Category": "Specific affect category (e.g., 'Content/Happy')" }, // Other categories as needed (B-O, B-S, C-O, C-S, D) }, "maladaptive_evidence": { // Same structure as adaptive_evidence }, "well_being_score": integer from 1-10, "reasoning": "brief explanation of your assessment" }

Important: 1. Include only EXACT text spans from the post, without any modifications. 2. Only include categories where you found clear evidence. 3. Be specific about the subcategory (e.g., "Content/Happy" not just "Affect"). 4. Make sure your well-being score aligns with the detailed scale provided. 5. If you find no clear evidence of any self-states, return empty objects for the evidence. 6. Your response should be ONLY the JSON. No other text before or after.

At runtime, the system retrieves k similar posts from our vector database using the Linq-Embed-Mistral embeddings and dynamically formats them as examples using the pattern shown above.

A.2 Task B Prompt: Post-level Summary of Self-state Dynamics

The following shows how the Task B prompt is augmented with in-context learning examples:

System Message: You are an expert in clinical psychology analyzing social media posts.

User Message:
You are analyzing a social media post for the CLPsych 2025 shared task, focusing on Task B - Post-level summary of self-state's inner dynamics.

Your task is to generate a summary of the interplay between adaptive and maladaptive self-states identified in the post. You need to:

1. Determine which self-state is dominant (adaptive/maladaptive) and describe it first. 2. For each self-state, identify the central organizing aspect (A, B, C, or D) that drives the state. 3. Structure the summary around this central aspect, describing how it influences the rest. 4. Emphasize potential causal relationships between the aspects. 5. Then, proceed to the second self-state and follow the same approach. 6. If the post contains only one self-state, summarize only that state.

Self-State Definitions
Self-states constitute identifiable units characterized by specific combinations of Affect, Behavior, Cognition, and Desire/Need (ABCD) that tend to be coactivated in a meaningful manner for limited periods of time.

- An adaptive self-state pertains to aspects of Affect, Behavior, and Cognition towards the self or others, which is conducive to the fulfillment of basic desires/needs (D). - A maladaptive self-state pertains to aspects of Affect, Behavior, and Cognition towards the self or others, that hinder the fulfillment of basic desires/needs (D).

Similar Examples for Reference
Here are some examples of similar posts with their evidence and summaries:

Example 1:

Post: "{example_post_1}"

Evidence:
- Adaptive:
A: "{adaptive_evidence_span}" ({adaptive_category})
...

- Maladaptive:

...

Summary:
"{example_summary_1}"

Example 2:

Post: "{example_post_2}"

Evidence:

...

Summary:
"{example_summary_2}"

[ADDITIONAL EXAMPLES UP TO k=5 OR k=10]

Please analyze the target post following a similar approach to these examples, but make your own assessment based on the specific content.

Post to Analyze
Here is the post: "{target_post}"

Evidence Identified in the Post
Adaptive evidence:
{formatted_adaptive_evidence}

Maladaptive evidence:
{formatted_maladaptive_evidence}

Response Instructions
Write a cohesive paragraph summary (200-300 words) that:

1. First describes the dominant self-state (whichever has more significant evidence). 2. Identifies which ABCD aspect (Affect, Behavior-Self, Behavior-Other, Cognition-Self, Cognition-Other, or Desire/Expectation) is central to each self-state. 3. Explains how this central aspect influences other aspects, focusing on causal relationships. 4. Naturally integrates ABCD elements into the description without explicitly highlighting them. 5. Uses clinical language appropriate for psychological assessment.

Do not use bulleted lists or headers in your summary. Write in a fluid, paragraph style.

At runtime, our system retrieves k similar posts through vector similarity, prioritizing examples that have both evidence annotations and existing high-quality summaries. This selective filtering ensures that the examples provided to the model demonstrate appropriate summary creation. Unlike Task A, which only requires evidence identification, Task B examples must showcase how evidence is integrated into coherent summaries that identify central organizing aspects and causal relationships.

A.3 Task C Prompt: Timeline-level Summary of Self-state Dynamics

The following shows how the Task C prompt is augmented with in-context learning examples, specifically using timeline-level representations:

System Message: You are an expert in clinical psychology analyzing social media posts.

User Message:
You are analyzing a social media timeline for the CLPsych 2025 shared task, focusing on Task C - Timeline-level summary of self-state's dynamics.

Your task is to generate a summary focusing on the interplay between adaptive and maladaptive self-states along the timeline. You need to:

1. Emphasize temporal dynamics focusing on concepts such as flexibility, rigidity, improvement, and deterioration. 2. Describe the extent to which the dominance of the self-states changes over time. 3. Explain how changes in aspects (Affect, Behavior, Cognition, and Desire) contribute to these transitions.

Timeline to Analyze

This timeline contains {len(posts_with_evidence)} posts spanning from {posts_with_evidence[0]["date"]} to {posts_with_evidence[-1]["date"]}.

Here are the posts with their Well-being scores and evidence:

{chronological_post_listing}

Post-level Summaries (if available)

{post_level_summaries}

Similar Timelines for Reference
Here are some examples of similar timelines with their summaries:

Example 1 (Timeline ID: {timeline_id_1}):

Timeline Characteristics:
ID: {timeline_id_1}
Time span: {duration_text}
Post count: {post_count}
Average well-being: {avg_well_being}
Well-being range: {min_well_being} to {max_well_being}
Well-being trend: {trend}
Self-state pattern: {state_dynamics}

Timeline Summary:
"{example_summary_1}"

Example 2 (Timeline ID: {timeline_id_2}):

Timeline Characteristics:

...

Timeline Summary:
"{example_summary_2}"

[ADDITIONAL EXAMPLES UP TO k=5 OR k=10]

Please analyze the target timeline following a similar approach to these examples, but make your own assessment based on the specific content.

Response Instructions

Write a cohesive paragraph summary (200-400 words) that:

1. Describes the overall pattern of self-states across the timeline (e.g., predominantly adaptive, predominantly maladaptive, fluctuating). 2. Identifies any shifts or transitions between dominance of adaptive and maladaptive states. 3. Analyzes temporal dynamics using concepts like flexibility, rigidity, improvement, and deterioration. 4. Explains how changes in ABCD aspects (Affect, Behavior, Cognition, Desire) contribute to these transitions. 5. Uses clinical language appropriate for psychological assessment.

Emphasize how the individual's self-states evolve over time. Do not use bulleted lists or headers in your summary. Write in a fluid, paragraph style.

For Task C, our system uses a specialized timeline-level vector database different from the post-level database used in Tasks A and B. The system creates a comprehensive timeline representation that includes:

- Chronological timeline metadata (duration, post frequency, date range)
- Well-being statistics (average scores, range, trends)
- Self-state pattern analysis (predominantly adaptive, maladaptive, mixed, or balanced)
- Sequence of individual posts with their dominant self-states

This timeline representation is embedded using Linq-Embed-Mistral and used to retrieve similar timelines for in-context learning. The system specifically filters for examples that have existing high-quality timeline summaries, providing effective demonstrations of temporal analysis. Task C builds upon the outputs of Tasks A and B, using the evidence spans, well-being scores, and post-level summaries as inputs for the timeline-level analysis.

B Example System Outputs

B.1 Task B Example Output

The following is an example of a post-level summary generated by our system for Task B:

> "The dominant self-state is maladaptive, primarily organized around feelings of anxiety and tension (Affect). The individual expresses significant worry about an upcoming presentation, manifested through rumination and catastrophic thinking. This anxiety drives avoidant behavior, as they mention repeatedly postponing preparation and distracting themselves with less important tasks. Their cognition reveals self-criticism, particularly regarding their perceived incompetence and inability to handle pressure. Despite these maladaptive elements, there is a minor adaptive self-state present, characterized by moments of self-awareness when they acknowledge the need to address their procrastination and apply coping strategies they've learned previously."

B.2 Task C Example Output

The following is an example of a timeline-level summary generated by our system for Task C:

> "This timeline demonstrates a fluctuating pattern between maladaptive and adaptive self-states over a three-month period, with a gradual improvement trend. Initially dominated by maladaptive states characterized by anxiety and depressive affect, the individual's posts show rigid negative cognitions about themselves and their future. Around the midpoint, a significant transition occurs following engagement with therapy, where adaptive self-states begin emerging with greater frequency. This shift is marked by increasing cognitive flexibility, with the individual demonstrating greater self-compassion and employing healthier coping behaviors. While maladaptive states still recur during stressful periods, they become less intense and persistent. The timeline reveals a dynamic interplay between affect and cognition as primary drivers of state transitions, with improvements in cognitive patterns (reduced self-criticism, increased perspective-taking) typically preceding positive affect changes."

References

Tom B. Brown, Benjamin Mann, Nick Ryder, Melanie Subbiah, Jared Kaplan, Prafulla Dhariwal, Arvind Neelakantan, Pranav Shyam, Girish Sastry, Amanda Askell, et al. 2020. Language models are few-shot learners. *Advances in Neural Information Processing Systems*, 33:1877–1901.

Ilias Chalkidis, Ion Androutsopoulos, and Nikolaos Aletras. 2022. An empirical study on neural methods for legal judgment prediction. In *Proceedings of the 2022 Conference on Empirical Methods in Natural Language Processing*, pages 4783–4798. Association for Computational Linguistics.

Jihoon Kwon Sangmo Gu Yejin Kim Minkyung Cho Jy-yong Sohn Chanyeol Choi Junseong Kim, Seolhwa Lee. 2024. Linq-embed-mistral:elevating text retrieval with improved gpt data through task-specific control and quality refinement. Linq AI Research Blog.

Patrick Lewis, Ethan Perez, Aleksandra Piktus, Fabio Petroni, Vladimir Karpukhin, Naman Goyal, Heinrich Küttler, Mike Lewis, Wen-tau Yih, Tim Rocktäschel, Sebastian Riedel, and Douwe Kiela. 2020. Retrieval-augmented generation for knowledge-intensive nlp tasks. In *Advances in Neural Information Processing Systems*, volume 33, pages 9459–9474.

Jiawei Liu, Zhiwei Tu, William Yang Wang, Dongkuan Zhang, Yiquan Cui, and Philip S. Yu. 2023. Retrieval-augmented generation for knowledge-intensive nlp tasks: A survey. In *Proceedings of the 2023 Conference of the North American Chapter of the Association for Computational Linguistics: Human Language Technologies*, pages 10616–10649. Association for Computational Linguistics.

Yury A. Malkov and Dmitry A. Yashunin. 2018. Efficient and robust approximate nearest neighbor search using hierarchical navigable small world graphs. *IEEE Transactions on Pattern Analysis and Machine Intelligence*, 42(4):824–836.

Niklas Muennighoff, Nouamane Tazi, Loïc Magne, and Nils Reimers. 2023. Mteb: Massive text embedding benchmark. *Preprint*, arXiv:2210.07316.

Han-Chin Shing, Suraj Nair, Ayah Zirikly, Meir Friedenberg, Hal Daumé III, and Philip Resnik. 2018. Expert, crowdsourced, and machine assessment of suicide risk via online postings. In *Proceedings of the Fifth Workshop on Computational Linguistics and Clinical Psychology: From Keyboard to Clinic*, pages 25–36, New Orleans, LA. Association for Computational Linguistics.

Adam Tsakalidis, Jenny Chim, Iman Munire Bilal, Ayah Zirikly, Dana Atzil-Slonim, Federico Nanni, Philip Resnik, Manas Gaur, Kaushik Roy, Becky Inkster, Jeff Leintz, and Maria Liakata. 2022. Overview of the CLPsych 2022 shared task: Capturing moments of change in longitudinal user posts. In *Proceedings of the Eighth Workshop on Computational Linguistics and Clinical Psychology*, pages 184–198, Seattle, USA. Association for Computational Linguistics.

Talia Tseriotou, Jenny Chim, Ayal Klein, Aya Shamir, Guy Dvir, Iqra Ali, Cian Kennedy, Guneet Singh Kohli, Anthony Hills, Ayah Zirikly, Dana Atzil-Slonim, and Maria Liakata. 2025. Overview of the clpsych 2025 shared task: Capturing mental health dynamics from social media timelines. In *Proceedings of the 10th Workshop on Computational Linguistics and Clinical Psychology*. Association for Computational Linguistics.

Ahmet Yavuz Uluslu, Andrianos Michail, and Simon Clematide. 2024. Utilizing large language models to identify evidence of suicidality risk through analysis of emotionally charged posts. In *Proceedings of the 9th Workshop on Computational Linguistics and Clinical Psychology (CLPsych 2024)*, pages 264–269, St. Julians, Malta. Association for Computational Linguistics.

Jingqing Wang, Rupert He, Amos Koker, Zhihong Ren, and Percy Liang. 2023. A survey on retrieval-augmented text generation. *ACM Computing Surveys*, 55(12):1–36.

World Health Organization. 2022. Mental disorders. *World Health Organization Fact Sheets*. Retrieved from https://www.who.int/news-room/fact-sheets/detail/mental-disorders.

Ayah Zirikly, Philip Resnik, Özlem Uzuner, and Kristy Hollingshead. 2019. CLPsych 2019 shared task: Predicting the degree of suicide risk in Reddit posts. In *Proceedings of the Sixth Workshop on Computational Linguistics and Clinical Psychology*, pages 24–33, Minneapolis, Minnesota. Association for Computational Linguistics.

Self-State Evidence Extraction and Well-Being Prediction from Social Media Timelines

Suchandra Chakraborty **Sudeshna Jana** **Manjira Sinha** **Tirthankar Dasgupta**
TCS Research, Kolkata
suchandrac2001@gmail.com,
{sudeshna.jana, sinha.manjira, dasgupta.tirthankar}@tcs.com

Abstract

This study explores the application of Large Language Models (LLMs) and supervised learning to analyze social media posts from Reddit users, addressing two key objectives: first, to extract adaptive and maladaptive self-state evidence that supports psychological assessment (Task A1); and second, to predict a well-being score that reflects the user's mental state (Task A2). We propose i) a fine-tuned RoBERTa (Liu et al., 2019) model for Task A1 to identify self-state evidence spans and ii) evaluate two approaches for Task A2: a retrieval-augmented DeepSeek-7B (DeepSeek-AI et al., 2025) model and a Random Forest regression model trained on sentence embeddings. While LLM-based prompting utilizes contextual reasoning, our findings indicate that supervised learning provides more reliable numerical predictions. The RoBERTa model achieves the highest recall (0.602) for Task A1, and Random Forest regression outperforms DeepSeek-7B for Task A2 (MSE: 2.994 vs. 6.610). These results highlight the strengths and limitations of generative vs. supervised methods in mental health NLP, contributing to the development of privacy-conscious, resource-efficient approaches for psychological assessment. This work is part of the CLPsych 2025 shared task (Tseriotou et al., 2025).

1 Introduction

Mental health assessment using natural language processing (NLP) has evolved from static risk classification to longitudinal modeling of self-states and psychological well-being. The CLPsych Shared Task has progressively introduced more nuanced challenges, moving beyond binary risk assessment to capture dynamic shifts in mental health. The CLPsych 2022 Shared Task (Tsakalidis et al., 2022) was the first to introduce longitudinal modeling, focusing on detecting "Moments of Change" in a user's mood over time and exploring its connection to suicidality risk. The CLPsych 2024 Shared Task (Chim et al., 2024) expanded on this by requiring models to find textual evidence that supports suicide risk levels.

The CLPsych 2025 Shared Task (Tseriotou et al., 2025) extends this research by combining longitudinal modeling with evidence extraction, promoting models that generate human-interpretable rationales while recognizing mental states as they evolve. The shared task consists of four subtasks:

- **Task A1** (Self-State Evidence Extraction): Identifying spans of text that provide evidence for adaptive and maladaptive self-states in a given post.
- **Task A2** (Well-Being Score Prediction): Assigning a well-being score (1–10) to measure the user's psychological state.
- **Task B** (Post-Level Summarization): Generating a summary of the interaction between adaptive and maladaptive states identified in the post.
- **Task C** (Timeline-Level Summarization): Producing longitudinal summaries that capture the trajectory of a user's mental state across multiple posts.

This work focuses on Tasks A1 and A2, which require precise extraction of self-state evidence and structured estimation of well-being scores from the given Reddit post.

Two main approaches exist for these tasks: supervised learning and generative modeling. Supervised methods leverage annotated datasets for structured predictions, using transformer-based models such as BERT (Devlin et al., 2019) and RoBERTa (Liu et al., 2019). Generative models, particularly Large Language Models (LLMs), offer contextual reasoning but require carefully designed prompts to ensure reliable outputs.

Proceedings of the 10th Workshop on Computational Linguistics and Clinical Psychology (CLPsych 2025), pages 279–286
May 3, 2025 ©2025 Association for Computational Linguistics

In this work, we make the following contributions:

1. **Span-Based Evidence Extraction**: We fine-tune a RoBERTa model to extract adaptive and maladaptive self-state evidence (Task A1), achieving a high recall of 0.602. This demonstrates the effectiveness of structured supervision in identifying psychological markers.

2. **Comparative Study of Well-Being Score Prediction**: We evaluate two distinct approaches for Task A2:

   (a) A **retrieval-augmented DeepSeek-7B** model for contextualized well-being estimation.

   (b) A **Random Forest regression** model trained on sentence embeddings for structured numerical prediction.

Our results indicate that supervised learning outperforms LLM-based approaches for numerical well-being regression, while LLMs capture nuanced mental health signals but introduce high variance in predictions. This study contributes to the ongoing development of interpretable, data-driven methods for mental health NLP. The following sections outline our methodology, experiments, and findings.

2 Task Description and Dataset

This study focuses on **Task A1**(Self-State Evidence Extraction) and **Task A2** (Well-Being Score Prediction).

2.1 Task A1: Self-State Evidence Extraction

Given a Reddit post p_j, Task A1 requires identifying spans of text within the post that indicate **adaptive** or **maladaptive** self-states. We define the task as learning a function $f_{\text{A1}} : X_j \rightarrow \{S_{\text{adaptive}}, S_{\text{maladaptive}}\}$, where X_j represents the text of post p_j, and $S_{\text{adaptive}}, S_{\text{maladaptive}}$ are sets of non-overlapping spans belonging to X_j that reflect positive coping mechanisms or distress-driven thought patterns.

2.2 Task A2: Well-Being Score Prediction

Task A2 involves assigning a **well-being score** y_j to each post p_j, where scores range from **1 (severe distress) to 10 (minimal impairment)**, aligning with the Global Assessment of Functioning (GAF) scale. This is framed as a regression problem $f_{\text{A2}} : p_j \rightarrow y_j, \quad y_j \in \{1, 2, ..., 10\}$.

2.3 Dataset Overview

The dataset (Shing et al., 2018; Zirikly et al., 2019; Tsakalidis et al., 2022) consists of 30 user timeline JSON files (343 posts) in the training set and 10 user timeline JSON files (94 posts) in the test set. Each training JSON file contains a timeline ID, a list of posts, and a timeline summary. Each post includes a post index, post ID, timestamp, post text, post summary, well-being score, and evidence annotations. The evidence annotations consist of adaptive and maladaptive states, each with categories and highlighted evidence spans. Each test JSON consists of a timeline ID and a list of posts.

The evidence spans (`adaptive-state` and `maladaptive-state`) are **substrings of the given post text**. The dataset follows the **MIND framework** (Slonim, 2024), modeling mental health as a **dynamic fluctuation of self-states** over time.

3 Methodology

3.1 Task A1: Self-State Evidence Extraction

We frame Task A1 as a **token classification problem**, where each token in a Reddit post is labeled as **adaptive (1)**, **maladaptive (2)**, or **non-evidence (0)**.

3.1.1 Data Preprocessing and Augmentation:

The training data was extracted from annotated JSON files and converted into a CSV format containing Timeline ID, Post, Adaptive Evidence, and Maladaptive Evidence. Posts without any evidence spans were removed, resulting in 199 posts. To enhance robustness, we generated **50 additional posts** using the `nlpaug` (Ma, 2019) library, which provides various NLP-based augmentation methods. Specifically, we applied synonym replacement using the **SynonymAug** augmenter and explicitly configured it to use WordNet as the synonym source. We also applied random word swapping using the **RandomWordAug** augmenter, which randomly exchanges the positions of words within a sentence. This introduced lexical and structural variations while preserving the meaning of the posts, thereby enhancing the overall diversity of the dataset.

3.1.2 Tokenization and Labeling:

We used the **RoBERTa tokenizer** with `add_prefix_space=True` to preserve subword alignment. Evidence spans were mapped to token positions using a rule-based matching algorithm. Labels were assigned directly at the word level by matching evidence spans; each word was initially labeled as non-evidence (0) and then relabeled as adaptive (1) or maladaptive (2) if it was part of the corresponding evidence spans.

3.1.3 Model Architecture and Training:

We fine-tuned a **RoBERTaForTokenClassification** model with three output labels corresponding to evidence categories. The model was trained using Cross-Entropy Loss, AdamW (Loshchilov and Hutter, 2019) optimizer (learning rate = $2e^{-5}$), batch size = 16, and 3 epochs with early stopping based on validation loss to prevent overfitting. We used mixed precision training (fp16) to enhance GPU utilization and speed up training.

3.1.4 Post-Processing and Inference:

During inference, each post is initially split into sentences using a sentence tokenizer. The model generates token-level predictions for each sentence, and the predicted label that is most frequent in that sentence is used as its overall classification.

3.2 Task A2: Well-Being Score Prediction

Task A2 involves assigning each Reddit post a **well-being score** ranging from 1 (severe distress) to 10 (minimal impairment). The training data was extracted from annotated JSON files and converted into a CSV format containing Timeline ID, Post, and Well-being Score. Rows with missing well-being scores were removed.

3.2.1 LLM-Based Approach (DeepSeek-7B)

For this method, we use a retrieval-augmented prompting strategy using DeepSeek-7B, an instruction-tuned causal language model. An overview of this method is shown in Figure 1. The training data is used to generate sentence embeddings via all-MiniLM-L6-v2 (Wang et al., 2020). For each test post, the embedding is computed and compared against the training embeddings using cosine similarity to retrieve the top-k most similar examples. These retrieved examples, along with their well-being scores, are incorporated into a detailed few-shot prompt that begins with a description of the well-being scale based on GAF criteria, followed by instructions to produce a justification sentence and a predicted well-being score. The prompt is tokenized using DeepSeek-7B's tokenizer, and the model generates an output (with parameters such as max_new_tokens set to 50 and temperature to 0.1) from which the numerical score is parsed. Finally, this entire inference pipeline iterates over the test set, and the predicted scores are saved for evaluation. An example prompt used in our approach is provided in the appendix A for reference.

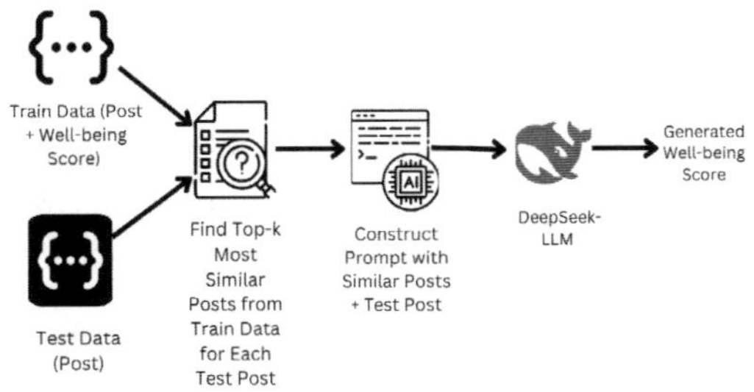

Figure 1: LLM-based Well-being Score Prediction

3.2.2 Supervised Learning Approach (Random Forest Regression)

We also experimented with a supervised regression approach using a Random Forest model trained on sentence embeddings. Sentence representations are generated using all-MiniLM-L6-v2 (Wang et al., 2020), a compact transformer-based embedding model. The feature matrix consists of the embeddings, while the well-being scores serve as the target variable. An 80-20 train-validation split is applied, and a Random Forest Regressor with 200 estimators and a fixed random state is trained on the dataset. Predictions are made on the validation set, and post-processing ensures that outputs are rounded and clipped to integer values within the 1–10 range. For inference, embeddings are generated for the test posts and passed through the trained model. The predicted well-being scores are then stored in a CSV file alongside their corresponding Timeline_ID and Post. Validation performance is assessed using Mean Absolute Error (MAE) and accuracy.

3.2.3 Post-processing:

For DeepSeek-7B, any non-numeric outputs were filtered, and scores exceeding 1–10 were discarded. For Random Forest Regression, predictions were clipped and rounded to ensure numerical consistency.

4 Evaluation Metrics

Task A.1: Evidence of Adaptive and Maladaptive Self-States

- **Recall**: Average of maximum BERTScore for gold spans:

$$\text{Recall} = \frac{1}{|E|}\sum_{e\in E}\max_{h\in H}\text{BERTScore}(e,h)$$

- **Weighted Recall**: Adjusted for predicted span lengths:

$$w = \begin{cases} \frac{n_{\text{gold}}}{n_{\text{pred}}} & \text{if } n_{\text{pred}} > n_{\text{gold}} \\ 1 & \text{otherwise} \end{cases}$$

- **Null Handling**: Defaults to 0 if no spans are submitted.

Task A.2: Well-being Score Prediction

- **Mean Squared Error (MSE)**: Averaged over timelines, computed for:
  - Serious impairment (scores 1-4)
  - Impaired functioning (scores 5-6)
  - Minimal impairment (scores 7-10)
- **Null Handling**: Ignored if no gold score; penalized by max error if no prediction.

5 Results

In Tables 1 and 2, we present the test set results for Task A1 and Task A2. The performance of our methods is compared against baseline models.

5.1 Task A1: Self-State Evidence Extraction

Table 1 presents the results for self-state evidence extraction. Our RoBERTa-based model (MMKA RoBERTa) achieves the **second-highest performance** in the shared task, with an **overall recall of 0.602**. The model shows stronger performance in detecting maladaptive spans (0.681 recall) compared to adaptive spans (0.522 recall), suggesting that distress-related expressions were more easily identifiable by the model. The weighted recall is lower, indicating some level of over-extraction. For a detailed analysis of common misclassification patterns, refer to Appendix B.

Model	Overall		Adaptive		Maladaptive	
	R	W	R	W	R	W
Llama 3.1	0.358	0.337	0.306	0.293	0.382	**0.411**
w/ Window	0.496	0.262	0.365	0.252	0.627	0.272
BART	0.404	**0.382**	0.473	**0.464**	0.336	0.299
w/ Window	0.260	0.258	0.282	0.279	0.238	0.237
MRoBERTa (Ours)	**0.602**	0.343	**0.522**	0.374	**0.681**	0.313

Table 1: Results of our proposed method against baselines on Task A1. "R" and "W" denote recall and weighted recall; *w/ Window* represents the incorporation of post windows.

Model	MSE↓	M-S	M-I	M-M	F1
Llama 3.1	4.22	4.67	3.66	3.20	0.255
w/ Window	4.46	**1.67**	3.20	7.06	**0.274**
BERT	**2.90**	3.39	2.32	2.81	0.139
w/ Window	4.56	5.68	1.01	5.34	0.135
MMKA DS (Ours)	6.61	4.22	11.76	4.95	0.257
MMKA RF	2.99	4.25	**0.78**	**2.60**	0.197

Table 2: Results of our proposed method against baselines on Task A2. "M-S", "M-I", and "M-M" denote MSE across serious impairment, impaired, and minimal impairment. MMKA DS is our Deepseek approach, and MMKA RF is our Random Forest approach which was not a part of our initial submission.

5.2 Task A2: Well-Being Score Prediction

Table 2 presents the results for well-being score prediction. Our Random Forest Regression model (MMKA Random Forest) achieves the second lowest overall **MSE of 2.994**, outperforming both our submission model DeepSeek-7B and most baselines. However, this approach was not part of our official submission. The DeepSeek-7B model exhibited higher variance and struggled, particularly in severe distress cases, yielding an MSE of 6.610. The results indicate that while LLM-based methods (DeepSeek-7B) capture contextual information, they struggle with numerical stability, often generating inconsistent well-being scores. Additionally, while using LLM-based methods for Task A2, we have faced hallucination issues of LLMs, which is a major drawback of this method. Random Forest Regression, by contrast, provides more stable predictions but lacks interpretability compared to LLM-generated justifications.

5.2.1 Performance Comparison for Task TA2: DeepSeek-7B vs Random Forest

For Task A2 (Well-being Score Prediction), the **Random Forest model** outperformed **DeepSeek-7B**, highlighting key differences between structured machine learning and large language models

(LLMs) for numerical prediction.

Key Factors for Random Forest's Superior Performance

- **Structured Learning:** Random Forest utilizes explicit numerical features and supervised training, which helps the model predict well-being scores precisely. DeepSeek-7B relies on retrieval-augmented prompting, which lacks direct optimization for numerical regression.
- **Stability Interpretability:** Random Forest provides consistent predictions and feature importance insights, while DeepSeek-7B's blackbox nature leads to variability and reduced interpretability.
- **Efficiency:** Random Forest makes deterministic predictions efficiently, whereas DeepSeek-7B is computationally expensive and sensitive to retrieval quality.

Future Improvements Enhancing DeepSeek-7B's performance could involve fine-tuning it on domain-specific data, improving retrieval mechanisms, and constraining numerical outputs. Exploring hybrid models combining structured learning with LLM-based contextual reasoning is a promising direction.

6 Conclusion

In this work, we explored approaches for self-state evidence extraction (Task A1) and well-being score prediction (Task A2) as part of the CLPsych 2025 Shared Task. Our RoBERTa-based token classification model achieved the second-best recall (0.602) for Task A1, demonstrating strong performance in detecting maladaptive self-state evidence. For Task A2, we compared a retrieval-augmented LLM (DeepSeek-7B) and a Random Forest regression model. While the DeepSeek-7B model captured contextual information, it exhibited numerical instability. Our Random Forest model outperformed all baselines (MSE = 2.994) except for BERT, but this approach was not part of the official submission.

7 Future Work

For Task A1, future work can focus on **span-level** annotation rather than **sentence-level** classification, allowing the model to distinguish adaptive and maladaptive cues within the same sentence. Future work can also explore **data augmentation using LLMs** for Task A1, which could improve self-state extraction by generating additional diverse training instances. This was not attempted due to computational constraints but presents a promising avenue for enhancing model generalization. Additionally, incorporating **stylistic features** such as sentiment shifts, discourse markers, and writing patterns could provide deeper contextual insights, improving both evidence extraction and well-being prediction. Further, hybrid models that combine the contextual reasoning of LLMs with the numerical stability of regression-based approaches could lead to more robust well-being assessments. Finally, extending models to **capture temporal trends** in user well-being may provide deeper insights into longitudinal mental health assessment.

8 Limitations

Our study has several limitations. First, initial experiments using prompting-based approaches with models such as Mistral-7B (Jiang et al., 2023) and LLaMA (Touvron et al., 2023) for Task A1 resulted in poor performance, with frequent hallucinations and unreliable evidence extraction. As a result, we opted for a RoBERTa-based token classification model, which demonstrated improved robustness. Second, data scarcity remains a significant challenge for both Task A1 and Task A2. Although we applied basic data augmentation techniques to increase the number of training instances, these methods are limited in their ability to capture the full variability of mental health expressions. More advanced data augmentation using LLMs, coupled with a BERT-based model, could potentially yield better performance. Finally, capturing the nuanced and inherently subjective aspects of self-state evidence and well-being scores proved difficult. While we initially anticipated that LLMs would excel in both tasks, they often failed to provide consistent and interpretable results. This suggests that larger models, which might better capture these subtleties, are computationally expensive and present a trade-off between performance and resource requirements.

9 Ethics

The data used in this study consists of sensitive, real user posts collected from Reddit. Although the data are publicly available, we have ensured that all processing is conducted within a secure environment, and no personally identifiable information is shared externally. We strictly adhere to ethical guidelines for data usage and privacy, ensuring that our findings are reported responsibly and without stigmatizing individuals. All analyses and results are derived solely for research purposes and to advance our understanding of mental health dynamics in social media.

References

Jenny Chim, Adam Tsakalidis, Dimitris Gkoumas, Dana Atzil-Slonim, Yaakov Ophir, Ayah Zirikly, Philip Resnik, and Maria Liakata. 2024. Overview of the CLPsych 2024 shared task: Leveraging large language models to identify evidence of suicidality risk in online posts. In *Proceedings of the 9th Workshop on Computational Linguistics and Clinical Psychology (CLPsych 2024)*, pages 177–190, St. Julians, Malta. Association for Computational Linguistics.

DeepSeek-AI, Daya Guo, Dejian Yang, Haowei Zhang, Junxiao Song, Ruoyu Zhang, Runxin Xu, Qihao Zhu, Shirong Ma, Peiyi Wang, Xiao Bi, Xiaokang Zhang, Xingkai Yu, Yu Wu, Z. F. Wu, Zhibin Gou, Zhihong Shao, Zhuoshu Li, Ziyi Gao, and 181 others. 2025. Deepseek-r1: Incentivizing reasoning capability in llms via reinforcement learning. *Preprint*, arXiv:2501.12948.

Jacob Devlin, Ming-Wei Chang, Kenton Lee, and Kristina Toutanova. 2019. Bert: Pre-training of deep bidirectional transformers for language understanding. *Preprint*, arXiv:1810.04805.

Albert Q. Jiang, Alexandre Sablayrolles, Arthur Mensch, Chris Bamford, Devendra Singh Chaplot, Diego de las Casas, Florian Bressand, Gianna Lengyel, Guillaume Lample, Lucile Saulnier, Lélio Renard Lavaud, Marie-Anne Lachaux, Pierre Stock, Teven Le Scao, Thibaut Lavril, Thomas Wang, Timothée Lacroix, and William El Sayed. 2023. Mistral 7b. *Preprint*, arXiv:2310.06825.

Yinhan Liu, Myle Ott, Naman Goyal, Jingfei Du, Mandar Joshi, Danqi Chen, Omer Levy, Mike Lewis, Luke Zettlemoyer, and Veselin Stoyanov. 2019. Roberta: A robustly optimized bert pretraining approach. *Preprint*, arXiv:1907.11692.

Ilya Loshchilov and Frank Hutter. 2019. Decoupled weight decay regularization. *Preprint*, arXiv:1711.05101.

Edward Ma. 2019. Nlp augmentation. https://github.com/makcedward/nlpaug.

Han-Chin Shing, Suraj Nair, Ayah Zirikly, Meir Friedenberg, Hal Daumé III, and Philip Resnik. 2018. Expert, crowdsourced, and machine assessment of suicide risk via online postings. In *Proceedings of the Fifth Workshop on Computational Linguistics and Clinical Psychology: From Keyboard to Clinic*, pages 25–36, New Orleans, LA. Association for Computational Linguistics.

Dana Atzil Slonim. 2024. Self-other dynamics (sod): A transtheoretical coding manual.

Hugo Touvron, Thibaut Lavril, Gautier Izacard, Xavier Martinet, Marie-Anne Lachaux, Timothée Lacroix, Baptiste Rozière, Naman Goyal, Eric Hambro, Faisal Azhar, Aurelien Rodriguez, Armand Joulin, Edouard Grave, and Guillaume Lample. 2023. Llama: Open and efficient foundation language models. *Preprint*, arXiv:2302.13971.

Adam Tsakalidis, Jenny Chim, Iman Munire Bilal, Ayah Zirikly, Dana Atzil-Slonim, Federico Nanni, Philip Resnik, Manas Gaur, Kaushik Roy, Becky Inkster, Jeff Leintz, and Maria Liakata. 2022. Overview of the clpsych 2022 shared task: Capturing moments of change in longitudinal user posts. In *Proceedings of the Eighth Workshop on Computational Linguistics and Clinical Psychology*, pages 184–198, Seattle, USA. Association for Computational Linguistics.

Talia Tseriotou, Jenny Chim, Ayal Klein, Aya Shamir, Guy Dvir, Iqra Ali, Cian Kennedy, Guneet Singh Kohli, Anthony Hills, Ayah Zirikly, Dana Atzil-Slonim, and Maria Liakata. 2025. Overview of the clpsych 2025 shared task: Capturing mental health dynamics from social media timelines. In *Proceedings of the 10th Workshop on Computational Linguistics and Clinical Psychology*.

Wenhui Wang, Furu Wei, Li Dong, Hangbo Bao, Nan Yang, and Ming Zhou. 2020. Minilm: Deep self-attention distillation for task-agnostic compression of pre-trained transformers. *Preprint*, arXiv:2002.10957.

Ayah Zirikly, Philip Resnik, Özlem Uzuner, and Kristy Hollingshead. 2019. Clpsych 2019 shared task: Predicting the degree of suicide risk in reddit posts. In *Proceedings of the Sixth Workshop on Computational Linguistics and Clinical Psychology*, pages 24–33, Minneapolis, Minnesota. Association for Computational Linguistics.

A Example Prompt for DeepSeek-7B for Task A2.

To illustrate the retrieval-augmented prompting strategy used for well-being score prediction, we provide the following dummy example prompt.

You are an advanced language model tasked with rating the overall well-being presented in a given post on a scale from 1 (low well-being) to 10 (high well-being). The score is based on GAF (American Psychiatric Association, 2000).
The well-being scale is given below:
1 – The person is in persistent danger of severely hurting self or has attempted a serious suicidal act with a clear expectation of death.
2 – In danger of hurting self or others (e.g., suicide attempts; frequently violent; manic excitement) or significant impairment in communication (e.g., incoherent or mute)
.
.
.
10 – No symptoms and superior functioning in a wide range of activities

Examples:
Post: "I've been feeling extremely overwhelmed with work, but I'm trying to manage it by taking breaks."
Well-being score: 7
Post: "Nothing feels enjoyable anymore, and I don't see the point in getting up most days."
Well-being score: 3

Now, read the following post and predict the well-being score. Use the above scale and examples to predict the well-being score. Before predicting the score, justify the predicted score in one full sentence.
Post: "I feel exhausted every day, but I still push through to meet my responsibilities."
Well-being score:

This example demonstrates how the model is guided to score well-being by utilising top-2 retrieved posts with their corresponding labels as examples.

B Error Analysis for Task A1 (Self-State Evidence Extraction)

In our analysis of misclassified instances, we identified several recurring patterns where the model failed to correctly classify evidence spans. Since gold labels were not available for the test data, the analysis was primarily conducted manually. Below, we categorize these errors and provide examples similar to the ones from the test dataset, which the model failed to classify. Note that the current RoBERTa based model has been fine-tuned for **sentence classification** rather than **span classification**. This means that instead of identifying specific spans within a sentence, the model assigns a label to the entire sentence. As a result, it struggles with cases where both adaptive and maladaptive evidence co-exist in a single sentence, leading to ambiguous predictions. In addition, posts with no adaptive or maladaptive evidences from the training data were excluded during the fine-tuning of RoBERTa. This likely contributed to the lower weighted recall compared to recall, as the model struggled to classify sentences as "none" and instead attempted to assign them to one of the predefined classes, even when they did not belong to either.

Mixed Sentiment

Example: "I feel really down, but I know things will get better soon."
Possible Cause: The model struggles to decide whether the sentence leans more positive or negative.

Negation Handling

Example: "I don't think I'm actually sad, just a bit tired."
Possible Cause: The presence of negation ("don't think") may confuse the model into classifying incorrectly.

Ambiguous Language

Example: "Why is everything like this?"
Possible Cause: Without context, the model might not distinguish uncertainty from definitive negative sentiment.

Strong Emotional Words

Example: "I'm completely exhausted and drained, I wish it was not like this."
Possible Cause: The model might overemphasize strong words like "exhausted" and "drained," ignoring the broader context.

The model appears to struggle with mixed sentiments, ambiguous language, emotionally charged words, and multiple ideas within a single post. It seems biased towards classifying strongly emotional statements as maladaptive, even when they contain adaptive elements. Additionally, it might not effectively handle negations or contextual shifts within a sentence, leading to inconsistent classifications. Further analysis could explore the influence of specific keywords and sentence structures in model errors.

These findings suggest that improving contextual understanding and refining the handling of ambiguity in language could enhance model performance in Task A1.

Team ISM at CLPsych 2025: Capturing Mental Health Dynamics from Social Media Timelines using A Pretrained Large Language Model with In-Context Learning

Vu Tran
The Institute of Statistical Mathematics
Tokyo, Japan
vutran@ism.ac.jp

Tomoko Matsui
The Institute of Statistical Mathematics
Tokyo, Japan
tmatsui@ism.ac.jp

Abstract

We tackle the task by using a pretrained large language model (LLM) and in-context learning with template-based instructions to guide the LLM. To improve generation quality, we employ a two-step procedure: sampling and selection. For the sampling step, we randomly sample a subset of the provided training data for the context of LLM prompting. Next, for the selection step, we map the LLM generated outputs into a vector space and employ the Gaussian kernel density estimation to select the most likely output. The results show that the approach can achieve a certain degree of performance and there is still room for improvement.

1 Introduction

The CLPsych 2025 shared task (Tseriotou et al., 2025) combines longitudinal modeling in social media timelines with evidence generation (Chim et al., 2024), promoting the generation of humanly understandable rationales that support recognizing mental states as they dynamically change over time.

The task is structured around the MIND framework (Slonim, 2024), a pan-theoretical scheme for capturing self-states as combinations of Affect, Behavior, Cognition, and Desire (ABCD) components, and identifying mental fluctuations over time.

The shared task's provided dataset contains annotations of evidence aligned with the ABCD paradigm, well-being score and expert summaries at post-level and timeline-level (Shing et al., 2018; Zirikly et al., 2019; Tsakalidis et al., 2022).

Particularly, the shared task is organized into 4 tasks namely A.1, A.2, B, and C, focusing on different aspects of analyzing a given user's mental health state. Task A.1 focuses on extracting evidence of adaptive and maladaptive mental state from user posts. Task A.2 focuses on scoring the well-being of a user within the context of a given user post. Task B focuses on writing a summary of the user's mental health state within the context of a given user post. Task C focuses on writing a summary of the user's mental health state within the context of a given user timeline consisting of a series of posts.

We tackle the task by utilizing a pretrained large language model (LLM) and in-context learning (Dong et al., 2024) with template-based instructions to guide the LLM. Since we approach with a pretrained model without further fine-tuning and in-context learning is limited to the number of in-context examples, to improve generation quality, we employ a two-step procedure: sampling and selection. For the sampling step, we repeatedly randomly sample a subset of the provided training data for the context of LLM prompting. For the selection step, we map the LLM generated outputs into a vector space and employ the Gaussian kernel density estimation (Scott, 2015; Silverman, 2018) to select the most likely output. Details of our method is described in the next section.

2 Method

2.1 Overview

We design our framework consisting of an LLM and utilize in-context learning with a two-step procedure: sampling and selection.

Sampling We randomly sample a subset of the provided training data for the context of LLM prompting, and repeat for a number of rounds. We used meta-llama/Meta-Llama-3-8B-Instruct[1] as the LLM and set the sample size to 225. The temperature of LLM generation is set to 0.1.

Selection We map the LLM generated outputs into a vector space and employ the Gaussian kernel density estimation (Scott, 2015; Silverman, 2018) with the Scott's Rule for bandwidth selec-

[1]https://huggingface.co/meta-llama/Meta-Llama-3-8B-Instruct

Proceedings of the 10th Workshop on Computational Linguistics and Clinical Psychology (CLPsych 2025), pages 287–291
May 3, 2025

```
You are a mental health expert and analyzing
a patient's social media post to determine
their well-being, their dominant self-state
of either adaptive or maladaptive.  The fol-
lowing is your past analysis.

Analysis 1:
<patient post contents>
Adaptive post segments:
* &lt;segment 1&gt;
* ...
Maladaptive post segments:
* &lt;segment 1&gt;
* ...
Well-being:  <well-being score>
Assessment:
<post summary>
...

Analysis i:  ...

Now analyze the following patent post.
<patient post>
Adaptive post segments:
<fill only post segments here, no analysis>
Maladaptive post segments:
<fill only post segments here, no analysis>
Well-being:  <give your score here>
Assessment:
<fill your assessment here>
```

Figure 1: Template for tasks A, B.

tion (Turlach, 1993; Bashtannyk and Hyndman, 2001) to select the most likely output. We used sentence-transformers/all-MiniLM-L6-v2[2] as the sentence embedder model.

2.2 Tasks A & B

Since the evidence of adaptive and maladaptive states is the key for generating the summary of the given user post, we jointly tackle the two tasks A and B in one single flow. We design a prompting template (Figure 1) that instructs the LLM to extract evidence and summarize a given user post jointly. Specifically, we set the number of past analyses to 5, i.e. giving the LLM 5 past user posts with annotations as in-context learning examples.

After performing the sampling step, we collected a set of candidates for each post. We, then, proceed to the selection step. For each candidate, we map a triplet of ⟨adaptive-evidence, maladaptive-evidence, summary⟩ to a triplet of vectors ⟨vector(adaptive-evidence), vector(maladaptive-evidence), vector(summary)⟩. The concatenation of the 3 vectors in the triplet forms the representative vector of the candidate. The set of candidates' vectors are put through the Gaussian kernel density estimation, and the candidate whose vector has the highest density is selected as the final output for the given user post.

[2]https://huggingface.co/sentence-transformers/all-MiniLM-L6-v2

```
You are a mental health expert and analyzing
a patient's social media post to determine
their well-being, their dominant self-state
of either adaptive or maladaptive.  The fol-
lowing is your past analysis.
Past patient 1:
<patient post 1>
<patient post 2>
...
Final Assessment:

...

Past patient i:  ...

Now analyze the following patient.
<patient post 1>
<patient post 2>
...
Final Assessment:  <fill your assessment
here; it should be concise, and focus on
change of self-state in the beginning, middle,
and end of the post timeline; no need to men-
tion detailed post contents; must start with
Final Assessment:>
```

Figure 2: Template for tasks C.

2.3 Task C

Since a timeline may contain a lot of posts, and our resource is limited, even though we believe that the evidence and post-summary are valuable for making the timeline summary, we had to abandon the information and only use the timeline posts as the sole input. That leads to our designed prompting template shown in Figure 2. We set the number of past example timelines to 3. In our observation, a number of past timelines greater than 3 often resulted in junk responses, indicating that the selected LLM cannot handle such a long context.

The selection step is performed as described in Subsection 2.1, where each candidate is a summary generated.

3 Results

As shown in Table 1, our method achieved relatively good performance overall. Particularly, our system performs relatively better in evidence extraction than well-being scoring and summary generation.

For the results of Task A.1 (Table 2), our system, also similar to some other systems, did put more focus on extracting evidence related to maladaptive state than adaptive state. In one perspective, it is a sign that our system did put more alert on negative contents when doing analysis, which is understandable since many public LLMs, including the LLM used in this work, are aligned to recognize negative inputs for the purpose of safeguarding.

For the results of Task A.2 (Table 3), our system also did put more focus on problematic well-being

Team	Task A1 Recall	Task A2 MSE	Task B Mean Consistency	Task C Mean Consistency
Aquarius	0.507	2.010	0.880	0.915
BLUE	0.555	2.260	**0.910**	**0.946**
BULUSI	0.433	**1.920**	0.868	0.941
CIOL	0.246	3.990	0.612	0.610
CSIRO-LT	0.460	2.040	-	-
EAIonFlux	0.517	2.080	0.888	0.913
MMKA	0.602	6.610	-	-
NoviceTrio	-0.028	13.830	0.686	0.855
PsyMetric	0.168	3.230	0.698	0.926
ResBin	0.470	8.020	0.764	0.898
Seq2Psych	0.276	3.270	-	-
uOttawa	**0.637**	2.620	0.860	0.943
Zissou	0.579	3.140	0.846	-
ISM (ours)	0.561	2.760	0.859	0.852
our rank	4	7	6	9

Table 1: Official test results of participants.

	overall		adaptive		maladaptive	
Teams	Recall	Weighed Recall	Recall	Weighed Recall	Recall	Weighed Recall
Aquarius	0.507	0.456	0.499	0.465	0.516	0.446
BLUE	0.555	0.392	0.472	0.400	0.639	0.384
BULUSI	0.433	0.370	0.339	0.339	0.526	0.402
CIOL	0.246	0.174	0.230	0.151	0.262	0.198
CSIRO-LT	0.460	0.427	0.384	0.377	0.537	**0.478**
EAIonFlux	0.517	0.471	0.517	0.480	0.518	0.462
MMKA	0.602	0.343	0.522	0.374	0.681	0.313
NoviceTrio	-0.028	-0.028	-0.104	-0.104	0.047	0.047
PsyMetric	0.168	0.168	0.152	0.152	0.184	0.184
ResBin	0.470	0.302	0.258	0.255	0.682	0.350
Seq2Psych	0.276	0.236	0.245	0.238	0.308	0.235
uOttawa	**0.637**	**0.498**	**0.594**	**0.542**	0.681	0.455
Zissou	0.579	0.320	0.445	0.305	**0.713**	0.335
ISM (ours)	0.561	0.452	0.488	0.460	0.633	0.444
our rank	4	4	5	4	6	5

Table 2: Test results for task A.1.

Teams	MSE	MSE serious	MSE impaired	MSE minimal	F1 Macro
Aquarius	2.010	2.160	3.110	1.250	0.366
BLUE	2.260	**1.410**	3.690	2.060	**0.393**
BULUSI	**1.920**	3.040	1.190	**0.650**	0.351
CIOL	3.990	7.310	**0.490**	2.890	0.119
CSIRO-LT	2.040	1.820	3.680	1.080	0.344
EAIonFlux	2.080	1.770	3.710	2.110	0.321
MMKA	6.610	4.220	11.760	4.950	0.257
NoviceTrio	13.830	3.160	11.590	18.620	0.135
PsyMetric	3.230	2.520	6.630	3.280	0.300
ResBin	8.020	20.260	3.710	1.890	0.192
Seq2Psych	3.270	4.980	1.380	2.630	0.191
uOttawa	2.620	2.280	4.030	2.910	0.302
Zissou	3.140	2.910	4.320	3.090	0.344
ISM (ours)	2.760	1.930	5.000	2.740	0.319
our rank	7	4	11	8	8

Table 3: Test results for task A.2.

Teams	Mean Consistency	Max Contradiction
Aquarius	0.880	0.781
BLUE	**0.910**	**0.533**
BULUSI	0.868	0.805
CIOL	0.612	0.966
CSIRO-LT	-	-
EAIonFlux	0.888	0.782
MMKA	-	-
NoviceTrio	0.686	0.885
PsyMetric	0.698	0.563
ResBin	0.764	0.835
Seq2Psych	-	-
uOttawa	0.860	0.832
Zissou	0.846	0.772
ISM (ours)	0.859	0.777
our rank	6	4

Table 4: Test results for task B.

Teams	Mean Consistency	Max contradiction
Aquarius	0.915	0.876
BLUE	**0.946**	0.540
BULUSI	0.941	0.714
CIOL	0.610	1.000
CSIRO-LT	-	-
EAIonFlux	0.913	0.760
MMKA	-	-
NoviceTrio	0.855	0.596
PsyMetric	0.926	**0.354**
ResBin	0.898	0.816
Seq2Psych	-	-
uOttawa	0.943	0.714
Zissou	-	-
ISM (ours)	0.852	0.833
our rank	9	8

Table 5: Test results for task C.

state as can be seen that MSE serious is relatively better than other categories.

For the results of Tasks B, and C (Tables 4, and 5), our system can generate relatively good summaries highly consistent with the expert annotated summaries. However, max contradiction metric results show that our system added contradictory analysis in the output summaries, which raises the concern of hallucination, a critical problem often found with LLMs (Huang et al., 2025).

4 Conclusion

We have presented our approach for the task by using a pretrained large language model (LLM) and in-context learning with template-based instructions to guide the LLM and designing a two-step procedure, namely sampling and selection, to improve system response quality. We achieved promising results even though the method is simple and requires manageable resources for processing. There is still room for improvement in several directions including choosing stronger LLMs, or fine-tuning with domain knowledge.

Limitations

- No guarantee of adequate domain knowledge. The LLM used in this paper was pre-trained on data extracted from the open Web, which means the model is not guaranteed to be trained on high-quality professional data needed to understand the domain data in this task. Finetuning the model with high-quality professional data may improve the limitation.
- No guarantee of adequate domain context understanding. Though in-context learning is an effective method for guiding an LLM to deal with a new task, the LLM may not understand fully the context, especially since there is no guarantee of adequate domain knowledge in the pre-trained model.

Ethics Statement

Secure access to the shared task dataset was provided with IRB approval under University of Maryland, College Park protocol 1642625 and approval by the Biomedical and Scientific Research Ethics Committee (BSREC) at the University of Warwick (ethical application reference BSREC 40/19-20).

Acknowledgments

This work was supported by "Strategic Research Projects" grant from ROIS (Research Organization of Information and Systems), Japan and JSPS KAKENHI Grant Number JP23K16954. Any opinions, findings, and conclusions or recommendations expressed in this material are those of the author(s) and do not necessarily reflect the views of the author(s)' organization, JSPS or MEXT.

References

David M Bashtannyk and Rob J Hyndman. 2001. Bandwidth selection for kernel conditional density estimation. *Computational Statistics & Data Analysis*, 36(3):279–298.

Jenny Chim, Adam Tsakalidis, Dimitris Gkoumas, Dana Atzil-Slonim, Yaakov Ophir, Ayah Zirikly, Philip Resnik, and Maria Liakata. 2024. Overview of the clpsych 2024 shared task: Leveraging large language models to identify evidence of suicidality risk in online posts. In *Proceedings of the 9th Workshop on Computational Linguistics and Clinical Psychology (CLPsych 2024)*, pages 177–190.

Qingxiu Dong, Lei Li, Damai Dai, Ce Zheng, Jingyuan Ma, Rui Li, Heming Xia, Jingjing Xu, Zhiyong Wu, Baobao Chang, and 1 others. 2024. A survey on in-context learning. In *Proceedings of the 2024 Conference on Empirical Methods in Natural Language Processing*, pages 1107–1128.

Lei Huang, Weijiang Yu, Weitao Ma, Weihong Zhong, Zhangyin Feng, Haotian Wang, Qianglong Chen, Weihua Peng, Xiaocheng Feng, Bing Qin, and Ting Liu. 2025. A survey on hallucination in large language models: Principles, taxonomy, challenges, and open questions. *ACM Trans. Inf. Syst.*, 43(2).

David W Scott. 2015. *Multivariate density estimation: theory, practice, and visualization*. John Wiley & Sons.

Han-Chin Shing, Suraj Nair, Ayah Zirikly, Meir Friedenberg, Hal Daumé III, and Philip Resnik. 2018. Expert, crowdsourced, and machine assessment of suicide risk via online postings. In *Proceedings of the fifth workshop on computational linguistics and clinical psychology: from keyboard to clinic*, pages 25–36.

Bernard W Silverman. 2018. *Density estimation for statistics and data analysis*. Routledge.

Dana Atzil Slonim. 2024. Self-other dynamics (sod): A transtheoretical coding manual.

Adam Tsakalidis, Jenny Chim, Iman Munire Bilal, Ayah Zirikly, Dana Atzil-Slonim, Federico Nanni, Philip Resnik, Manas Gaur, Kaushik Roy, Becky Inkster, and 1 others. 2022. Overview of the clpsych 2022 shared task: Capturing moments of change in longitudinal user posts. In *Proceedings of the Eighth Workshop on Computational Linguistics and Clinical Psychology*, pages 184–198.

Talia Tseriotou, Jenny Chim, Ayal Klein, Aya Shamir, Guy Dvir, Iqra Ali, Cian Kennedy, Guneet Singh Kohli, Anthony Hills, Ayah Zirikly, Dana Atzil-Slonim, and Maria Liakata. 2025. Overview of the clpsych 2025 shared task: Capturing mental health dynamics from social media timelines. In *Proceedings of the 10th Workshop on Computational Linguistics and Clinical Psychology*. Association for Computational Linguistics.

Berwin A Turlach. 1993. Bandwidth selection in kernel density estimation: a rewiew. Technical report, Humboldt Universitaet Berlin.

Ayah Zirikly, Philip Resnik, Ozlem Uzuner, and Kristy Hollingshead. 2019. Clpsych 2019 shared task: Predicting the degree of suicide risk in reddit posts. In *Proceedings of the sixth workshop on computational linguistics and clinical psychology*, pages 24–33.

Transformer-Based Analysis of Adaptive and Maladaptive Self-States in Longitudinal Social Media Data

Abhin B and **Renukasakshi V Patil**
Department of Information Technology
National Institute of Technology Karnataka, Surathkal
Mangalore, India
{abhinb.211ai003@nitk.edu.in, renukasakshivpatil.211ai030@nitk.edu.in}

Abstract

The CLPsych workshop, held annually since 2014, promotes the application of computational linguistics to behavioral analysis and neurological health assessment. The CLPsych 2025 shared task, extending the framework of the 2022 iteration, leverages the MIND framework to model temporal fluctuations in mental states. This shared task comprises three subtasks, each presenting substantial challenges to natural language processing (NLP) systems, requiring sensitive and precise outcomes in analyzing adaptive and maladaptive behaviors. In this study, we employed a range of modeling strategies tailored to the requirements and expected outputs of each subtask. Our approach mostly utilized traditional language models like BERT, LongFormer and Pegasus diverging from the prevalent trend of prompt-tuned large language models. We achieved an overall ranking of 13th, with subtask rankings of 8th in Task 1a, 13th in Task 1b, 8th in Task 2, and 7th in Task 3. These results highlight the efficacy of our methods while underscoring areas for further refinement in handling complex behavioral data.

1 Introduction

Understanding mental health through digital footprints has become a critical area of research, with social media providing a unique lens into users' psychological states over time. The CLPsych 2025 Shared Task builds upon prior research efforts by integrating longitudinal modeling with evidence generation The CLPsych 2025 Shared Task builds upon prior research efforts (Tsakalidis et al., 2022; Tseriotou et al., 2025; Zirikly et al., 2019) by integrating longitudinal modeling with evidence generation..., focusing on adaptive and maladaptive self-states in user timelines. Our work in this shared task contributes to the growing field of computational mental health assessment by employing **state-of-the-art transformer-based models** across various subtasks. A key preprocessing step in our approach was **removing posts with null values**, ensuring cleaner and more informative datasets for analysis. Unlike previous studies that retained all posts for completeness, our decision aimed to enhance the **signal-to-noise ratio**, thereby improving model performance. By applying a combination of specialized NLP models, we effectively extracted relevant psychological markers, assigned well-being scores, and summarized self-state dynamics in both individual posts and entire timelines.

For well-being score prediction, we fine-tuned MentalBERT, a model specifically trained for mental health applications, to enhance accuracy in assessing psychological functioning. For post-level summarization, we leveraged Longformer, which excels in capturing contextual dependencies in lengthy texts, ensuring comprehensive self-state summaries. At the timeline level, we utilized Pegasus-X-Large, a model optimized for abstractive summarization, to generate coherent narratives capturing self-state transitions. Lastly, for evidence extraction, we employed Mistral, a robust transformer model capable of identifying relevant spans with high precision. Our results indicate that MentalBERT achieved state-of-the-art performance in well-being score prediction, Longformer provided detailed and context-aware post summaries, while Pegasus-X effectively distilled timeline-level insights, and Mistral demonstrated high recall in extracting meaningful evidence spans. These findings reinforce the potential of advanced NLP techniques in modeling dynamic mental health patterns and offer promising directions for future clinical and computational research.

2 Background

Toxicity detection in NLP has focused on span-level identification of harmful content. SemEval-

Proceedings of the 10th Workshop on Computational Linguistics and Clinical Psychology (CLPsych 2025), pages 292–299
May 3, 2025 ©2025 Association for Computational Linguistics

2021 Task 5 (Ji et al., 2021a) highlighted challenges in detecting toxic spans using token classification and span prediction. Transformer-based models like BERT, RoBERTa, and SpanBERT improved performance by combining these approaches. In mental health, domain-specific models like MentalBERT (Chhablani et al., 2021) have enhanced social media text analysis for early detection of mental disorders and suicidal ideation, demonstrating the value of contextualized representations in both toxicity detection and mental health assessment (Shing et al., 2018; Zirikly et al., 2019).

Timeline summarization distills event evolution from timestamped documents, requiring coherence and diversity. Evolutionary Timeline Summarization (ETS) (Yan et al., 2011) optimizes relevance and coverage, while graph-based methods enhance abstractive and extractive summarization. Handling long-context dependencies remains a challenge for transformer models (Qin, 2024), with research on dynamic-resolution encoding (e.g., Nugget) improving efficiency.

Zero-shot prompting in LLMs faces challenges in generating concise, coherent summaries. Chain-of-event (CoE) prompting (Wei et al., 2022) structures summarization into four steps, improving abstraction and coherence. Models like Pegasus (Zhang et al., 2020) further enhance contextual understanding. We also reference the Self-Other Dynamics (SOD) framework (Slonim, 2024) for analyzing adaptive and maladaptive self-states.

3 Methodology

Our methodology focuses on exploiting pre-trained language models and fine-tuning them effectively with the provided data. We employ various language models like BERT, Longformer, and Pegasus across multiple tasks.

3.1 Task A.1: Evidence Extraction for Adaptive and Maladaptive Self-States

Task A.1 involves extracting text spans indicating adaptive or maladaptive self-states. Adaptive self-states support fundamental needs fulfillment, while maladaptive self-states obstruct these needs(Slonim, 2024). The objective is identifying spans evidencing adaptive, maladaptive or neither self-states.

Initially, we considered BERT for token classification (Devlin et al., 2019), labeling each token as part of an adaptive, maladaptive, or neutral span. However, this risked producing fragmented spans with misclassified portions, compromising contextual integrity.

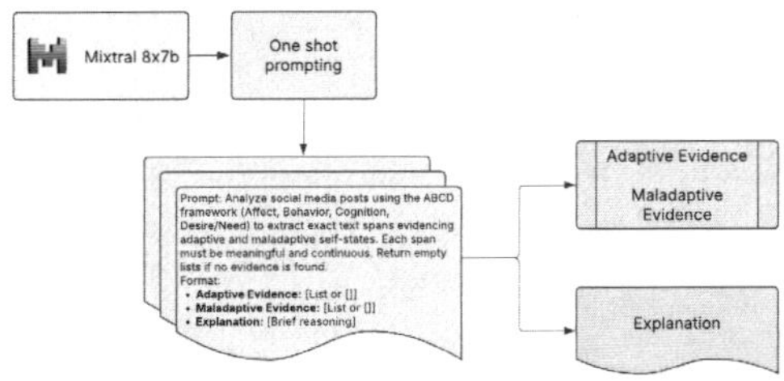

Figure 1: (Task 1a) Methodology diagram showing the prompt tuning of Mistral 8x7b for the purpose of accurate extraction of adaptive and maladaptive spans with addition of explanation component.

Consequently, we adopted Mixtral 8x7B (Jiang et al., 2024), a large language model renowned for robust natural language understanding. This MoE model was selected for its proficiency in interpreting complex natural language and extracting coherent spans, overcoming BERT's fragmentation issues.

The model was guided by a carefully designed prompt leveraging one-shot learning to capitalize on Mistral 8x7B's pre-trained knowledge. The prompt structure can be found in A which was finalized after a careful effort in experimentation based on general research of what adaptive and maladaptive behaviors are.

To enhance performance, we integrated the "explain then annotate" strategy (Lee et al., 2020). This technique requires providing a rationale before finalizing annotations, improving accuracy through deeper contextual understanding. In our implementation, Mixtral 8x7B generated brief explanations for each classification, refining the span extraction process, as illustrated in Figure 1.

3.2 Task 1b: Well-Being Score Prediction

Task 1b involves predicting a well-being score for each post as a classification task. We leveraged embeddings from transformer-based encoder models as input features for a Random Forest Classifier (Breiman, 2001).

For contextual embeddings extraction, we employed various models including BERT (Devlin et al., 2019), MiniLM (Wang et al., 2020), RoBERTa (Liu et al., 2019), ClinicalBERT (Huang et al., 2020), and MentalBERT (Ji et al., 2021b).

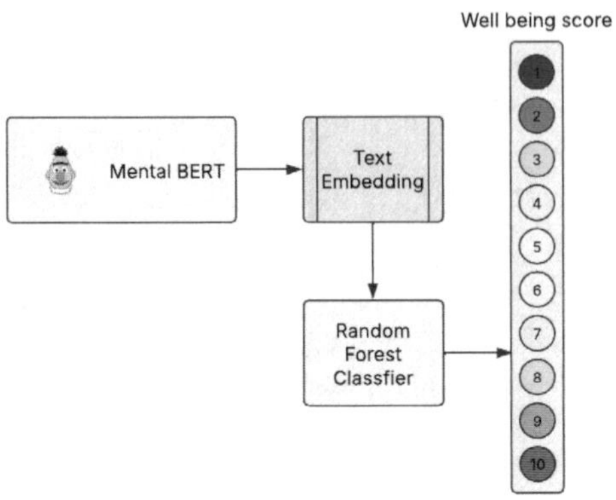

Figure 2: (Task 1b) Methodology of well-being score prediction using Mental BERT and Random Forest Classifier

The [CLS] token embedding from each model's final layer served as post representation, feeding into a Random Forest Classifier for multi-class classification.

Performance evaluation using accuracy as the primary metric revealed MentalBERT as the best performer. This superior performance likely stems from MentalBERT's pre-training on mental health-related text, enabling better capture of domain-specific patterns relevant to well-being assessment, as shown in Figure 2.

3.3 Task 2: Post-Level Summaries of Self-State Dynamics

Task 2 involves generating summaries describing the interplay between adaptive and maladaptive self-states, focusing on the dominant state and its driving ABCD aspect. The summary must outline how this central aspect influences other components and address any complementary self-state.

Initially, we utilized BART (Lewis et al., 2019) for summarization. However, BART's 1024 token context limitation proved insufficient for longer posts. We switched to Longformer (Beltagy et al., 2020) with its 4096 token capacity, allowing effective processing of extended posts with their evidence spans.

We leveraged outputs from Task 1a—specifically, the extracted evidence spans—as additional input to inform summarization. The data was structured as:

```
post: {post}
adaptive_evi: [list of spans]
maladaptive_evi: [list of spans]
```

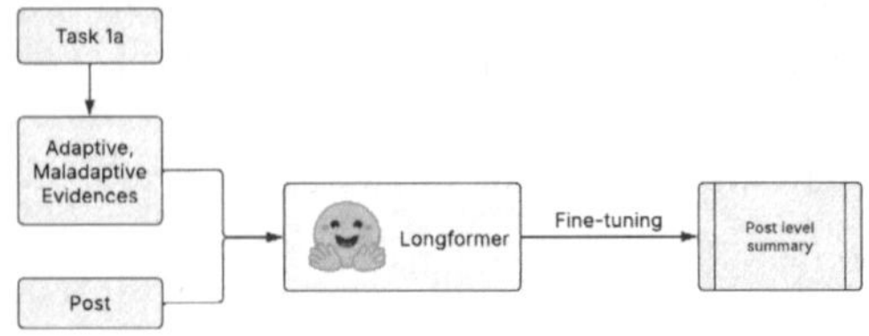

Figure 3: (Task 2) Methodology diagram depicting the finetuning process of LongFormer for the task of post-level summary generation.

Figure 4: (Task 3) Methodology diagram for the task 3 showing the finetuning of pegasus-X-large for timeline summary generation.

This structured input enabled Longformer to contextualize posts with identified self-state evidence, generating summaries accurately reflecting self-state interplay, as depicted in Figure 3.

3.4 Task 3: Timeline-Level Summaries

Task 3 required generating timeline-level summaries reflecting an individual's self-state dynamics over time. We combined multiple posts from the same timeline into cohesive text, preserving chronological order to maintain timeline integrity for effective self-state progression analysis.

Analysis revealed a mean word count of 1224 words, with the longest text reaching 5555 words. This necessitated a model capable of processing long sequences. We selected Pegasus-X-large (Zhang et al., 2020), which accommodates up to 16,000 tokens, eliminating truncation concerns.

Pegasus-X-large was fine-tuned using combined posts as input and expected timeline summaries as targets, enabling the generation of summaries effectively capturing temporal dynamics and transitions, as shown in Figure 4.

4 Experimental Results

In this section, we present results for each task in the CLPsych 2025 shared task. Experiments were conducted using Kaggle's free 2xT4 GPUs, providing sufficient computational resources.

The initial dataset contained 343 rows. After removing null values from well-being scores and post summaries, 199 rows remained for Tasks 1b and 3. For Tasks 1a and 2, we utilized the full dataset as these tasks were unaffected by the missing data.

4.1 Task 1a: Evidence Extraction

For Task 1a, we employed Mixtral-8x7b (Jiang et al., 2024), a mixture of experts model quantized to 4-bit to optimize memory usage. A temperature of 0.2 was selected after experimentation, balancing creativity and precision for effective extraction of adaptive and maladaptive self-state indicators.

4.2 Task 1b: Well-Being Score Prediction

In Task 1b, we tested various models including BERT (Devlin et al., 2019) and LLaMA (Touvron et al., 2023) for sequence classification. Optimal performance was achieved by combining Mental-BERT (Ji et al., 2021b) embeddings with a Random Forest Classifier (Breiman, 2001) (100 estimators). Embeddings were extracted using the Sentence Transformers library, capturing domain-specific nuances relevant to well-being assessment.

4.3 Task 2: Post-Level Summaries

For Task 2, we utilized Longformer (Beltagy et al., 2020), chosen for its long-sequence handling capability. Input data combined post content with Task 1a outputs (adaptive/maladaptive evidence spans). The model was fine-tuned using the Seq2SeqTrainer from Transformers (Wolf et al., 2019) for 10 epochs, generating coherent summaries reflecting self-state interplay.

4.4 Task 3: Timeline-Level Summaries

Task 3 involved timeline-level summaries by concatenating posts from the same timeline. Posts were separated chronologically using a delimiter (`\n\n\—-\n\n`). Pegasus-X-large (Zhang et al., 2020), with its 16,000-token context window, was fine-tuned for 10 epochs using Seq2SeqTrainer.

The performance across all tasks is summarized in Table 1, presenting our team's (ResBin) scores.

Table 1: Experimental Results for Team ResBin Across Different Tasks

Task	Metric	Score	Rank
Task 1a	Recall	0.470	8
Task 1b	Mean Consistency	0.764	13
Task 2	MSE	8.020	8
Task 3	Mean Consistency	0.898	7

5 Limitations

Our approach faces several constraints despite promising results. Using traditional language models rather than prompt-tuned LLMs may have limited our performance on tasks requiring nuanced psychological inferences. Effective pre-processing of the data was missing and no sort of data augmentation was carried out. Because of limited data the pre-trained transformer based models could not generalize to the extent LLMs are capable of generalizing. Furthermore fine-tuning of LLMs was not carried out in our work which could have possibly given even better results with extreme level of natural language understanding. Additionally, our models may inadequately capture long-term temporal dependencies in user posts, potentially missing subtle shifts in mental states.

6 Ethics

Our research adheres to strict ethical guidelines protecting data privacy and dignity. We store data exclusively on local machines and google drives with team-restricted access and will delete all dataset files and derived models after the CLPsych 2025 workshop. We commit to not redistributing the dataset and to not submitting any part of it to platforms that might use it as training data. We acknowledge that models developed in this research are for computational research purposes only, not for direct clinical application.

7 Conclusions

In this work, we focused on fine-tuning natural language models like BERT, Longformer, and Pegasus (Tseriotou et al., 2025) for the CLPsych 2025 shared task, building on prior methodologies (Tsakalidis et al., 2022; Zirikly et al., 2019). Our approach effectively addressed evidence span extraction, well-being score prediction, and summary generation at both post and timeline levels. While we prioritized task-specific models over large language models (LLMs) for interpretability, future work could explore supervised LLM fine-tuning with sufficient computational resources to enhance prediction and summarization capabilities. This direction may further bridge the gap between general-purpose language models and domain-specific mental health analysis tasks.

References

Iz Beltagy, Matthew E. Peters, and Arman Cohan. 2020. Longformer: The long-document transformer. *Preprint*, arXiv:2004.05150.

Leo Breiman. 2001. Random forests. *Machine learning*, 45:5–32.

Gunjan Chhablani, Abheesht Sharma, Harshit Pandey, Yash Bhartia, and Shan Suthaharan. 2021. Nlrg at semeval-2021 task 5: Toxic spans detection leveraging bert-based token classification and span prediction techniques. *arXiv preprint arXiv:2102.12254*.

Jacob Devlin, Ming-Wei Chang, Kenton Lee, and Kristina Toutanova. 2019. Bert: Pre-training of deep bidirectional transformers for language understanding. *Preprint*, arXiv:1810.04805.

Kexin Huang, Jaan Altosaar, and Rajesh Ranganath. 2020. Clinicalbert: Modeling clinical notes and predicting hospital readmission. *Preprint*, arXiv:1904.05342.

Shaoxiong Ji, Tianlin Zhang, Luna Ansari, Jie Fu, Prayag Tiwari, and Erik Cambria. 2021a. Mentalbert: Publicly available pretrained language models for mental healthcare. *arXiv preprint arXiv:2110.15621*.

Shaoxiong Ji, Tianlin Zhang, Luna Ansari, Jie Fu, Prayag Tiwari, and Erik Cambria. 2021b. Mentalbert: Publicly available pretrained language models for mental healthcare. *Preprint*, arXiv:2110.15621.

Albert Q Jiang, Alexandre Sablayrolles, Antoine Roux, Arthur Mensch, Blanche Savary, Chris Bamford, Devendra Singh Chaplot, Diego de las Casas, Emma Bou Hanna, Florian Bressand, and 1 others. 2024. Mixtral of experts. *arXiv preprint arXiv:2401.04088*.

Dong-Ho Lee, Rahul Khanna, Bill Yuchen Lin, Jamin Chen, Seyeon Lee, Qinyuan Ye, Elizabeth Boschee, Leonardo Neves, and Xiang Ren. 2020. Lean-life: A label-efficient annotation framework towards learning from explanation. *arXiv preprint arXiv:2004.07499*.

Mike Lewis, Yinhan Liu, Naman Goyal, Marjan Ghazvininejad, Abdelrahman Mohamed, Omer Levy, Ves Stoyanov, and Luke Zettlemoyer. 2019. Bart: Denoising sequence-to-sequence pre-training for natural language generation, translation, and comprehension. *Preprint*, arXiv:1910.13461.

Yinhan Liu, Myle Ott, Naman Goyal, Jingfei Du, Mandar Joshi, Danqi Chen, Omer Levy, Mike Lewis, Luke Zettlemoyer, and Veselin Stoyanov. 2019. Roberta: A robustly optimized bert pretraining approach. *Preprint*, arXiv:1907.11692.

Guanghui Qin. 2024. *Towards Efficient Long-Context Natural Language Processing*. Ph.D. thesis, Johns Hopkins University.

Han-Chin Shing, Suraj Nair, Ayah Zirikly, Meir Friedenberg, Hal Daumé III, and Philip Resnik. 2018. Expert, crowdsourced, and machine assessment of suicide risk via online postings. In *Proceedings of the fifth workshop on computational linguistics and clinical psychology: from keyboard to clinic*, pages 25–36.

Dana Atzil Slonim. 2024. Self-other dynamics (sod): A transtheoretical coding manual.

Hugo Touvron, Thibaut Lavril, Gautier Izacard, Xavier Martinet, Marie-Anne Lachaux, Timothée Lacroix, Baptiste Rozière, Naman Goyal, Eric Hambro, Faisal Azhar, and 1 others. 2023. Llama: Open and efficient foundation language models. *arXiv preprint arXiv:2302.13971*.

Adam Tsakalidis, Jenny Chim, Iman Munire Bilal, Ayah Zirikly, Dana Atzil-Slonim, Federico Nanni, Philip Resnik, Manas Gaur, Kaushik Roy, Becky Inkster, and 1 others. 2022. Overview of the clpsych 2022 shared task: Capturing moments of change in longitudinal user posts. In *Proceedings of the Eighth Workshop on Computational Linguistics and Clinical Psychology*, pages 184–198.

Talia Tseriotou, Jenny Chim, Ayal Klein, Aya Shamir, Guy Dvir, Iqra Ali, Cian Kennedy, Guneet Singh Kohli, Anthony Hills, Ayah Zirikly, Dana Atzil-Slonim, and Maria Liakata. 2025. Overview of the clpsych 2025 shared task: Capturing mental health dynamics from social media timelines. In *Proceedings of the 10th Workshop on Computational Linguistics and Clinical Psychology*. Association for Computational Linguistics.

Wenhui Wang, Furu Wei, Li Dong, Hangbo Bao, Nan Yang, and Ming Zhou. 2020. Minilm: Deep self-attention distillation for task-agnostic compression of pre-trained transformers. *Preprint*, arXiv:2002.10957.

Jason Wei, Xuezhi Wang, Dale Schuurmans, Maarten Bosma, Fei Xia, Ed Chi, Quoc V Le, Denny Zhou, and 1 others. 2022. Chain-of-thought prompting elicits reasoning in large language models. *Advances in neural information processing systems*, 35:24824–24837.

Thomas Wolf, Lysandre Debut, Victor Sanh, Julien Chaumond, Clement Delangue, Anthony Moi, Pierric Cistac, Tim Rault, Rémi Louf, Morgan Funtowicz, and 1 others. 2019. Huggingface's transformers: State-of-the-art natural language processing. *arXiv preprint arXiv:1910.03771*.

Rui Yan, Xiaojun Wan, Jahna Otterbacher, Liang Kong, Xiaoming Li, and Yan Zhang. 2011. Evolutionary timeline summarization: a balanced optimization framework via iterative substitution. In *Proceedings of the 34th international ACM SIGIR conference on Research and development in Information Retrieval*, pages 745–754.

Jingqing Zhang, Yao Zhao, Mohammad Saleh, and Peter J. Liu. 2020. Pegasus: Pre-training with extracted gap-sentences for abstractive summarization. *Preprint*, arXiv:1912.08777.

Ayah Zirikly, Philip Resnik, Ozlem Uzuner, and Kristy Hollingshead. 2019. Clpsych 2019 shared task: Predicting the degree of suicide risk in reddit posts. In *Proceedings of the sixth workshop on computational linguistics and clinical psychology*, pages 24–33.

A Appendix

A.1 Prompt used in Task 1a

You are a mental health expert analyzing social media posts from the given posts to identify evidence of adaptive and maladaptive self-states. Each self-state is characterized by **Affect (A)**, **Behavior (B)**, **Cognition (C)**, and **Desire/Need (D)** (ABCD framework). Your task is to extract continuous spans of text from the post that directly evidence these self-states.

Definitions

Adaptive Self-State

Aspects of Affect, Behavior, Cognition, and Desire/Need that support the fulfillment of basic needs like relatedness (connection with others), autonomy (independence), and competence (feeling capable). These are healthy coping mechanisms or positive mental states. Examples include:

- **Affect (A)**: Positive emotions like happiness, hope, or pride (e.g., "I'm feeling really hopeful about my future").
- **Behavior (B)**: Healthy actions towards self or others, such as:
  - Support: Seeking or appreciating help from others (e.g., "I talked to my friend about my stress").
  - Physical Activity: Engaging in exercise like walking or yoga (e.g., "I went for a run to clear my mind").
  - Relaxation: Calming activities like meditation or listening to music (e.g., "I listened to soft music to relax").
  - Problem-Solving: Actively addressing issues (e.g., "I made a plan to tackle my workload").
- **Cognition (C)**: Positive perceptions of self or others, like self-acceptance or viewing others as supportive (e.g., "I believe I can improve with help").
- **Desire/Need (D)**: Healthy intentions or needs, like seeking relatedness or autonomy (e.g., "I want to connect with others to feel supported").

Maladaptive Self-State

Aspects of Affect, Behavior, Cognition, and Desire/Need that hinder the fulfillment of basic needs, reflecting unhealthy coping mechanisms or negative mental states. Examples include:

- **Affect (A)**: Negative emotions like depression, anxiety, shame, or hopelessness (e.g., "I feel so hopeless and sad").
- **Behavior (B)**: Unhealthy actions towards self or others, such as:
  - Escape: Withdrawing socially or over-engaging in solitary activities (e.g., "I stayed in my room all day scrolling online").
  - Unhealthy Self-Soothing: Overeating, binge drinking, or excessive internet use (e.g., "I binged on snacks to feel better").

- Numbing: Using substances to numb emotions (e.g., "I drank to forget my problems").
- Self-Harm: Engaging in self-injury (e.g., "I want to hurt myself").
- Compulsions/Risk-Taking: Seeking adrenaline through risky behaviors (e.g., "I drove recklessly to feel something").

- **Cognition (C)**: Negative perceptions, like self-criticism or expecting rejection (e.g., "I think I'm a failure and no one cares").
- **Desire/Need (D)**: Unhealthy expectations or fears, like expecting failure or rejection (e.g., "I feel like I'll never be good enough").

Task A.1 - States' Evidence

- Identify continuous spans of text in the post that directly provide evidence of adaptive self-states and maladaptive self-states.
- Each span should be a complete, meaningful segment of text (e.g., a full sentence or a phrase) that clearly reflects an adaptive or maladaptive self-state.
- A post may contain evidence for one self-state (adaptive or maladaptive), both, or neither.
- If no evidence is found for a self-state, return an empty list ([]) for that category.
- Ensure the spans are exact substrings of the post, preserving the original wording and punctuation.

Post to Analyze

{post}

Output Format

Provide the following in a structured format:

- **Adaptive Evidence**: [List of continuous text spans (strings) showing adaptive self-states, or [] if none]
- **Maladaptive Evidence**: [List of continuous text spans (strings) showing maladaptive self-states, or [] if none]
- **Explanation**: [Brief explanation of your evidence selection, referencing the ABCD framework and coping mechanisms]

Example

Post: "My friend went for a gym session to relieve stress, but he sometimes gets dissapointed and feels hopeless about his situation."
Output:

- **Adaptive Evidence**: ["My friend went for a gym session to relieve stress"]
- **Maladaptive Evidence**: ["he sometimes gets dissapointed and feels hopeless about his situation"]
- **Explanation**: The span "My friend went for a gym session to relieve stress" reflects an adaptive self-state through Behavior (B) - Physical Activity, as walking is a healthy coping mechanism to relieve stress. The span "he sometimes gets dissapointed and feels hopeless about his situation" indicates a maladaptive self-state through Affect (A) - hopelessness, a negative emotion that hinders well-being.

Disclaimer

Paraphrased representative data is being used in the prompt and not real data from training data provided during the shared task.

Who We Are, Where We Are: Mental Health at the Intersection of Person, Situation, and Large Language Models

Nikita Soni[1], **August Håkan Nilsson**[2], **Syeda Mahwish**[1],
Vasudha Varadarajan[1], **H. Andrew Schwartz**[1], **Ryan L. Boyd**[3]
[1]Department of Computer Science, Stony Brook University
[2]Oslo Metropolitan University, Oslo Business School
[3]University of Texas at Dallas, Dept. of Psychology
{nisoni, smahwish, vvaradarajan, has}@cs.stonybrook.edu,
august.nilsson1907@gmail.com, boyd@utdallas.edu

Abstract

Mental health is not a fixed trait but a dynamic process shaped by the interplay between individual dispositions and situational contexts. Building on interactionist and constructionist psychological theories, we develop interpretable models to predict well-being and identify adaptive and maladaptive self-states in longitudinal social media data. Our approach integrates person-level psychological traits (e.g., resilience, cognitive distortions, implicit motives) with language-inferred situational features derived from the Situational 8 DIAMONDS framework. We compare these theory-grounded features to embeddings from a psychometrically-informed language model that captures temporal and individual-specific patterns. Results show that our principled, theory-driven features provide competitive performance while offering greater interpretability. Qualitative analyses further highlight the psychological coherence of features most predictive of well-being. These findings underscore the value of integrating computational modeling with psychological theory to assess dynamic mental states in contextually sensitive and human-understandable ways.

1 Introduction

Understanding mental health through language has long been a foundational goal in clinical psychology and computational social science. Human expression — especially as manifested through digital communication — provides a unique window into internal states, social interactions, and psychological well-being. The CLPsych 2025 shared task builds on prior work in computational linguistics and clinical psychology, extending the analysis of mental health from static assessments to a dynamic, temporally anchored perspective. Seq2Psych, our interdisciplinary team, approaches this challenge by integrating psychological theory with computational methods, ensuring that our models are both empirically grounded and practically applicable. Our work emphasizes not only predictive accuracy but also a principled, theory-grounded approach to interpretability, facilitating a nuanced explanation of how self-states fluctuate over time.

Primary Contributions made in this work include: (1) proposal of a theory-driven baseline that combines language-inferred person-level traits (e.g., well-being, cognitive distortions, resilience) with situational context features derived from the Situational 8 DIAMONDS framework (Rauthmann et al., 2014); (2) use of a human-centered language model (Soni et al., 2022, 2024c) trained on temporal user histories, to generate person-contextualized embeddings aligned with psychometric theory; (3) evaluation of these representations — individually and in hybrid configurations — for predicting well-being and identifying adaptive/maladaptive self-states in longitudinal text data; (4) an analysis of the most predictive psychological features to highlight interpretable connections between language, context, and mental health outcomes.

2 Background

Traditional models of psychological assessment often rely on static categories — diagnostic labels that imply stable traits or enduring conditions. However, integrative psychological theories emphasize that mental states are inherently dynamic, shaped by a complex interplay between individual dispositions and situational contexts (Buss, 1987; Ekehammar, 1974; Boyd and Markowitz, 2024). The constructionist view of emotions, for example, posits that emotional experiences emerge from interactions between an individual's traits, cognitive processes, and environmental affordances (Barrett, 2017). Likewise, interactionist approaches in personality psychology highlight that adaptiveness or maladaptiveness of a given behavior is highly contingent upon situational fit (Mischel and Shoda,

Proceedings of the 10th Workshop on Computational Linguistics and Clinical Psychology (CLPsych 2025), pages 300–313
May 3, 2025

1995; Fleeson, 2004).

The notion of situational fit is central to understanding mental health dynamics. Psychological well-being is not merely an individual trait but a function of how well a person's responses align with their context. A behavior that is adaptive in one situation may be maladaptive in another. For instance, hypervigilant behaviors may be adaptive in some contexts (e.g., military personnel in combat situations), but highly *maladaptive* in another (e.g., a classroom or casual social gathering; Vyas et al. 2023). This perspective aligns with the broader movement in psychology that views well-being as a dynamic process rather than a fixed state (see, e.g., Hollenstein, 2015)

2.1 A Principled Baseline: Integration of Person-Level Traits and Situational Context

To model mental health dynamics in a principled manner, our approach combines person-level traits with psychological dimensions of the situation. Specifically, we leverage:

Psychological Characteristics of the Situation Using a large language model, we annotated each post for the psychological characteristics of its context, based on the Situational 8 DIAMONDS (S8D) framework (Rauthmann et al., 2014). This framework captures eight psychosocial aspects of a given situation — **D**uty, **I**ntellect, **A**dversity, **M**ating, p**O**sitivity, **N**egativity, **D**eception, and **S**ociality — that shape the meaning and context of person-environment transactions.

Person-Level Traits (PLT) We employed existing models and methods to assess and estimate individual differences in implicit motives (Nilsson et al., 2024), depression and anxiety (Gu et al.), harmony in life and satisfaction with life (Kjell et al., 2022a) valence (Eijsbroek, 2023), cognitive distortions (Varadarajan et al., 2025), and resilience (Mahwish et al.). These traits serve as stable psychological anchors that interact dynamically with situational contexts in well-established fashions (Mejía and Hooker, 2015; Ungar, 2013; Joiner Jr. and Timmons, 2009).

By combining these features, we constructed a baseline model that aligns with psychological theory, providing an interpretable reference point against which more data-driven approaches can be evaluated. Our method represents a true interdisciplinary effort in computational social science, bridging insights from personality psychology, emotion theory, and NLP to advance the study of mental health dynamics in digital contexts.

2.2 Human Language Modeling: HaRT for Person-Contextual Embeddings

Our principled baseline approach offers a clear explanatory mechanism for predicting well-being and distinguishing between adaptive and maladaptive self-states. However, we anticipate that more advanced language models will enhance predictive accuracy and provide a richer representation of language and individuals. HaRT, trained on the Human Language Modeling (HuLM) task — which predicts the next word based on prior words, incorporating a latent user representation derived from their temporal historical language — enables a person-contextualized understanding of language (Soni et al., 2022). Grounded in psychometric theory on the stability of psychological traits (Watson, 2004) HuLM processes an author's language collectively, recognizing that linguistic patterns are best understood within the context of the individual themselves, over time (Soni et al., 2024b; Ganesan et al., 2024). This approach is particularly well-suited for our tasks, given the dataset's longitudinal structure, where language is nested within individuals, and has proven to be effective in mental health assessments (Ganesan et al., 2022; Varadarajan et al., 2024b), psychological assessments (Soni et al., 2025), and user attributes assessments (Soni et al., 2024a).

3 Data & Tasks

Dataset. The CLPsych 2025 shared task (Tseriotou et al., 2025) provided annotated evidence for adaptive and maladaptive self-states (Slonim, 2024) as spans of texts from posts written by individuals historically in addition to a score representing the overall well-being in a post. The data consists of 30 users (timelines) with a total of 343 posts of which 199 posts were annotated.

Shared Tasks. The shared tasks focus on the longitudinal modeling of changes in individual's mood and states (Shing et al., 2018; Zirikly et al., 2019; Tsakalidis et al., 2022). We participate in 2 subtasks targeted at post-level judgments: a) predicting the overall well-being, and b) identifying evidence for adaptive and maladaptive self-states.

4 Methods

We extracted two categories of features: a theory-informed baseline — comprising Situational 8 DIAMONDS (S8D) and Person-Level Traits (PLT; see Section 2.1) — and person-contextualized embeddings. While PLT features were computed at both sentence and post levels, S8D were limited to post-level annotations due to their reliance on broader context.

Situational 8 DIAMONDS (S8D). We used Deepseek-R1 (DeepSeek-AI et al., 2025) with few-shot prompting to infer scores for each of the eight situational dimensions at the post level. Each dimension was prompted separately using two manually annotated exemplars tailored to its psychological construct (see § A.2). Scores ranged from 1 (not present) to 9 (highly present), reflecting the inferred prominence of each situational characteristic.

Person-Level Traits (PLT). We extracted 19 features across four subdomains:
Implicit Motives. Following Nilsson et al. (b), we applied fine-tuned RoBERTa-Large models to estimate three subconscious motives — achievement, affiliation, and power — at the sentence level. These predictions were averaged and adjusted for word count to yield post-level scores (details in Appendix A.3).
Mental Health. Using the Language-Based Assessment Model Library (Nilsson et al., a), we inferred six psychological dimensions: valence (Eijsbroek, 2023), harmony in life, satisfaction with life (Kjell et al., 2022b), anxiety, and two depression indices (Gu et al., 2024). Features were extracted at the sentence level and averaged to generate post-level estimates.
Resilience. We implemented the Resilience through Language Modeling (ReLM) framework (Mahwish et al.) to compute scores for nine resilience-related facets (e.g., optimism, coping toolkit) at both sentence and post levels. See Appendix A.4 for details.
Cognitive Distortions. Drawing on prior work (Varadarajan et al., 2025), we used pretrained models to estimate levels of cognitive distortion, a known correlate of maladaptive emotional states (Mann et al., 2002; Bathina et al., 2021), at both sentence and post levels.

Person-Contextual Embeddings: HaRT. We fine-tuned HaRT (Human-aware Recurrent Transformer) (Soni et al., 2022) to predict continuous well-being scores at the post level and binary adaptive/maladaptive labels at the sentence level. To do so, we split the CLPsych training data into internal training and validation sets. HaRT processes users' historical posts in sequence, enabling the generation of temporally informed, person-specific embeddings at both the sentence and post levels.

We evaluated these embeddings across three tasks using 5-fold nested cross-validation: (a) continuous well-being prediction via ridge regression, (b) adaptive label prediction using logistic regression, and (c) maladaptive label prediction using logistic regression. We chose 5-fold CV to mitigate overfitting given the small sample size. For all classifiers, we used a penalty range of [10, 0, -1, -0.10, 0.10]. For span identification, we predicted label probabilities and applied thresholds of 0.45 (adaptive) and 0.4 (maladaptive) to extract evidence-level annotations.

	5-fold Ridge CV	
	$r \uparrow$	$MSE \downarrow$
S8D	0.528	2.556
Dist	0.365	3.059
ReLM	0.533	2.538
PLT	**0.629**	**2.149**
S8D + ReLM + Dist	0.623	2.178
S8D + PLT	0.622	2.174

Table 1: Pearson correlation (r) and Mean Squared Error (MSE) when training a ridge regression model using different "principled" baseline features to predict continuous Well-being scores using nested 5-fold cross-validation.

5 Results & Discussion

Well-being Scores. Situational characteristics (S8D) inferred from posts were predictive of annotated well-being scores (see Table 1). When combined with PLT features, our "principled" baseline — grounded in interactionist theory — yielded improved performance. The psychometric theory-inspired HaRT model outperformed baselines on the internal validation set, although we observed signs of overfitting in 5-fold CV (see Table 2). Nonetheless, official results showed similar trends (Table A2), with the theory-driven S8D +PLT baseline outperforming the theory-agnostic $\text{HaRT}_{\text{WB-FT + Ridge}}$ variant.

Adaptive and Maladaptive States. HaRT models performed well in the binary classification of adaptive and maladaptive self-states (Table A4),

	Internal Val Set		5-fold Ridge CV	
	$r\uparrow$	$MSE\downarrow$	$r\uparrow$	$MSE\downarrow$
HaRT$_{WB\text{-}FT}$	0.684	1.828	0.876	0.828
HaRT$_{WB\text{-}FT}$ + S8D + ReLM + Dist	-	-	0.883	0.787
HaRT$_{WB\text{-}FT}$ + S8D + PLT	-	-	0.884	0.783

Table 2: Pearson correlation (r) and Mean Squared Error (MSE) when fine-tuning HaRT using internal train and validation splits and further training a ridge regression model using resulting embeddings and principled baseline features. Note: Pearson r values for the 5-fold ridge CV numbers are likely inflated due to partial data contamination across the fine-tuning and cross-validation datasets. The internal validation set was used while finetuning the HaRT model, after which the weights were separately used as inputs for well-being task. The internal validation numbers have been omitted for the post-finetuned model.

	5-fold Log.reg. CV	
	$F1_{macro}$	AUC
Dist adaptive	**0.54**	0.75
HaRT$_{WB\text{-}FT}$	0.50	0.74
HaRT$_{WB\text{-}FT}$ + ReLM + Dist adaptive	0.53	**0.76**
PLT adaptive	0.48	0.66
HaRT$_{WB\text{-}FT}$ + PLT$_{adaptive}$	0.52	**0.76**
Dist maladaptive	0.56	0.73
HaRT$_{WB\text{-}FT}$	0.56	0.73
HaRT$_{WB\text{-}FT}$ + ReLM + Dist maladaptive	0.57	0.76
PLT maladaptive	0.49	0.70
HaRT$_{WB\text{-}FT}$ + PLT$_{maladaptive}$	**0.58**	**0.77**

Table 3: Macro F1 and AUC results when training a logistic regression model to predict binary adaptive and maladaptive labels separately over sentences split from posts.

with additional gains observed when combined with PLT features (Table 3). While Dist and PLT features alone showed reasonable performance, they produced minimal variation in predicted probabilities across examples (Figure 1). In contrast, HaRT-based models exhibited greater sensitivity to language variation and were more effective at identifying adaptive and maladaptive evidence spans. Additional supporting results and probability distributions can be found in Appendix § A.6.

5.1 Discussion.

Our interactionist, theory-based "principled" baseline approach effectively predicts annotated well-being scores. However, it struggled to capture evidence of adaptive and maladaptive states within posts, highlighting the challenge of disentangling self-states from their situational context — an issue that even human observers can find difficult to assess accurately (Uleman et al., 1996; Nisbett et al., 1973; Ross, 1977).

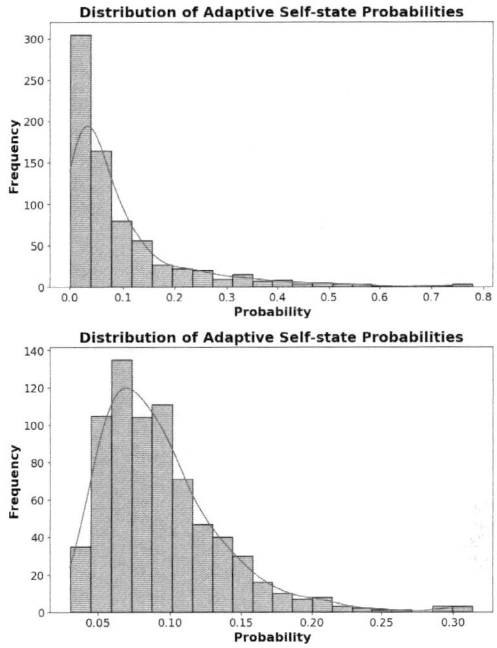

Figure 1: Distribution of probabilities to predict adaptive state for a given sentence. On the top is using HaRT$_{WB\text{-}FT}$ + PLT features, and the bottom is using PLT features in Logistic Regression models.

To further explore the predictive power of our principled baseline features, we conduct a qualitative analysis of well-being correlations. As shown in Figure 2, the top three features positively associated with well-being scores are: 'satisfaction with life' (from mental health in PLT), 'positivity' in the situation (from S8D), and 'harmony in life' (from mental health in PLT). Conversely, the top three features negatively correlated with well-being scores include: 'higher power belief' (resilience from PLT), 'depression scale' (mental health from PLT), and overall 'resilience score' (from PLT).

Additionally, Figure 2 shows that the ridge regression model, leveraging our principled baseline features, assigns positive importance to 'sociality', 'positivity', and 'intellect' (all from S8D), while attributing negative importance to 'daily lived experience' (from resilience in PLT), 'need for affiliation'

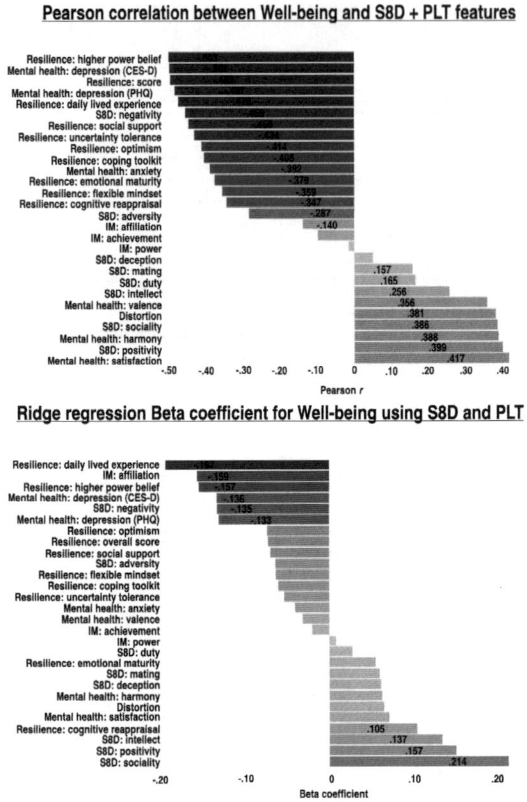

Figure 2: Qualitative analysis of features in our principled baseline consisting of psychological characteristics of the situation and person-level traits. Left: Pearson correlation coefficients; Right: Ridge regression beta coefficients for predicting well-being with the S8D and PLT features.

(from implicit motives in PLT)[1], and 'belief in a higher power' (from resilience in PLT).

These findings align with prior research indicating that well-being is closely tied to life satisfaction and positive social interactions (Diener et al., 1999; Cacioppo and Cacioppo, 2014). Similarly, negative associations with depression, present-focus, and certain aspects of religiosity are consistent with existing psychological literature on mental health dynamics (Beck, 1967; Himmelstein et al., 2018; McCullough and Larson, 1999; Braam and Koenig, 2019). A more detailed discussion of these results can be found in the Appendix § A.5.

[1]High expression of affiliation-related language may indicate a frustrated, rather than fulfilled, need for social connection. Classic motivation theories (McClelland, 1987) suggest that individuals who frequently verbalize their affiliation motive may be experiencing social deprivation or unmet interpersonal needs. This aligns with research on compensatory behaviors, where socially disconnected individuals often amplify affiliative overtures to seek connection (Baumeister and Leary, 1995; Richman and Leary, 2009).

6 Conclusion

Mental health is not a static trait but a dynamic outcome shaped by ongoing interactions between person and context. In this work, we operationalized this psychological insight by combining person-level traits and situational characteristics — core tenets of interactionist and constructionist theory — to model well-being and adaptive self-states in language.

Our theory-driven baseline, built from the Situational 8 DIAMONDS and language-inferred psychological traits, demonstrated strong performance while offering interpretable, psychologically grounded predictions. Features like positivity, satisfaction with life, and harmony in life emerged as key indicators of well-being, while markers of cognitive distortion and unmet affiliation needs were linked to maladaptive patterns. HaRT's person-contextualized embeddings added value in modeling temporal variation, particularly for adaptive and maladaptive evidence detection.

These findings highlight the value of bridging computational models with psychological theory — not only to improve prediction, but to ensure outputs are meaningful and human-understandable. Future work should explore how different contexts modulate trait adaptiveness, and how language-based systems might support more flexible, resilient self-states over time.

By integrating theory and computation, we move toward systems that understand individuals not as fixed entities, but as contextually situated and dynamically evolving.

Limitations

While this work offers a psychologically grounded and interpretable approach to modeling mental health from language, several limitations must be acknowledged — both technical and conceptual.

First, our analyses are constrained by the scale and structure of the CLPsych 2025 dataset. With only 30 users and fewer than 200 annotated posts, the generalizability of our findings is limited. Although we used robust cross-validation and avoided tuning on the test set, future work should evaluate these models on larger, more diverse, and demographically representative datasets.

Second, the ground truth labels themselves are inherently interpretive, reflecting human judgments of well-being and self-states based on textual evidence. This raises a broader epistemological question: to what extent can self-states be reliably inferred from language alone? Our work assumes that linguistic expressions are sufficient proxies for psychological states — a premise that, while useful for modeling, must be critically examined in clinical and applied settings.

Third, our feature extraction relies on pretrained models and heuristics that may carry latent biases or be insensitive to cultural or contextual nuance. For example, expressions of distress or resilience may vary across communities, and models trained on general corpora may fail to capture such variation meaningfully.

Finally, although we draw from psychological theory, our models remain correlational. They can identify linguistic markers of mental health but in no way are able to definitively speak to underlying mechanisms, causal relationships, or interventions. Future work should incorporate longitudinal clinical assessments to validate language-based features against real-world outcomes.

These limitations do not undermine the value of this work, but rather highlight the need for computational psychology to remain grounded in balancing predictive power with interpretive care, and data-driven modeling with theoretical accountability.

Ethical Considerations

Modeling mental health from language data presents profound ethical challenges. While the tools developed in this work aim to advance understanding of mental states in contextually sensitive and interpretable ways, their misuse — or even well-intentioned use without adequate safeguards — poses real risks to privacy, autonomy, and well-being.

First and foremost is the issue of consent. Although the data used in this study were shared with participant permission as part of a structured research challenge, this controlled environment does not reflect broader real-world settings in which language-based models might be applied. Any future deployment must ensure that individuals are aware of — and have control over — how their language data are interpreted, stored, and acted upon.

Second, language-based inferences about mental health are probabilistic and inherently contain some degree of uncertainty. Over-reliance on model outputs — particularly in clinical, legal, or surveillance contexts — could lead to misdiagnoses, stigmatization, or unwarranted interventions. The interpretability of our features helps mitigate this risk, but human oversight and psychological expertise remain essential in any applied use.

Third, there are critical concerns around representation and bias. Our models are trained on English-language data from social media forums, which may reflect particular cultural, demographic, and socioeconomic perspectives. As a result, model outputs may not generalize across populations and could even reinforce existing inequities if deployed indiscriminately. Expanding the diversity of training data and engaging with cultural psychology are necessary steps forward.

Finally, as researchers in computational social science, we must remain vigilant about the institutional and commercial pressures that can shape how mental health technologies are built and used. The potential to infer mental states from language at scale invites both promise and peril. Ethical research in this space demands more than compliance — it requires an ongoing commitment to transparency, self-reflection, and the prioritization of human dignity.

Acknowledgments

We are grateful to the organizers of the CLPsych 2025 Shared Task for their efforts in curating the dataset, designing an impactful task, and fostering interdisciplinary collaboration.
This research is supported in part by the Office of the Director of National Intelligence (ODNI), Intelligence Advanced Research Projects Activity (IARPA), via the HIATUS Program contract #2022-22072200005, a grant from the NIH-NIAAA (R01 AA028032), and a grant from the CDC/NIOSH (U01 OH012476), Developing and Evaluating Artificial Intelligence-based Longitudinal Assessments of PTSD in 9/11 Responders. The views and conclusions contained herein are those of the authors and should not be interpreted as necessarily representing the official policies, either expressed or implied, of ODNI, IARPA, any other government organization, or the U.S. Government. The U.S. Government is authorized to reproduce and distribute reprints for governmental purposes notwithstanding any copyright annotation therein.

References

Lisa Feldman Barrett. 2017. *How emotions are made: The secret life of the brain.* How emotions are made: The secret life of the brain. Houghton Mifflin Harcourt, Boston, MA.

Krishna C Bathina, Marijn Ten Thij, Lorenzo Lorenzo-Luaces, Lauren A Rutter, and Johan Bollen. 2021. Individuals with depression express more distorted thinking on social media. *Nature human behaviour*, 5(4):458–466.

Roy F. Baumeister and Mark R. Leary. 1995. The need to belong: Desire for interpersonal attachments as a fundamental human motivation. *Psychological Bulletin*, 117(3):497–529. Place: US Publisher: American Psychological Association.

Aaron T. Beck. 1967. *Depression: Clinical, experimental, and theoretical aspects*, no additional printings listed edition edition. see notes for publisher info.

Ryan L. Boyd and David M. Markowitz. 2024. Verbal behavior and the future of social science. *American Psychologist*, pages 1–23. Place: US Publisher: American Psychological Association.

Arjan W. Braam and Harold G. Koenig. 2019. Religion, spirituality and depression in prospective studies: A systematic review. *Journal of Affective Disorders*, 257:428–438.

David M. Buss. 1987. Selection, evocation, and manipulation. *Journal of Personality and Social Psychology*, 53(6):1214–1221.

John T. Cacioppo and Stephanie Cacioppo. 2014. Social relationships and health: The toxic effects of perceived social isolation. *Social and Personality Psychology Compass*, 8(2):58–72. _eprint: https://onlinelibrary.wiley.com/doi/pdf/10.1111/spc3.12087.

DeepSeek-AI, Daya Guo, Dejian Yang, Haowei Zhang, Junxiao Song, Ruoyu Zhang, Runxin Xu, Qihao Zhu, Shirong Ma, Peiyi Wang, Xiao Bi, Xiaokang Zhang, Xingkai Yu, Yu Wu, Z. F. Wu, Zhibin Gou, Zhihong Shao, Zhuoshu Li, Ziyi Gao, Aixin Liu, Bing Xue, Bingxuan Wang, Bochao Wu, Bei Feng, Chengda Lu, Chenggang Zhao, Chengqi Deng, Chenyu Zhang, Chong Ruan, Damai Dai, Deli Chen, Dongjie Ji, Erhang Li, Fangyun Lin, Fucong Dai, Fuli Luo, Guangbo Hao, Guanting Chen, Guowei Li, H. Zhang, Han Bao, Hanwei Xu, Haocheng Wang, Honghui Ding, Huajian Xin, Huazuo Gao, Hui Qu, Hui Li, Jianzhong Guo, Jiashi Li, Jiawei Wang, Jingchang Chen, Jingyang Yuan, Junjie Qiu, Junlong Li, J. L. Cai, Jiaqi Ni, Jian Liang, Jin Chen, Kai Dong, Kai Hu, Kaige Gao, Kang Guan, Kexin Huang, Kuai Yu, Lean Wang, Lecong Zhang, Liang Zhao, Litong Wang, Liyue Zhang, Lei Xu, Leyi Xia, Mingchuan Zhang, Minghua Zhang, Minghui Tang, Meng Li, Miaojun Wang, Mingming Li, Ning Tian, Panpan Huang, Peng Zhang, Qiancheng Wang, Qinyu Chen, Qiushi Du, Ruiqi Ge, Ruisong Zhang, Ruizhe Pan, Runji Wang, R. J. Chen, R. L. Jin, Ruyi Chen, Shanghao Lu, Shangyan Zhou, Shanhuang Chen, Shengfeng Ye, Shiyu Wang, Shuiping Yu, Shunfeng Zhou, Shuting Pan, S. S. Li, Shuang Zhou, Shaoqing Wu, Shengfeng Ye, Tao Yun, Tian Pei, Tianyu Sun, T. Wang, Wangding Zeng, Wanjia Zhao, Wen Liu, Wenfeng Liang, Wenjun Gao, Wenqin Yu, Wentao Zhang, W. L. Xiao, Wei An, Xiaodong Liu, Xiaohan Wang, Xiaokang Chen, Xiaotao Nie, Xin Cheng, Xin Liu, Xin Xie, Xingchao Liu, Xinyu Yang, Xinyuan Li, Xuecheng Su, Xuheng Lin, X. Q. Li, Xiangyue Jin, Xiaojin Shen, Xiaosha Chen, Xiaowen Sun, Xiaoxiang Wang, Xinnan Song, Xinyi Zhou, Xianzu Wang, Xinxia Shan, Y. K. Li, Y. Q. Wang, Y. X. Wei, Yang Zhang, Yanhong Xu, Yao Li, Yao Zhao, Yaofeng Sun, Yaohui Wang, Yi Yu, Yichao Zhang, Yifan Shi, Yiliang Xiong, Ying He, Yishi Piao, Yisong Wang, Yixuan Tan, Yiyang Ma, Yiyuan Liu, Yongqiang Guo, Yuan Ou, Yuduan Wang, Yue Gong, Yuheng Zou, Yujia He, Yunfan Xiong, Yuxiang Luo, Yuxiang You, Yuxuan Liu, Yuyang Zhou, Y. X. Zhu, Yanhong Xu, Yanping Huang, Yaohui Li, Yi Zheng, Yuchen Zhu, Yunxian Ma, Ying Tang, Yukun Zha, Yuting Yan, Z. Z. Ren, Zehui Ren, Zhangli Sha, Zhe Fu, Zhean Xu, Zhenda Xie, Zhengyan Zhang, Zhewen Hao, Zhicheng Ma, Zhigang Yan, Zhiyu Wu, Zihui Gu, Zijia Zhu, Zijun Liu, Zilin Li, Ziwei Xie, Ziyang Song, Zizheng Pan, Zhen Huang, Zhipeng Xu, Zhongyu Zhang, and Zhen Zhang. 2025. DeepSeek-R1: Incentivizing reasoning capability in LLMs via reinforcement learning. *arXiv preprint*. ArXiv:2501.12948 [cs].

Ed Diener, Eunkook M. Suh, Richard E. Lucas, and Heidi L. Smith. 1999. Subjective well-being: Three decades of progress. *Psychological Bulletin*, 125(2):276–302. Place: US Publisher: American Psychological Association.

Veerle C Eijsbroek. 2023. A comparison of the assessments from a person and the person's AI–model using computational language assessments.

Bo Ekehammar. 1974. Interactionism in personality from a historical perspective. *Psychological Bulletin*, 81(12):1026–1048. Place: US Publisher: American Psychological Association.

William Fleeson. 2004. Moving personality beyond the person-situation debate: The challenge and the opportunity of within-person variability. *Current Directions in Psychological Science*, 13(2):83–87. Publisher: SAGE Publications Inc.

Adithya V Ganesan, Siddharth Mangalik, Vasudha Varadarajan, Nikita Soni, Swanie Juhng, João Sedoc, H Andrew Schwartz, Salvatore Giorgi, and Ryan L Boyd. 2024. From text to context: Contextualizing language with humans, groups, and communities for socially aware nlp. In *Proceedings of the 2024 Conference of the North American Chapter of the Association for Computational Linguistics: Human Language Technologies (Volume 5: Tutorial Abstracts)*, pages 26–33.

Adithya V Ganesan, Vasudha Varadarajan, Juhi Mittal, Shashanka Subrahmanya, Matthew Matero, Nikita

Soni, Sharath Chandra Guntuku, Johannes Eichstaedt, and H Andrew Schwartz. 2022. Wwbp-sqt-lite: Multi-level models and difference embeddings for moments of change identification in mental health forums. In *Proceedings of the Eighth Workshop on Computational Linguistics and Clinical Psychology*, pages 251–258.

Zhuojun Gu, Katarina Kjell, H. Andrew Schwartz, and Oscar Kjell. Natural language response formats for assessing depression and worry with large language models: A sequential evaluation with model preregistration. *Assessment*.

Zhuojun Gu, Katarina Kjell, H Andrew Schwartz, Oscar Kjell, et al. 2024. Natural language response formats for assessing depression and worry with large language models: A sequential evaluation with model pre-registration.

Philip Himmelstein, Scott Barb, Mark A. Finlayson, and Kymberly D. Young. 2018. Linguistic analysis of the autobiographical memories of individuals with major depressive disorder. *PLOS ONE*, 13(11):e0207814.

Tom Hollenstein. 2015. This time, it's real: Affective flexibility, time scales, feedback loops, and the regulation of emotion. *Emotion Review*, 7(4):308–315. Publisher: SAGE Publications.

Thomas E. Joiner Jr. and Katherine A. Timmons. 2009. Depression in its interpersonal context. In *Handbook of depression, 2nd ed*, pages 322–339. The Guilford Press, New York, NY, US.

Oscar N. E. Kjell, Sverker Sikström, Katarina Kjell, and H. Andrew Schwartz. 2022a. Natural language analyzed with AI-based transformers predict traditional subjective well-being measures approaching the theoretical upper limits in accuracy. *Scientific Reports*, 12(1):3918.

Oscar NE Kjell, Sverker Sikström, Katarina Kjell, and H Andrew Schwartz. 2022b. Natural language analyzed with ai-based transformers predict traditional subjective well-being measures approaching the theoretical upper limits in accuracy. *Scientific reports*, 12(1):3918.

Kurt Kroenke, Robert L. Spitzer, and Janet B. W. Williams. 2011. Patient Health Questionnaire-9. Institution: American Psychological Association.

Syeda Mahwish, Ryan L. Boyd, Vasudha Varadarajan, Roman Kotov, Benjamin J. Luft, H. Andrew Schwartz, and Sean A. P. Clouston. Measuring resilience using language modeling (ReLM): A computational approach to observing resilience.

Ruth E Mann, Anthony R Beech, T Ward, DR Laws, and SM Hudson. 2002. Cognitive distortions, schemas, and implicit theories. *Sexual Deviance: Issues, Theories and Treatment*, pages 135–153.

David C. McClelland. 1987. *Human motivation*. CUP Archive.

Michael E. McCullough and David B. Larson. 1999. Religion and depression: A review of the literature. *Twin Research and Human Genetics*, 2(2):126–136.

Shannon T. Mejía and Karen Hooker. 2015. Emotional well-being and interactions with older adults' close social partners: Daily variation in social context matters. *Psychology and Aging*, 30(3):517–528. Place: US Publisher: American Psychological Association.

Walter Mischel and Yuichi Shoda. 1995. A cognitive-affective system theory of personality: Reconceptualizing situations, dispositions, dynamics, and invariance in personality structure. *Psychological Review*, 102(2):246–268. Place: US Publisher: American Psychological Association.

August Nilsson, Zhuojun Gu, Veerle C Eijsbroek, Katarina Kjell, Salvatore Giorgi, Roman Kotov, H. Andrew Schwartz, and Oscar N. E. Kjell. a. The Language-Based Assessment Model (L-BAM) library: Open model sharing for independent validation and broader application.

August Nilsson, J Malte Runge, Oscar Kjell, Adithya V Ganesan, Carl Viggo Nilsson, et al. 2024. Automatic implicit motives codings are as accurate as humans' and 99% faster.

August Nilsson, J. Malte Runge, Oscar Kjell, Nikita Soni, Adithya V. Ganesan, and Carl Viggo Nilsson. b. Automatic implicit motives codings are as accurate as humans' and 99% faster. *Journal of Personality and Social Psychology: Personality Processes and Individual Differences*.

Richard E. Nisbett, Craig Caputo, Patricia Legant, and Jeanne Marecek. 1973. Behavior as seen by the actor and as seen by the observer. *Journal of Personality and Social Psychology*, 27(2):154–164. Place: US Publisher: American Psychological Association.

Lenore S. Radloff. 1977. The CES-D Scale: A self-report depression scale for research in the general population. *Applied Psychological Measurement*, 1(3):385–401. Place: US Publisher: Sage Publications.

John F. Rauthmann, David Gallardo-Pujol, Esther M. Guillaume, Elysia Todd, Christopher S. Nave, Ryne A. Sherman, Matthias Ziegler, Ashley Bell Jones, and David C. Funder. 2014. The Situational Eight DIAMONDS: A taxonomy of major dimensions of situation characteristics. *Journal of Personality and Social Psychology*, 107(4):677–718.

Laura Smart Richman and Mark R. Leary. 2009. Reactions to discrimination, stigmatization, ostracism, and other forms of interpersonal rejection. *Psychological Review*, 116(2):365–383.

Lee Ross. 1977. The intuitive psychologist and his shortcomings: Distortions in the attribution Process. In Leonard Berkowitz, editor, *Advances in Experimental Social Psychology*, volume 10, pages 173–220. Academic Press.

Han-Chin Shing, Suraj Nair, Ayah Zirikly, Meir Friedenberg, Hal Daumé III, and Philip Resnik. 2018. Expert, crowdsourced, and machine assessment of suicide risk via online postings. In *Proceedings of the Fifth Workshop on Computational Linguistics and Clinical Psychology: From Keyboard to Clinic*, pages 25–36, New Orleans, LA. Association for Computational Linguistics.

Dana Atzil Slonim. 2024. Self-other dynamics (sod): A transtheoretical coding manual.

Nikita Soni, Niranjan Balasubramanian, H Schwartz, and Dirk Hovy. 2024a. Comparing pre-trained human language models: Is it better with human context as groups, individual traits, or both? In *Proceedings of the 14th Workshop on Computational Approaches to Subjectivity, Sentiment, & Social Media Analysis*, pages 316–328.

Nikita Soni, Pranav Chitale, Khushboo Singh, Niranjan Balasubramanian, and H. Andrew Schwartz. 2025. Evaluation of llms-based hidden states as author representations for psychological human-centered nlp tasks. *Preprint*, arXiv:2503.00124.

Nikita Soni, Lucie Flek, Ashish Sharma, Diyi Yang, Sara Hooker, and H Andrew Schwartz. 2024b. Proceedings of the 1st human-centered large language modeling workshop. In *Proceedings of the 1st Human-Centered Large Language Modeling Workshop*.

Nikita Soni, Matthew Matero, Niranjan Balasubramanian, and H. Andrew Schwartz. 2022. Human language modeling. In *Findings of the Association for Computational Linguistics: ACL 2022*, pages 622–636, Dublin, Ireland. Association for Computational Linguistics.

Nikita Soni, H Schwartz, João Sedoc, and Niranjan Balasubramanian. 2024c. Large human language models: A need and the challenges. In *Proceedings of the 2024 Conference of the North American Chapter of the Association for Computational Linguistics: Human Language Technologies (Volume 1: Long Papers)*, pages 8623–8638.

Adam Tsakalidis, Jenny Chim, Iman Munire Bilal, Ayah Zirikly, Dana Atzil-Slonim, Federico Nanni, Philip Resnik, Manas Gaur, Kaushik Roy, Becky Inkster, Jeff Leintz, and Maria Liakata. 2022. Overview of the CLPsych 2022 shared task: Capturing moments of change in longitudinal user posts. In *Proceedings of the Eighth Workshop on Computational Linguistics and Clinical Psychology*, pages 184–198, Seattle, USA. Association for Computational Linguistics.

Talia Tseriotou, Jenny Chim, Ayal Klein, Aya Shamir, Guy Dvir, Iqra Ali, Cian Kennedy, Guneet Singh Kohli, Anthony Hills, Ayah Zirikly, Dana Atzil-Slonim, and Maria Liakata. 2025. Overview of the clpsych 2025 shared task: Capturing mental health dynamics from social media timelines. In *Proceedings of the 10th Workshop on Computational Linguistics and Clinical Psychology*. Association for Computational Linguistics.

James S. Uleman, Leonard S. Newman, and Gordon B. Moskowitz. 1996. People as flexible interpreters: Evidence and issues from spontaneous trait inference. In Mark P. Zanna, editor, *Advances in Experimental Social Psychology*, volume 28, pages 211–279. Academic Press.

Michael Ungar. 2013. Resilience, trauma, context, and culture. *Trauma, Violence, & Abuse*, 14(3):255–266. Publisher: SAGE Publications.

Vasudha Varadarajan, Allison Lahnala, Adithya V. Ganesan, Gourab Dey, Siddharth Mangalik, Ana-Maria Bucur, Nikita Soni, Rajath Rao, Kevin Lanning, Isabella Vallejo, Lucie Flek, H. Andrew Schwartz, Charles Welch, and Ryan L. Boyd. 2024a. Archetypes and entropy: Theory-driven extraction of evidence for suicide risk. In *Proceedings of the 9th Workshop on Computational Linguistics and Clinical Psychology (CLPsych 2024)*, pages 278–291, St. Julians, Malta. Association for Computational Linguistics.

Vasudha Varadarajan, Allison Lahnala, Adithya V Ganesan, Gourab Dey, Siddharth Mangalik, Ana-Maria Bucur, Nikita Soni, Rajath Rao, Kevin Lanning, Isabella Vallejo, et al. 2024b. Archetypes and entropy: Theory-driven extraction of evidence for suicide risk. In *Proceedings of the 9th Workshop on Computational Linguistics and Clinical Psychology (CLPsych 2024)*, pages 278–291.

Vasudha Varadarajan, Allison C. Lahnala, Sujeeth Vankudari, Syeda Mahwish, Akshay Raghavan, Scott Feltman, Camilo Ruggero, Roman Kotov, and H. Andrew Schwartz. 2025. Linking language-based distortion detection to mental health outcomes. In *Proceedings of The 10th Workshop on Computational Linguistics and Clinical Psychology*, Albuquerque, USA. Association of Computational Linguistics.

Karishma Vyas, Dominic Murphy, and Neil Greenberg. 2023. Cognitive biases in military personnel with and without PTSD: a systematic review. *Journal of Mental Health*, 32(1):248–259. Publisher: Routledge _eprint: https://doi.org/10.1080/09638237.2020.1766000.

David Watson. 2004. Stability versus change, dependability versus error: Issues in the assessment of personality over time. *Journal of Research in Personality*, 38(4):319–350.

Ayah Zirikly, Philip Resnik, Özlem Uzuner, and Kristy Hollingshead. 2019. CLPsych 2019 shared task: Predicting the degree of suicide risk in Reddit posts. In *Proceedings of the Sixth Workshop on Computational Linguistics and Clinical Psychology*, pages 24–33, Minneapolis, Minnesota. Association for Computational Linguistics.

A Appendix

A.1 Official Submissions and results

Table A1 shows the official shared task results for Task A.1 (evidence extraction), and Table A2 shows the official results for Task A.2 (wellbeing prediction).

A.2 Situational 8 DIAMONDS Prompts

We used the items associated with each psychosocial situation from Rauthmann et al. (2014) to define each of the S8D to Deepseek-R1. We individually prompted for each of the S8D and provided two personalized exemplars and annotations for each of the exemplars (2-shot). Some few-shot examples were manually created and some were picked from the CLPsych data. The prompts template would read as follows:

> *Instruction: You are an expert in situational perception and psychological analysis. Your task is to evaluate a given block of text for the (insert situation) dimension from the Situational 8 DIAMONDS taxonomy. Individuals who score higher in the (insert situation) dimension relate to the following situations:*
>
> *{{S8D items corresponding to relevant dimension were inserted here}}*
>
> *Your task is to provide the following in a structured JSON format: Rating: Assign a numerical rating for the mating dimension on a scale of 1 to 9 (where 1 = Not at all present and 9 = Highly present).*
>
> *Reasoning: Provide a justification for the rating based on the text. Span Extraction: Identify specific phrases in the text that support your rating.*
>
> *Below are two examples with respective input texts and corresponding outputs to illustrate the task:*
>
> *{{Example texts with corresponding annotations were inserted here}}*
>
> *Now, evaluate the following input text:*
>
> *{{Text requiring annotation was inserted here}}*

We then curated two examples catered towards each situation and hand-annotated the example to provide as a guideline for output. We provided two messages from the CLPsych 2025 dataset for the situations Duty and Intellect, although for the remaining six S8D, we curated our own examples. An example annotation for the situation Adversity (including manually creates few-shot examples) goes as follows:

> "Example 1: "I can't believe how unfair my manager is being. I worked overtime all last week, skipped my breaks, and still got blamed for a project delay that wasn't even my fault. He called me out in front of the entire team, making it sound like I was slacking off. I tried to defend myself, but he just dismissed me and moved on. It's exhausting constantly feeling like I have to prove myself, only to be treated like I'm incompetent."
>
> **Output:** "adversity": 8, "reasoning": "The individual is experiencing direct blame from their manager for a project delay that was not their fault. They describe being publicly criticized in front of colleagues and dismissed when attempting to defend themselves. The tone reflects frustration and exhaustion from repeated unfair treatment, which strongly aligns with the Adversity dimension.","supporting spans": "I worked overtime all last week, skipped my breaks, and still got blamed for a project delay that wasn't even my fault.","He called me out in front of the entire team, making it sound like I was slacking off.","I tried to defend myself, but he just dismissed me and moved on.","It's exhausting constantly feeling like I have to prove myself, only to be treated like I'm incompetent."
>
> Example 2: "Every time I try to express my opinion, my older brother just shuts me down. He talks over me, mocks what I say, and makes me feel like I'm too stupid to contribute. It's like my thoughts don't matter in my own family. Even when I call him out on it, he just laughs and says I'm being too sensitive. I don't know how to get him to take me seriously."
>
> **Output:** "adversity": 7, "reasoning":

	A.1 Adaptive/Maladaptive					
	recall			**weighted recall**		
	overall	adaptive	maladaptive	overall	adaptive	maladaptive
HaRT$_{\text{WB-FT}}$ + ReLM + Dist (LinSVC)	0.108	0.099	0.116	0.103	0.097	0.109
HaRT$_{\text{WB-FT + Log.reg.}}$	0.105	0.077	0.132	0.102	0.075	0.129
HaRT$_{\text{adaptive-FT}}$	**0.276**	**0.245**	**0.308**	-	-	-
HaRT$_{\text{maladaptive-FT}}$	-	-	-	**0.236**	**0.238**	**0.235**

Table A1: Task A.1 Adaptive and Maladaptive Evidence task. We found that finetuning HaRT for independent adaptive and maladaptive evidence classification can yield significant boosts over traditional principled baselines.

	A.2 Well-being				
	timeline MSE ↓				**F1**
	overall	minimal impairment	impaired	serious impairment	macro
HaRT$_{\text{WB-FT}}$	3.73	2.76	1.95	5.92	0.15
HaRT$_{\text{WB-FT + Ridge}}$	3.42	2.95	**1.33**	5.1	0.17
HaRT$_{\text{WB-FT}}$ + S8D + ReLM + Dist (Ridge)	3.27	2.63	1.38	4.98	0.19
HaRT$_{\text{WB-FT}}$ + S8D + PLT (Ridge)	3.22	1.40	2.60	4.86	0.19
S8D + ReLM + Dist (Ridge)	3.02	**1.21**	1.79	5.25	0.17
S8D + PLT (Ridge)	**2.78**	1.84	2.14	**3.89**	0.19

Table A2: Task A.2 Wellbeing task. We found that unlike Task A.1, finetuning HaRT to the wellbeing prediction task need not consistently offer boosts, instead, principled and theoretical methods can offer significant advantages, with a small number of interpretable dimensions without compromising on the accuracy. The first 3 rows in this table were our official submissions while others are presented for additional analysis. We note that using HaRt fine-tuned for the respective adaptive and maladaptive binary classifications may provide benefits over using HaRT$_{\text{WB-FT}}$ in all combinations, however, due to time constraints we do not have empirical results for the same.

"The individual describes repeated experiences of being dismissed, mocked, and dominated by their older brother. The situation involves verbal criticism, a power imbalance, and an inability to be taken seriously, all of which strongly align with the Adversity dimension.", "supporting spans": "Every time I try to express my opinion, my older brother just shuts me down.", "He talks over me, mocks what I say, and makes me feel like I'm too stupid to contribute.", "Even when I call him out on it, he just laughs and says I'm being too sensitive.""

A.3 Implicit Motives and Mental Health in PLT

To construct our person-level trait (PLT) features, we extracted both implicit motivational needs and core mental health dimensions from participants' language using pre-existing, validated models.

Implicit Motives. Following classic motivational theory (McClelland, 1987), we define three core implicit motives reflected in language: (1) the need for achievement, indicated by references to striving for excellence; (2) the need for affiliation, reflected in efforts to initiate or maintain friendly relationships; and (3) the need for power, expressed as influence or control over others or institutions. We used RoBERTa-based models from prior work (Nilsson et al., 2024), trained on expert-coded Picture Story Exercises, to infer these motives at the sentence level. Sentence-level predictions were then aggregated to post level using word count-adjusted averaging procedures.

Mental Health Dimensions. We further extracted six features representing key aspects of mental health using models from the Language-Based Assessment Model Library (Nilsson et al., a). These include:

- **Valence:** Trained on annotated Facebook posts rated for emotional positivity or negativity. The model's out-of-sample correlation with human ratings was $r = .81$.

- **Harmony in Life & Satisfaction with Life:** Trained on open-text responses rated using validated scales (Kjell et al., 2022a). The models achieved out-of-sample correlations of $r = .73$ and $r = .71$, respectively.

- **Depression:** Two separate models were used — one trained to the Patient Health Questionnaire-9 (PHQ-9; Kroenke et al., 2011), and the other to the Center for Epidemiologic Studies Depression Scale (CES-D; Radloff, 1977). These models yielded correlations of $r = .66$ and $r = .73$, respectively.

- **Anxiety:** Trained on worry-based language mapped to the Generalized Anxiety Disorder 7-item scale (GAD-7), with a correlation of $r = .63$.

All models were previously pre-registered for their respective source projects, evaluated using nested cross-validation, and applied out-of-sample to the present dataset. These features form part of our psychologically interpretable PLT baseline.

A.4 Resilience in PLT

Traditional views of resilience often reduce it to the absence of psychopathology or the ability to recover from stress. However, contemporary psychological frameworks emphasize a broader understanding: resilience as a multidimensional capacity for adaptive functioning in the face of adversity. To operationalize this richer perspective, we used the ReLM (Resilience using Language Modeling) framework (Mahwish et al.), which integrates an archetype-based approach to assess resilience from language.

ReLM captures nine core facets of resilience: optimism, flexibility mindset, sense of social support (SoS), continued activities of daily living (CADL), cognitive reappraisal, emotional maturity, uncertainty tolerance, belief in a higher power, and coping toolkit. Each facet is represented by four prototype statements — brief exemplar sentences derived from a synthesis of resilience literature and analysis of archival interviews with individuals who have demonstrated stability in the face of trauma. For example, a prototype for flexibility mindset reads: "I always try new things because I'm open to exploring."

To assess individual alignment with each facet, ReLM embeds both prototype statements and participant text using Sentence RoBERTa and computes their semantic similarity (see: Varadarajan et al., 2024a). The resulting scores quantify how strongly a participant's language reflects each dimension of resilience. A higher score indicates greater expression of that facet.

Finally, a composite resilience score is computed by applying exploratory factor analysis to the nine facet scores. Across multiple datasets, a single-factor solution consistently explained 49–56% of the variance (Mahwish et al.), supporting the use of a unified resilience metric. This composite score provides a theoretically grounded and interpretable estimate of an individual's language-based resilience profile.

A.5 Continued Discussion

The 'daily lived experience' facet in this dataset exhibits a strong negative association with well-being. This relationship likely stems from the facet's focus on individuals persisting through routine tasks despite ongoing stressors. Participant statements such as "I've adjusted to my new environment—it was hard at first, but I'm improving" and "I've been leaving the house more" (rewritten for anonymity) illustrate gradual adaptation and effort. However, because well-being in this dataset is framed in terms of symptom absence and unimpaired functioning, the 'daily lived experience' facet presents a paradox. While it reflects resilience — people continuing daily tasks despite struggles — it also signals underlying difficulty. The very act of pushing forward in the face of these challenges may indicate diminished well-being, as it suggests persistent symptomatology masked by forced functionality.

The Belief in a Higher Power facet reflects trust in external forces during times of struggle, as illustrated by participant statements like "I accept things as they are… I trust that things will get better over time" and "I know that life will work itself out" (anonymized). These responses suggest a surrender of personal control to fate or higher being—a mechanism that may offer emotional relief. However, the model's negative weighting of this facet stems from a tension between its definition of well-being (rooted in agency, engagement, and lack of symptom) and a resilience strategy that relies on external control. While faith can provide comfort, the framework may be interpreting passive reliance on higher powers as maladaptive in contexts where well-being is tied to active mastery of one's circumstances.

A.6 Supplementary results and figures

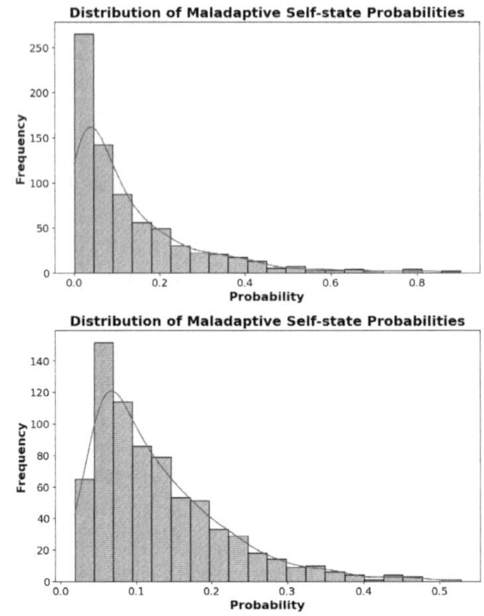

Figure A1: Distribution of probabilities to predict maladaptive state for a given sentence. On the top is using $\text{HaRT}_{\text{WB-FT}}$ + PLT features, and the bottom is using PLT features in Logistic Regression models.

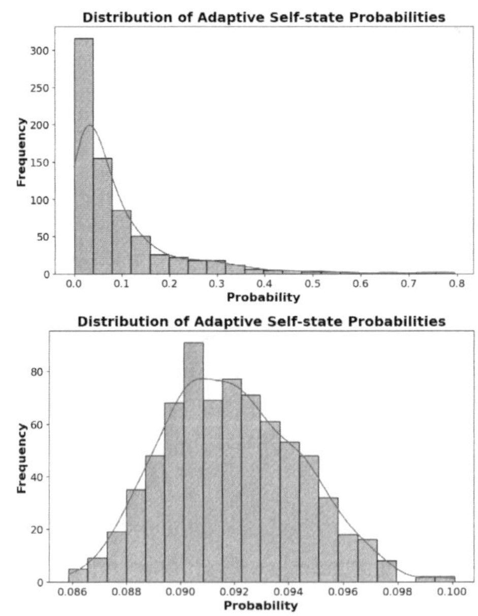

Figure A2: Distribution of probabilities to predicting adaptive state for a given sentence. On the top is using $\text{HaRT}_{\text{WB-FT}}$ + ReLM + Dist$_{\text{adaptive}}$ features, and the bottom is using ReLM + Dist$_{\text{adaptive}}$ features in Logistic Regression models.

	Internal Val Set			
	$F1_{macro}$	$F1_{wtd}$	AUC	Acc
Dist$_{\text{adaptive}}$	0.53	0.85	0.75	0.89
HaRT$_{\text{adaptive-FT}}$	0.67	0.89	0.66	0.89
Dist$_{\text{maladaptive}}$	0.56	0.82	0.76	0.84
HaRT$_{\text{maladaptive-FT}}$	0.60	0.82	0.60	0.82

Table A3: Task A.1 Additional results on internal validation set when predicting binary adaptive and maladaptive labels separately over sentences split from posts.

	5-fold Log.reg. CV			
	$F1_{macro}$	$F1_{wtd}$	AUC	Acc
Dist$_{\text{adaptive}}$	0.54	0.87	0.75	0.90
HaRT$_{\text{WB-FT}}$	0.50	0.87	0.74	0.90
HaRT$_{\text{WB-FT}}$ + ReLM + Dist$_{\text{adaptive}}$	0.53	0.87	0.76	0.91
PLT$_{\text{adaptive}}$	0.48	0.86	0.66	0.91
HaRT$_{\text{WB-FT}}$ + PLT$_{\text{adaptive}}$	0.52	0.87	0.76	0.90
Dist$_{\text{maladaptive}}$	0.56	0.81	0.73	0.85
HaRT$_{\text{WB-FT}}$	0.56	0.81	0.73	0.85
HaRT$_{\text{WB-FT}}$ + ReLM + Dist$_{\text{maladaptive}}$	0.57	0.82	0.76	0.86
PLT$_{\text{maladaptive}}$	0.49	0.80	0.70	0.86
HaRT$_{\text{WB-FT}}$ + PLT$_{\text{maladaptive}}$	0.58	0.83	0.77	0.86

Table A4: Additional results when training a logistic regression model to predict binary adaptive and maladaptive labels separately over sentences split from posts.

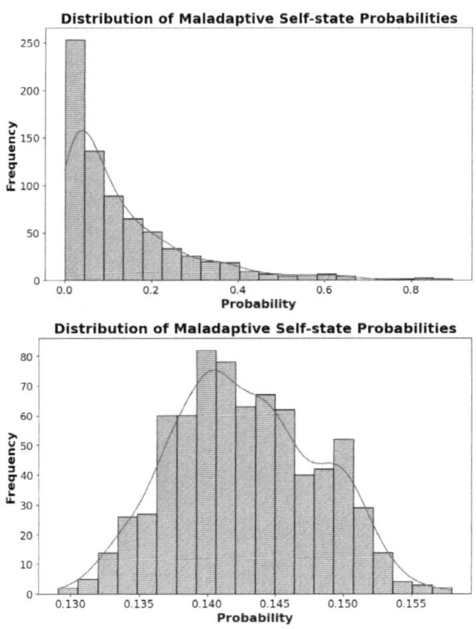

Figure A3: Distribution of probabilities to predicting maladaptive state for a given sentence. On the top is using HaRT$_{\text{WB-FT}}$ + ReLM + Dist$_{\text{maladaptive}}$ features, and the bottom is using ReLM + Dist$_{\text{maladaptive}}$ features in Logistic Regression models.

Author Index

Association for Computational Linguistics
209 N. Eighth Street
Stroudsburg, Pennsylvania 18360

ISBN 979-8-3313-1915-1

CURRAN ASSOCIATES INC.
proceedings.com